T0276080

The Geology of the Canary Islands

The Geology of the Canary Islands

Juan Carlos Carracedo

Honorary Emeritus Professor, University of Las Palmas de Gran Canaria

Valentin R. Troll

Chair in Petrology and Geochemistry, Uppsala University, Sweden

ELSEVIER

AMSTERDAM • BOSTON • HEIDELBERG • LONDON • NEW YORK • OXFORD
PARIS • SAN DIEGO • SAN FRANCISCO • SINGAPORE • SYDNEY • TOKYO

Elsevier
Radarweg 29, PO Box 211, 1000 AE Amsterdam, Netherlands
The Boulevard, Langford Lane, Kidlington, Oxford OX5 1GB, UK
50 Hampshire Street, 5th Floor, Cambridge, MA 02139, USA

British Library Cataloguing-in-Publication Data
A catalogue record for this book is available from the British Library.

Library of Congress Cataloging-in-Publication Data
A catalog record for this book is available from the Library of Congress.

ISBN: 978-0-12-809663-5

For Information on all Elsevier publications
visit our website at http://www.elsevier.com/

Working together
to grow libraries in
developing countries

www.elsevier.com • www.bookaid.org

Publisher: Candice Janco
Acquisition Editor: Louisa Hutchins
Editorial Project Manager: Marisa LaFleur
Production Project Manager: Mohanapriyan Rajendran
Designer: Mark Rogers

Typeset by MPS Limited, Chennai, India

Contents

Foreword

The Canary Islands are highly popular with tourists, yet relatively few venture away from the resorts. However, for those who do venture inland, the superb volcanic scenery can stir an interest and provoke a desire to learn more about it. Today, some of the fascinating volcanological sites are marked by notice boards and interpretive centers, but there is clearly a need for more comprehensive descriptions. The relevance is even greater for visiting geologists and university field excursions, and this book satisfies this need.

A clearly written and extremely well-illustrated book, it brings up to date and extends the earlier "Canary Islands" publication by Carracedo and Day, which appeared in the Classic Geology of Europe series published by Terra Publishing in 2002, with much new data, especially from chemical and isotopic analyses, allowing the advancement of interpretations regarding formation, growth, and erosion of these fascinating volcanic islands.

The opening chapter provides an overview of the archipelago and advances an up-to-date analysis of its tectonic and magmatic history in the context of its plate tectonic setting. The authors make a very clear, well-supported case for the islands to have developed on the edge of a continental plate moving across a stationary Mantle Plume, and for the occurrence of recent volcanism on the most easterly island, Lanzarote, to be due to the interaction of small-scale upper mantle convection at the edge of the African craton that interacts with the upwelling material of the Canary mantle plume. They support this interpretation rather than the alternative model in which crustal fractures, such as a potential fracture propagating from the Atlas fault, gave rise to volcanism upon cutting through the lithosphere, thereby possibly generating fusion by decompression.

The authors are well-positioned to publish this book because both have performed extensive research in the islands. Both have led field excursion programs with the result that the account of every island is supported by geological excursion itineraries that reveal the essential aspects of the geology, petrology, and volcano-tectonics of each island.

There are a number of geology texts already available regarding individual islands and literally hundreds of articles published regarding specific features, including newspaper headline reports of potential tsunamis resulting from landslides caused by recent volcanism on La Palma and of the sub-marine eruption in 2011 a few kilometres south of El Hierro, which produced floating pumice. However, this book is different. It is a comprehensive description of the whole archipelago. The authors have used recently developed methods and techniques to sort out the various tectonic and volcanological concepts concerning the formation and history of each island and have applied these to the evolution of the whole archipelago. Interestingly, they have also been able to link this to the volcanoes of the Madeira volcanic province and explain, in some detail, the origin of many of the volcanic features seen in both groups of islands. This publication therefore sets a benchmark for geological and volcanological efforts in the region.

By, Prof. C.J. Stillman
Department of Geology, Trinity College Dublin, Ireland

Acknowledgments

The geological information that has accumulated on the Canary Islands since the first descriptions by the great naturalists in the 18th and 19th centuries is by now extensive and continues to grow, making the Canary Islands one of the best studied oceanic archipelagos on the globe. To condense this mountain of information into a single book is a momentous task, which accounts for the few attempts brought to completion (eg, the *Geological Evolution of the Canary Islands*, compiled by Schmincke and Sumita in 2010 and the first volume of *Geología de Canarias I* published by Carracedo in 2011). An inevitable challenge is to select the most relevant geological topics without excluding significant details, while staying within the naturally limited space of such a publication. For obvious reasons (time), the introductory text and the offered routes can only include a fraction of the geology of the Canary Islands, but are aimed to achieve a representative overview of the main geological features and events. Our approach to this restriction is to provide as many illustrations as possibly feasible, following the aphorism "a picture is worth a thousand words." We are extremely grateful to Elsevier for accommodating this approach.

We are furthermore indebted to C.J. Stillman for writing the foreword and to S. Wiesmaier for help with some of the initial translations of earlier Spanish publications for the Introduction chapter and for providing several field photographs, especially from Fuerteventura. L.M. Schwarzkopf is thanked for many photos from Gran Canaria and also for many hours of driving there, and S. Socorro kindly contributed a series of spectacular panorama photographs. Further photographic contributions were made by S. Burchardt, R. Casillas, H. Castro, F.M. Deegan, A.M. Díaz Rodríguez([†]), A. Hansen, C. Karsten, S. Krastel, M.-A. Longpré, C. Moreno, and L.K. Samrock. We are also indebted for access to their photographic materials from the archives of the Gobierno de Canarias, the Cabildos of all the main islands, the Parque Nacional del Teide, the Archivo General de Simancas, in Valladolid, and the Spanish Archivo Histórico Nacional de España, Madrid, GRAFCAN, Guardia Civil, NASA, and the private collection of the Marqués de Vilaflor. We are furthermore indebted to C. Karsten, P. Agnew, and L.K. Samrock, who provided countless hours of organizational effort and are sincerely thanked for greatly improving our manuscript. We also thank the student assistants L. Barke, T. Mattsson, M. Jensen, and K. Thomaidis for assisting us during the final write-up, especially for generously giving off their time whenever help was needed. We greatly acknowledge institutional support too, in particular from the University of Las Palmas de Gran Canaria (Spain) and from Uppsala University (Sweden). Those researchers who have inspired our thinking on the geology of the Canary Islands shall be acknowledged here also, especially J.M. Fuster([†]) and H.-U. Schmincke, the respective PhD supervisors and mentors to the two authors, but also G. Ablay, A.K. Barker, M. Canals, R. Cas, H. Clarke, J. Day, S. Day, F.M. Deegan, A. Delcamp, A. Demeney, E. Donoghue, J. Geldmacher, H. Guillou, A. Gurenko, T. H. Hansteen, A. Hansen, H. Hausen, K. Hoernle, E. Ibarrola, A. Klügel, S. Krastel, M.-A. Longpré, J. Martí, D.G. Masson, J.M. Navarro([†]), R. Paris, F. Perez-Torrado, M. Petronis, E. Rodriguez-Badiola, A. Rodriguez-Gonzalez, P. Rothe, L.K. Samrock, C. Solana, C.J. Stillman, R.I. Tilling, J.L. Turiel, B. van Wyk de Vries, T.R. Walter, N.D. Watkins, A.B. Watts, S. Wiesmaier, J.A. Wolff, and K. Zazcek.

THE CANARY ISLANDS: AN INTRODUCTION

The Geology of the Canary Islands. DOI: http://dx.doi.org/10.1016/B978-0-12-809663-5.00001-3

THE CANARY ISLANDS

The Canary Islands are an archipelago of volcanic origin located off northwest Africa (Fig. 1.1A). The archipelago comprises seven major islands: Fuerteventura, Lanzarote, Gran Canaria, Tenerife, La Gomera, La Palma, and El Hierro, along with the four smaller islets of La Graciosa, Alegranza, Isla de Lobos, and Montaña Clara located north of Lanzarote (Fig. 1.1B). The total surface area of the islands is approximately 7490 km² and the easternmost islands, Lanzarote and Fuerteventura, are approximately 100 km from the African coast. Measured from the easternmost tip of Alegranza to the westernmost point of El Hierro, the entire archipelago stretches over approximately 500 km in total. The Canary Islands are considered part of the ecoregion of Macaronesia, which also includes the Azores, Madeira, Cape Verde, and the Selvagem islands.

The present population of the Canary Islands is approximately 2,118,700, with Tenerife hosting (in 2015) 888,000 inhabitants, Gran Canaria 847,000, Lanzarote 143,000, Fuerteventura 107,000, La Palma 82,000, La Gomera 21,000, and El Hierro 11,000. In addition, between 9 and 12 million tourists visit the islands annually. The different islands receive variable tourist attention, with approximately 400,000 visitors to the Taburiente Caldera on La Palma, but approximately 3.5 million to Teide volcano and the Las Cañadas National Park on Tenerife.

The Canary Islands are also one of the greatest natural laboratories for volcanology on our planet. In the 19th century, many key concepts in volcanology have been coined here by famous

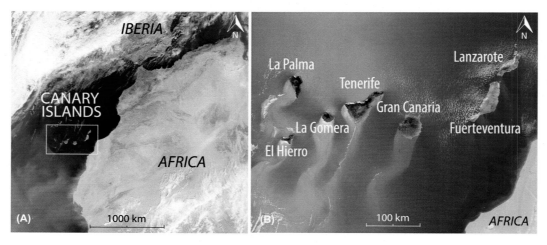

FIGURE 1.1

(A) Geographic setting of the Canary Islands. The eastern islands are close to the African continent (~100 km) and are located on the passive continental margin. All the islands rest on oceanic crust, but the central and western islands (~200 to 600 km off Africa) lie on crust of increasingly oceanic character (ie, with progressively thinner sediment cover). (B) The seven main islands of the Canary archipelago (images from NASA).

naturalists such as Alexander von Humboldt, Charles Lyell, and Leopold von Buch. The wide variety of volcanic features combined with the mild and agreeable climate, plus excellent rock exposures, make the "Canaries" the ideal location for both beginners and advanced enthusiasts of ocean island volcanology.

The variety of exposed volcanic features in the Canaries ranges from flow structures of fresh lava flows to the plutonic hearts of volcanoes exposed in deeply incised valleys and through giant landslides. Here, we aim to offer the reader an in-depth view of the geology and volcanology of the Canary Islands. We present not only a detailed overview of traditional and the latest research but also 27 day-long field itineraries that include all of the seven islands. This will bring the reader to the essential outcrops and so affords the opportunity to experience geological evidence of the key concepts first-hand. The itineraries can be combined or shortened at will to fit the complex daily schedules of travelers, but are designed to fill an entire day each. In this chapter, however, we first outline regional tectonic and volcanological building blocks of the Canary archipelago before offering a detailed overview of the volcanic features and processes recorded on the individual islands.

The Canary archipelago is volcanically active and all the islands, with the exception of La Gomera, display signs of Holocene volcanism. Lanzarote, Tenerife, and La Palma had historical eruptions (ie, during the past 500 years). The last onshore eruption occurred on La Palma in 1971 (Fig. 1.2), and the last Canary eruption was a submarine event off El Hierro between late 2011 and spring 2012.

The Canary archipelago formed progressively and from a long-lived magma source over some 60 million years (Ma). Movement of the African plate successively shifted the plate over a stationary mantle plume, which separated the islands; eventually, an island will becomes disconnected from the mantle source and a new island begins to form. This is a matter of repetitive and similar volcanic processes, and the archipelago is indeed the result of a reiteration of the same type of island-forming processes with only some local variations. The noteworthy differences in altitude and scenery between the islands are a consequence of the state of development of each island, that is, of the present stage of island evolution.

Therefore, to understand the geology of the Canary archipelago, each island has to be individually studied and a synthesis of all islands should provide a picture of the complete course of evolution of an individual island, as well as the archipelago as a whole. This approach can be compared to acquiring knowledge about the different growth stages of a human (eg, childhood, adolescence, youth, maturity, and old age) that are seen represented in a community. Transferred to the Canaries, this relates to the mature old islands in the East and childlike ones in the West.

In most groups of oceanic volcanic islands, such as the Hawaiian archipelago, the youngest islands were among the first to be studied because it was logical to assume that the geological evidence would be clearer due to the better preservation of the volcanic features and formations. Conversely, when the study of the Canary Islands commenced in modern geological times, the decision to start in the eastern and central islands exacted lasting complications and required a greater effort and a longer study period before erroneous concepts were identified. Initial association with the geology and tectonics of the nearby African continent was favored (eg, Rothe and Schmincke, 1968) but was later abandoned after decades of scientific debate.

FIGURE 1.2

Teneguía volcano, the last on-shore eruption recorded in the Canaries, occurred on La Palma in 1971. The eruption started at 15:00 on October 26, 1971, from a volcanic fracture, and it eventually formed several overlapping strombolian vents. The photograph was taken 8 hours after the eruption onset from a pre-1971 cone, looking SE (picture courtesy of A.M. Días Rodríguez).

The relatively recent progress in seafloor mapping and geomagnetic research eventually promoted the understanding of the geology of the western islands, which has been a crucial factor that led to an overall view of the Canary archipelago as a geologically closed entity. In this book we aim to provide insight into the key aspects of their origin, stages of evolution, structural changes, temporal and spatial distribution of volcanism, and past and current eruptive hazards.

The western islands, La Palma and El Hierro, lie on a deeper and younger oceanic basement than the remaining islands of the archipelago. However, the single alignment of the Canaries may become a "dual line," with these two youngest islands potentially setting a new trend of simultaneous island formation (see Fig. 1.1B).

A decisive factor in the reconstruction of the geological history of the Canaries was increasing the availability and accuracy of radioisotopic ages (eg, Abdel Monem et al., 1971, 1972; McDougall and Schmincke, 1976; Carracedo, 1979; Cantagrel et al., 1984; Ancochea et al., 1990, 1994, 1996, 2006; Guillou et al., 1996; 1998, 2004a,b; van den Bogaard and Schmincke, 1998; Carracedo et al., 2007a,b, 2011b). A further decisive factor has been comparison to other groups of oceanic volcanic islands, such as the Hawaiian Islands (Carracedo, 1999) or the Cape Verdes (Carracedo et al., 2015b), which share close resemblances due to their common origin from mantle hot spots. In particular, stages of island formation and structural features are similar, including large rift structures and associated volcanism and giant lateral landslides. In this respect, the benefit has been double: just as the accumulation of knowledge of this type of islands in the Pacific Ocean has served to better understand the Atlantic islands, so has the geological research and advancement of knowledge of the Canaries been helping in acquiring a more universal picture of ocean island volcanism, where the Canaries contributed significantly to the understanding of processes recorded in volcanic ocean islands around the world.

The spectacular increase in information regarding the geology of the Canary Islands has also led to an increase in the quality of the applied concepts. The most popular one at present is that of a fixed magmatic source or "hot spot" capable of sending magma to the surface through a slow-moving plate above. In this geological setting, a single pattern of island formation is systematically repeated and differences arise due to their different age and associated state of evolution.

THE PRE-CANARIES GEOLOGICAL SETTING; CREATION OF OCEANIC CRUST AND THE OPENING OF THE CENTRAL ATLANTIC

OCEANIC CRUST IN THE CANARY REGION

For an oceanic archipelago to form there must be an ocean, but, at the end of the Triassic, 200 Ma ago, the Atlantic Ocean did not yet exist (Fig. 1.3A). In fact, there was only one single continent (Pangea) surrounded by an all-encompassing gigantic ocean (Panthalassa). At that time, in the central part of Pangea, in the area where the Canaries would develop later, a vast tholeiitic flood basalt province became active, extending over more than 10 million km^2 (Fig. 1.3A). This large igneous province (LIP), known as the Central Atlantic Magmatic Province (CAMP), evolved into a mid-oceanic rift that first produced ocean crust in the Early Jurassic. This process continued ever since (Fig. 1.3B), widening the Atlantic steadily to form the present-day Atlantic Ocean (see, eg, McHone, 2000).

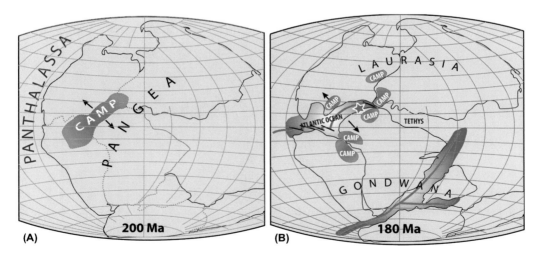

FIGURE 1.3

(A) Prior to the breakup of the largest and most recent supercontinent, Pangea, volcanism formed a large igneous province, the Central Atlantic Magmatic Province (CAMP). This volcanic region surrounded the initial Pangean rift zone over an area of approximately 10 million km^2. Recent high-quality radiometric dates point toward a brief time period for most of the CAMP magmatism, likely less than 2 Ma, at approximately 200 Ma. (B) Initial stages of the opening of the Central Atlantic. The area where the Canaries will develop approximately 150 million years later is indicated with a red star (see text for details).

The formation of the Canary Islands and the magmatic activity of the Mid-Atlantic ridge appear completely unrelated, because the composition of their respective basaltic materials differs fundamentally. Moreover, the crust on which the present Canaries reside is oceanic and approximately 150−170 million years old (which constitutes some of the oldest oceanic crust on the planet), and it formed at least 120 Ma before the onset of Canary Island volcanism during the early phase of the opening of the Atlantic Ocean (Fig. 1.3B). The present-day Canary Islands began to form on this oceanic crust, but only when it was already cold and dense and covered with a thick sedimentary sequence, up to 10 km near Africa but only 1 km or less below El Hierro (Bosshard and MacFarlane, 1970; Martínez del Olmo and Buitrago, 2002).

Although at present there is a general consensus that the entire alignment of the Canaries developed on oceanic crust, earlier interpretations postulated a basement of continental crust for the islands, suggesting a block of the African continent detached in the initial stage of continental rifting (Dietz and Sproll, 1979) (Figs. 1.4 and 1.5). Rothe and Schmincke (1968), for example, suggested that at least Fuerteventura and Lanzarote were underlain by continental crust. Additional evidence favoring a continental origin for the Canaries was indicated by Rothe (1974) on the basis of the discovery of an egg of a "nonflying" fossil ostrich found within Late Tertiary sediments intercalated in a basaltic sequence on Lanzarote. This was thought to confirm the existence of a land bridge with the African continent. However, newer observations suggest that these likely were large oceanic birds, probably resembling albatrosses, except that they were much larger than any modern albatross (García Talavera, 1990) (see chapter: The Geology of Lanzarote), thus removing the need for a "paleontological" land bridge.

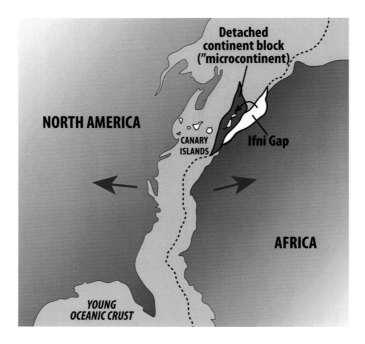

FIGURE 1.4.

Dietz and Sproll (1979) interpreted a gap (the "Ifni gap") in the fit between Africa and North America as a continental block that detached and broke loose during the initial stage of continental rifting. It was initially proposed that continental crust forms the basement to the Canary Islands, especially the eastern ones.

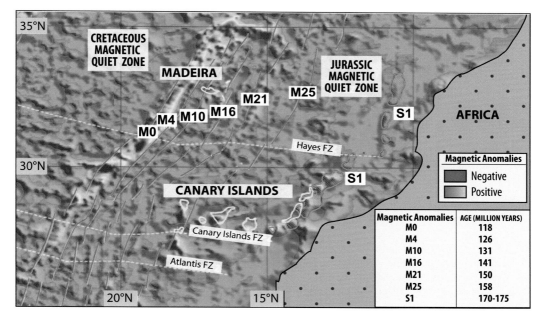

FIGURE 1.5

Magnetic anomaly map of the Eastern Central Atlantic (courtesy of the US National Geophysical Data Center, 1993). The Canary Islands lie on oceanic crust created between anomalies S1 (175 Ma) and M25 (158 Ma), indicating that the drift between Africa and North America was initiated approximately 180 million years ago. The polarity of the anomalies and their ages are shown in the inset. The islands of the Canary Volcanic Province themselves were created much later by an upwelling mantle plume at 65 Ma or less, a process that is effectively independent of the dynamics of the oceanic crust on which the islands rest.

A corollary of the described geological setting is the probable existence of oil off the eastern Canary Islands, in the basin between the eastern islands and the coast of Morocco, where up to 10 km of sediments have accumulated (Fig. 1.6). Portions of these were deposited in a very shallow sea at the initial stages of rifting; thus, they are likely to contain organic components. After decades of exploration in the area, the phase of prospecting and drilling has recently begun. The topic forms the core of a heated public debate on the islands at the time of writing, and likely for many more years to come.

THE CANARY VOLCANIC PROVINCE
SEAMOUNTS

The early history of the Central Atlantic Ocean basin is receiving new interest in the form of recent ^{40}Ar/^{39}Ar dates of seamounts scattered in the Central Atlantic (eg, south of the Canaries) (Fig. 1.7A). These older submarine cones and ridges of Cretaceous ages include Henry seamount

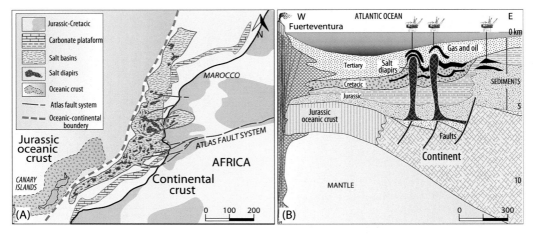

FIGURE 1.6

(A) Tectonic reconstruction of the African continental margin in the area of the Canary Islands during the initial stages of the opening of the Central Atlantic. (B) Cross-section showing the sources of shallow oil and gas deposits between the Canary Islands and Morocco, which are likely caused by salt and evaporate diapirs (after Davidson, 2005).

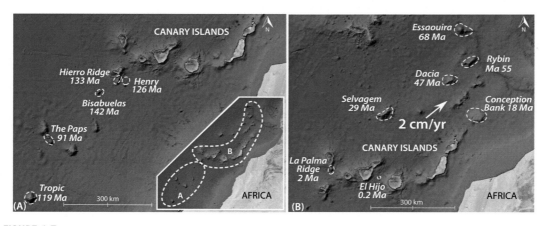

FIGURE 1.7

(A) Seamounts located southwest of the Canary archipelago were recently dated by van den Bogaard (2013) and give apparent ages from 91 to 142 Ma (images from Google Earth). The Canary Islands and their associated seamounts extend to the northeast and show different ages with a systematic distribution over the past approximately 65 Ma (farther away=older). In contrast, the Cretaceous seamounts to the southwest are scattered with respect to their ages (B) and likely follow an ancient oceanic fracture, which would also explain their seemingly random age distribution (see eg, Feraud et al., 1980).

and part of the southern ridge of El Hierro (Klügel et al., 2011; van den Bogaard, 2013). These much older seamounts do not directly belong to the Canary Volcanic Province because their ages are scattered and their broad trend is at an angle to that of the Canary Islands (Fig. 1.7A) but parallels the magnetic ocean floor anomaly M25 (Fig. 1.5).

There are many volcanoes in the Central Atlantic that remain hidden under the sea surface. These so-called seamounts are older islands that have been eroded back to below the sea surface, or cones that were arrested during their growth and never reached the sea surface, or young volcanoes currently in an active phase of growth, where only future activity will decide whether they will emerge. In this geodynamic context, the Cretaceous alkaline seamounts of van den Bogaard (2013) and Klügel et al. (2011) seem unrelated to CAMP tholeiitic volcanism and, most likely, they also developed independently of the seamounts and islands that form the present-day Canarian Volcanic Province (CVP).

For example, the cluster of seamounts southwest of the archipelago (Fig. 1.7A) were dated by van den Bogaard (2013) and found to range from 91 to 142 million years in age. While the seamounts to the northeast of the Canary Islands show a systematic distribution of their ages (those farther away being older), the seamounts to the southwest are scattered with respect to their age distribution, characteristic of fracture-controlled volcanism. So far, no isotopic fingerprinting has been performed on samples from these seamounts; therefore, it remains unclear for the moment whether these Cretaceous seamounts represent an early magmatic manifestation of the Canary Islands. However, an assumption that volcanic seamounts and islands that are located in the greater area of the Canaries must have originated from the same source through the past 140 Ma is likely oversimplified. It may well be that an earlier episode of magmatism produced older seamounts in this area, whereas later Canary Island magmatism was then superimposed (eg, Zazcek et al., 2015).

In the Canaries, there exist at least five large seamounts that are located northeast of the archipelago and are complemented by many smaller ones in between (Figs. 1.7B and 1.8). In this cluster of seamounts, those that are farther away from the archipelago are systematically older than the closer ones, up to an age of 68 Ma for the Lars/Essaouira seamount (Geldmacher et al., 2005; van den Bogaard, 2013). Isotopic fingerprinting, a technique used for identifying the original source of a magmatic rock in the Earth's mantle, yields very similar data for the older seamounts along the Canary trend (to the North) to the fingerprints of the rocks from the emerged Canary Islands (Geldmacher et al., 2005). This means that these particular seamounts have most likely originated from the same source in the Earth's mantle that feeds the Canary Islands today. As a result, they are thought to be a part of the CVP, along with the Canary Islands themselves.

Two seamounts located in and around the Canary Islands are repeatedly reported as being young and active, and thus are potential candidates for the formation of future subaerial islands. The first one is *Las Hijas* (The Daughters) to the southwest of El Hierro (Fig. 1.9A). Rihm et al. (1998) found only little sediment deposited on the surface of this seamount and therefore interpreted Las Hijas to have a young age. In contrast, a trachyte sample dredged from the flanks of Las Hijas gave a radiometric age of 133 Ma (van den Bogaard, 2013), which prompted the author to suggest a new name, *Las Bisabuelas* (the great-grandmothers), for this seamount. Because sampling at great depths is difficult, the origin of *Las Hijas* is not well constrained at the moment. It may well be that both studies complement each other, with one showing a structure continuously resurfaced by new activity (hence, little sediment cover), whereas the other happened to sample very old portions

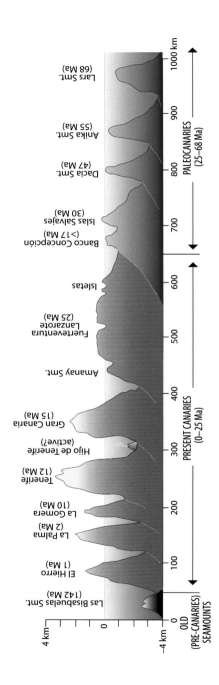

FIGURE 1.8

Schematic diagram showing the chain of islands and seamounts that form the Canary Volcanic Province (CVP). The ages of the oldest available rocks from each site show a systematic increase from El Hierro toward the northeast. Compiled after Geldmacher et al. (2001), Guillou et al. (2004a), and Zaczek et al. (2015).

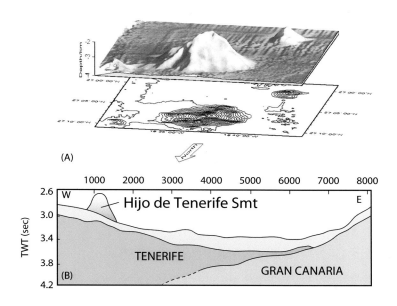

FIGURE 1.9

(A) Three-dimensional shaded image of the main *Las Hijas* seamount viewed from the north-northeast, derived from a combination of bathymetric data with shading from GLORIA sidescan. This small group of seamounts, located 70 km southwest of El Hierro, was named *Las Hijas* (the daughters) because it was suspected to represent the next upcoming island of the archipelago (Rihm et al., 1998). However, the seamount group was dated by van den Bogaard (2013) at 133 Ma, and so they range among the oldest seamounts in the Atlantic Ocean, and van den Bogaard suggested the alternative name *Bisabuelas* (the great-grandmothers) as a new name for this seamount group (image courtesy of S. Krastel). (B) Seismic line P117 along the channel between Gran Canaria and Tenerife. The younger volcanic flank of Tenerife onlaps the older and steeper flank of Gran Canaria. *Hijo de Tenerife* seamount is growing onto these overlapping aprons. No indication of a major oceanic fault is visible in this profile (after Krastel and Schmincke, 2002b), whereas seismicity is very active in this area, suggesting that *Hijo de Tenerife* is an actively growing submarine volcano.

from the same edifice or, more likely, a much older edifice located in the same area (as also seen at El Hierro for example). To assume that the dated formations belong to an active volcano, in turn, would require an abnormally long life for any volcanic edifice.

The more frequently mentioned other seamount is *El Hijo de Tenerife* (son of Tenerife), which is located between Tenerife and Gran Canaria (Fig. 1.9B) and dated at 0.2 Ma (van den Bogaard, 2013).

Whether either of the two seamounts will break the sea surface to form an eighth Canary Island is unclear. Using previously determined growth rates of, for example, the shield stage of Gran Canaria (Schmincke and Sumita, 1998) and assuming these apply to *Las Hijas* as well, the seamount may be able to form a new subaerial island within the next 500,000 years (Rihm et al., 1998).

Age of the islands

Although there are distinct processes involved when comparing submarine eruptions to subaerial ones, by and large, eruptions produce material that is deposited around an eruptive vent, thereby adding layers on layers of material onto older eruptive products and, by doing so, increasingly adding height to a volcano. Mature islands are usually the result of several smaller volcanoes or volcanic episodes superimposed on one another in space and time.

The progressive west-to-east age increase in the archipelago has been determined by means of radioisotopic and paleomagnetic dating of the oldest emerged volcanic rocks (Fig. 1.10A). Marine geology studies have shown that each island is surrounded by loose fragmented volcanic material produced in slumps and landslides derived from the flanks of the insular edifices that form extensive aprons around the islands. These aprons are progressively onlapping onto each other in a northeast-southwest direction (Fig. 1.10B), unequivocally corroborating the age progression in the archipelago determined by radiometric dating and implying that a hot spot has generated the islands (eg, Carracedo et al., 1998; Urgelés et al., 1998).

GENESIS OF THE CANARY VOLCANIC PROVINCES

Volcanism in this 800×400 km volcanic belt progressively decreased in age from the 68-million-year-old Lars Seamount in the Northeast to the 1.2-million-year-old island of El Hierro in the Southeast (Fig. 1.10A). This sequence is widely interpreted as originating from a hot spot (Carracedo et al., 1998; Carracedo, 1999; Geldmacher et al., 2005).

The CVP rests on oceanic crust of Jurassic age (150–170 million years) that formed during the initial stages of the opening of the Central Atlantic and that represents some of the oldest crusts in the oceanic basins of the globe with magnetic anomalies that parallel the continental margin. The Canary alignment, in turn, follows a trend parallel to that of the Volcanic Province of Madeira (Fig. 1.11). Although the hot spots are likely unrelated because their different isotopic compositions indicate independent magmatic sources, the parallel island chains demonstrate the relative movement of the African Plate during this time interval. The parallel curved path of both volcanic alignments is notably unrelated to the Atlantic fracture zones, and the simultaneous volcanism over the past 70 million years, coupled with a similar age progression for both archipelagos, can only be adequately explained by the hot spot or fixed mantle plume model (eg, Troll et al., 2015).

THE CANARIAN HOT SPOT

A wide variety of models have been proposed for the origin of the Canary Islands, including fractures that give rise to volcanism on cutting through the lithosphere, generating fusion by decompression, such as a potential fracture propagating from the Atlas fault (Anguita and Hernán, 1975, 2000). The propagating fracture model opposes the hot spot model and has been maintained by some scientists for a considerable time. However, the fracture has never been located, even with modern detailed geophysical studies (eg, due to extensive hydrocarbon prospection). It is therefore compelling that the parallel trend and the coinciding ages of the

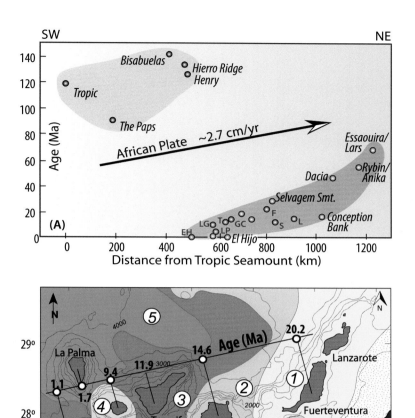

FIGURE 1.10

(A) Distribution of ages versus distance from Tropic Seamount along a southwest-northeast transect through the Cretaceous seamount group (black dots), the Canary Islands (red), and the Canary seamounts (blue). The arrow indicates the direction and average movement of the African plate during the past 100 Ma (modified after van den Bogaard, 2013). (B) Progressive east-west age variation of oldest exposed rocks in the Canary Islands is commonly interpreted to reflect a mantle plume as the underlying reason for the island's volcanism. Ages are after Guillou et al. (2004a,b), and aprons (in colour) are modified after Urgelés et al. (1998).

volcanic provinces of Madeira and the Canaries likely bear little connection to the Atlas Mountains. In fact, the most northeastern edge of the Canarian alignment commences more than 300 km northwest of the Atlas range and within oceanic, not continental, crust (eg, Geldmacher et al., 2005).

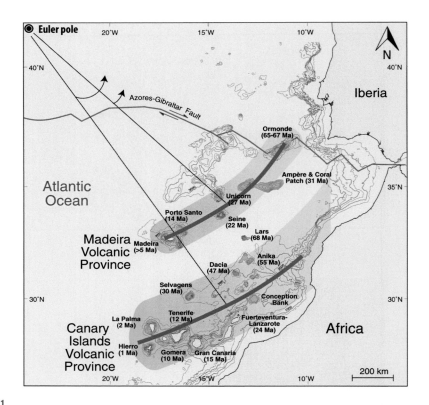

FIGURE 1.11

The Canary and Madeira Volcanic Provinces, including the main islands and the associated seamounts. The two volcanic chains follow a parallel and curved trend as the islands formed approximately at the same time and at a similar rate, consistent with the displacement of the Africa plate above two stationary melting anomalies. Modified after Geldmacher et al. (2005).

However, fractures cutting through such an old (Mesozoic) section of the lithosphere could not generate significant volumes of magma by decompression alone (eg, McKenzie and Bickle, 1988). By contrast, the hot spot or mantle-plume model (Wilson, 1963) is independent of the lithosphere and volcanism is generated as a consequence of the existence of a fixed and long-living thermal mantle anomaly. As Wilson reasoned, oceanic island alignments originate by relatively focused, long-lasting, and exceptionally hot mantle regions called hot spots or mantle plumes that provide localized volcanism. Wilson suggested that once an island has formed, continuing plate movement eventually carries the island beyond the "hot spot," thus cutting it off from the magma source, causing volcanism to cease. As one island volcano becomes detached from the magma source, another develops over the still active hot spot, and the cycle is repeated. Therefore, an age progression of the successive islands along the chain is critical evidence for a fixed-plume origin (Fig. 1.12).

Although several aspects of this issue have yet to be resolved, the model relating the genesis of the Canary Islands to a hot spot or fixed mantle plume is also important to fully explain the erosion

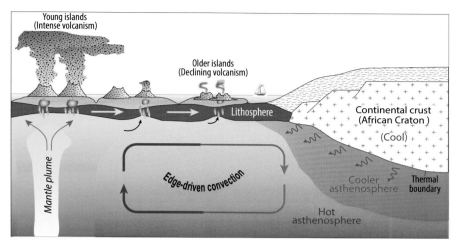

FIGURE 1.12

A mantle plume can explain the linear younging direction along a northeast-southwest—oriented path for the Canary Islands (Carracedo et al., 1998), although the conventional hot spot model cannot readily explain the occurrence of recent volcanism in Lanzarote, opposite to the inferred location of the present hot spot. A possible explanation may be the small-scale upper mantle convection at the edge of the African craton that is interacting with the Canary mantle plume, which may lead to local eruptive anomalies. Synthesized after Carracedo (1999), Geldmacher et al. (2005), King (2007), Gurenko et al. (2010).

levels and spatial distribution of the Canary Islands that can only be reasonably explained with a hot spot type of model for the archipelago (eg, Carracedo et al., 1998).

Arguments against a simple Hawaiian-type hot spot model, such as the long volcanic history of the islands and the fact that volcanism persists even in the older islands of the archipelago (eg, Lanzarote), have recently been explained by edge-driven mantle convection (King, 2007; Gurenko et al., 2010), which creates a contact of hot asthenosphere with colder passive sub-continental mantle domains, in this case from the African craton. The convection cells generated would move hot and rising plume material toward the east and northeast also, thus reaching Lanzarote and hence producing the sporadic eruptions that have taken place in the Eastern Canaries in recent times. Moreover, Hoernle and Schmincke (1993) proposed a "blob model" for the Canarian hot spot. According to these authors, the multicycle evolution of island volcanism and the temporal variations in chemistry and melt production within each cycle represent dynamic decompression melting of discrete mantle "blobs" of plume material beneath each island.

New finite-frequency tomographic images from seismic wave velocities now confirm the existence of deep mantle plumes below a large number of known island clusters and chains, including the Canaries (Fig. 1.13). The three Macaronesian plumes (Canaries, Azores, and Cape Verde) are robust deep mantle features appearing as isolated anomalies down to >1000 km depth, and thus they are likely sourced from the very deep mantle off the coast of Africa (Montelli et al., 2004, 2006).

Notably, additional evidence in favor of a mantle plume comes from calcareous nannofossils recently recovered from xeno-pumice erupted during the 2011 submarine events off El Hierro

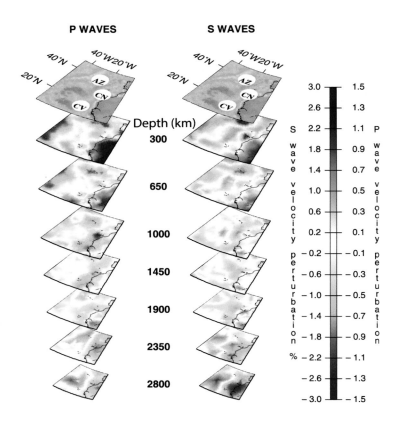

FIGURE 1.13

Three-dimensional view of the melting anomalies (plumes) beneath the AZ (Azores), CN (Canary), and CV (Cape Verde) archipelagos in both (left) P-wave and (right) S-wave tomographic models (from Montelli et al., 2004). Note the three anomalies are traceable down to the core−mantle boundary.

(see chapter: The Geology of El Hierro). These nannofossils define the sub-island sedimentary rocks under El Hierro as Cretaceous to Pliocene in age. These pre-El Hierro sedimentary rocks reach to substantially younger ages than the Miocene sedimentary strata under the older eastern islands (Fig. 1.14), and therefore support an age progression among the islands and hence a mantle-plume as the most probable driver for Canary volcanism (Carracedo et al., 2015a; Troll et al., 2015; Zazcek et al., 2015).

VOLCANIC GROWTH VERSUS GRAVITATIONAL SUBSIDENCE IN THE CANARY ISLANDS

Wilson (1963) defined his hot spot model for the Hawaiian-Emperor volcanic chain, where the islands that move off the hot spot begin to subside to become seamounts (guyots). The flexible

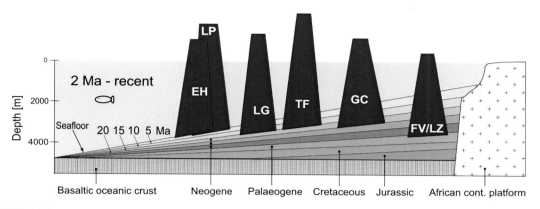

FIGURE 1.14

Schematic cross-section through the Canary archipelago and the African continental margin (thicknesses of sedimentary layers not to scale). Nannofossils in El Hierro eruptives now demonstrate, in agreement with available radiometric ages of the oldest subaerial lavas, that progressively younger pre-volcanic sediments are present in the west of the archipelago, which supports the previously established onshore age progression and thus provides further evidence in favor of the mantle plume hypothesis (from Zaczek et al., 2015).

oceanic crust and the large volume (weight) of the islands causes them to subside into the substrate. For example, Mauna Loa is estimated to comprise approximately 80,000 km^3 of basalt, a mass so great that it will depress the underlying and comparatively young Pacific crust. Thus, volcanic growth due to lava accumulation in the Hawaiian Islands is compensated by subsidence and landsliding. This process accounts for the fact that the oldest island of the 600-km-long Hawaiian archipelago (Kauai) is only 5.1 million years old, whereas Fuerteventura is more than 20 Ma. Most of the subsidence in the Hawaiian Islands is due to an increase in density of the rocks as they cool and the concentration of high-density material at their depth, and because the island edifice moves off the swell caused by the hot spot, and therefore the plume's dynamic support comes to an end.

One crucial factor in the Canaries is the velocity of plate motion, which is considerably higher for the Pacific (approximately 7 cm/year) than for the Atlantic (approximately 2 cm/year) plate. These velocities are related to plate movement around a pole of rotation (Euler pole), which is approximately 13,000 km from the Hawaiian islands and within the Pacific plate, but only 3,800 km from the Canaries, and located on the Atlantic plate (eg, Troll et al., 2015).

For example, in Hawaii, islands are elevated by approximately 1000 m during their active volcanic period. This swell is caused by the buoyancy of the less dense material of the hot spot underneath the oceanic crust. Additionally, and importantly, the oceanic crust underneath the Hawaiian archipelago is approximately 95 Ma and, thus, younger than the ocean crust beneath the eastern Canaries (>170 Ma). Furthermore, it is also hotter than the old oceanic crust of Jurassic age underneath the Canaries. The relatively young oceanic crust under Hawaii is thus much thinner and more flexible than the crust under the Canaries and flexes approximately

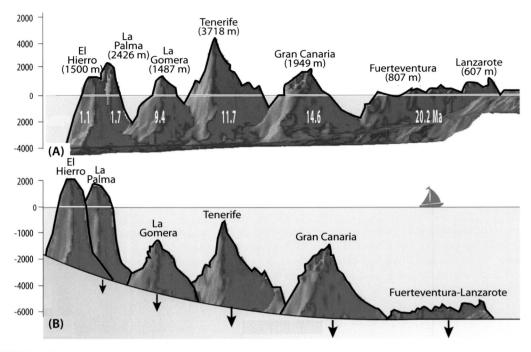

FIGURE 1.15

(A) West-East profile showing the present Canary Islands along the African continental margin. (B) Simulated position of the Canary Islands if the subsidence rates in the archipelago were similar to that of the Hawaiian Islands (based on Carracedo, 2011). Likely explanations for the differences between the Canaries and Hawaii may be the distance to a continental margin and the age and thickness of the underlying oceanic crust.

10-times more than the sub-Canary crust. For this reason, the Canary Islands remain emerged for a much longer period (>20 Ma) because the oceanic crust on which the Canary Islands rest has a thickness of approximately 7 km for the igneous part, that is, the part that formed at the mid-ocean ridge in Jurassic to Cretaceous times. Over millions of years, sedimentation, mostly from the continental margin of Africa, deposited large thicknesses of sediments onto this basaltic oceanic crust. Thus, the Canary Islands formed over crust of up to 18 km in thickness close to the African coast, but only 8 km or less at the western end of the archipelago (see Fig. 1.6B). If the Canaries were on an oceanic crust similar to that of the Hawaiian Islands and had a comparable rate of plate movement, then all the islands except La Palma and El Hierro would likely be sunken seamounts by now, even when accounting for the lower volume and weight of the Canarian archipelago relative to Hawaii (Fig. 1.15).

Two other factors cause the older Canary Islands to stay emerged longer than islands in other oceanic archipelagos. Besides its great thickness, the crust underneath the Canaries is also significantly more rigid than under other ocean islands. The oceanic crust underneath the Canaries ranges among the oldest worldwide and, moreover, the proximity to the African continent implies that the oceanic crust underneath the Canaries is likely laterally attached to buoyant continental crust. Continental crust has an average density of approximately 2.65 g/cm^3, which is much lower than the dense basaltic oceanic crust (2.8−3 g/cm^3). Continental crust is thus more buoyant than the denser oceanic crust and, in this sense, the continental margin of Africa potentially acts as a buoyancy anchor for the Jurassic oceanic crust attached to it.

This absence of significant subsidence leaves the typical Canary Island exposed to erosion above sea level for much longer periods than in archipelagos of comparable origin, such as Hawaii. Because of this much longer period of emergence, it is possible to reconstruct the evolution of a Canary island in much greater detail and much older stages. For example, because of its age, Fuerteventura has been eroded to much deeper levels, exposing rock sequences in the very heart of the volcanic island, which sheds light on processes of island formation (eg, Javoy et al., 1986; Le Bas et al., 1986; Hobson et al., 1998; Stillman, 1999; Allibon et al., 2011). This cannot be observed in archipelagos where subsidence is much more pronounced.

Finally, the very long period of active volcanism and low rates of magma production likely allow for magmatic evolution toward more differentiated compositions, explaining why evolved magmas (trachytes, phonolites, rhyolites) abound in the Canaries relative to, for example, Hawaii, and a variety of crustal differentiation processes are likely at work to cause felsic magma generation in the Canaries (Schmincke, 1969; Stillman et al., 1975; Freundt and Schmincke, 1995; Wolff et al., 2000; Klügel et al., 2000; Troll and Schmincke, 2002; Wiesmaier et al., 2012).

ISLAND GROWTH STAGES IN THE CANARIES

The youngest islands (La Palma and El Hierro) appear to be in their juvenile stage, in which submarine and subsequently subaerial basaltic eruptions take place relatively frequently in a geological sense, forming the greater part of the volume of the islands in only a few million years (Fig. 1.16A). Volcanism usually subsides at the end of the shield stage and the islands begin a period of eruptive repose, which can last for a few million years (Fig. 1.16B). All the Canary Islands have passed through this cycle with the exception of La Palma and El Hierro, which will probably reach this stage in the geological future. The evolution of an island then progresses to a stage of posterosive rejuvenation, the duration of which may be several million years again (Fig. 1.16C). During this stage, the eruptions are spaced further apart in time than during the shield stage and the volcanic emission rates are much lower. Also, rejuvenation activity tends to generate either highly alkaline magmas (eg, on Gran Canaria) or more evolved ones (more siliceous and viscous). The latter phenomenon can give rise to an increase in height of silicic volcanoes as, for instance, seen at Mount Teide on Tenerife.

These main stages in the development of the Canaries are very similar to those of oceanic volcanoes in general and were first defined in the Hawaiian Islands (ie, Walker, 1990, 1999). For the Canaries, Fúster and coworkers (1968a−d) initiated the first modern and comprehensive geological study on the islands, commencing with Lanzarote and Fuerteventura, the oldest, posterosional islands in the East of the archipelago. There, the authors defined "Old" and "Recent" Series. However, the application of these volcano-stratigraphic units proved unfeasible in the western

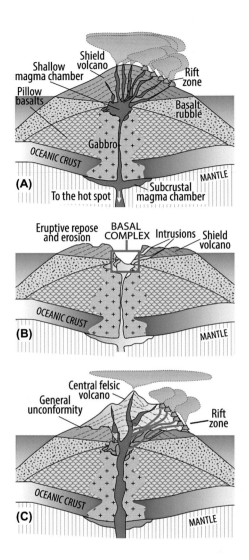

FIGURE 1.16

Cartoon illustrating the main stages of growth of a Canary island. (A) A short but high-productivity basaltic shield stage is followed by (B) a phase of eruptive repose, in which erosion may exhume the submarine part of the edifice and (C) a long posterosive stage, which is interrupted by sporadic pulses of small-scale volcanic rejuvenations that can produce felsic and more explosive volcanism in the form of stratovolcano-type edifices.

Canaries, that is, the "Old" Series of La Palma or El Hierro were found to be considerably younger than some units of the "Recent" Series of Fuerteventura, Lanzarote, or Gran Canaria. This problem was resolved once the main growth stages of oceanic volcanoes, as defined in the Hawaiian Islands, were applied to the Canaries also (see, eg, Carracedo et al., 1998 and references therein).

THE SEAMOUNT STAGE

Note that seamounts are not only very young volcanoes that are growing to reach the sea surface but also old ones, which either never made it above the sea surface or have already been eroded back or sunken to below sea level (Fig. 1.17). However, for the description of island evolution, the term seamount is used here to represent a young and growing submarine volcano.

The seamount stage combines all magmatic activity that occurs before the growing submarine island breaks the surface of the sea (see chapters: The Geology of La Palma and The Geology of Fuerteventura). From erupting the first magmatic material onto the seafloor to reaching the sea surface, a seamount may reach a vertical height similar to large stratovolcanoes on land. In the Canaries, the seafloor ranges from a depth of approximately 1000 m for the easternmost islands, Lanzarote and Fuerteventura, to approximately 3500 m for the westernmost islands, La Palma and El Hierro (Canales and Dañobeitia, 1998). By volume, rocks from the seamount stage thus represent the bulk of any volcanic island, even though these rocks remain largely inaccessible (Fig. 1.18).

The seamount stage, usually covered by subsequent volcanism, only crops out on Fuerteventura and La Palma, two islands that have been deeply eroded through major incisions caused by giant landslides and subsequent erosion. On the island of La Palma, one has the rare opportunity to inspect these rocks in the outcrop, because the seamount has been exposed in the Caldera de Taburiente, where the uplifted and tilted submarine volcano reaches 1500 m above sea level (Fig. 1.19). Here, a sequence of hyaloclastic rocks and pillow lavas is exposed, together with a

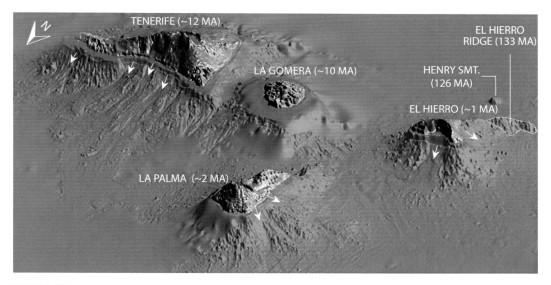

FIGURE 1.17

Bathymetry and topography of La Palma, La Gomera, and El Hierro viewed from the north. The main landslides are indicated with arrows in this shaded relief image (from Masson et al., 2002). Note the large volumes of the island edifices beneath sea level and the extensive sedimentary aprons to all the islands.

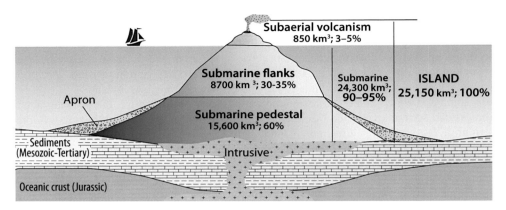

FIGURE 1.18

Estimates of the volumes of the different parts of an ocean island (using Gran Canaria as an example). Note the small subaerial island volume compared with the considerably more voluminous submarine part (modified after Schmincke and Sumita, 1998).

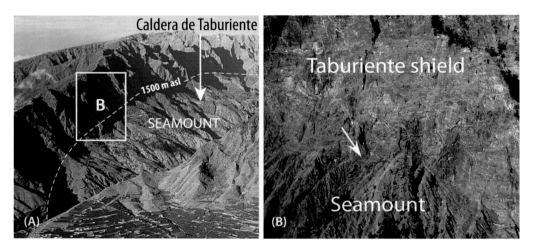

FIGURE 1.19

(A) View of the western wall of the Caldera de Taburiente showing the contact of the uplifted and tilted submarine volcano that reaches 1500 m above sea level. The submarine complex is overlain by the lava flows of the Taburiente shield. The square indicates the enlargement in (B), which shows a close-up of the unconformity (white arrow).

high concentration of mafic and felsic plutonics and dykes (Staudigel and Schmincke, 1984). The majority of material erupted in the seamount stage is of mafic composition, but occasionally trachytes and syenites do occur, as is also evident in the "Basal Complex" of Fuerteventura (eg, Stillman et al., 1975; Javoy et al., 1986; Hobson et al., 1998; Holloway et al., 2008).

CONSTRUCTIVE PROCESSES THAT BUILD VOLCANOES: POSITIVE VOLCANIC LANDFORMS

Volcanism constructs two main types of volcanoes, those that result from a single period of activity during their history and will not erupt again (eg, Teneguía volcano), and those that are built from many eruptive phases separated by periods of repose (eg, Teide volcano). While the former type is referred to as "monogenetic," the latter type is usually labeled "polygenetic."

POLYGENETIC VOLCANOES

In this category we include volcanic shields, composite volcanoes, and stratovolcanoes, and also the long-lived rift zones in the Canary Islands.

Volcanic shields

The shield stage represents the major phase of submarine and subaerial volcanic activity in a volcanic island. Currently, two islands in the Canaries are thought to be in the shield stage, La Palma and El Hierro. Both feature a single shield volcano on first glance; however, in terms of island evolution, we must recognize that the shield stage of both islands comprises more than one "volcanic system."

As can be seen in many outcrops of shield volcanism across the Canary Islands, this stage of a volcanic island formation is mainly composed of basaltic lava flows and dykes. The lavas are stacked on top of one another and usually dip gently toward the sea (Fig. 1.20). The eruptive behavior of shield stages can thus be reconstructed to be predominantly effusive, and explosive eruptions tend to be rare throughout this stage. This is consistent with observations from other active shield volcanoes on Earth, for example, Mauna Loa on Big Island, Hawaii. There, a quasi-continuous supply of fresh magma leads to layer after layer of basaltic material deposited onto the surface.

The predominantly basaltic composition of the lava flows in a shield volcano notably influences the morphology of such volcanoes. Basaltic lava flows possess low viscosity and often high eruptive rates, and thus run off the slope of a volcano much more rapidly. The resulting lava flows

FIGURE 1.20

Comparison of the size and aspect ratio of two oceanic shield volcanoes: (A) Mauna Loa (Hawaiian Islands) and (B) La Gomera (Canary Islands).

FIGURE 1.21

Famara cliff, Lanzarote. The internal structure of shield volcanoes can be observed in places where erosion has deeply incised the volcano, or by inspection in galerías. Erosional valleys and high cliffs permit excellent sections of the inner structure of such island shields that are frequently formed by repetitive stacks of lava flows and interbedded pyroclastics, as here in the cliff of Famara on Lanzarote.

form a comparably widespread and thin blanket (usually ≤1 m), not adding much thickness to the central part of the volcanic edifice. The name shield volcano results from this behavior; shield volcanoes are wide but have a low profile, resembling a warrior's shield (see, eg, chapter: The Geology of La Gomera) as much of the outpouring lava runs "off" the central high of the shield.

Notably, such a shield volcano is hence not a single volcanic cone, but rather a complex edifice that may include many individual and superimposed eruptive centers, fissures, and vents. Old shield volcanoes, such as the Teno massif on Tenerife, feature deeply incised erosional valleys that permit insight into the inner structure of such shield volcanoes. The walls of these valleys are commonly riddled with dykes that have cut through the stacked lava flows, testifying to the complexity of a shield volcano (Fig. 1.21). Again, this is consistent with the behavior of currently active shield volcanoes. To return to the previous example, the edifice of Mauna Loa is riddled with small cones and fissures across its flanks, each of which corresponds to a small eruption in the past and a feeder dyke or dyke complex at depth. A larger, central vent at the peak, often in the form of a caldera, is the result of one or several larger shallow magma pockets that emptied during individual eruptions, thus causing a collapse of the overlying roof rock in the summit region. Moreover, Mauna Loa illustrates why the shield stage comprises the major volume of material erupted above sea level on an island. Eruption rates are high there, with up to 6,000,000 m³/day, and in the period from 1832 to 1984, a total of 39 eruptions have been recorded (Decker et al., 1995). The total volume of the island is now estimated at 80,000 km³ of predominantly basaltic material.

For the shield stage of Gran Canaria, McDougall and Schmincke (1976) and van den Bogaard and Schmincke (1998) performed age determinations, which provided precise estimates of the age of Gran Canaria's shield stage (see chapter: The Geology of Gran Canaria). The data confirmed not only that the exposed Gran Canaria shield stage rocks built up in a geologically short time (less than 1 million years) but also that the shield volcanoes form the bulk of the erupted volume above sea level. Deposits, which were confirmed to have erupted much later, comprise only a fraction of what erupted during the subaerial phases of Gran Canaria. Similar relationships of initially rapid and voluminous construction were found in El Hierro, La Palma, and La Gomera; therefore, similar shield stages can be assumed to apply to the remaining Canary Islands also.

The cessation of the shield stage in an island may coincide with the beginning of a period of extended erosion. For example, the central shield on Tenerife entered a period of repose of approximately 4.5 million years until volcanic activity resumed in that part of the island. Contrary to the Tenerife shield, on Gran Canaria the shield stage did not end abruptly, but rather transgressed into a period of collapsed caldera volcanism with gradually declining eruption rates over approximately 4 to 5 million years.

STRATOVOLCANOES

Composite volcanoes or stratovolcanoes are the most common type of noted volcanic edifices, and they occur in all regions of volcanic activity and throughout the world. However, they are particularly frequent in subduction zones, for example, along the circum-Pacific belt of convergent plate margins. Conversely, composite volcanoes are less frequent at divergent plate margins, but they exist there, too (eg, Hekla on Iceland).

In intraplate settings, as in the Canaries, composite volcanoes are not uncommon and because of higher heat flux and magma differentiation processes within these volcanoes, composite intraplate volcanoes tend to be dominated by basaltic and intermediate eruptions.

Felsic stratovolcanoes (Fig. 1.22) are exceptional in intraplate oceanic islands, likely because rapid island subsidence commonly inhibits the development of long-term differentiation processes required. The apparent lack of subsidence and the long periods of volcanic activity in the Canaries (>10 Ma in the central and eastern islands) may account for the felsic stratovolcanoes that were constructed in these islands at different stages of their evolution, such as Teide volcano on Tenerife (Carracedo et al., 2007b), Roque Nublo on Gran Canaria (Pérez-Torrado et al., 1995), Vallehermoso volcano on La Gomera (Ancochea et al., 2003; Rodríguez-Losada et al., 2004; Paris et al., 2005b), and possibly a large and long-lost volcano on Fuerteventura (Stillman, 1999). In some cases in the Canaries, extreme differentiation led to very explosive (Plinian) caldera-forming eruptions, such as by the Las Cañadas volcano on Tenerife (see chapter: The Geology of Tenerife) or the Tejeda caldera on Gran Canaria (see chapter: The Geology of Gran Canaria).

The outstanding geological values of the Teide volcanic complex among the ocean volcanoes, particularly the features derived from its evolution to felsic compositions of the Ocean Island Alkali series, complement the basaltic volcanoes of Hawaii, which also belong, in part, to a

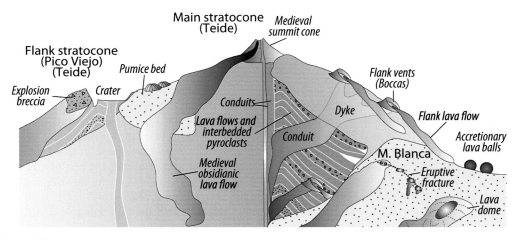

FIGURE 1.22

Anatomy of the most emblematic stratovolcano complex in the Canary Islands, the Teide and Pico Viejo volcanoes on Tenerife, and their associated volcanic features (after Carracedo and Troll, 2013).

different type of magma series, mainly the Tholeiitic magma series (OIT). These complementary differences were recently recognized through the successful World Heritage nomination of Mt. Teide in 2007 (see, eg, Carracedo, 2008a).

RIFT ZONES

Rift zones constitute some of the most prominent and persistent structures in the development of many oceanic volcanic islands. These structures control the construction of the insular edifices, possibly from the initial stages, and often form the main relief features that define the shape and topography of several of the islands. Rifts zones often concentrate eruptive vents (Fig. 1.23); therefore, they are crucial elements in the assessment of volcanic hazards in the Canaries. Finally, they frequently play a key role in the generation of flank collapses and the catastrophic disruption of well-established volcano plumbing systems, thus adding to the geometrical and compositional diversity of the Canary volcanic systems (eg, Carracedo et al., 2011a,b; Delcamp et al., 2010, 2012, Deegan et al., 2012).

Rift zones were initially recognized on the Hawaiian Islands (eg, Fiske and Jackson, 1972; Swanson et al., 1976; Walker, 1999), although a good part of the later progress made in understanding their genesis and structure has been achieved through their study in the Canaries (eg, Carracedo, 1994, Carracedo et al., 2007b, 2011a,b; Klügel et al., 1999; Walter and Schmincke, 2002; Walter and Troll, 2003; Delcamp et al., 2010, 2014; Deegan et al., 2012; Wiesmaier et al., 2011, 2012).

Compared with those of the Hawaiian Islands, the rifts of the Canaries are considerably longer-lasting. The lower magmatic activity of the mantle hot spot that has generated the Canaries probably resulted in much lower eruptive rates, favoring higher-aspect-ratio rift zones by accumulation

FIGURE 1.23

Satellite view of the Cumbre Vieja rift zone, La Palma. Note the concentration of the eruptive vents in the crest of the rift forming an arc that is likely caused by flank stresses of the steep ridge (image courtesy of NASA).

of relatively short flows, thus likely promoting the growth of prominent ridges as a main relief feature in the Canaries.

Rift zones can be inspected in El Hierro, Tenerife, and La Palma. Perhaps the best example of a rift zone is found on La Palma island, which consists mainly of a large, slightly older shield volcano in the North (Taburiente volcano), with a much younger rift zone onlapping to the South (the Cumbre Vieja volcano, Fig. 1.23, see also chapter: The Geology of La Palma). Historically, however, it was from gravity and seismic surveys by researchers such as MacFarlane and Ridley (1969) who detected conspicuous gravity ridges in the Bouguer anomaly map of Tenerife, which they interpreted as a high concentration of dykes that form a rift zone (Fig. 1.24). According to these workers, the growth of the island (both subaerial and submarine parts) is largely controlled by dyke injection along these three major rift zones, with angles of approximately 120° between them (see also Macdonald, 1972).

Plotting the vent distribution in the islands with the greater Quaternary activity (Tenerife, La Palma, and El Hierro), it is apparent that vents concentrate along such triple-armed rift zones (Fig. 1.25), which, together with observations from Hawaii (eg, Walker, 1999), led to the definition of rift zones in the Canaries as a linear clustering of surface vents fed by dyke swarms from the depth (Fig. 1.26). Walker argued that these structures may be an invariable characteristic of ocean volcanoes, and this appears to hold true, at least for the Canary Islands. There, significant advancements in oceanic rift research were made by taking advantage of the many hundreds of water tunnels on Tenerife and La Palma that are used for groundwater mining (locally called "*galerías*"). These tunnels facilitate access to the deeper structure of the rift zones, providing a unique opportunity for direct observations and sampling (Carracedo, 1994; Carracedo et al., 2007a,b, 2011a,b; Delcamp et al., 2010, 2012, 2014; Deegan et al., 2012).

To explain the long-lived concentration of dykes and vents, an initial model proposed by Luongo et al. (1991) postulated the opening of fracture systems along three branches, often

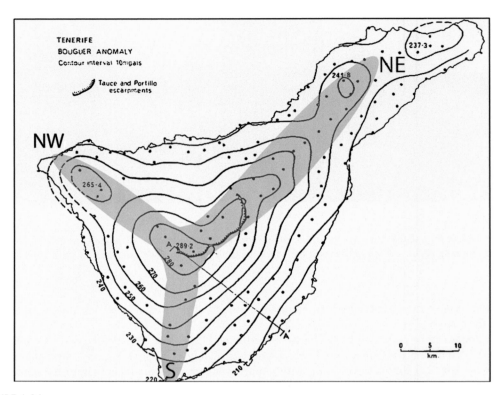

FIGURE 1.24

Original Bouguer anomaly map of Tenerife (from MacFarlane and Ridley, 1969) contoured at 10-mgal intervals showing a three-armed gravity anomaly, interpreted by the authors as underlying dyke complexes.

separated by a 120° angle (the least-effort geometry), in response to magma vertical upward loading (eg, by pushing of a plume or plume finger). This model was applied to the Canaries by Carracedo (1994) to explain the aligned concentration of eruptive sites, the longevity and direction of rift zones, and the genesis of volcano sector collapses (Fig. 1.27). In this model, the rift zones are thought to have initiated early in the history of the islands to formed their deep inner structure or backbone. However, some recent additions to this model appear necessary.

If triaxial rift zones developed simultaneously on particular islands (eg, Tenerife, Hawaii) the location of the centers of those rift systems should be sufficiently distant from one another considering the highly viscous relaxation behavior of the upper mantle and flexure wavelengths of the crust (Watts and Masson, 2001). If Tenerife shield volcanoes (Teno, Anaga, and Central shields) are thought to be triaxial structures, then they are probably located too close to one another to meet those conditions (Walter and Troll, 2003). Additionally, changes in rift orientation and the opening of new arms in mature edifices cannot be explained by deep mantle-induced fractures, but require an element of "edifice control" (eg, Walter and Troll, 2003; Walter et al., 2005). An alternative

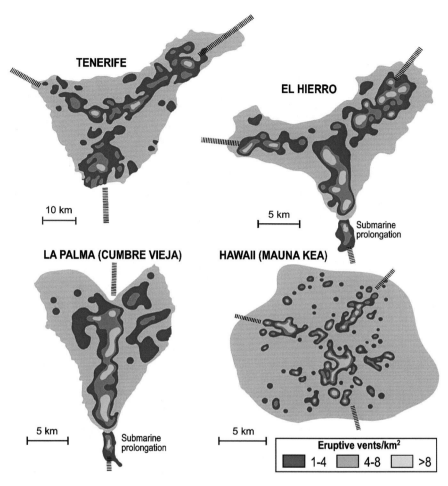

FIGURE 1.25

Concentration of Quaternary eruptive vents on Tenerife, El Hierro, the Cumbre Vieja on La Palma (after Carracedo, 1994), and on Mauna Kea, Hawaii (after Porter, 1972). Note that the Canary rift zones appear more clearly defined than the Hawaiian ones.

process proposed by these authors implies that flank deformation, once a volcano becomes sufficiently unstable, and dyke intrusions force the flanks of the volcano to spread and creep seaward (Walter and Troll, 2003; Walter et al., 2005; Delcamp et al., 2010, 2012), allowing for dynamic changes in the geometry of rifts, thereby eliminating the difficulty for superimposed and nearby volcanic rift systems (see Carracedo and Troll, 2013, for a more comprehensive discussion).

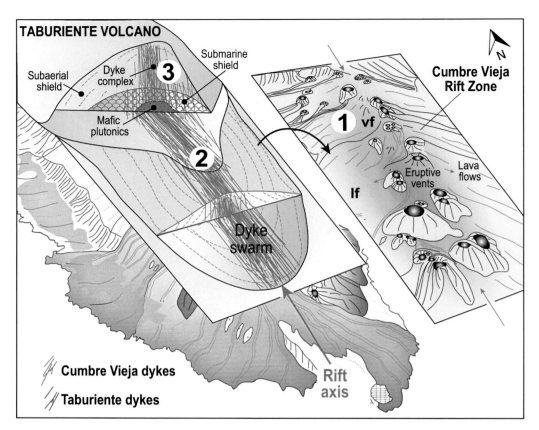

FIGURE 1.26

Anatomy of a characteristic Canary rift zone: the Cumbre Vieja, La Palma. The upper layer of the rift zone has been "detached" to show the internal structure of the rift: (1) Tight cluster of eruptive vents at the crest of the ridge; (2) Cumbre Vieja dyke swarm; and (3) dyke swarm of the late stages of the Taburiente shield volcano that dissect cumulate and plutonic rocks in the deeper part of the ridge structure (from Carracedo and Troll, 2013).

MONOGENETIC VOLCANOES
CINDER CONES

A characteristic of Canarian cinder cones is their frequent relationship with fissures and fissure eruptions in which vents often form alignments of several hundred meters or even kilometers in length (see chapters: The Geology of Lanzarote and The Geology of Tenerife).

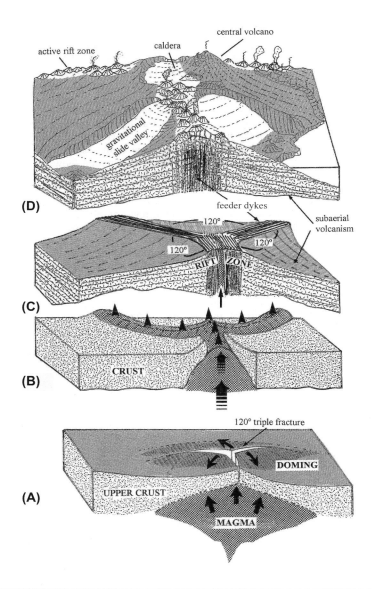

FIGURE 1.27

Original model proposed by Carracedo (1994) to account for the formation of ocean island rift zones. Plumes typically cause uplift and doming that eventually may rupture the rigid oceanic crust along a system of triple fractures, spaced at 120°. This seems to be the naturally preferred configuration and is thought to be a response to least-effort fracturing. Dykes increasingly concentrate along these fractures and feed the vents at the crest of the volcanic ridges that shape the topography of the islands. This model was also the first to connect the development of volcanic rift zones and large landslides in the Canary Islands.

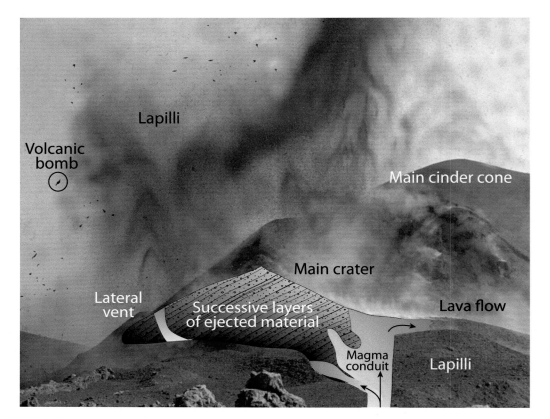

FIGURE 1.28

Volcanic products and schematic internal structure of a characteristic cinder cone (in this case, Teneguía volcano, 1971, La Palma).

Cinder cones are the most common volcanic landforms in the Canary Islands and in oceanic islands in general. They are relatively simple structures built of lapilli and scattered larger scoria fragments, bombs, and some lava flows (Fig. 1.28). Usually of basaltic composition, they construct the familiar, small, cone-shaped mountains with summit craters and are generally black, although high-temperature oxidation and alteration during and just after the eruption may cause the lapilli to develop a bright red to brown color (eg, by palagonization; Fig. 1.29). The inner structure of cinder cones can be inspected in many places in the Canaries, at times because they are partially eroded, but also because lapilli are traditionally quarried for construction and agricultural use (see, eg, Fig. 5.55).

In contrast to the polygenetic composite volcanoes that are usually long-living and that may grow very high (thousands of meters through many successive eruptions), monogenetic cinder cones are usually the result of single eruption episodes (eg, weeks to several months) and do not attain great heights (typically less than a few hundred meters).

FIGURE 1.29

El Corazoncito vent, Mñas. del Fuego, Lanzarote. Cinder cones in the Canaries are usually of basaltic composition, form the familiar cone-shaped small mountains with summit craters, and are generally black, although high-temperature oxidation during the eruption (or weathering) may give the lapilli a bright red to yellow color. These cones frequently form during a single eruption and are thus referred to as monogenetic cones.

The shape of cinder cones is controlled by the angle of repose, which is approximately $30-35°$ (ie, when a particle stops rolling down the flanks but remains in place). However, many factors can modify these simple proportions, such as temperature, particle size, explosivity, and interaction with water (alteration). Porter (1972) established a classification for cinder cones using their main morphometric parameters. Basal diameter (Wco), cone height (Hco), and crater diameter (Wcr) define a clear distinction between cones from strombolian eruptions and those corresponding to phreatomagmatic events, with the latter being generally shorter and considerably wider (see, eg, Figs. 3.40 and 5.20).

NEGATIVE VOLCANIC LANDFORMS

Apart from craters, which are generally small features, large negative landforms such as giant landslide depressions and calderas are also well represented in the Canaries.

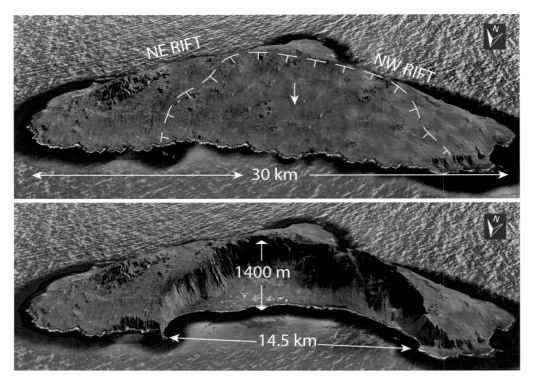

FIGURE 1.30

El Hierro island before and after the most recent giant landslide that formed the El Golfo embayment (images modified from Google Earth). Note the significant loss of mass that the island experienced, likely in a single catastrophic event (Manconi et al., 2009).

GIANT LANDSLIDES

In contrast to the slow and continuous processes of erosion, giant landslides are thought to happen almost instantaneously. Landslides may occur both in times of strong eruptive activity and during strong erosive phases; however, none has yet been witnessed in the Canaries. It is only from careful studies of island and seabed morphologies that scientists came to conclude that these big landslides are able to radically change an island's shape and volume (Fig. 1.30). They are assumed to do so in an extremely short time, possibly as little as a few hours (Ward and Day, 2001), although slow-motion creep may precede such a flank collapse for some time (eg, Walter et al., 2005; Delcamp et al., 2012).

Several causes for triggering giant landslides have been proposed and most are probably valid, but they may depend on the circumstances soon before an individual landslide occurs. For highly active volcanoes, continuous eruptions may oversteepen the slopes of a volcanic edifice to such a degree that the volcano becomes unstable under its own weight. Also, continuous injection of magma from below has been advocated to increase tension within the edifice and thus could destabilize it.

Repetitive injection of blade-like dykes will progressively increase the anisotropy of a volcano, forcing new dykes to wedge their path parallel to previous intrusions, like a knife between the pages of a book (see Fig. 5.23). If this process is sustained and if injections are sufficiently frequent, then parts of the rift zones may remain hot (thermal memory) and preferentially guide the path of even more successive intrusions (eg, Vogt and Smoot, 1984). However, intrusion can only progress in a dyke complex if the structure can accommodate fresh injections. Because repetitive intrusion would progressively increase compressive stresses, new injections can only occur if at least one flank of a rift zone is free to move apart (see Fig. 5.29). Therefore, extensional forces add up in growing rift zones and eventually reach a critical rupture threshold that can lead to massive landslides.

This would point to giant landslides as the only significant erosive force during times of high eruptive activity on an island. Note, however, that the extent of eruptive activity is not the only factor for a landslide to occur, and a generally valid correlation between highly active periods and giant landslides can probably not be made. For example, times of pronounced erosion or hydrothermal alteration may also favor the occurrence of giant landslides. Over long periods of time, continuous percolation of water through the sub-surface may alter formerly competent layers of rock and would weaken the solid rock material through hydration and partial decay to clays (hydrothermal overprint; eg, Donoghue et al., 2008, 2010). Residual magmatic heat from a slowly cooling magma reservoir deep inside the island may accelerate these alteration processes. Extensive formation of such clay minerals may then provide slippage horizons, so that overlying parcels of rock may start moving despite a possibly modest slope angle (see also Delcamp et al., 2010, 2012).

Approximately one dozen giant landslides have been documented in the Canary Islands. Several additional ones have been inferred and, likely, some others are obliterated by volcanism and sedimentation. As shown in Fig. 1.31, the volumes and area of the collapses in the Canaries are considerably lower than those on Hawaii, at least by an order of magnitude (Carracedo, 1999). One relevant factor here is probably the angle and depth of the sliding plane (decollement), because the failure planes of the structural collapses in the Canary Islands are comparatively shallow relative to those of the Hawaiian Islands (see inset in Fig. 1.31A).

Giant collapses are more frequent in the early stages of development of islands, when eruptive rates are greatest, leading to high edifices flanked by oversteepened slopes. This explains the conspicuous evidence of giant collapses on the youngest, western Canaries that are in their shield stage (Urgelés et al., 2001; Masson et al., 2002, Manconi et al., 2009), and on Tenerife, which at present is in its early rejuvenation stage (Watts and Masson, 1995; Carracedo et al., 2007a,b, 2011a,b). It is likely that the number and frequency of large landslides were similar on Gran Canaria (Funck and Schmincke, 1998; Krastel et al., 2001; Schmincke and Sumita, 1998) and in the eastern Canaries (Stillman, 1999), but evidence there is likely less complete due to the higher degree of erosional decay.

Direct onshore evidence of giant landslides is rare in the Canaries, but the San Andrés fault system on El Hierro runs along the axis of the northeast rift zone (see chapter: The Geology of El

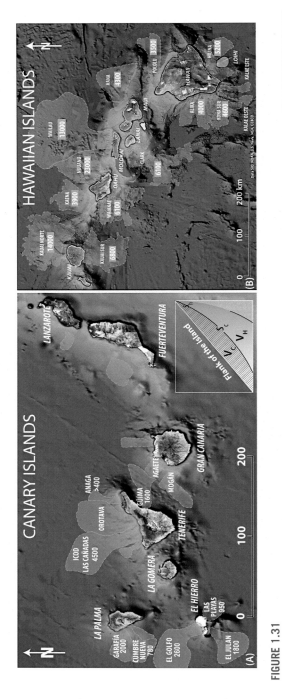

FIGURE 1.31

Mega-landslides in the Canaries (A) and the Hawaiian Islands (B) modified from Urgelés et al. (2001) and Moore et al. (1989). Inset in (A) shows volume versus slide scar angle for the Canaries (V_C and S_c) and for Hawaii (V_H and S_h) (after Moore, 1964). All numbers are in km^2. Base maps are from Google Earth.

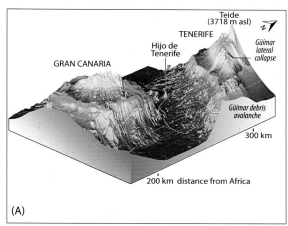
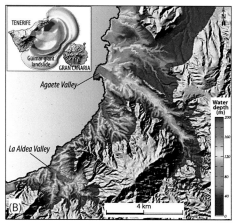

FIGURE 1.32

(A) Submarine morphology of Gran Canaria and Tenerife showing the Güímar lateral collapse scar and the associated submarine debris avalanche (modified after Krastel et al., 2001). (B) Maximum water column recorded inland in northwest Gran Canaria for a single event failure scenario of the Güímar slide on Tenerife. Inset shows the extent of the Güímar mega-landslide tsunami after 520 seconds for a single collapse event and a constant restricting stress of 150 kPa (after Giachetti et al., 2011).

Hierro), a feature that has been interpreted as a scar related to a partially arrested landslide (Carracedo, 1999; Carracedo et al., 2001; Gee et al., 2001). It is equally difficult to locate tsunami deposits produced by these mega-landslides, although the marine conglomerates that crop out at 41 to 188 meters above sea level (masl) in Barranco de Agaete on Gran Canaria are a noteworthy exception, comprising polymictic, rounded to angular volcanic clasts and fossil rhodolites, and marine shells (Paris et al., 2005a; Pérez-Torrado et al., 2006). The Agaete valley provides very favorable conditions for the deposition and preservation of tsunami deposits due to shallow slopes and many terraces. This particular deposit is most probably related to a catastrophic failure on the northeastern flank of the nearby island of Tenerife, directly facing the Agaete valley (the 0.8 Ma Güímar collapse; see Fig. 1.32).

Ward and Day (2001) inferred from geological evidence, such as a fault that opened during the 1949 eruption in the Cumbre Vieja on the island of La Palma, that during a future eruption of the Cumbre Vieja volcano, a catastrophic flank collapse may occur and trigger a tsunami that will reach the coasts of Florida (6100 km from La Palma), with waves as high as 120 m. Eruptions may occur any time in the near future on the Cumbre Vieja, likely even with limited warning. Although no movement was detected on this fault during the 1971 eruption, a failure of the flank in the geological future is conceivable. Therefore, the safest measure would be the evacuation of several hundred thousand people from the Canaries and the eastern coast of America in the event of any precursory sign of eruptive reactivation on La Palma, which is likely a logistical impossibility, however.

Recent studies on the same subject, however, now rebut this model as unrealistic and point out that media publicity regarding those predictions has created unnecessary public anxiety among Atlantic coastal communities (Pararas-Carayannis, 2002). For example, Mader (2001) used finite-volume Navier-Stokes modeling to predict that in the case of a giant landslide from La Palma, the maximum wave amplitude off the east coast of the United States would be approximately 1 m, concluding that even with shoaling, the wave would not constitute a significant hazard to the coastal population in the eastern United States.

In addition, there is no direct evidence that the fault that developed during the 1949 eruption on the Cumbre Vieja indicates detachment of the entire western flank of La Palma or just a near-surface volcanic fracture that formed during this eruption from an underlying dyke (see chapter: The Geology of La Palma). The suggestion that the entire western flank of Cumbre Vieja forms a detached block of rock (25 km long, 2−3 km deep, and 15−20 km wide) that sank 4 m in 1949 is certainly unrealistic because several coastal villages (*Puerto Naos*, *Tazacorte*, *El Remo*) would have been affected by a 4-m drop and would have disappeared under the sea. Thus, the 1949 faults must be predominantly of local relevance only.

CALDERAS

The term "caldera" can have several meanings, depending on the geographic area, and still leads to considerable confusion at times, especially in the Canary Islands.

Leopold von Buch was the first to introduce the word caldera during his 1815 visit to La Palma, where he applied it to the *Caldera de Taburiente*. The term caldera (meaning "pot") was originally proposed by von Buch to define what he believed to be a prototypical example of an "elevation crater."

In the Hawaiian Islands, the term caldera is used to describe large volcanic craters, generally many kilometers wide, formed by collapse of the volcano summit above shallow magma chambers, such as Mauna Loa and Kilauea calderas. In subduction zone systems, calderas refer to large collapsed magma chamber roofs of silicic systems that generally produce large and devastating eruptions.

In the Canaries, most of the calderas are erosive features that originated from gravitational landslides and subsequent widening by wind and rain erosion, such as *Caldera de Taburiente* and the *Las Cañadas caldera*. However, the *Caldera de Tejeda*, on Gran Canaria, is a classic example of a large silicic collapse caldera (vertical collapse) that formed during the emission of a series of voluminous silica-rich ignimbrites, with each between 30 and 100 km^3 in volume (Schmincke, 1969, 1974; Troll et al., 2002).

On Lanzarote, the term caldera is applied to craters that are wider than the characteristic cinder cones, frequently of phreatomagmatic origin, such as *Caldera Blanca* (Fig. 7.8). In this book, we come across all three types of calderas, and we aim to distinguish them following these definitions here.

THE ISLAND OF "LA CANARIA"

The different islands of the Canarian archipelago all appear to display a similar history if we compare their successive stages of evolution. As the islands repeat the same general scheme of events, their differences exist because they were formed and evolved at different times, rather than being synchronous. Thus, they represent different stages of growth, like a naturally mixed population where children, teenagers, adults, and the elderly live together. There, individuals go through the same cycle of growth, maturation, and decay, but offset from one another. As volcanoes grow, they also go through a constructive phase of evolution in which growth of the edifice through volcanic activity outpaces its destruction through mass wasting (eg, landsliding) and background erosion. At this stage, eruptive rates must be high and islands grow to considerable heights (eg, La Palma). Late cycles of volcanism are generally separated from the shield stage by extended periods of inactivity and usually show eruptive rates that are drastically diminished relative to the shield stage. Magmas can become more evolved (siliceous, and thus more viscous and often explosive), contributing to an increase in volcano height (eg, Tenerife). Finally, during the destructive phase of evolution, mass wasting and erosion outpace volcanic growth (eg, Gran Canaria), and the islands steadily decrease in size until they are eroded down to sea level and below (eg, Lanzarote and Fuerteventura).

In the early shield stage of evolution, an island shows "juvenile" morphology, characterized by great elevations (above 1500 masl) and deep barrancos (ravines) (eg, the central and western Canaries). At the "mature" stage, however, erosive and denudational landforms are predominant, island elevation decreases to 800 m or less, and relief is smooth (eg, in the eastern islands). This advanced denudation exhumed the deeper plutonic formations in the eastern Canaries, revealing the inner structure of the island edifices, which are inaccessible in the younger western islands due to a high rate of volcanic activity at present.

Notably, the closing of the isthmus of Panama initiated the cold water and clockwise circulating ocean current of the North Atlantic, thereby setting up the trade winds. This reduced the heating effect of the Sahara to the east and changed the previous tropical climate in the area of the Canaries to a more moderate one (Meco et al., 2003, 2007). Only the islands taller than approximately 1200 masl can intercept the humid trade winds, causing notably different climatic conditions on different islands and at different altitudes. The trade winds flow past the low-altitude islands, which consequently have an extremely arid climate, similar to that of the Sahara desert (eg, the 126 mm average yearly rainfall on Lanzarote and 199 mm on Fuerteventura), whereas the taller islands also have lush regions, especially in the northern and northwestern parts of the islands. Another consequence of the Pliocene climatic change was the end of the tropical fauna in the Canaries. Biogenic white sands only occur on beaches in the older Miocene islands, an aspect of likely limited geological consequence, but a crucial factor for the tourism-based economy of the archipelago (eg, on Gran Canaria, Lanzarote, and Fuerteventura).

Type of magma (OIB series)	Volcanic rocks	Volcanic features		Canary Islands	Hawaiian Islands
Initial terms (mafic)	Tholeiites Basanites Basalts	Lava flows and related features	lava flows, lava tubes and lava lakes	Yes	Yes
		Pyroclasts	Lapilli and scoria beds	Yes	Yes
		Volcanic cones	Shield volcanoes Lapilli and scoria (strombolian) cones Phreatomagmatic tuff cones and tuff rings	Yes	Yes
Intermediate and magma mixing	Trachybasalts Tephrites Hawaiites	Lava flows and related features	lava flows, lava tubes and lava lakes	Yes	Rare
		Pyroclasts	Lapilli and scoria beds	Yes	Rare
		Volcanic cones	Lapilli and scoria (strombolian) cones	Yes	Rare
		Magma mixing	Intermediate and evolved lavas in a single eruption	Yes	Rare
Evolved terms (felsic)	Phonolites Benmoreites Trachytes	Lava flows and related features	Lava flows, lava tubes and lava lakes	Yes	Very rare or none
		Pyroclasts	Pumice and phonolitic scoria beds Alternating basaltic lapilli and pumice beds Ignimbrites	Yes	Very rare or none
		Volcanic cones	Pumice and phonolitic strombolian cones Lava domes and coulees	Yes	Very rare or none
			Composite felsic volcanoes (stratocones)	Yes	None
		Calderas	Collapse caldera Cone sheet	Yes	None

FIGURE 1.33

Main differences in volcanic features between the Canary and Hawaiian archipelagos. The contrasting features are dominantly the consequence of felsic volcanism. The characteristic eruptive mechanisms, volcanic landforms, and products of felsic volcanism are common in the Canary Islands but are very scant in the Hawaiian archipelago. These differences were the main reason for the inclusion of the Teide National Park into the UNESCO World Heritage List in 2007. After Carracedo (2008b).

The wealth of geological diversity in relation to the seven islands is thus caused by their long life (more than three times that of the Hawaiian Islands), which is due to the lack of subsidence. This not only allows for lasting periods of magmatic evolution, which can lead to abundant felsic eruptions, but also permits tracing the different life stages of a typical Canary island. The abundant felsic magmatism in the Canaries results in a unique variety of eruptive mechanisms, products, and structures relative to other ocean island systems such as Hawaii (Fig. 1.33). A sketch is shown in Fig. 1.34, where the imaginary island of *La Canaria* is illustrated, combining the main features of the different Canary Islands into a schematic prototype Canary volcanic island near the end of a long and geologically eventful life.

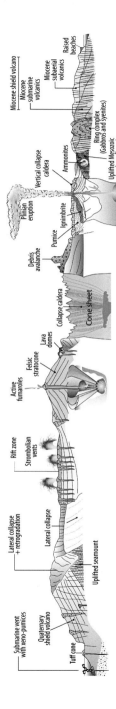

FIGURE 1.34

The most significant geological features of the Canary Islands are cataloged here in the form of the fictional island of *La Canaria*. Although no island that exposes all these features exists at present, many of the phenomena that have occurred on the older islands are likely to occur on the younger ones also. For example, features that derived from felsic magmas are rare or absent in the young volcanoes (eg, El Hierro), whereas the deeply eroded eastern islands (eg, Fuerteventura and Gran Canaria) allow inspection of the inner workings of felsic ocean island volcanoes, and the mid-level erosional stage of La Gomera, for example, presents evidence for the feeder levels from the magma reservoirs to the surface. It is therefore reasonable to take lessons from the older systems and apply them to the younger islands, as well as taking those from the younger ones to help reconstruct the older systems. Applying this concept to El Hierro, for example, felsic magmatism should soon become more relevant there, as seen on the slightly older island of La Palma, where phonolite domes occur already with reasonable frequency.

THE GEOLOGY OF EL HIERRO

The Geology of the Canary Islands. DOI: http://dx.doi.org/10.1016/B978-0-12-809663-5.00002-5

43

THE ISLAND OF EL HIERRO

The island of El Hierro is the youngest (1.12 Ma), smallest (268 km^2), and westernmost of the Canaries. The most characteristic feature of the island is its truncated trihedral shape, with three convergent ridges of volcanic cones that are separated by wide, horseshoe-shaped embayments (Fig. 2.1). El Hierro is also the least populated island of the Canaries (10,500 inhabitants) and 58% of its surface area is protected. Since 2000, El Hierro has been a UNESCO Biosphere Reserve. The geology of El Hierro is representative of an oceanic island in the early stage of shield development. Most of the island is mantled by recent cinder cones and lavas, and it is barely incised by incipient erosional barrancos. The internal structure of the island, in turn, has been exhumed by gigantic landslides, revealing overlapping volcanic edifices that successively developed and collapsed. These features provide one of the best possible geological scenarios to investigate the relationship between rift zones and giant lateral collapses, which are apparently a common process in the earlier stages of oceanic island growth (see, eg, Carracedo and Troll, 2013).

FIGURE 2.1

Triple-armed geometry of El Hierro. Three linear rift arms and three horseshoe-shaped landslide depressions in between the rift arms are observed, and define the characteristic shape of El Hierro (image from *IDE Canarias visor 3.0*, GRAFCAN).

THE SUBMARINE EDIFICE

El Hierro is the emergent summit of a volcanic shield that rises from a 3700-m deep seafloor. The ocean floor comprises approximately 1 km of sedimentary rocks that overlie Jurassic igneous oceanic crust of almost 150 million years beneath this part of the archipelago (Fig. 2.2). Marine research in the surroundings of the Canaries has been intense in recent years and includes sidescan sonar imaging of the submerged part of the island edifices. The images of the island of El Hierro reveal a triple-armed edifice, with amphitheater-like embayments and steeply sloping flanks. The total height of the edifice is approximately 5000 m (from the 1500-m altitude recorded at *Malpaso* to the base of the island at a depth of 3500 m below sea level). A characteristic feature of the submarine edifice is the long (more than 20 km) and curved prolongation of the South rift, which was also the location of the 2011 to 2012 submarine eruption. The final tip of this rift, the El Hierro submarine Ridge, dated at 133 million years (Ma) (van den Bogaard, 2013), is probably a much older feature unrelated to the Canary Volcanic Province (eg, Zazcek et al., 2015; Troll et al., 2015).

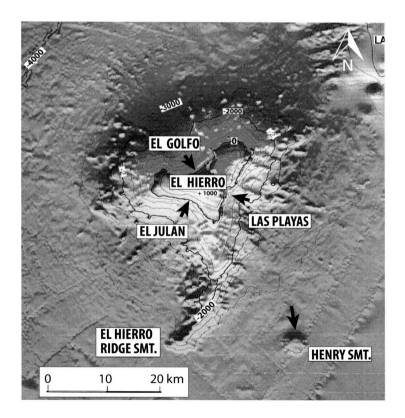

FIGURE 2.2

Color-shaded relief bathymetry and topography of El Hierro (image from Masson et al., 2002) revealing a triple-armed edifice with amphitheater-like embayments that extend below sea level. Note the submarine prolongation of the southern rift arm that extends for another approximately 20 km toward the southeast.

AGE OF VOLCANISM: RADIOMETRIC AGES AND GEOMAGNETIC REVERSALS

Abundant radiometric ages have recently become available that now allow defining the temporal sequence of the different volcanic formations that constructed El Hierro island in the Quaternary (Abdel Monem et al., 1971; Guillou et al., 1996; Pérez-Torrado et al., 2011). Geomagnetic reversals are a powerful tool for this period of growth and complement radiometric age determinations. Polarity of the Earth's magnetic field, recorded in a lava flow upon cooling, allows us to pinpoint the age of the different volcanic formations by correlation of the magnetic polarity of the El Hierro lavas with the established global scale of palaeomagnetic inversions (Fig. 2.3). For the period of time covering the construction of El Hierro, there are two inversions: the *Brunhes* epoch, or normal polarity, that covers the past 780,000 years, and the previous *Matuyama* inverse polarity epoch. A short normal polarity event, known as the *Jaramillo* event, took place within the *Matuyama* epoch at approximately 0.99 to 1.05 Ma.

MAIN STRATIGRAPHIC UNITS

The period of construction of the island of El Hierro, which comprises effectively the last 1 million years, makes it particularly suitable for the application of the Geomagnetic Polarity Timescale (GPTS), which can hence assist in defining the volcanic stratigraphy and the dating of volcanic formations. Few changes in magnetic polarity took place during that period and they are relatively easy to identify and to correlate on El Hierro. The geochronological and palaeomagnetic data in sections provided by the collapse scars and the deep barrancos and via wells, boreholes, and water-mining tunnels (galerías) on the island define three large superimposed volcanoes: the Tiñor, the El Golfo, and the recent Rift volcano (Figs. 2.3, 2.4).

TIÑOR VOLCANO

The Tiñor volcano forms the first stage of subaerial growth of El Hierro, and its outcrops are confined to the northeast flank of the island and to the interior of the Las Playas embayment (see Figs. 2.3 to 2.5). The Tiñor volcano developed very rapidly from approximately 1.12 Ma to 1.03 Ma, and there is no consistent compositional variation with time that can be mapped in the field.

However, there are several differences between units that may reflect the morphological evolution of the developing edifice: (1) a basal unit of relatively thin, steeply dipping flows probably corresponds to the initial stages of subaerial growth of the volcano; (2) an intermediate unit of thicker lavas that progressively trend to sub-horizontal flows in the center of the edifice; and (3) a group of emission vents with well-preserved wide craters (the Ventejís volcano group; see Fig. 2.5B) and associated lavas, which occupy the valleys and canyons that were carved into the older underlying rocks. The flows of the Ventejís unit are very easily identified by their significant peridotite mantle xenolith content, likely from a terminal explosive stage associated with the collapse of the northwest flank of the Tiñor volcano (c.f. Carracedo et al., 2001; Manconi et al., 2009).

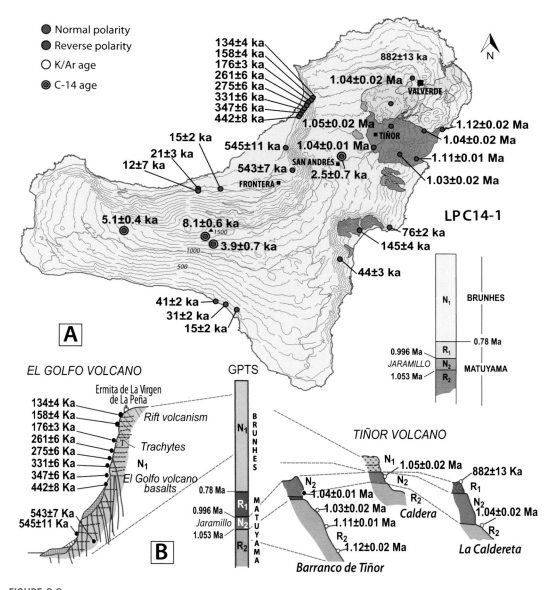

FIGURE 2.3

(A) Radioisotopic ages (K/Ar) and polarity reversals of the Earth's magnetic field determined on El Hierro and correlated with the international Geomagnetic Polarity Time Scale (GPTS). Most of the lavas on the island correspond to the Brunhes epoch of normal polarity. Older volcanic formations, of reverse (Matuyama) polarity, are located at the northeast part of the island (after Guillou et al., 1996). (B) Volcanic sequence in the El Golfo cliff. The K/Ar ages record rapid growth of the El Golfo volcano during the Upper Brunhes period. To the right, sections in the northeast of the island. There, ages and polarities show the presence of an older volcano (Tiñor volcano) that formed already during the Upper Matuyama period (modified after Guillou et al., 1996).

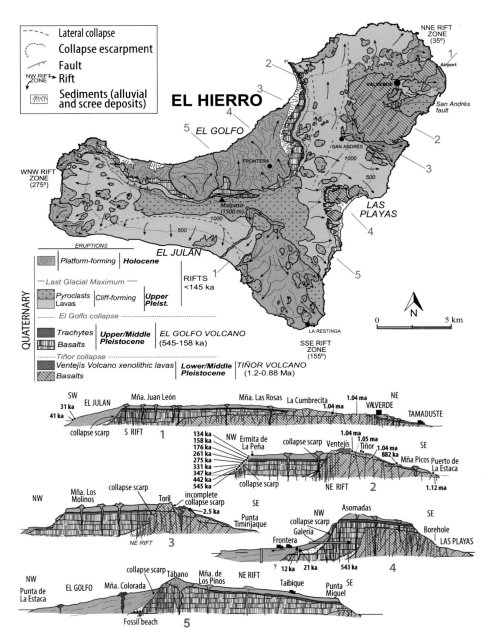

FIGURE 2.4

Simplified geological map and representative cross-sections of El Hierro. The cross-sections show the island to be constructed of three overlapping volcanoes (Tiñor, El Golfo, and the recent rift activity). These successive volcanic formations are separated by major unconformities that resulted from giant lateral collapses (after Carracedo et al., 2001).

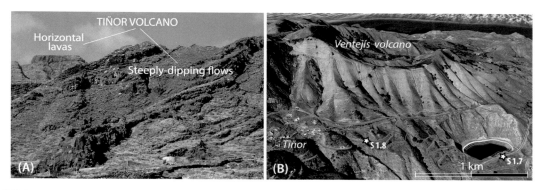

FIGURE 2.5

The Tiñor volcano records the initial stage of the subaerial development of El Hierro from approximately 1.11 Ma to 0.88 Ma. (A) The oldest flows dip steeply, indicating fast growth of the Tiñor shield. The younger Tiñor lavas form a plateau of horizontal lavas at the top of the former volcanic edifice. (B) Cluster of cinder cones, including the Ventejís volcano, to the northwest of *Valverde* (image from Google Earth).

EL GOLFO VOLCANO

The Tiñor volcano eventually collapsed at approximately 0.88 Ma, and a new volcanic edifice, the El Golfo volcano, developed and eventually filled the northwest-facing collapse embayment of the Tiñor volcano. Finally, the El Golfo lavas spilled toward the east coast also (Fig. 2.6).

The El Golfo volcano developed entirely in the Brunhes epoch, from approximately 545 ka onward, and the age of the lowest flow indicates a semi-radial dip in the lava flows, illustrating that the El Golfo edifice grew nested inside the Tiñor lateral collapse embayment that was centered near the present day town of Frontera.

Two sub-units may be identified in the El Golfo volcano from morphological differences and from local unconformities: (1) a basal unit mainly composed of strombolian and surtseyan pyroclastics (cinder cones and tuff rings) with subordinate lava flows and (2) an upper unit predominantly composed of lava flows. The lower unit is cut by numerous dykes that form northeast-, east-southeast-, and west-northwest-trending swarms that match the present volcanic vent systems and indicate that the triple-arm rift system was also an important feature for the older El Golfo edifice. The upper El Golfo units, in turn, are topped by several differentiated lava flows and block-and-ash deposits (trachybasalts, trachytes), likely representing the terminal stages of activity of the El Golfo volcano.

The duration of the growth of El Golfo volcano can be estimated to be approximately 360 to 380 ka, as indicated by the age of 545 ka for the basal lavas and the age of the trachytic lavas in the collapse scarp section (176 ka). Notably, the association of giant landslides with postcollapse felsic volcanism (eg, phonolites and trachytes) is a frequent feature in the Canary Islands (see also chapters: The Geology of La Palma and The Geology of Tenerife; and Carracedo et al., 2011b).

FIGURE 2.6

Eastern scarp of the El Golfo embayment. The section represents approximately 400 kyrs, from 545 ky to 130 ky. Most of the sequence corresponds to the El Golfo volcano, and only the top lavas and the cinder cone in the center belong to the recent rift volcanism. The thicker and light color units in the upper El Golfo sequence (indicated with arrows) are rare trachyte flows from the final stages of El Golfo volcanism.

RECENT RIFT VOLCANISM

The recent rift volcanism represents the latest stage of island growth, with the three arms of the rift being simultaneously active. The rift series lavas rest broadly conformable on the El Golfo sequences over much of the island, apart from in the El Golfo embayment itself. The distribution of vents (Figs. 2.6, 2.7) implies that a relatively thin layer of basic lavas covered much of the island recently, and these lavas have largely filled the El Julan collapse embayment and, in part, also the El Golfo and Las Playas embayments. The rift eruptions commenced after the differentiated (felsic) emissions of the El Golfo volcano ceased at approximately 158 ka (Guillou et al., 1996), and will be inspected on day 1 of the field itinerary.

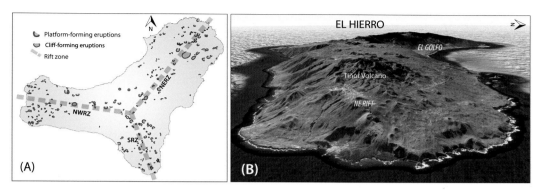

FIGURE 2.7

(A) Distribution of eruptive vents on El Hierro, outlining a well-defined regular triple-armed rift system.
(B) *Google Earth* view from the east onto the northeast rift zone of El Hierro.

COMPOSITIONAL VARIATION

Rocks from volcanic oceanic islands are generated by partial mantle fusion due to the action of fixed hot spots that give rise to Oceanic Island Basalts (OIB). Within this series there are two trends depending fundamentally on the degree of mantle fusion: the Hawaiian-type Tholeiites (OIT) are typical of islands originated by vigorous hot spot magmatism, with high partial mantle fusion rates such as displayed by the Hawaiian Islands, the Galápagos Islands, and, in part, also by the Canary Islands. In turn, the alkaline Ocean Island series is characteristic of hot spot islands with lower mantle fusion rates, such as the Canary Islands (for most parts) or the Cape Verdes. This is because large alkali elements are the first to enter a partial mantle melt and, therefore, small melt fractions enrich these elements in the resulting mantle magma.

The TAS diagrams of lavas from the westernmost Canary Islands (Figs. 2.8 and 3.4) show that silica-saturated rocks with a high alkali content clearly dominate, corresponding to the Canarian ocean island alkaline trend (OIB series). However, some volcanoes (El Golfo volcano on El Hierro and the submarine part and Garafía volcanoes on La Palma) present Hawaiian tendencies, and merge into the OIT Series. The conclusion reached is that fusion rates were greater during earlier stages of volcanism of La Palma, whereas on El Hierro this stage of larger melt production was only reached during the evolution of the El Golfo volcano. Alternative ways to make OITs in the Canary Islands have recently been contemplated by Aparicio et al. (2006, 2010), who consider the 1736 tholeiites of Lanzarote to have formed originally from OIB magmas also, but then assimilated silicic crustal sediments to give them a tholeiitic character, a process that was highlighted again "in action" during the 2011 El Hierro eruptive events (eg, Troll et al., 2011, 2012; Carracedo et al., 2015a).

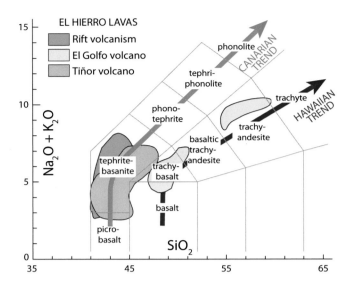

FIGURE 2.8

Total alkali versus silica (TAS) diagram comparing lavas from El Hierro with the general Canary and Hawaiian magmatic evolutionary trends. The early El Hierro geochemical variations of the Tiñor volcano differ from the remaining islands in the archipelago, showing a trend somewhat similar to that of the tholeiites of the Hawaiian Islands and likely reflecting very high degrees of mantle melting during the Tiñor eruptive episode (after Carracedo et al., 2001).

MAIN STRUCTURAL ELEMENTS OF EL HIERRO

The most significant structural elements that condition the growth, shape, and relief of the Canary Islands are the '*dorsales*' or rifts and the associated large gravitational landslides (eg, Carracedo, 1996a, b; Walter and Troll, 2003; Manconi et al., 2009; Carracedo and Troll, 2013).

The role of these structural elements is particularly noteworthy and evident on El Hierro, because of its earlier stage of development relative to the other islands of the archipelago. These are either in more advanced stages of evolution and erosion has dismantled a large part of the edifices or are affected by rejuvenation volcanism that is often less organized.

RIFT ZONES

The origin, geometry, and disposition of the rifts of the Canary Islands have been reviewed in the case of Tenerife (see chapter: The Geology of Tenerife, Figs. 5.21 and 5.23). On El Hierro, these structures are also very pronounced and the majority of the eruptive vents group along the axes of the rifts (Fig. 2.9). Moreover, rift zones give El Hierro and Tenerife their characteristic triple-armed shape because collapse embayments are consistently located in-between the rift axes of these islands.

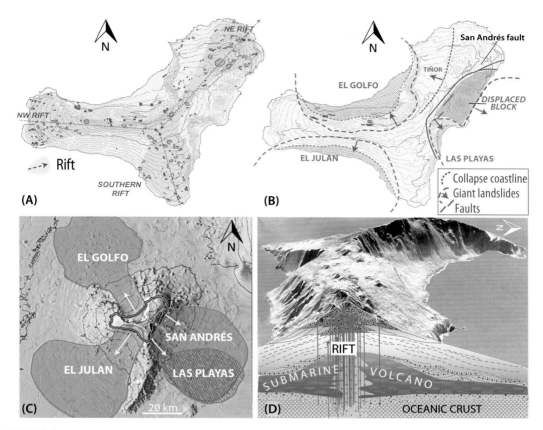

FIGURE 2.9

Recent rift zones of El Hierro. (A and B) Geometry and disposition of the rifts and giant landslides of El Hierro.
(C) Debris avalanche deposits from lateral collapses off the flanks of El Hierro (after Masson et al., 2002).
(D) Schematic cross-section through the El Hierro northeast rift arm (image in part from Google Earth).

THE GIANT LANDSLIDES OF EL HIERRO

The existence of a triple rift system on El Hierro has favored the concentration of eruptions at the center of the island, increasing its height and progressively leading to gravitational instability. Dyke injections into the axes of the rifts probably help trigger lateral collapses (Fig. 2.9). Notably, large landslides may not only be extraordinarily important for the development of oceanic volcanic islands, but also for the present living conditions on the El Hierro, because the location of underground water reserves is controlled by the island's geometry also.

The island of El Hierro, together with the Hawaiian island of Molokai, is well suited to analyze how, when, and why catastrophic landslides take place. In addition to the large number of these structures on El Hierro island (at least four, with very different characteristics; Fig. 2.9), the El Golfo collapse embayment is the most recent and best preserved of all such embayments in the Canary archipelago.

The knowledge on El Hierro's giant landslides results from a combination of onshore studies, defining the ages of the formations involved and the character of the embayments (eg, Guillou et al., 1996; Day et al., 1997), with offshore studies performed by oceanographic vessels using side-scan sonar equipment that allowed the nature and extent of the avalanche deposits to be defined (Urgelés et al., 1998; Collier and Watts, 2001; Gee et al., 2001).

TIÑOR LATERAL COLLAPSE

The Tiñor landslide, calculated to have occurred approximately 880,000 years ago, is the oldest giant landslide of El Hierro for which evidence has been found. This landslide probably removed more than half of the Tiñor volcanic edifice, which was a substantial part of the island of El Hierro at that time (Fig. 2.10A).

The resultant collapse embayment was then filled by subsequent eruptive activity (El Golfo volcano and the incipient volcanism at the rifts). However, the fossilized collapse scarp can be traced by the brusque change from relatively old volcanic formations (>0.88 Ma) on one side of the scarp to considerably younger ones (<0.545 Ma) on the other side (Fig. 2.10B).

However, this landslide has not been identified in the offshore sonar images obtained. This is perhaps because the avalanche deposits that accumulated on the northwest flank of the island have been covered by sediments and by material from the much more recent El Golfo lateral collapse.

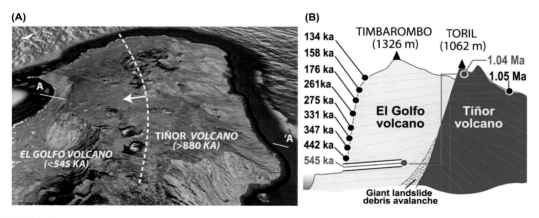

FIGURE 2.10

(A) *Google Earth* image of the Tiñor landslide, calculated to have occurred approximately 0.88 Ma ago. This is the oldest giant landslide of El Hierro for which evidence has been found (Guillou et al., 1996). (B) Geological evidence for the Tiñor giant landslide includes radioisotopic dating, which shows the entire volcanic sequence of the western sector of the island to be younger than the formations that make up the eastern part (from Guillou et al., 1996; Carracedo et al., 2001).

As illustrated in Fig. 2.10B, the roof of the sequence that remained after the Tiñor collapse has been dated at 1.04 million years, whereas a water tunnel crossing the El Golfo sequence into this old formation gave an age at the end of the tunnel of 0.545 Ma for the oldest postcollapse volcanism. This relationship can only be explained if we consider the western part to represent a more recent fill, implying the prior collapse of the older formation.

EL JULAN LANDSLIDE

The El Julan landslide depression has also been almost completely filled, in this case by recent eruptive activity of the northwest and south rifts. The shape of the collapse embayment is still evident, although the scarp itself is by now completely covered (Figs. 2.11, 2.12A).

Sediments from the edge of the African continent and detritus from the island surround the island of El Hierro and in part cover the avalanche deposits of the El Julan landslide (Masson et al., 2002). The thickness of the sediments that cover the El Julan avalanche is 10−12 m, indicating that this landslide must be relatively old, probably older than 0.6 Ma, and therefore it would pre-date the El Golfo landslide. The material dispersed into the ocean by the El Julan landslide has been estimated at a volume of approximately 130 km^3 (Carracedo et al., 2001).

LAS PLAYAS LANDSLIDE AND THE SAN ANDRÉS FAULT

The more complex gravitational collapse of the southeast flank of El Hierro involved a large landslide that formed the Las Playas depression (Masson et al., 2002). The corresponding offshore blocky debris avalanche deposit may have been caused by a series of events, whereas another part

FIGURE 2.11

The El Julan embayment is located in the leeward southeast flank of the island, that is, the side that is sheltered from the trade winds (the Sea of Calms). It has been designated the status of a *Spanish National Marine Reserve* because of its clear waters, the high concentration of different species of fish and dolphins, pilot whales, and beaked whales, not to mention innumerable species of marine flora. The Sea of Calms was the site of the last eruption on the island in 2011, which caused large losses in marine life in this area at the time (photo courtesy of S. Socorro).

of the collapsed block slipped but remained anchored (Fig. 2.12B). The apparent lack of postslump displacement of the formerly moving block, particularly during the younger El Golfo collapse, suggests its present stability.

The scar of this aborted collapse is readily observed as a fault that runs along the southeast flank of the island, from *La Caleta* to the headwall of *Las Playas*, passing through the village of *San Andrés* (Fig. 2.12B). Some stretches of the fault are overlain by lava flows emplaced after the El Golfo landslide, several of which are dated at 145,000 years, thus providing a minimum age for the Las Playas collapse.

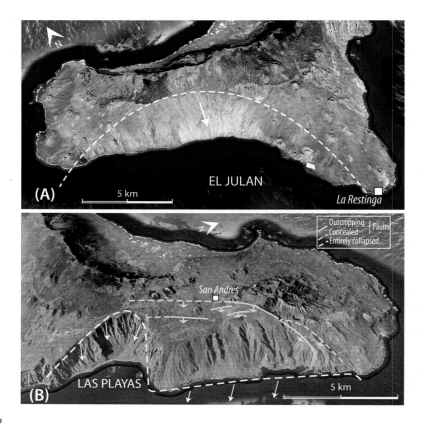

FIGURE 2.12

(A) Oblique aerial view from the south onto the El Julan embayment. The original fault scarp is covered with younger lavas and is no longer exposed. (B) Oblique aerial view of the northeast part of El Hierro showing the San Andrés fault system enclosing a partially collapsed block. The collapsed block slipped but remained in part anchored, likely representing an aborted giant landslide. This situation provides a unique scientific opportunity to study the processes involved in these large lateral collapses (images modified from Google Earth).

Where the San Andres fault is visible, it appears as a northeast-trending normal fault zone, dipping at an angle of 60 to 70° (Day et al., 1997). Traces of the fault are exposed for approximately 10 km. Its slip plane is locally marked by pseudo-tachylite and due to ongoing road improvements, larger segments can be studied in detail.

THE EL GOLFO GIANT COLLAPSE

This is the largest and most recent collapse of the island of El Hierro and indeed the most readily recognizable giant collapse of the entire archipelago (Fig. 2.13). The collapse scar runs from an altitude of 1500 m (*Malpaso*) to a depth of 3200 meters above sea level (masl) (Fig. 2.14A). The scree next to the island forms a wide fan, which has been smoothed by the action of the loose and migrating collapse materials. The friction of the passing rocks incised a submarine ravine with walls up to 600 m high on either side (see Fig. 2.14A).

The mobilized materials apparently reached high velocities and transport energy, fanning out over an area of 1500 km^2. The sonar images show materials that once were part of the island and are now deposited over distances of up to 65 km from the coast (Fig. 2.14B). These deposits are formed by randomly distributed angular blocks that can reach 1200 m in diameter and 300 m in height and are composed of sequences of pyroclastics and lava flows that are readily identifiable (Masson et al., 2002). Mass wasting of 150−180 km^3 of rocks by the collapse left a horseshoe-type depression of 15 × 5 km. The floor of the depression is covered with recent postcollapse lavas that form a coastal platform and have started to extend the precollapse coastline again. The vertical collapse scarp reaches 1200 m in the area of *Jinama*, providing one of the most spectacular panoramic views of the entire archipelago (Fig. 2.13).

The age of this lateral collapse is still the subject of debate due to the apparently conflicting information deduced from both onshore and offshore observations. The evidence obtained in marine geology studies, fundamentally from the age of turbidites associated with the avalanche deposits

FIGURE 2.13

Panoramic view over the El Golfo landslide embayment from the eastern edge of the collapse scarp (*Mirador de La Peña*). The cliff on the left is approximately 1200 m high. Note the lavas from nested eruptive vents near the collapse scarp have begun to fill the embayment (photo courtesy of S. Socorro).

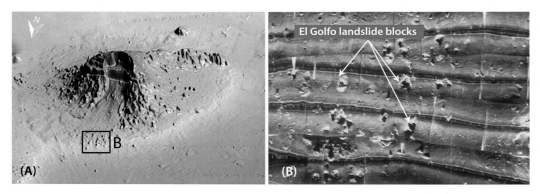

FIGURE 2.14

(A) Color-shaded relief bathymetry and topography of El Hierro showing the subaerial and submarine parts of the El Golfo lateral collapse embayment and its associated off-shore debris avalanche deposits. (B) The sonar image shows randomly distributed angular blocks that reach 1200 m in diameter and 300 m in height that once formed a part of the island. The blocks are composed of piled-up heaps of pyroclasts and lava flows that are now deposited more than 65 km from their former position on El Hierro (see box in A; images from Masson et al., 2002).

of this giant landslide, points to an age between 13,000 and 17,000 years (Masson et al., 2002), notably coinciding with the last glaciation.

In contrast, observation of the scarp reveals a complex succession of processes that would appear to require a much longer period of time to take place. The El Golfo scarp, north of *Fuga de Gorreta*, shows remains of apparently older lava flows (although later than the collapse), intercalated between altered and compacted slope deposits (Fig. 2.15A). These deposits appear to adapt to an old curved profile that overlies a vertical cliff. The formation of this 180-m cliff required considerable time and must have taken place before the lavas filling the collapse depression were emitted and "fossilized" the cliff, thus isolating it from marine erosion (Carracedo and Day, 2002). At least two generations of such *"piedmonts"* exist, one covered by the lava flows forming the recent platform and the other overlying the flows (Fig. 2.15B).

To reconcile onshore and offshore information, we may need to assume that two collapse events have taken place. This would imply one collapse with an age similar to that of the lava flows that form the ceiling of the scarp (approximately 133,000 years), which coincides with the Riss glacial stage, and another more recent collapse, simultaneous with the Würm last glacial maximum (Fig. 2.15C). This time interval would be sufficient to generate all the structures described. Longpré et al. (2011) recently obtained new $^{40}Ar/^{39}Ar$ ages using both furnace-heating and laser-heating dating techniques for the altitudinally highest lava flows cut by the El Golfo landslide headwall as well as a postcollapse lava flow erupted from within the embayment. The authors' data imply the formation of the El Golfo debris avalanche between 87 ± 8 ka and 39 ± 13 ka ago, thus placing further constraints on the onset of the El Golfo collapse.

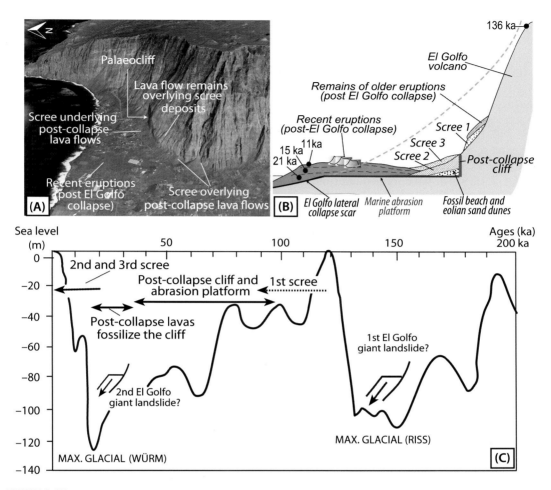

FIGURE 2.15

Geomorphological features that provide information about the relative age and processes involved in the development of the present El Golfo scarp. (A) Main geomorphological features of the eastern part of the scarp (*Google Earth* image). (B) Section indicating the required sequence of processes involved in the generation of these features. (C) Correlations of ages of formation of the different geomorphological features with the low sea level stand in the last glacial maximum and the cliff-forming and platform-forming eruptions recorded here. The data support the formation of the El Golfo scarp in two successive giant landslides that seem to have occurred in connection with the two last glaciations (Riss, 180,000−130,000 years and Würm, 70,000−10,000 years BP). The initial El Golfo giant landslide removed part of the superstructure of the El Golfo volcano, whereas the second slide may have been confined to the submarine part of the island. Recently, Longpré et al. (2011) obtained radiometric ages that constrain the main El Golfo landslide to between 87 ± 8 ka and 39 ± 13 ka ago.

RECENT VOLCANISM ON EL HIERRO

La Palma and Tenerife were the most volcanically active islands during the Holocene (the past 10,000 years). La Gomera has had no volcanic activity for at least 2 million years, whereas volcanic activity has been relatively scant on El Hierro in the Holocene, or at least much lower than would be expected for one of the youngest islands of the archipelago. A possible explanation is an "on−off" alternation of volcanism between El Hierro and La Palma during the past 120−130,000 years, with La Palma being perhaps favored by the tectonic situation because volcanism was much more extensive on the island of La Palma during this period.

As also seen on La Palma, the recent eruptions of El Hierro frequently form coastal platforms (see Fig. 2.4), which have formed after the last glacial maximum approximately 18,000 years ago. The best-developed coastal platforms on El Hierro are located at the extreme ends of the rifts and in the interior of the El Golfo depression (Fig. 2.16A). Small platforms of this type are also found at the end of the northeast rift, in the *Tamaduste* area, forming a plateau at sea level on which the airport runway was eventually built. Another quite well-developed platform lies at the end of the South rift, forming the *Lajial del Julan* and the *Llano de La Restinga*.

However, the most recent coastal lava plains are located at the western end of the northwest rift, forming the most westerly point of the island, of the Canaries and thereby of Spain and Europe as well. Several of these platform-forming eruptions of the northwest-rift may be very recent, but none is historic. The Lomo Negro eruption, in *Hoya del Verodal*, has been related to a series of earthquakes recorded on the island in 1793, but there is no report or indication that the said seismic crisis culminated in an eruption. Naturally, a submarine eruption similar to the 2011/2012 events could have taken place near the coast and possibly went unnoticed or was swiftly forgotten.

Radioisotopic ages obtained from rocks recovered from boreholes drilled into the El Golfo collapse basin show that the fill is composed of lava from predominantly pre-Holocene eruptions (Fig. 2.16B). However, it appears evident that the eruptive activity on El Hierro has been less intense than, for example, on La Palma over the past few thousand years, although El Hierro is presently believed to be in the juvenile and, hence, most vigorous stage of growth. Therefore, it must be assumed that, although on a short-term time scale (thousands of years), these islands grow in discrete, volcanic pulses separated by reasonably long interruptions. Geologically speaking, however, this type of activity must nevertheless be considered to be semi-continuous in a geological sense, that is, over longer periods of time (more than thousands of years).

After the 2011 submarine eruption south of El Hierro, an alternative to consider is that El Hierro is much more active than the scant Holocene eruption record seems to indicate, but most of the eruptive activity was possibly submarine. The 2011 submarine eruption was barely noticed, except by the seismic stations deployed by the Spanish *Instituto Geográfico Nacional* (IGN). Although the eruptive vents were relatively shallow (~90 mbsl), the only visible manifestation of the eruption was a discolored and bubbling area on the sea surface (locally known as *La Mancha* = the stain, or as *the Jacuzzi*) near the fishing village of *La Restinga* (Figs. 2.17A,B and 2.18). For several weeks, lava bombs that reached the ocean surface as lava balloons (Fig. 2.17C) and also xenoliths of pre-island oceanic sediment early during the eruption (Fig. 2.17D) were the only manifestations in terms of accessible eruption deposits, which would have almost no long-term preservation potential in the onshore geological record.

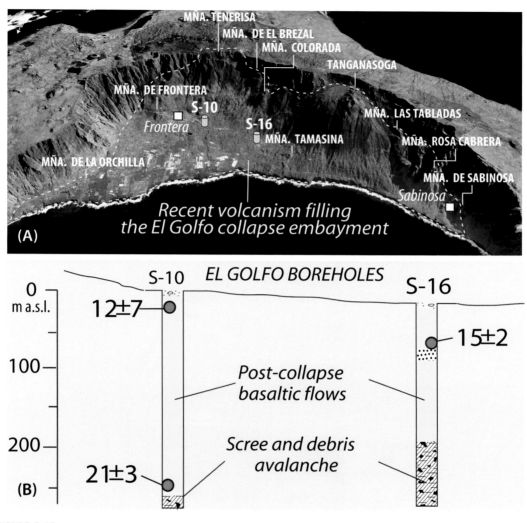

FIGURE 2.16

(A) Eruptive vents nested inside the El Golfo embayment and location of the two boreholes that were drilled through this sequence (image from *Google Earth*). (B) In accordance with radiometric ages obtained from these borehole samples, the oldest lavas to overlie the likely collapse-related debris avalanche deposit are approximately 21 ka old (after Guillou et al., 1996).

Only three earthquakes reached a magnitude between M 4.3 and M 4.5 in 2011 (Fig. 2.18) and were only felt on part of the island. The eruption had great media impact because the IGN provided online seismic real-time data with thousands of seismic events of very low magnitude that were only noticeable instrumentally. Therefore, it is possible that a number

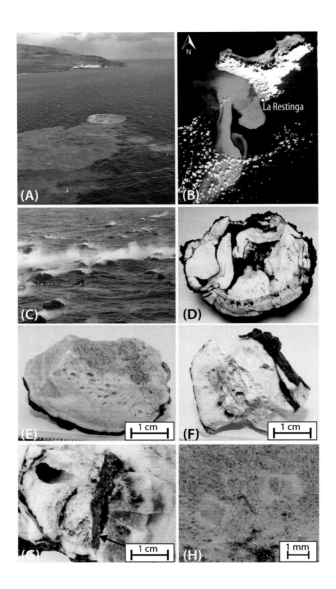

FIGURE 2.17

(A) The 2011 "*Jakuzzi*" of dissolved magmatic gases and suspended matter produced green and bright discoloration of seawater and was locally known as *La mancha* (the stain) (photos in A and C courtesy of *Guardia Civil*, taken during monitoring helicopter flights). (B) The submarine eruption commenced on October 10, 2011 and produced a plume that was distributed for several kilometers to the southwest before drifting off into the Atlantic (*Satellite image by RapidEye*). (C) Phases of strong degassing occurred with abundant rock fragments bursting above the water surface approximately 2.5 km off the coast from the village of *La Restinga*. Some of the large water bubbles

(Continued)

of submarine eruptions had taken place on El Hierro even in historic times but were previously unnoticed or unrecorded, especially before seismic monitoring was available in the archipelago.

PREHISTORIC ERUPTIONS

The most recent eruption to take place on the island of El Hierro (prior to the 2011 events) appears to be that of *Montaña Chamuscada* (2500 ± 70 years BP; Guillou et al., 1996), a group of volcanic cones situated immediately north of *San Andrés*. Lava flows emitted from these cones reach the south coast at *Punta de Timirijaque* (see cross-section 3 in Fig. 2.4).

The Tanganasoga volcano, a large volcano emplaced in the scarp of the El Golfo depression (Fig. 2.16A), may have been the main vent to emit lava within this depression during the Holocene. The Tanganasoga volcano yields a ^{14}C age of 6700 years (Pellicer, 1977). This age is compatible with the 12,000 ± 7000 years obtained by K/Ar in a borehole (S-10) that actually cuts through the upper lavas of Tanganasoga volcano (see Fig. 2.16B). A K/Ar age obtained from the lava at the bottom of that same borehole near the contact with the collapse breccia gave an age of 21,000 ± 3000 years (Guillou et al., 1996).

The majority of the recent volcanism is in fact sourced from the northwest rift, where several eruptions, including those of the more prominent Tanganasoga volcano, have been vented. However, no historical eruptions occurred on El Hierro until 2011, which is in contrast to the intense volcanism on La Palma that took place during the same time (see chapter: The Geology of La Palma, Figs. 3.21A and 3.23A).

◀ were 15 m high (photo from November 8, 2011). (D–H) In the early stage of the eruption, floating bombs with white frothy cores were erupted, which have been termed 'xeno-pumice'. Since their brief appearance, the nature and origin of these "floating stones" have been vigorously debated among researchers, with important implications for the interpretation of the hazard potential of this type of eruption. The xeno-pumice rocks have been proposed to be: (i) juvenile high-silica magma (eg, rhyolite); (ii) remelted magmatic material (trachyte); (iii) melted oceanic sediments; or (iv) altered volcanic rock, such as reheated hyaloclastites or zeolite from the submarine slopes of El Hierro. On the basis of mineralogy, whole rock and mineral compositions, and isotope geochemistry, Troll et al. (2012) concluded that the El Hierro xeno-pumice specimens are in fact xenoliths from pre-island sedimentary layers that were picked up and heated by the ascending magma, causing them to partially melt and vesiculate, similar to occurrences of frothy sedimentary xenoliths on several other Canary Islands (see, eg, chapter: The Geology of La Palma, The Geology of Gran Canaria, and The Geology of Lanzarote). The discovery of Cretaceous to Pliocene fossil assemblages in the xeno-pumice samples from El Hierro now seem confirm this conclusion (eg, Zaczek et al., 2015). (E) The core of this xeno-pumice contains clay-type sedimentary material surrounded by vesicles up to 2 mm in size and a finer vesicular layer near the margin. (F) Xeno-pumice with a highly vesicular inner domain around a sedimentary fragment that was "caught in the act" of degassing and inflation upon melting and pre-eruptive decompression. (G) Jasper fragment, partially dissolved in xeno-pumice (note the large vesicle in the upper left part of the sample, which indicates intense degassing). (H) Large relict detrital quartz grains in a xeno-pumice sample (images in D-H from Troll et al., 2015).

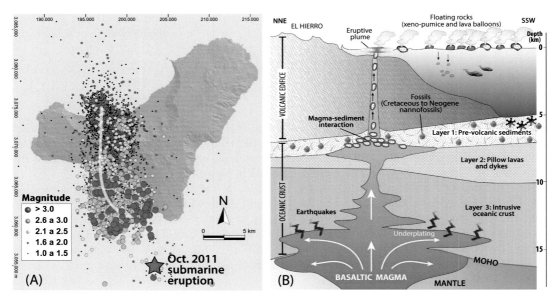

FIGURE 2.18

(A) Distribution of epicenters under the island of El Hierro between July 19 and October 8, 2011 (data from *Instituto Geográfico Nacional, IGN*). (B) Sketch showing the structure of El Hierro Island and the 2011 events. The ascending magma, according to the distribution of seismic events prior to eruption, moved initially in a sub-horizontal fashion from north to south at the depth of the oceanic crust, likely interacting with pre-volcanic sedimentary rocks at that point. The xeno-pumice rocks found during the first week of the 2011 eruption at El Hierro are thus probably dominantly the products of magma−sediment interaction beneath the volcano. Sedimentary rock fragments were picked up by migrating magma and melted and vesiculated while immersed in magma during transport to the vent site. Once erupted onto the ocean floor, they separated from the erupting lava and floated on the sea surface due to their high vesicularity and low density (modified after Troll et al., 2011, 2012).

HISTORIC ERUPTIONS: THE 2011 SUBMARINE ERUPTION

The only historic reference to a possible submarine eruption off El Hierro was in 1793, which appeared in the works of Bory de St. Vincent (1803), von Humboldt (1814), and Darias y Padrón (1929); the latter provided a detailed account from eye-witness reports of a seismic crisis on the island in that period. From March 27 to June 15, 1793, strong earthquakes shook the island, damaging buildings and raising public alarm. The prospect of a volcanic eruption prompted the authorities to prepare the first known general evacuation plan in the Canaries (Bethéncourt-Massieu, 1982).

The seismicity, initially felt only around El Golfo but later on the entire island, declined in intensity in midsummer and finally ceased without any mention of manifestations of subaerial volcanic activity in any of the reports (see also Stop 2.7).

After 218 years of quiescence since the 1793 unrest, a submarine eruption commenced offshore from El Hierro on October 10, 2011 (Carracedo et al., 2012a, b; Lopéz et al., 2012; Pérez-Torrado et al., 2012; González et al., 2013). Seismic precursors allowed early detection of the activity and its approximate location, indicating the impending eruption to be submarine. Prior to the first visible manifestations, such as the floating plume and stained waters and, later, voluminous gas exhalations and floating lava bombs (Fig. 2.17), the only indication of the upcoming eruption was the geophysical data (seismic shocks and ground deformation) recorded by the Spanish *Instituto Geográfico Nacional* (IGN). The majority of seismic events were insignificant from a hazard point of view and were not felt by the population. However, they provided a useful precursory phenomenon that heralded the eventual volcanic event, allowing time to prepare for the likely emergency (eg, Carracedo et al., 2015a).

Just prior to the eruption, seismicity increased in frequency and magnitude (up to M_L 4.3) and hypocenters became shallower (12−14 km) and progressed toward the south tip of the island (Fig. 2.18A). Finally, on October 10, 2011, a harmonic tremor started to be recorded by all of the seismic stations on the island, with the highest amplitudes registered offshore from *La Restinga* (López et al., 2012). The harmonic tremor marked the start of the submarine eruption and continued with variations in amplitude until the end of the eruption on March 5, 2012.

Magmatic underplating lifted the entire island edifice by >12 cm (Fig. 2.18B). In turn, the increases in total volume and height of El Hierro caused by the volcanic material expelled during the 2011−2012 submarine event were modest and localized (329×10^6 m^3; Rivera et al., 2013, 2014). Periods of unrest and underplating may be much more frequent and even more important than eruptions, as was observed in this case. This implies that a significant part of oceanic island growth may be related to intense magmatic underplating (Hansteen et al., 1998; Klügel, 1998; Klügel et al., 2005; Longpré et al., 2008a; Stroncik et al., 2009; Barker et al., 2015; Carracedo et al., 2015a).

The nature and origin of some of the floating lava bombs, particularly those with a glassy, dark basanite crust enveloping light-colored, highly vesicular, white glassy interiors (Fig. 2.17D), were initially explained by a number of hypotheses. These included: (a) the white cores are sedimentary xenoliths with varying degrees of melting (Troll et al., 2012); (b) they are remobilized hydrothermally altered trachytic/rhyolitic volcanic rocks (Meletlidis et al., 2012); and (c) these cores are remobilized trachyte magma that assimilated approximately 10% sedimentary quartz to form a rhyolite composition (Sigmarsson et al., 2013).

Several lines of evidence support the cores being sedimentary xenoliths. The cores include large, clear, and rounded quartz crystals, whereas igneous rocks on El Hierro do not contain any free (primary) magmatic quartz crystals. Furthermore, they contain sedimentary relicts of gypsum, jasper, and clay fragments and have high $\delta^{18}O$ values (9 to $\geq 12\%$) that overlap with the lower end of global sedimentary reference values (9 to 30%), but not with regular magmatic $\delta^{18}O$ isotope compositions (5.5−7.5%; eg, Hoefs, 1997). Moreover, some of the clay-rich relicts in the El Hierro xeno-pumice samples contain Cretaceous to Pliocene nannofossils (Zaczek et al., 2015; Troll et al., 2015). This notably defines that the youngest sedimentary strata in the archipelago reside under El Hierro (up to Pliocene), thus confirming the El Hierro edifice as the youngest in the archipelago, which, in turn, strengthens the plume hypothesis (see Fig. 1.14). Finally, the white cores show dissimilar incompatible trace element concentrations compared to trachyte and rhyolite from any known Canary compositions (eg, Carracedo et al., 2015a). These pieces of evidence likely support the idea that the core samples are dominantly xenoliths from originally pre-island sedimentary layers. These xenoliths were probably picked up and heated by the ascending magma, causing them to partially melt and vesiculate on decompression (Fig. 2.18B). Thus, the term "xeno-pumice" was coined during the El Hierro 2011 events (eg, Troll et al., 2011, 2012) to accurately characterize this type of rock, and was linked to previous descriptions of the phenomenon on eg, Lanzarote (Aparicio et al., 2006, 2010) or Gran Canaria (Hansteen and Troll, 2003).

The progress of the 2011/2012 eruption was monitored by surveys from onboard the Spanish research vessel Ramón Margalef, focusing on changes in seabed elevation and the water column. By using acoustic techniques, they showed that at the end of the eruption the submarine volcano had grown from 350 m to approximately 88 m below the ocean surface (Fig. 2.19). A key factor in the assessment of the potential for explosive (surtseyan-type) submarine eruptions is the depth of the vent, with the volume ratio of steam reaching critical explosive levels at depths of 100 mbsl or less (Schmincke, 2004). This implies that factors such as waning magma supply have likely also played a role in the El Hierro eruption in 2012, where surtseyan activity did not occur at any stage.

The final submarine cone, nested in a valley (Fig. 2.20), consisted of at least four vents along a north-northwest to south-southeast lineation. The total accumulated volume was 329×10^6 m^3. Bathymetric data revealed that the eruption constructed a double cone, with lavas filling part of the surrounding valley and forming an extended apron down to a depth of more than 2000 mbsl (Rivera et al., 2013, 2014). A relevant additional lesson from the 2011/2012 submarine eruption is the observation that intrusions caused significant inflation and accumulated a permanent vertical uplift of approximately 12 to 15 cm, whereas eruptive volumes (329×10^6 m^3; Rivera et al.,

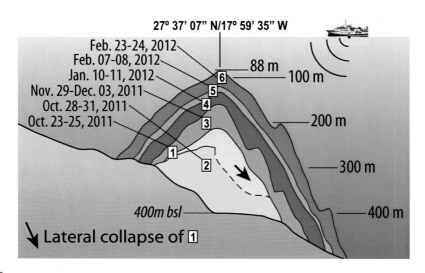

FIGURE 2.19

Development of the 2011 eruption vent site from data obtained in successive bathymetries of the RV Ramon Margalef (data from *Instituto Español de Oceanografía, IEO*). The decrease in height of the cone from the first to the second bathymetry was caused by a flank collapse of the rapidly growing cone (after Rivera et al., 2013).

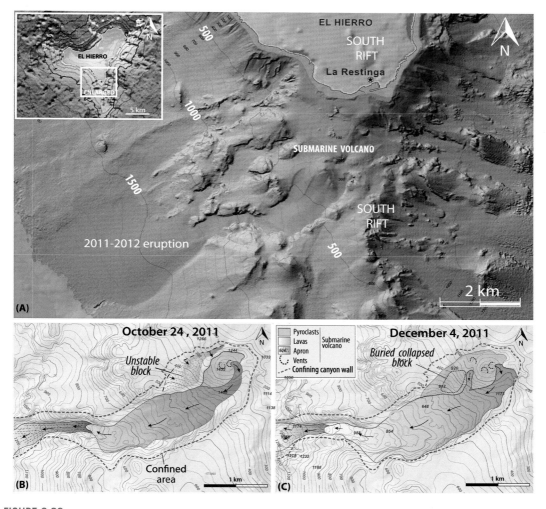

FIGURE 2.20

(A) Bathymetric map of the underwater volcano, which began to grow on October 10, 2011 on the south ridge of El Hierro Island. The map covers 21,000 hectares (210 million m²). (B) Geological map of the submarine eruption using the first bathymetry obtained on October 24, 2011 by the *RV Ramon Margalef*. (C) Geological map of the same area for December 4, 2011 (following a subsequent bathymetric survey) (from Carracedo et al., 2012, 2015a and Rivera et al., 2013).

2013) are almost negligible in comparison to this uplift volume. Magmatic underplating thus considerably raised the entire island edifice in 2011 and 2012, and we must realize that underplating represents a significant aspect of island growth (eg, González et al., 2013; Carracedo et al., 2015a).

GEOLOGICAL ROUTES

El Hierro is a relatively small island with a population of approximately 12,000 (Fig. 2.21). Economically, it is one of the weaker islands of the Canary archipelago and, consequently, infrastructure is not always as well developed as on the larger islands. For this reason, a number of the trips we suggest are on well laid out dirt tracks. Hopefully, these will gradually become tarmac roads within the next few years.

EL HIERRO DAY 1: TIÑOR VOLCANO AND SAN ANDRÉS LANDSLIDE

Start the day in *Valverde* (Fig. 2.22). Make your way to *Plazoleta del Consejo* in the city center and park the car there on the open parking space.

Stop 1.1. Plazoleta Del Consejo (N 27.8092 W 17.9158)

Columnar jointed lavas of reverse magnetic polarity are exposed here. The lavas here are basaltic in character, with abundant crystals of olivine and pyroxene, plus xenoliths fragments of pyroxenite and peridotite. The reverse magnetic polarity previously measured here implies them to be older than 0.78 Ma (see Fig. 2.3). Thus, they are some of the oldest lavas exposed on El Hierro.

Continue in a southwesterly direction and, toward the end of *Valverde*, notice that a small road (unmarked) goes off to the left just before a sports ground (Fig. 2.22). Follow this simple road for

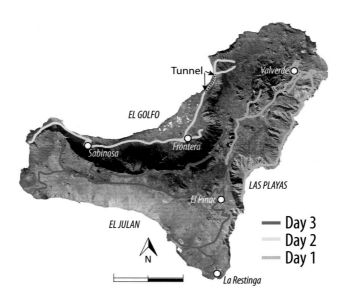

FIGURE 2.21

Geological itineraries on El Hierro. See also Figs. 2.24, 2.33, and 2.44 for further details. Map modified from IDE Canarias visor 3.0, GRAFCAN.

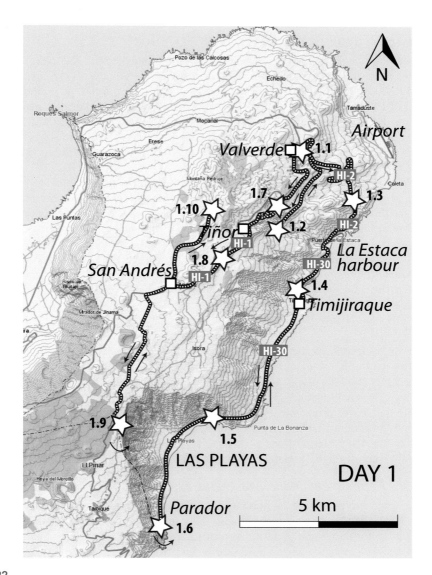

FIGURE 2.22

Detailed itinerary for *Day 1*, marking the stops and other relevant features in and around the Tiñor volcano and the San Andrés fault and landslide system. Map modified from IDE Canarias visor 3.0, GRAFCAN.

approximately 4 km to inspect the spectacular San Andrés fault that, after some time, will appear on your right in a superb and relatively new road cut.

Stop 1.2. San Andrés Fault (N 27.7884 W 17.9221)

The San Andrés fault system marks the site of a large landslide in the old Tiñor volcano (Fig. 2.23A). However, this collapse did not fail completely (Guillou et al., 1996; Day et al., 1997; Carracedo et al., 2001). The fault plane dips steeply and is, in fact, found at a number of places around this part of El Hierro (Fig. 2.23B,C). The white staining in the lower part marks where the fault was covered before the road was put in place (Fig. 2.23D). Behind the fault, the outcrop shows younger lava on top of the fault rock assemblage. This canyon-filling lava is from the current rift and was dated at approximately 0.1 Ma (Guillou et al., 1996), thus defining a rough age for the activity of the San Andrés fault system.

The fault plane exposed here records rapid movement because pseudotachylite glass (friction melt) is still found in places that can be used to estimate speeds of fault movement. For the production of melts via direct high-speed frictional sliding, a slip rate of 1 m/s or more is required (Fig. 2.23D) (Day et al., 1997). The fault also shows a damage zone of several meters behind the fault plane. Slickensides are still locally preserved and the downthrown side of the fault system can be seen as well, with steep dipping lavas that reveal a "drag" against the fault plane. The estimated offset is approximately 300 m on this fault, but it may have varied locally along the fault system and likely involved several fault splays. Also, note the thick cream-colored fault gouge next to the fault plane, which is made up of finely crushed material that got caught during the fault movement. Also note the IGN station that measures seismic activity here, which was put in place for good reason.

Make your way back to *Valverde* on the small road on which you came. Note the windmills and pumping pipelines of the hydro-electric plant on the way back, which we explore in more detail later. Wind energy is used to pump water to higher elevations, which can then be released on demand to produce electricity (see, eg, Stop 1.7).

Once back in *Valverde*, drive into town and follow signs for *"aeropuerto."* Leave *Valverde* toward the airport, and right after the airport turn off from EH-911; the San Andrés fault becomes visible again. Pass the km 3 signpost and park in a large lay-by on your right that is covered with black lapilli. Walk south for approximately 100 m to inspect the fault (Fig. 2.23E).

Stop 1.3. Curved Northern End of the San Andrés Fault (N 27.7941 W 17.8973)

The San Andrés fault curves toward its northern end and fault displacement is likely smaller here (eg, Day et al., 1997), defining a potential collapse mass inside the curved and likely listric outline of the destabilized sector in the east and southeast of El Hierro (Fig. 2.23A–E).

Continue on EH-911 toward *Puerto de la Estaca*, and from there follow signs (brown) for *Las Playas*. Pass lavas, scoria, and volcanoclastic sedimentary rocks of the Tiñor volcano on your right.

Immediately after the first tunnel on EH-911, park the car in a lay-by on your right.

Stop 1.4. Pillow Lavas at Timijiraque (N 27.7709 W 17.9142)

Walk up the old coastal road on the opposite side of EH-911 for approximately 300 m, where you will encounter rounded boulders under the Tiñor lavas. A few steps further, a small number of lava pillows are exposed (Fig. 2.24). This former beach is approximately 10 m above the current beach,

FIGURE 2.23

(A) The San Andrés fault delimits a collapsed block that slumped by approximately 300 m but remained anchored, apart from in the Las Playas collapse embayment (image from *Google Earth*). The squares indicate the different views of the fault. (B) Well-preserved fault plane at the middle exposure level of the fault. The white carbonate stained part was only recently exhumed by the opening of the track (image courtesy of *C. E. Karsten*). (C) Partly eroded fault plane in the upper exposure of the fault. (D) Close-up view of the mid-level fault plane, showing down-dip grooves. (E) View of the southeast edge of the San Andrés fault.

FIGURE 2.24

Pillow lavas at *Timijiraque* beach. Pillow lavas must be volumetrically abundant because they likely comprise large portions of the submarine part of the island, but outcrops are relatively scant in the Canary Islands. This former beach is approximately 10 m above the current beach, documenting that El Hierro is in a state of uplift since the days of the Tiñor volcano that produced these pillows. Note that the accumulated, and now largely permanent, uplift during the 2011/2012 eruption was more than 12 cm in total.

testifying that El Hierro has been in a state of uplift since the days of the Tiñor volcano. Note the lasting uplift during the 2011/2012 events was 12 cm. El Hierro thus shows the typical signs of a growing ocean island in which island subsidence is not a major phenomenon.

Return to your car and continue toward *Las Playas*. You will pass more Tiñor lavas on the coastal road, but now several younger (recent) flows can also be seen to have cascaded over the steep barranco walls of the old Tiñor shield. Soon you will come to a second tunnel. Again, right after the tunnel, park the car in a lay-by, but this time on the left.

Stop 1.5. Las Playas Embayment (N 27.7308 W 17.9423)

The bay in front of you is *Las Playas*, the part of the destabilized rock mass of the San Andrés system that actually slumped into the sea (Fig. 2.25A). Seismic and sonar data show two different landslides: the San Andrés and the younger Las Playas landslide (Gee et al., 2001). These authors interpret the San Andrés as a slump, probably involving several separate sliding events, whereas the Las Playas event is a debris avalanche, smaller than, but similar to, the El Golfo landslide (see Fig. 2.9B,C).

The Las Playas landslide scar exhumed the deeper structure of this part of the island. The cliff section is topped by thick trachyte flows overlying a sequence of basaltic lavas that is intensely intruded by dykes. This stratigraphic sequence is similar to that found in the *El Golfo* cliff, but here in *Las Playas* the older Tiñor lavas form the lower part of the section (see Fig. 2.4).

You will also see steeply dipping lavas in the foreground, which are recent infillings from the rift zone above. Note the different dip angles of the deposits at the front. The conical scree fans are at the angle of repose ($\sim 33°$). Only lavas can exceed this angle (see Fig. 2.25A) because of the internal cohesion of magma (Newtonian vs Bingham fluids).

Continue into *Las Playas* bay. Stop at the *Parador de Turismo* (a state-owned hotel) at the very end of the road.

FIGURE 2.25

(A) *Las Playas* landslide embayment. The volcanic sequence comprises eruptive rocks of the Tiñor volcano that are overlain by deposits from the El Golfo volcano. In the foreground, the steeply dipping rift lavas contrast with the more gentle screes. (B) *Roque de la Bonanza*, in the *Las Playas* bay, is a sea stack that has resisted erosion because it is crossed by a dyke. Bonanza (prosperity) refers to the rich fishing grounds in the bay here.

Stop 1.6. Parador de Turismo (N 27.7172 W 17.9583)

If open, this might be a nice refreshment stop. Behind the *Parador de Turismo*, Tiñor lavas and dykes are exposed. These are among the oldest rocks exposed on El Hierro (≥1.1 Ma).

The San Andrés collapse has obviously removed considerable amounts of material in this part of the collapse system. Thus, the Las Playas embayment most likely reflects a partly aborted landslide; that is, a sliver of the collapsed mass simply stayed behind and is now present as the slipped mass of arrested rock (Stop 1.2).

Return on EH-911 toward *Valverde*. Just before the tunnel at the end of Las Playas bay, you will have a good view of *Roque de la Bonanza*, a sea stack that has resisted erosion because it was intruded by a dyke and is hence more weathering-resistant (Fig. 2.25B). The term *Bonanza* here refers to the calm bay that used to be rich in fish, ie, a place of plentiful catches. Continue toward *Valverde*. Return to the main road that led to Stop 1.2 (*Calle San Juan*), but continue on the main road uphill. The road climbs steadily and you will get to a major crossing after a short drive. Right at the crossing is a mirador. Drive there and park the car.

Stop 1.7. Planta Hidroelectrica (N 27.7945 W 17.9223)

Here, you can view the water reservoir used by the hydroelectric power plant. This volume of water can be released to produce power when electricity is in demand (eg, during peak tourist season). The system is effectively renewable and self-sustained, providing a powerful example for alternative energy concepts of the future. The power plant uses the reservoir to store water that is pumped there by means of wind energy. Another reservoir further uphill (700 masl) inside a phreatomagmatic caldera also exists, and water can be transferred between them (Fig. 2.26). When released,

FIGURE 2.26

(A) An explosion crater, probably of phreatomagmatic origin, was converted to a water reservoir for the El Hierro Wind and Hydro-Power Station. This project integrates a wind farm, a pumping station, and a hydro-electrical plant (B and C). The wind farm supplies electrical energy directly to the network while simultaneously feeding a pump that raises water to an elevated reservoir as a form of power storage. When wind power is poor, the hydro-electrical plant uses the stored potential energy of the water in the elevated reservoir to generate additional electricity, thus guaranteeing a steady and sustainable power supply.

water creates hydroelectric power, thus supplying electricity independently of wind strength at time of power demand.

Note that the rocks in the surrounding hillside are of the oldest Tiñor volcano and are dated at just over 1 Ma in this area.

Return to the crossing at the water reservoir and drive uphill toward *Tiñor* village using the old road (TF9111/HI-104). Just a short stretch further, you will pass the upper exposure of the San Andrés fault again.

Stop 1.8. San Andrés Fault (N 27.7792 W 17.9384)

This old road (TF-911) has been abandoned since the construction of HI-1 and is now unused. Thus, it is possible to stop anywhere on the road to inspect part of the San Andrés fault (see Fig. 2.23C). Here, slickenslides and the damage zone of the fault are visible again, although the surface has lost its high polish due to weathering.

Continue on the old TF-911 road to the next roundabout and join HI-1 in a westward direction (left). Pass the lapilli deposits of *Mt. Chamuscada* (dated by ^{14}C at 2500 years BP) before entering the village of *San Andrés*. Continue toward *El Golfo* and *La Restinga* and after another 800 m, turn left onto HI-4 (orange) toward *El Pinar* and *La Restinga*. After 4.5 km on HI-4, turn left into HI-402 (white) and park at the upcoming mirador.

Stop 1.9. Mirador de las Playas (N 27.7318 W 17.9716)

From here, you can view the "dent" in the San Andrés landslide mass from above (ie, the section that collapsed completely) (Fig. 2.27A). Note, submarine ocean floor mapping documents a debris avalanche deposit offshore that links with the surface evidence for a large collapse in this part of the island. Also note the unconformity that separates the El Golfo and Tiñor volcanoes in the valley wall, and the extensive dyke swarm in the older Tiñor volcanics (Fig. 2.27B).

Take a walk around the head of the embayment along the walking trail that starts at the mirador. Once you have enjoyed the views and the breath-taking heights sufficiently, return to the car and return to HI-1 toward *San Andrés*. Then, take HI-10; after 300 m, take a track to the right posted *El Garoé*. Follow the indicators until you reach the entrance to the *El Garoé* site. An entrance fee of €1.50 per person applies.

Stop 1.10. The Holy Tree (Garoé) (N 27.7959 W 17.9427)

The Canary Islands lack extrusive surface water (eg, rivers and permanent lakes). Freshwater in the archipelago is found mainly as groundwater from wells and *galerías*, particularly in the central and western Canaries. Impermeable layers (eg, palaeosols, baked soils, dykes) help to retain groundwater reserves (see, eg, chapter: The Geology of La Gomera, and chapter: The Geology of Tenerife, eg, Fig. 5.107). El Hierro, however, is formed by dominantly permeable volcanic sequences that are unable to retain groundwater to a significant degree.

The absence of springs on the island of El Hierro has obliged the inhabitants until recently to store rainwater for both domestic purposes and for irrigation. One can only imagine how difficult life must have been for the aboriginal inhabitants, the Bimbaches. However, legend has it that in ancient times there was a tree on the island from which water was issued (Fig. 2.28A,B). That tree was the *Garoé*, which is said to mean "lagoon"; it was one of the Bimbaches' greatest treasures because at times there was no other source of water on the island.

FIGURE 2.27

(A) Las Playas collapse embayment viewed from the *Mirador de las Playas*. (B) Stratigraphic sequence in the wall of the embayment. Note that the deposits of the El Golfo volcano overlie those of the Tiñor volcano in an erosive unconformity, and the rift lavas unconformably rest on top of both.

The existence of the *Garoé* is historically documented. It would appear to have been a laurel tree (*Ocotea foetens*), endemic to the islands of Madeira and the Canaries, where an individual tree trunk can have a diameter of up to 1.5 m. This tree remained standing until 1610, when it collapsed during a hurricane. The historian Abreu Galindo asserted that the sacred tree was located in "a gully running up a valley from the sea to the front of a steep rock face."

However, the historian Viera y Clavijo expounds that "the Garoé" neither "gave in a single night twenty thousand barrels of water nor did it extract from the aridity of the ground the copious moisture

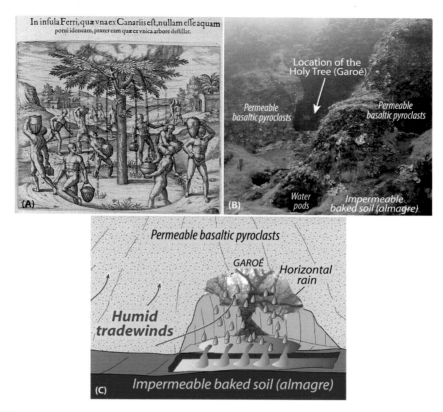

In infula Ferri, quæ vna ex Canariis eft, nullam effe aquam potui idoneam, præter eam quæ ex vnica arbore deftillat.

(A)

Location of the Holy Tree (Garoé)

Permeable basaltic pyroclasts

Permeable basaltic pyroclasts

(B)

Water pods

Impermeable baked soil (almagre)

Permeable basaltic pyroclasts

GAROÉ

Horizontal rain

Humid tradewinds

Impermeable baked soil (almagre)

(C)

FIGURE 2.28

(A) The "Holy Tree" (*El Garoé*) in a 1748 copy from an illustration by *T. de Bry (1596)* depicting how the water was collected around the Garoé (image courtesy of *Cabildo Insular de El Hierro*). (B) A sequence of porous basaltic scoria beds absorb water from the humid trade wind. A perched aquifer retains the water through an impermeable baked palaeosol (*almagre*), and waterholes carved into the *almagre* collect the water. (C) The presence of large trees like the Garoé improved the interception and precipitation of water from the humid trade winds and provided a steady supply of water to the bimbaches.

later distilled". The scientific explanation is simple and is based on well-known concepts. The trade winds, saturated with water vapor from their path over the ocean, ascend the northeast flanks of El Hierro, cooling progressively as they gain altitude. On reaching a characteristic temperature (dew point), the water vapor dissolved in the air condenses to form fine droplets that are visible to the naked eye. The fog or stratified cloud is referred to in the Canaries as the sea of clouds (*mar de nubes*) (see also chapter: The Geology of Tenerife, Fig. 5.46). In the condensation zones, if the wind pushes the fog against a plant cover, turbulence is created that favors tiny droplets to attach to leaves and branches and coalesce to form larger drops of water that fall to the ground because of their weight (Fig. 2.28C). Thus, a "rain" from the Garoé was created that formed pools of water underneath the tree, whereas the surrounding ground was dry (indicating a locally impermeable substrate).

In tall isolated trees, this process demonstrates its maximum efficiency. The phenomenon requires the tree to be situated at a suitable altitude for maximum condensation (in the area of the Garoé, 1.020 masl) and in the path of the tradewinds. It also requires the existence of a thick impermeable layer of *almagre*, a baked red clay, under the tree to create a basin that prevents water migration from the ground.

The Garoé tree is located in an area called *Los Lomos*, which is part of the island's oldest rock formation. However, there is a thick layer of volcanic scoria at this site from recent subaerial volcanism. Beneath this layer of scoria, there is an impermeable layer of an argillaceous paleosol (Fig. 2.28B,C). The distinctive slope of the scoria layer guides water that infiltrates by condensation or rainfall to underneath the scoria and directly onto the layer of fine yellow lapilli that was baked by the overflowing scoria, thus making this underlying horizon impermeable. The Bimbaches dug holes into this impermeable bed, thus creating sinkholes to catch the water of this "perched aquifer."

Today, a new well (*Pozo de Los Padrones*) supplies all the water requirements of the island. This well, located near the base of the El Golfo scarp, is 54 m deep and continues into a 760-m-long galeria that is penetrating the El Golfo volcano (Fig. 2.29). This galeria crosses relatively recent lavas (<0.5 Ma) that fill the El Golfo landslide scarp and an altered debris avalanche deposit, which, in turn, rests on the older (>1.04 Ma) Tiñor lavas (Figs. 2.10, 2.29). The geological setting of the El Golfo embayment (tomorrow's trip) has thus replaced the Garoé tree, allowing modern society to draw on fossil water rather than on rainwater only.

Return to *Valverde* via HI-1.

END OF DAY 1.

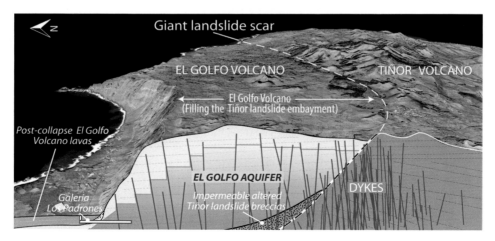

FIGURE 2.29

The more than 1000-m-thick sequence of the El Golfo volcano, of dominantly basaltic flows, provides a porous rock mass capable of storing important reserves of groundwater (background image by *Google Earth*). The basal debris avalanche from the giant lateral collapse was likely transformed by alteration and compaction into an impermeable clay layer, thus preventing draining of the aquifer. This aquifer yields a sufficient volume of water to cover current demand on the island but is not replenished at the same rate (after Carracedo, 2008b).

EL HIERRO DAY 2: THE EL GOLFO COLLAPSE SCARP

We start once more in *Valverde*, and today we explore the northern side of the island (Fig. 2.30). There are many opportunities to find refreshments along the way, which means not much needs to be packed. However, if you would like to swim in one of the natural pools along the El Golfo coast or the beach at the last stop of Day 2, then bring your swimming gear with you. Otherwise, sunscreen and a fleece jacket are required.

Drive out of *Valverde* on HI-5 toward *Frontera/El Golfo*. Pass lavas of the Tiñor volcano on your left. Right after passing through *El Mocanal*, turn toward *Guarazoca* and *La Peña* (HI-10). Park your car at *Mirador La Peña*.

Stop 2.1. Mirador de La Peña (N 27.8072 W 17.9807)

Here, you can walk along a paved path to view the El Golfo embayment, the youngest giant collapse of the Canaries (see Fig. 2.13). The headwall is up to 1200 m tall, making this one of the tallest semi-vertical cliffs in the world. Giant landslides in ocean volcanoes cause many of the highest cliffs in the world, such as Cabo Girao on Madeira (570 m), the Kalaupapa cliff (1010 m) on the island of Molokai, and of course the El Golfo scarp (1235 m) right here on El Hierro.

This mirador includes a restaurant that was designed by the famous Spanish artist and architect César Manrique. The 700-m-high mirador offers an "aerial" view of the entire El Golfo landslide scarp that is partially filled with postcollapse volcanism, and also bathroom facilities and strong coffee.

Recent eruptive activity on El Hierro has started to fill the depression that originated from the lateral collapse of the El Golfo volcano. Several eruptions, whose centers are located in the interior

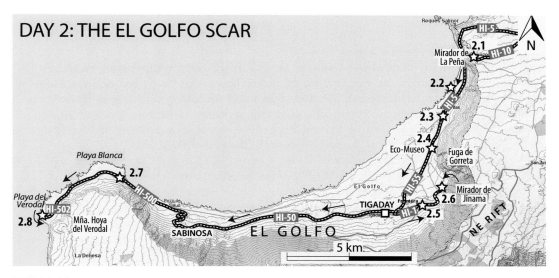

FIGURE 2.30

Detailed itinerary of *Day 2* with stops and other relevant features in and around the El Golfo embayment (map from IDE Canarias visor 3.0, GRAFCAN).

of the landslide basin and on the collapse scarp itself, can be observed. This particularly includes the Tanganasoga volcano, one of the most voluminous and youngest eruptions located inside the El Golfo landslide embayment (see Fig. 2.16A).

After having taken in some breathtaking views from *La Peña*, return to HI-5 and follow signs toward *El Golfo* and *Frontera*. This road will bring you to the El Golfo tunnel. Pass seaward-dipping lavas and volcaniclastics of the Tiñor volcano on your left before entering the tunnel a short while later. After the tunnel, the El Golfo embayment will open up in front of you when you enter the El Golfo valley. Not too long after leaving the tunnel, turn right onto HI-55 toward *Punta Grande*, where, according to the Guinness Book of Records, the smallest hotel in the world can be found. Drive there and park the car.

Stop 2.2. Punta Grande (N 27.7973 W 17.9910)

The cliffs to the northeast reveal two thicker gray bands—the two trachyte flows of the El Golfo volcano (Fig. 2.31A,B). The two sea stacks further out in the bay (*Roques de Salmor*) are made of this trachyte material and thus are likely more weathering-resistant than the predominantly basaltic material of the remaining El Golfo sequence (Fig. 2.31C).

A large boulder of this light gray material lies on the open pebbled viewing platform here at *Punta Grande*. There, several larger and smaller blocks of blackish basalt are also present. The trachyte carries very few crystals of sanidine and pyroxene, whereas one of the basaltic blocks carries olivine and pyroxene. One of the smaller blocks carries olivine only and another smaller basalt block contains predominantly reddish (oxidized) olivine, thus providing a visual difference between these different rock types here.

Return to your car and proceed on HI-5 into the El Golfo valley and toward *Tigaday*. A few hundred meters later, park in a lay-by on the right, immediately after you have passed the turn for *Pozo de la Salud* (HI-550) on your right.

Stop 2.3. HI-5/HI-550 Crossroads (N 27.7870 W 17.9937)

Here, you can inspect a morphological section through the El Golfo wall. A sharp secondary cliff occurs in the lower part of the wall, topped by postcollapse lavas and scree deposits, including recent scree (at the lower cliff) (Fig. 2.32). The sequence of formation was likely as follows: (1) El Golfo landslide forms the scarp; (2) postcollapse eruptions mantle the scarp; (3) marine erosion retrogrades the coast and generates the lower cliff; and (4) recent lavas fossilize the lower cliff to form a coastal platform that allows for scree deposits at the foot of the scarp.

This sequence of events likely required a longer period of time than the approximately 11,000 years attributed by Masson (1996) to the El Golfo landslide events (see Figs. 2.16 and 2.32). However, younger events that gave rise to submarine debris avalanche episodes are very conceivable also, hence reconciling the submarine debris layer dated at approximately 11,000 years by Masson (1996) with the onshore geological evidence that provides details for a maximum age for the main El Golfo collapse of between approximately 130 and 80 ky (Guillou et al., 1996; Carracedo et al., 2001) but a minimum age of at least 39 ka (Longpré et al., 2011).

Continue on HI-5 into the El Golfo valley. After a short drive (~3 km), the Guinea Ecomuseum appears on your left. Drive into their carpark to view current instability issues along the El Golfo headwall.

FIGURE 2.31

View from *Punta Grande*. (A) Volcanic stratigraphy of the eastern cliff of El Golfo. (B) Close-up view showing the El Golfo volcano sequence topped by trachyte flows and overlying rift lavas. (C) Detail of the *Roques de Salmor*, two erosive remnants of the trachyte lava flows.

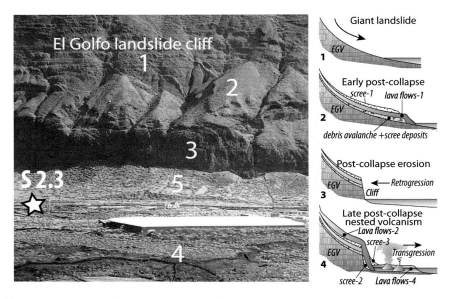

FIGURE 2.32

View of the El Golfo collapse escarpment (Stop 2.3) showing geomorphological features that suggest a pre-Holocene age for the main El Golfo collapse. The likely sequence of events is as follows: the initial formation of the collapse scarp (1); postcollapse eruptions mantling the scarp with lava flows and inter-bedded scree deposits (2); a basal cliff formed by retrogression of the coast (3); and recent lavas fossilized the cliff and scree deposits attached to the escarpment on top of the recent lavas (4). It would appear unreasonable that this entire sequence of events had been completed in 11,000 years or less (after Carracedo & Day, 2002).

Stop 2.4. The Guinea Ecomuseum (N 27.7744 W 17.9991)

The Guinea Ecomuseum shows part of the ancient village of *Guinea*. The village was established on a pre-Hispanic settlement and abandoned after part of it was buried many years ago because of frequent floods and landslides from an active stream with a source in the cliff of El Golfo (Fig. 2.33A). The entrance fee is 6.50 EUR and, if you wish, guided tours are hourly, starting at 10:30 am.

Ignoring the dangerous location on the delta of the stream, the village was restored and the ecomuseum was installed, largely with funds by the European Union. In addition, a laboratory devoted to the reproduction of the giant lizard (*gallotia simonyi*), which is endemic to the island and presently confined to this cliff, has been established here. However, precautions against future landslides and floods are surprisingly limited and the protective barrier erected in 2004 may be inadequate (Fig. 2.33B).

Indeed, torrential rains in 2007 resulted in the loss of the greater part of the museum facilities, the laboratory, and many animals undergoing study. Fortunately, there was no loss of human life. Notwithstanding, the village and the ecomuseum have been restored again and activities have resumed, despite the obvious dangers of the site.

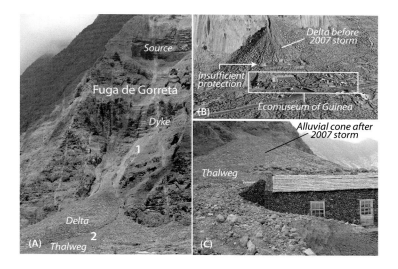

FIGURE 2.33

(A) Active "dejection cone" at the foot of the El Golfo cliff (*Fuga de Gorreta*) showing the area of the eroded fan (1) and the debris of the 2007 flood (2). (B) Location of the old and abandoned *Poblado de Guinea* and, recently, of the Guinea Ecomuseum. (C) The 2007 flood partly buried these facilities, and it remains to be seen if the rebuilt museum complex will fare any better, but probably not, at least not in the long run.

Walk to the eastern end of the village to inspect remnants of recent mass movements into the village (Fig. 2.33C).

Another notable but volcanological feature to inspect here in the outdoor museum is a number of lava tunnels in the recent (postcollapse) lavas, which were used for simple housing by the Guanches and as cool storage facilities by the Spanish settlers. It is possible to go inside some of these tunnels, at least under guidance.

There is also information available on the El Hierro ("giant") lizard and the Canary Island endemic flora and fauna, allowing a wider appreciation of El Hierro's natural world and cultural heritage.

Continue on HI-5 toward *Frontera* (ie, into the El Golfo embayment) until it rises uphill and signs for *Valverde* appear. HI-5 turns into HI-55 in *Frontera*. Turn left at the junction between HI-55 and HI-1 (toward *Valverde*). Drive uphill for a few minutes until you reach the church and tower of the *Iglesia de la Candelaria* (built in 1615). Park in front of the church.

Stop 2.5. Iglesia de la Candelaria (N 27.7551 W 18.0014)

Walk a few steps up the hill behind the church building toward the tower for some superb overview impressions. (Fig. 2.34). Note the recent black and reddish volcanic foundation on which the tower stands while walking up the path. The view here includes the entire El Golfo embayment, with the collapse scarp circling around the postcollapse coastal platform that is gently dipping toward the sea. Boreholes drilled in this area to monitor the depth of the groundwater level crossed these

FIGURE 2.34

Iglesia de la Candelaria. The bell tower tops one of the larger cinder cones inside the El Golfo embayment. The El Golfo cliff is visible in the background.

young lava flows that are overlying aeolian sand deposits and allowed dating the oldest postcollapse lavas to a K/Ar age of 21 ka (see Fig. 2.16B).

The church tower here sits on an eruptive vent inside the El Golfo embayment. This exemplifies the way the collapse embayments fill up as a result of postcollapse volcanic activity (see also the Güímar collapse on Tenerife). Be warned that the church bell rings quite loudly at the close of each hour.

Return to your car and turn into the small road right in front of the church that leads to *El Lunchón* and *Las Lapas*. Follow this road by car or on foot for approximately 1 km to inspect intra-embayment lava flows. Follow the small road until it becomes a dirt track for a short stretch. Eventually, it ends in an abandoned old quarry behind the last houses in that street.

Stop 2.6. Abandoned Quarry (N 27.7576 W 17.9938)

Here, deeply incised gullies offer the opportunity to investigate the postcollapse infill and provide an idea of the amount of erosion that has already occurred again. Obviously, volcanic refilling is not keeping up with erosion throughout the entire El Golfo embayment, although the ultimate result will be a filled landslide embayment comparable with the Tiñor and the El Julan landslides on El Hierro and those on other islands in the Canaries (eg, Carracedo et al., 2011b).

Steeply inclined intra-embayment lavas are exposed here that are interlayered with debris fan deposits (scree). Note how the lavas thicken toward your position when compared to the thickness of the lavas exposed above the quarry site (Fig. 2.35A). The lavas are locally very crystal-rich

FIGURE 2.35

(A) Steeply inclined intra-embayment lavas are interlayered with debris fan deposits (scree). These flows were erupted from vents located at the base of the El Golfo collapse scarp. (B) The scree increasingly coarsens at the foot of the scarp and up to house-sized blocks that show intense internal deformation. These may correspond to the debris avalanche that formed from the initial collapse.

(olivine and pyroxene) and are a part of the recent series that cascaded down the El Golfo headwall.

Walk down the small dirt track beneath the quarry where the scree becomes increasingly coarser and where house-sized blocks show intense internal deformation at the bottom of the track (Fig. 2.35B). These are some of the earliest deposits laid down inside the El Golfo embayment, and they may even correspond to the debris avalanche from the initial collapse.

Return to *Iglesia de la Candelaria* and from there to HI-55. Turn left there and drive into the old town center of *Frontera*. You will find refreshment opportunities, a post office, and a pharmacy in case you are in need of some supplies. When continuing on HI-55, you will soon meet HI-50 (green), where you turn left toward *Sabinosa*. Stay on HI-50 for several kilometers. Once houses become less frequent, you will pass the recent cinder cone of *Montaña Tamasina* (270 masl).

Soon you will drive through extensive postcollapse lava deposits on your left again for several kilometers along the road to *Sabinosa*. Many of these originate from vents high up in the collapse scarp, often very close to the headwall itself (see Fig. 2.16A). Continue into *Sabinosa*. This village is one of the oldest settlements on the island and was built high into the mountain face to spot pirate landings well in advance.

The road goes downhill in a bendy fashion after you leave *Sabinosa*, and you will pass extensive lapilli deposits on the way. After a while, HI-50 will meet HI-500. Turn left here toward *Ermita Virgen de los Reyes*. Follow this flat road for several kilometers. On the left, the El Golfo cliffs rise, while on the right, extended coastal lava platforms from several recent eruptive vents are visible.

Soon after, you will drive through a boulder field of basaltic and ankaramitic blocks, which indicates active erosion of the El Golfo cliff at this site. After another 2 km on this flat road, you will see a conspicuous orange − colored outcrop of part of a tuff ring.

Stop 2.7. El Verodal Tuff Ring (N 27.7610 W 18.1417)

On El Hierro, there are several spectacular examples of phreatomagmatic eruptions in which sea water or groundwater came into contact with magma to generate clearly identifiable hydro-volcanism.

The violent vaporization of water increases the explosivity of an eruption, fragmenting the lava much more efficiently than in nonphreatomagmatic eruptive events. In addition to being fragmented, lava from phreatomagmatic eruptions is quickly cooled and altered by the hot water and gases. The result is a very characteristic type of orange to reddish pyroclastic deposit (palagonitic tuffs) that may be highly vitreous and brightly colored in tones of orange to yellow, making them easily recognizable (Fig. 2.36A).

FIGURE 2.36

(A) Part of a tuff ring is exposed within the basaltic sequence at *La Hoya del Verodal*, at the northwest edge of the island. Note the distinct bright yellow to orange color, typical of palagonite that formed by the interaction between water and fresh basaltic glass. (B) Tuff rings are generally wider than strombolian cones, with a large open crater and a typically annular structure.

Image from Google Earth.

Water—magma interaction has further effects, such as modifying the shape and dimensions of the resulting cones that are usually wider than typical scoria cones, and usually show a large open crater (see chapter: The Geology of Lanzarote). These cones tend to form typically annular structures (tuff rings). One of these tuff rings with a distinct bright yellow color, typical of palagonite, outcrops in the cliff at *La Hoya del Verodal* (Fig. 2.36A). Two different sections of the tuff ring crop out on either side of the Lomo Negro volcano, which is itself a recent basaltic cone attached to the cliff (Fig. 2.36B). Both outcrops are part of the same tuff ring, the eastern section of which is in part covered by the volcanic sequence forming the cliff, whereas the western part has been dismantled by marine erosion. Accordingly, the diameter of the crater of this older phreatomagmatic volcano is approximately 820 m, considerably wider than the typical values for strombolian magmatic cones (see eg, Figs. 3.40 and 5.20).

Continue along this road and after ca. 1 km you will come to a junction (with HI-502) where the road ascends the Lomo Negro cone. Right next to the junction are the hornito-type vents that emitted the spectacular ankaramite flow exposed here (Fig. 2.37). Turn right at the junction and continue downhill for a few meters and then park. Take a walk back to inspect the hornitos. Hernández-Pacheco (1982) associated this eruption of Lomo Negro with the 1793 seismic crisis on El Hierro, based on a ^{14}C age of 1800 AD $\pm$ 50. However, the author deemed the reliability of the age to be low, because the dated sample consisted of unburned plant remains under a lava flow (and therefore can be considerably younger than the lavas).

From March 27 to June 15, 1793, strong earthquakes shook the island, damaging buildings and raising public alarm. The prospect of a volcanic eruption prompted the authorities to prepare the first known general evacuation plan in the Canaries. The seismicity, initially felt only at El Golfo but later on the entire island, declined in intensity and frequency and finally ceased, without any mention of manifestations of subaerial volcanic activity in any of the reports. On the contrary, as these reports pointed out, an evacuation would have been conducted *"had volcanism ruined the island."* Orders were given to report any sign of volcanic activity on the island, thus making it improbable that an onshore eruption occurred but went unnoticed. The reports therefore imply that this episode of unrest ended without an eruption or that the eruption was submarine, a plausible conclusion after the occurrence of the 2011 submarine events. The Lomo Negro ankaramites, in turn, are likely prehistoric.

FIGURE 2.37

Hornitos and lavas of the Lomo Negro eruption, exposed at the northwest end of the island. This eruption was linked to the 1793 seismic crisis on El Hierro by some researchers, but it is more likely prehistoric in age. The 1793 seismic activity was probably associated with a submarine eruption, perhaps similar to the one in 2011/2012.

FIGURE 2.38

The *El Vallito* cliff and *Verodal* beach at the far end of the northwest rift zone. The red sand of the beach is oxidized basaltic lapilli from *Mña. de los Charcos*, the recent cone exposed at the top of the cliff.

Continue in the direction of *Playa del Verodal* (approximately 1 km from here). Note the road soon turns into a reasonably well laid out dirt track. Continue until you reach the parking site at the beach.

Stop 2.8. Verodal Beach (N 27.7491 W 18.1523)

A steep cliff of older El Golfo lavas is topped by reddish-brown lapilli from the postcollapse rift zone activity that helped to form this distinctly red beach here (Fig. 2.38). Enjoy the beach! But beware of possible rockfalls below the cliff and avoid swimming in heavy swell.

For a speedy return to *Valverde*, drive along the lower coastal road (HI-550 yellow), which offers a fast-motion run through the entire El Golfo valley on the way home.

Enjoy the drive!
END OF DAY 2.

EL HIERRO DAY 3: THE RIFT ZONE

This day will involve some amazing landscapes but also a lot of driving (Fig. 2.39). Ensure your car has sufficient gasoline and bring a lunchbox and fleece with you because you will be outside at high elevations for most of the day. At the end of the day there is also the chance to go for a swim. The tour is best taken on a clear day with no clouds over the central rift zone.

Make your way from *Valverde* onto HI-1 with direction for *San Andrés* and *El Pinar* (eg, via *Calle San Juan*). Pass through *San Andrés* village and a few kilometers further, you will see a brown sign for *Jinama*. Turn right onto HI-120 (yellow) and park at *Jinama* viewpoint.

Stop 3.1. Mirador de Jinama (N 27.7629 W 17.9808)

Here, you will get a fantastic view of the El Golfo embayment (see Fig. 2.13). The viewpoint also marks the start of a hiking trail down the El Golfo cliff.

A few meters into the trail, a major dyke is seen standing proud, implying that the surrounding rocks were removed by backward erosion of the El Golfo cliff. Walk down the trail for a few steps to inspect the dyke. Note the full trail will need several hours and organized transport at the other end!

Return to your car and go to HI-1 in the direction of *El Golfo* (ie, turn right at the HI-120/HI-1 crossing). Follow HI-1 uphill for a while until you reach the flat part right behind the El Golfo cliff. Soon after, turn left and follow brown signs for *Cruz de los Reyes* and *Ermita Virgen de los Reyes* (HI-45 green).

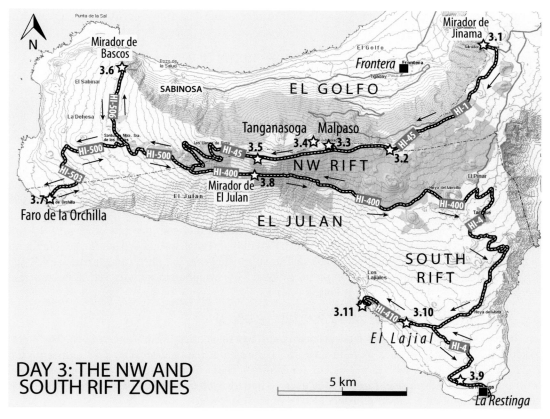

FIGURE 2.39

Detailed map of *Day 3* (northwest and south rift zones), with the stops and other relevant features indicated (map from IDE Canarias visor 3.0, GRAFCAN).

Drive down the not so wide HI-45 road and park at a lay-by on the right that comes up after a few kilometers.

Stop 3.2. Mirador de El Tomillar (N 27.7268 W 18.0173)

Here, you can look down the south rift and get an impression of the extensive recent rift volcanism that produced the widespread lapilli fields found on top of the island's main ridge. Continue on HI-45, but note that it will change into a well laid out dirt track a little further ahead. Follow the track in the direction of *Malpaso* (brown sign). After a few minutes on the dirt track, a sharp right turn brings you again onto a paved road and up to the next viewpoint.

Stop 3.3. Mirador Malpaso (N 27.7292 W 18.0404)

This site represents the highest elevation on the island (1500 masl). From here, you will see the El Golfo embayment and the islands of La Gomera and Tenerife. To the west, the Tanganasoga volcano, a post−El Golfo scarp eruptive mass, is visible (Fig. 2.40). From here, follow a dirt track (on foot) that takes you west for approximately 500 to 600 m and that runs approximately parallel to the cliff's outline.

Stop 3.4. Tanganosoga Volcano (N 27.7290 W 18.0447)

The large Tanganasoga volcano, which is nested in the El Golfo collapse scarp (see Fig. 2.16A), will now present itself in full beauty. The edifice has several vents and emitted large volumes of lavas and also ash (Fig. 2.40A,B). An interesting aspect of this volcano is the occurrence of violent phreatomagmatic episodes that mantled the area here with large blocks (remnants of blown out rock) and ashes (Fig. 2.40C,D). These ashes are very fine in places and very light in color, suggesting intense fragmentation and rather high explosivity (Pedrazzi et al., 2014).

Tanganasoga's explosive episodes can also be inspected when you return to the viewpoint and look along the stretch of road right below you (Fig. 2.40E). The gray ash layers and pumice-rich beds of the Tanganosoga volcano up here clearly bear witness to its explosive episodes of the last 6 to 12 ky.

From the viewpoint, return to the dirt track below (HI-45) and continue toward *Ermita Virgen de los Reyes*. After a few hundred meters, the light-colored Tanganosoga ashes are seen again on your right, generally underlying fresh black lapilli, suggesting that the phreatomagmatic explosions occurred in the early phases of the eruption, followed by strombolian activity (c.f. Clarke et al., 2009). Continue on HI-45 for a few minutes.

Stop 3.5. Tanganasoga Ashes (N 27.7227 W 18.0665)

Here (Fig. 2.41), you can inspect inter-bedded lapilli and finer ashes from Tanganosoga. The light-colored fine ash is present once more in the lower right area of the outcrop. The dark surface on top of the outcrop is littered with larger blocks (or bombs). These blocks are early Tanganasoga rocks blown out by a later explosive Tanganasoga event. The larger blocks that mark the sides of the track that you are driving on are most probably also from this explosion.

These Tanganasoga ashes are spread at least 3.5 km from their vents, underlining the explosivity of this volcano. Note, in this case, the word "ash" is suitable (grain size <2.54 mm), but not so in the common strombolian eruptions, where "lapilli" is the appropriate term (2.54 mm to 6.4 cm; bombs are >6.4 cm). Fragmentation efficiency depends on the degree of explosiveness of the

FIGURE 2.40

(A) *Google Earth* image of the multivent Tanganasoga volcano. Note the lava flows emitted from a hornito-type vent at the southern end of the chain of craters. (B) Close-up view of Tanganasoga lava flows. (C) Violent phreatomagmatic episodes of Tanganasoga volcano mantled the area with large ejected blocks and ashes. (D) Plan view of ash layers from an explosive phreatomagmatic episode of the Tanganasoga volcano. (E) A layer of probably trachytic, fine, and phreatomagmatic ashes mantles altered lapilli from older eruptions. These ashes are observed underlying basaltic pyroclasts, indicating that the felsic explosive events describe the initial phases of this particular Tanganasoga eruption. With progressive eruptive duration, however, the products became dominantly basaltic in composition.

eruption (the more explosive, the higher the fragmentation and the finer the resulting particles). This, in turn, depends on the ratio of lava (fuel) to water (coolant). If the proportion is adequate, then violent explosions occur that produce clouds of volcanic ash and steam that deposit the finely fragmented lava rocks onto the area surrounding the eruptive vent. In such situations of high levels of explosive water vaporization, larger blocks from already solidified rocks (eg, dykes and older lavas) are ripped out and hurled up together with the ash.

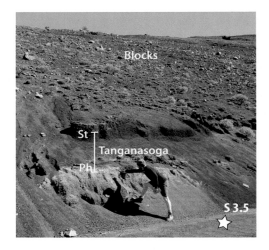

FIGURE 2.41

Tanganasoga ashes at Stop 3.5. The phreatomagmatic ash layer (Ph) overlies older altered lapilli and is topped by Tanganasoga strombolian lapilli (St). Note the basaltic blocks on the top of the sequence, indicating the occurrence of a series of explosive episodes during the construction of the Tanganasoga volcano.

Continue on the dirt track, passing more ash, lapilli, and block deposits on the way. Soon the track will become rather bendy and will bring you downhill. After a few more kilometers, HI-45 will meet HI-400 (yellow). Turn right onto HI-400 and follow signs for *Ermita Virgen de los Reyes* and now also *Faro de Orchilla*.

A short drive further, several vents will show up in the lower ground to your left. Stop for a few minutes to inspect the vents from a distance, then continue on HI-400. Drive past the *Ermita Virgen de los Reyes*. We are aiming for *Mirador de Bascos* before approaching *Faro de Orchilla* on the return.

Follow signs for *Mirador de Bascos*, but note you will be on a dirt track again soon, and there is a cattle grid where you will need to open and close the gate to drive to the mirador.

Stop 3.6. Mirador de Bascos (N 27.7551 W 18.1181)

At *Mirador de Bascos* you can inspect the El Golfo embayment from the west, with *Sabinosa* village in the foreground. The Tanganosoga massif is right in front of you, displaying all its might, and so is the entire El Golfo landslide embayment (Fig. 2.42).

Once you have taken in the views, return to your car and go to HI-400 in the direction of *Faro de Orchilla*. Just after passing the *Ermita Virgen de los Reyes*, a right turn will bring you to HI-500, and signs for the *Faro de Orchilla* will guide your route for the next little while. After some downhill driving on a narrow road and after passing the *Montaña Tenaca* cone, you will come to a triple crossing where one arm leads down to the lighthouse (the dirt track option). If traffic permits, consider briefly stopping here at the triple junction to inspect the hornito that gave rise to

FIGURE 2.42

Views of El Golfo collapse scarp from *Mirador de Bascos*, which provides a view over the western part of the El Golfo embayment. (A) The precollapse volcanic sequence can be seen on the right of the photograph. (B) The *Mirador de Bascos* provides a view from the west of the entire El Golfo embayment.

lava flows in the area here. Behind the hornito is another vent (*Montaña de las Calcosas*) that forms part of the rift alignment.

Drive down the dirt track and park the car at the lighthouse.

Stop 3.7. Faro de Orchilla (N 27.7067 W 18.1476)

The lighthouse is located right behind another rift-zone vent (*Montaña de la Orchilla*), and is also very close to the former 0-Meridian, which in 1795 was transferred to Greenwich in England.

From the lighthouse, look eastward to view the El Julan landslide scarp, another large collapse embayment that pre-dates the El Golfo collapse and the subsequent rift volcanism. Next to the Faro (and behind it), a stack of lava flow lobes from *Montaña de la Orchilla* are exposed in a vertical section (Fig. 2.43), allowing close examination of its physical and mineralogical make-up. Many of the lobes contain abundant olivine and pyroxene, and thus are ankaramitic to picritic in composition.

Ankaramites (pyroxene > olivine) are very dense and show crystal assemblages typical of great depth. Likely, these dense post—El Golfo lavas are, at least in part, a result of decompression caused by the collapse, which allows dense magmas from depth to ascend rapidly (Manconi et al., 2009). Ankaramites are the typical magmas to erupt after large-scale landslides, as can also be seen in the Icod slide and the Teno and Orotava collapses on Tenerife, and especially here at the El Golfo embayment (eg, Longpré et al., 2011; Carracedo et al., 2011b; Delcamp et al., 2012).

Return to HI-500 and turn right toward *Ermita de los Reyes*, aiming now for *La Restinga* in the very south of El Hierro. After a short drive you will meet a T-junction with HI-506 (white). Turn right here, toward *El Julan*. Soon, however, you will join HI-400 yellow again, in the direction of *El Julan* and *El Pinar*. Approximately 10 minutes later, you will pass *Mirador de El Julan*, which offers a vantage point to inspect the El Julan landslide embayment in more detail.

FIGURE 2.43

Mña. La Orchilla, a vent at the edge of the northwest rift zone. Behind the main cone is the lighthouse, the official reference of the former 0-Meridian. In the 2nd century AD, Ptolemy considered this zero meridian to be the westernmost position of the known world. In 1795, however, the 0-Meridian was transferred to Greenwich near London.

Stop 3.8. Mirador de El Julan [N 27.7169 W 18.0668]

The El Julan giant landslide dates back to approximately 160,000 years ago and is a major geometric control on the three-armed geometry of El Hierro. Notably, large slides appear to always occur in between the main rift axes, suggesting that intrusive activity in the rift zones (rift push) may be a key factor in pushing the flanks of these volcanoes out to a point where they collapse laterally, and likely in a catastrophic fashion (see Fig. 5.29; Carracedo, 1996 a,b; Carracedo and Troll, 2013).

The El Julan collapse embayment is completely filled with lavas from the northwest rift zone (Fig. 2.44), but the deep structure must be similar to that of the El Golfo collapse (see Fig. 2.10B). Possibly a deep water reservoir is hosted in the porous sequence of interbedded lavas and pyroclasts overlying the collapse debris avalanche deposits, which is presumably altered and by now highly impermeable (see Fig. 2.29).

Continue toward *El Pinar*. After another 10 minutes or so, you will slowly leave the pine forest and an impression of the south rift zone will open up and provide a good view of the alignment of vents that cluster all the way to the southern tip (and are known to continue below sea level, as exemplified by the 2011 events). After a few more minutes, a road posted for *El Pinar* takes you off HI-400. Stay on this road until you reach a T-junction with HI-4 (red). Turn right at the T-junction toward *El Pinar* and *La Restinga*. Now, follow the sign for *Tanajara* that will take you to the summit of *Mña. Tanajara*, which offers a panoramic view of the southern rift zone. When completed, return to *El Pinar* and continue downhill to *La Restinga*.

FIGURE 2.44

El Julan collapse embayment viewed from Stop 3.8. Note that the collapse basin has been almost completely filled by eruptions from the northwest rift zone. Ultimately, in the geological future, the El Golfo valley will probably experience a similar fate.

Stop 3.9. La Restinga Pahoehoe Lava Field (N 27.6438 W 17.9878)

Just before reaching *La Restinga*, take the time for a photo stop along this stretch of road, which shows one of the most beautiful pahoehoe lava textures that can be seen on the island. Cross the road for some amazing details! Note the pile up, the mini-lava tunnels, and the flow tongues on various scales (Fig. 2.45).

Lava flows are generally named or classified by the appearance of their surface. Because this classification was first developed in the Hawaiian Islands, the internationally accepted terms are Polynesian terms, and refer to the appearance of the "sea surface." Flows with a smooth surface are termed *pahoehoe*, the Hawaiian word for a calm sea. Repeating a word in Hawaiian, in this case "hoehoe," reinforces the concept. Rough-surface lavas relate to the term *'a'a*, which in Hawaiian means a windswept sea (Fig. 2.45A,B).

Pahoehoe flows may have smooth surfaces or the often more abundant folded surfaces that look like ropes or even heaps of entrails (Fig. 2.45C,D). If one observes an advancing pahoehoe flow, toes can be seen forming at the front that expand and inflate as they slowly fill with more lava (Fig. 2.45E,F). The elastic crust cools and quickly begins to solidify to the point that one can safely walk on the thin solid crust of lava on Hawaii, even if it is still incandescent. Moreover, pahoehoe flows often travel more slowly than 'a'a flows, thus representing only a limited risk to humans.

Continue into *La Restinga* and park near the harbor in the village for a refreshment stop in one of the beach cafés to contemplate the scenario of the 2011/2012 eruption (see Figs. 2.17–2.20 and Fig. 2.46).

The eruption ejected juvenile floating volcanic bombs and sedimentary xenoliths from the ocean crust (Fig. 2.46A). Volcanic ash and frothy bombs as well as gases formed a floating plume of

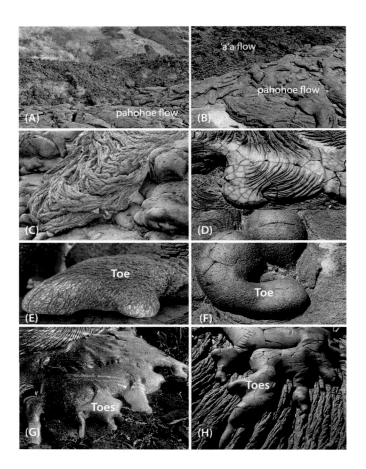

FIGURE 2.45

(A, C, E, G) Active pahoehoe and 'a'a lava flows on Hawaii and (B, D, F, H) pahoehoe and 'a'a lava flow textures at Stop 3.9. When rubbly 'a'a flows advance, they show a rough, jagged, spinose, and generally clinkery surface, whereas pahoehoe lava shows smooth top surfaces. The 'a'a flows advance much like the tread of a bulldozer (ie, by rollover) (see also Fig. 3.57D in chapter 3 The Geology of La Palma) while pahoehoe flows grow by internal flow, causing lobes to inflate and advance simultaneously. There is no difference in the composition of these two forms, and a single flow can start as pahoehoe and change to 'a'a as it loses heat and gas and starts to crystallize, which causes an increase in viscosity. Flowing down steep slopes or flowing very rapidly can generally accelerate the transition from pahoehoe to 'a'a (note 'a'a flows cannot become pahoehoe flows again, unless reheated). A ropy surface develops when the skin of cooling lava at the surface of a pahoehoe flow is pushed into folds by the moving lava just below the top skin.

FIGURE 2.46

Different views of the 2011 eruptive plume of stained water that produced floating basaltic lava balloons and xeno-pumice (photos courtesy of Guardia Civil, taken during monitoring helicopter flights).

discolored seawater at that time, which was visible in high-resolution satellite images, featuring a large stain (locally known as "La mancha") on the surface of the *Mar de Las Calmas* (Fig. 2.46B). The village was evacuated on two separate occasions during these events and occupied by troops of the Spanish army (Unidad Militar de Emergencias, UME). For a while, it was a ghost town patrolled solely by armed guards (Pérez-Torrado et al., 2012).

The eruption vents were at a depth of ∼100 m or deeper (except at the final declining stage of the eruption), which prevented explosivity and, therefore, never posed a real threat to the fishing village of *La Restinga* (600 inhabitants). Seismicity reached M 4.5 which led to road closures (eg, the tunnel connecting the capital *Valverde* with *Frontera* in the El Golfo embayment), adding to the desolation and despair of many of the inhabitants at the time (Fig. 2.47).

If you are keen to inspect more lavas and spectacular lava morphologies, return to HI-4 and turn left onto HI-410 toward *Tacorón*. After approximately 1 km, park the car near a white monolith.

Stop 3.10. El Julan Volcano and Lava Field (N 27.6643 W 18.0039)

Walk from here toward HI-4. This 1-km stretch crosses 'a'a and pahoehoe lavas from a relatively recent group of eruptive vents in the El Julan embayment (Fig. 2.48A,B), which erupted an extensive basaltic lava field that presents an extraordinary variety of forms and structures that is still in an excellent state of preservation. This is possibly the best representation of this type of lava in the Canary Islands and is equally spectacular as those of Kilauea, on the island of Hawaii.

You may want to go to the flat coastal platform 1 km below. Walking is easy on the pahoehoe lavas. There you can enjoy a superb display of 'a'a and pahoehoe lava features (Fig. 2.48C–F). Actually, this platform is almost a thematic park of volcanism and should be adequately protected, but the western part of the platform has already been occupied with a large greenhouse facility for banana cultivation.

FIGURE 2.47

Local newspaper front page from November 2011. The repeated evacuations of the fishing village of *La Restinga* on the southern tip of the island in 2011 was controversially discussed (see Pérez-Torrado et al., 2012). Some alarmist interpretations of the ongoing eruption caused fear, frustration, and distress among the population and initiated a lasting economic crisis on the island (image courtesy of Diario de Avisos).

Among these spectacular features are those formed in the slope (ancient sea cliff), where the lava flowed downstream, billowed, and folded, like the skin on cooling milk. Based on this appearance, terms such as corded or ropey lava have been used as a description.

On the coastal platform, high fluidity lava traveled over a smooth, flat surface, where it ponded and formed a smooth solidified crust. If the flow of lava continues below the already solidified lava crust (inflation), then hydrostatic pressure can elevate the crust, generating tumuli, where the pressure can break through and deform the crust through intruding lava wedges (see Fig. 2.48D).

Return to your vehicle and continue on HI-410 toward *Tacorón* beach. The road passes several cinder cones and lavas with conspicuous channels flowing from the Northwest rift zone higher on the slope. After a short distance, the road arrives at the top of the ancient sea cliff, now fossilized by a coastal platform of recent lavas. Park at the end of the road.

Stop 3.11. Tacorón Beach (N 27.6716 W 18.0263)

At *Tacorón* beach, an 'a'a lava pile can be seen, and another fine view of the El Julan collapse embayment presents itself.

You may wish to enjoy a swim at this spot. This is a good place to end the day (if you are lucky, with a spectacular sunset) and close the geological chapter of El Hierro.

Return to HI-410, and from there to your base.

END OF DAY 3.

FIGURE 2.48

(A) Vents to the lava field of the El Julan volcano (image from *Google Earth*). (B) Map outlining the vents and various types of lava features, which are particularly well preserved on the coastal lava platform below. (C) On gentle slopes, pahoehoe flows formed tumuli (pressure ridges). Uplifted sections of pahoehoe lava crust are caused by pressure from fluid lava that continue to migrate beneath the hardened crust. (D) Wedge of lava in a tumulus. The wedge forms as lava inside the tumulus extrudes from a crack in the hardened crust of the tumulus. (E) Folded pahoehoe in the flank of a tumulus. (F) Entrail lavas from a mini-hornito.

THE GEOLOGY OF LA PALMA

The Geology of the Canary Islands. DOI: http://dx.doi.org/10.1016/B978-0-12-809663-5.00003-7

101

THE ISLAND OF LA PALMA

La Palma is the fifth-largest (706 km^2) and second-highest island (2430 meters above sea level) of the Canary archipelago. It is elongated in a north–south direction (Fig. 3.1A) and comprises two

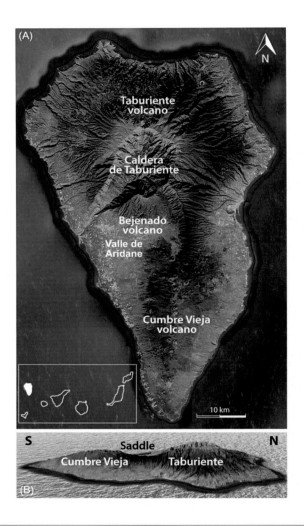

FIGURE 3.1

(A) *Google Earth* vertical satellite image of La Palma showing the main geological features: the Taburiente and Bejenado volcanoes and the Taburiente caldera, as well as the younger Cumbre Vieja rift zone. (B) The two main volcanoes (Taburiente and Cumbre Vieja) are separated by a saddle that is formed by the head of the lateral collapse embayment that formed the Caldera de Taburiente depression approximately 560 ka ago. After Guillou et al., 2001; image from Google Earth.

large volcanoes separated by a valley or saddle (which gives La Palma its characteristic profile): the extinct northern shield volcano and the southern Cumbre Vieja volcanic ridge (Fig. 3.1B), with the latter being currently the most active volcanic region in the Canaries.

Mean annual rainfall is relatively high (740 mm/year compared to the average 325 mm/year for the entire archipelago), and mean values in excess of 1400 mm/year are recorded on La Palma's northern highlands (eg, above the town of *Barlovento*). Erosion is highly active and has carved deep and locally very wide ravines ("barrancos") in the northern volcano, whereas the barrancos are still in their early stages of development in the younger southern Cumbre Vieja ridge.

Steep slopes are frequent on the island. The main depressions, the Valle de Aridane and the Caldera de Taburiente (see Fig. 3.1A), originated from lateral collapses and were subsequently enlarged by erosive retrogradation. Vertical cliffs hundreds of meters high are frequent in the northern shield, whereas on the Cumbre Vieja they are less developed. The beaches on La Palma are usually made of black basaltic sands and cobbles and are generally small and often with limited access.

RADIOMETRIC AGES AND GEOMAGNETIC REVERSALS

Radiometric ages (K/Ar and ^{40}Ar/^{39}Ar) have been essential to the reconstruction of the volcanic history of La Palma (Ancochea et al., 1994; Carracedo et al., 1999, 2001; Guillou et al., 1998, 2001). The period of construction of La Palma, approximately the past 1.72 million years (Ma), makes it particularly suitable for the application of the Geomagnetic Polarity Timescale (GPTS) (Guillou et al., 2001; Singer et al., 2002). Few changes in magnetic polarity took place during that period (the Cumbre Vieja volcano developed during the normal polarity Brunhes epoch) and they are relatively easy to identify and correlate (Figs. 3.2; 3.3A).

Most of the lavas on the island have ages younger than 780 ka and normal geomagnetic polarity, corresponding to the Brunhes normal polarity epoch (see Fig. 3.3A). Therefore, the Matuyama-to-Brunhes boundary is extremely useful because the simple determination of the geomagnetic polarity in the field allows us to identify older formations and is thus a powerful tool, especially when combined with radiometric ages (see Fig. 3.2).

Moreover, the use of short magnetic duration events (eg, Jaramillo, Cobb Mountain) offers the possibility to refine the stratigraphy of lava sequences.

Using these concepts, La Palma is dominantly covered by recent (Brunhes) lavas, although erosion has exposed older formations in the deep barrancos of the north and at the wall of the Caldera de Taburiente (Fig. 3.3A,B).

COMPOSITION OF ERUPTED ROCK SUITES

When using the Total Alkali versus Silica diagram (TAS) for the successive volcanoes of La Palma, the rocks consistently progress toward more differentiated lavas, particularly the Taburiente and Cumbre Vieja volcanoes in their terminal stages of growth (Fig. 3.4). Overall, volcanism of La Palma encompasses much of the Ocean Island Basalt (OIB) series and ranges from primitive

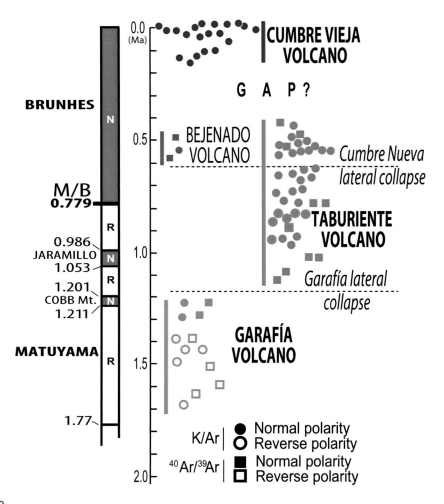

FIGURE 3.2

Successive volcanoes of La Palma as defined by K/Ar, ^{40}Ar/^{39}Ar, and ^{14}C chronostatigraphy and rock polarities relative to the geomagnetic polarity time scale (GPTS). The successive volcanoes are clearly separated in time, except the Bejenado volcano, which overlaps with the final stages of growth of the Taburiente volcano (after Carracedo et al., 2001 and Guillou et al., 2006).

basanites and basalts to highly evolved phonolites, with the latter being the common felsic rock-type in the western Canaries (eg, Rodríguez-Badiola et al., 2006).

For example, phonolitic domes of variable ages abound at the summit and on the flanks of the Cumbre Vieja rift. Although most of the eruptions of the Cumbre Vieja are basaltic lava emissions and strombolian events (some of them phreato-strombolian), the recognizable felsic eruptions show distinct evidence for dome collapse and block-and-ash deposits (*nuées ardentes*).

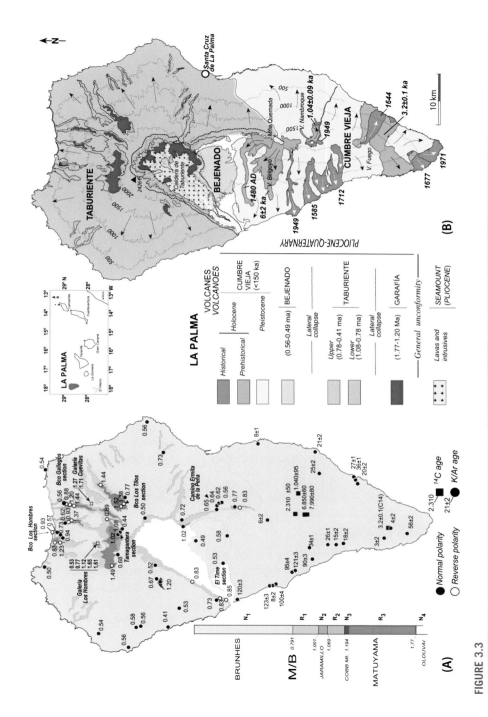

FIGURE 3.3

(A) Geographical distribution of available K/Ar, ^{40}Ar/^{39}Ar, ^{14}C ages and rock polarities for La Palma. The oldest formations are exposed in the windward flank of the Taburiente volcano, where landslides and erosion have incised deep barrancos. The geomagnetic reversal time scale is from Singer and Brown (2002). (B) Geological map of La Palma showing the successive volcanoes and volcanic formations (after Carracedo et al., 2001).

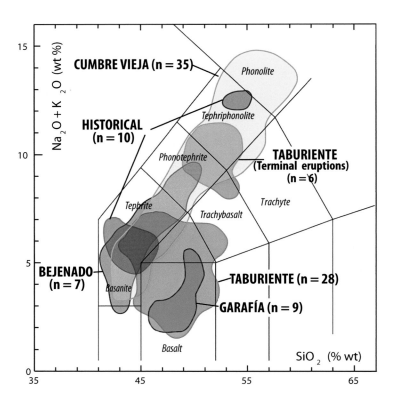

FIGURE 3.4

Total Alkali versus Silica diagram (TAS) for lavas from La Palma. The successive volcanoes progress toward more differentiated lava compositions, particularly the Taburiente and Cumbre Vieja volcanoes. Volcanism of La Palma encompasses the entire OIB series, from primitive basanites to highly evolved phonolites. Data from Rodríguez-Badiola et al., 2006.

GENERAL STRATIGRAPHY

The classic division into two main stratigraphic units, the Old Series and the Recent series, established by Fúster et al. (1968a,b) in the eastern Canaries, failed in the western Canaries. This is because the "Old" Series of La Palma and El Hierro were found to be considerably younger than, for example, the "Recent Series" of Fuerteventura, Lanzarote, Gran Canaria, or Tenerife. At present, the entire island of La Palma is in the shield stage, which is geologically equivalent to the oldest formations of the central and eastern Canaries, however.

An alternative and probably more efficient stratigraphic division is based on the stages of development of island volcanoes, such as the submarine, the shield, and the posterosional (or rejuvenated) stages of growth. The stratigraphy of La Palma, entirely in the shield stage, must therefore rely mainly on the distinction of the successive volcanic edifices (Fig. 3.5).

FIGURE 3.5

(A) The island of La Palma is formed by two main volcanoes (*DEM* from *GRAFCAN*). A circular shield volcano in the north, the Taburiente volcano, and an elongated ridge structure to the south, the Cumbre Vieja rift. (B) These volcanoes are progressively younger from north to south, indicating continuous migration of the emerged volcanism in a southerly direction (eg, Carracedo et al., 2001; Walter and Troll, 2003).

THE SUBMARINE EDIFICE

Swath bathymetry (sonar) coverage around La Palma (Fig. 3.6) shows important features related to the constructional and destructive events of the submarine part of the island. The shaded relief images outline the extension and basal perimeter of the island, which is 120 km in diameter and approximately 4000 m deep. The 6423-m-high volcanic edifice of La Palma is therefore one of the highest volcanic structures on Earth. The shaded relief image also shows the debris avalanche deposit that originated from gravitational slope failures and the submarine extension of the main southern rift zone.

Oceanographic vessels such as the RV Meteor or RV Poseidon (Germany) have dredged the flanks of the island of La Palma and obtained samples of recent volcanic activity from the submarine part of the edifice (Fig. 3.7). In some areas, such as the submarine prolongation of the Cumbre Vieja rift, the lavas appear to be very recent, closely resembling the historical subaerial lavas exposed onshore (eg, Schmincke and Graf, 2000). This geometry provides evidence that volcanic activity is also taking place in the submarine part of the islands, which is by far the more voluminous (see Figs. 1.17 and 1.18).

The greater part of these submarine eruptions may only manifest itself through seismic activity, because any increase in temperature or volcanic gas flux may not be observed on the island if the eruptions take place at greater depth in the ocean. Many periods of frequent and intense seismicity

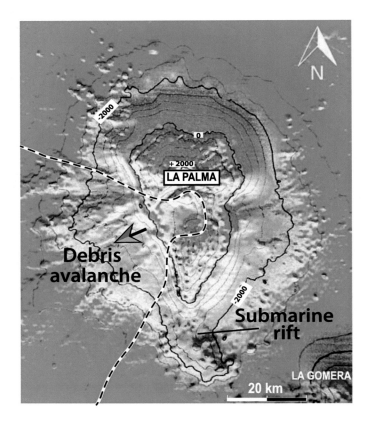

FIGURE 3.6

Color-shaded relief bathymetry and topography map of La Palma (from Masson et al., 2002). Note the shallow deposits off the Caldera de Taburiente to the southwest of the island, which represent the remains of a debris avalanche.

on the island, described in different accounts, did not culminate in eruptions on the island itself (eg, between 1936 and 1939), which may have been related to submarine eruptions, as has recently been witnessed off El Hierro in 2011 (see chapter: The Geology of El Hierro).

THE UPLIFTED SEAMOUNT

The circular Taburiente shield volcano is formed by three large overlapping volcanoes that show angular unconformities between them. The oldest is a submarine volcano formed in the Pliocene (possibly between 4 and 2 million years ago) that represents the stage of submarine construction through which all the islands must pass before emerging above sea level (Fig. 3.8).

The bottom of the Caldera de Taburiente offers the extraordinary opportunity to inspect the submarine stage of development of the island. The submarine volcano has been uplifted by

FIGURE 3.7

Submarine samples (recent basaltic rocks) dredged from the submarine part of the Cumbre Vieja rift during a cruise of the RV "Poseidon" in 2001 (in which both authors participated).

endogenous growth and intrusions to 1500 m above the present sea level. This elevation and a 50° tilt to the southeast, toward the mouth of the caldera, suggest that a traverse into the caldera progresses deeper and deeper into the submarine core of the volcano (see also daily routes).

The *"extrusive submarine series"* crops out in the Caldera de Taburiente. It has been dated as Pliocene (3−4 million years) by way of foraminiferous fossils (*Globorotalia crassaformis, Neogloboquadrina humerosa, Globoquadrina altispira*, and *Globorotalia puncticulata*) found in the palagonitic tuffs and pelagic sediments that are intercalated with the submarine lavas (Staudigel and Schmincke, 1981, 1984). This extrusive submarine series is made up of a sequence of pillow lavas and fragmentary volcanic materials that can be separated into a shallow-water and a deep-water facies. The former is exposed in Bco. de Las Angustias at 145 masl, and the latter at 365 masl.

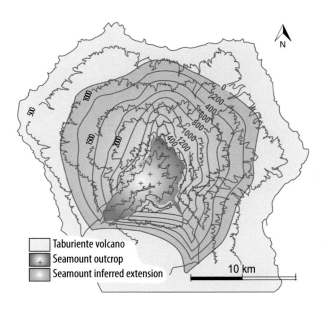

FIGURE 3.8

Actual outcrop extent and inferred extent of the unexposed submarine series of central La Palma (after Coello, 1987).

An interesting feature of this extrusive series is the progressive hydrothermal metamorphism that intensifies toward the interior of the caldera and ranges from low-temperature alteration (<10°C) at the upper part of the section (ie, in the lower course of the barranco) to greenschist facies metamorphism in the interior of the caldera (Fig. 3.9, Staudigel and Schmincke, 1981).

Mineral zoning implies a metamorphic gradient of 200−300°C per kilometer and the circulation of large volumes of hydrothermal fluids over a prolonged period of time. This process has resulted in the almost complete replacement of the original igneous minerals with clays (eg, montmorillonite, chlorite) and also epidote. Indeed, epidote is responsible for the bright pistachio green coloring of some of the rocks exposed within the Caldera de Taburiente.

In the "*intrusive submarine series*", the intrusions inside the Caldera de Taburiente correspond to both the submarine and the subsequent subaerial volcanism and can be separated into three main groups:

1. Trachytic and phonolitic domes of the Dos Aguas area. These crop out over a surface area of approximately 2 km^2.
2. Intrusive plutonic rocks (gabbros) occur in bodies of tens to hundreds of meters in thickness and crop out on the bottom of the Caldera de Taburiente over several square kilometers. These intrusions likely correspond to multiple events associated with submarine and probably also subaerial eruptions.
3. Dyke networks are highly developed in the rocks of the submarine series, reaching the highest density at the center of the caldera and along the Bco. de Las Angustias, with a gradual

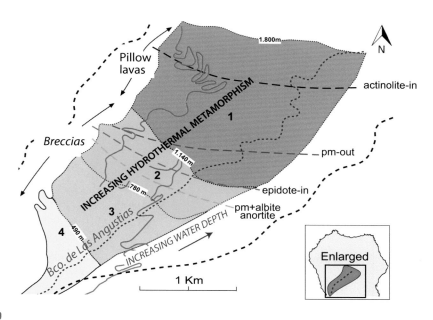

FIGURE 3.9

Greenschist facies metamorphism in lavas of the Caldera de Taburiente grades from low-grade (4-3) to medium-grade hydrothermal metamorphism (2-1) when traversing toward the interior of the barranco. Modified after Staudigel and Schmincke (1981, 1984).

decrease toward the mouth of this barranco (Fig. 3.10). Maximum intrusion density reaches almost 100% of the rock mass in certain areas and is more than 75% in the Taburiente shield (eg, in the barrancos Verduras de Alfonso and Los Cantos). Toward the area of *La Viña*, the dyke density decreases to only approximately 10% of the rock mass.

Two main types of dykes have been defined, with different age and orientation. The older dykes (green in Fig. 3.10) appear rotated by the tectonic uplift and tilt of the pre-emerged submarine volcano and, consequently, they likely represent feeder conduits to the submarine eruptions. The more recent, semi-vertical, nonrotated group (red in Fig. 3.10) are the feeder conduits of the eruptions of the later subaerial (emerged) stage (eg, Krumbholz et al., 2014).

THE NORTHERN SHIELD

The northern shield continued to grow following emergence from the submarine stage (Fig. 3.11) and formed the Garafía volcano. An angular unconformity on the north flank of the emerged seamount is present and the period of growth occurred between 1.8 and 1.2 million years ago (see Figs. 3.2 and 3.3). The Garafía volcano, completely covered now by the subsequent volcanism, only crops out at headwalls and the bottom of the deepest barrancos.

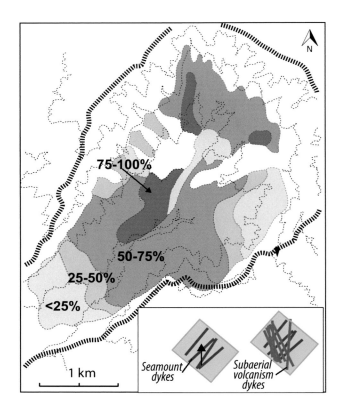

FIGURE 3.10

Density and orientation of dykes in the Caldera de Taburiente (after Staudigel et al., 1986; Carracedo et al., 2001; and Krumbholz et al., 2014). Dyke densities grade from less than 25% (gray shading) to approximately 100% in the interior of the barranco (dark red shading).

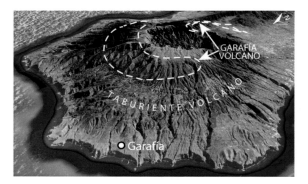

FIGURE 3.11

Oblique view of the northern shield of La Palma (image from Google Earth). The northern shield is formed by the superposition of three volcanoes: the emerged seamount, the Garafía edifice, and the overarching Taburiente volcano. The north of La Palma was exhumed by erosion and giant landslides, providing us with the geologically rare possibility to inspect a submarine volcanic edifice onshore.

The northern part of the island developed by the superposition of three volcanoes:

1. The submarine volcano, uncovered by large landslide and erosive processes and visible in the Caldera de Taburiente (see Fig. 3.8).
2. The Garafía volcano, which crops out in erosional windows at the bottom of the deepest barrancos in the north and west sectors of the island (Fig. 3.11).
3. The younger Taburiente volcano, which covers the greater part of the northern shield volcano.

Garafía volcano

The best exposures of the Garafía volcano are in the north of the island, in the area around the town of *Garafía*, and inside the Caldera de Taburiente (Figs. 3.11 and 3.12). The main outcrops are at the head and lower walls of the deeper barrancos (eg, Las Grajas, Barbudo, Los Hombres, Franceses, Jieque, and del Agua). The exposed parts form a 400-m-thick sequence of very steep (20–35°) and radially outward-dipping lava flows (Fig. 3.13). The flows are predominantly thin pahoehoe lavas, frequently with interbedded layers of basaltic lapilli. The formation is crossed by numerous radial dykes.

Garafía volcano may have reached a diameter of 23 km and a height of at least 2500 m on top of the older submarine volcano. Approximately 1.2 million years ago, the volcanic edifice collapsed toward the southwest. The debris avalanche deposits of this landslide have been identified in sonar images on the western flank of the island and as a fan-like deposit on the ocean floor (Masson et al., 2002).

Taburiente volcano

The partial destruction of Garafía volcano was followed by intense and continued volcanism that filled the depression and culminated in the formation of a new volcano, Taburiente, which eventually covered both preceding volcanoes (Figs. 3.13 and 3.14A). This stratigraphic setting can be

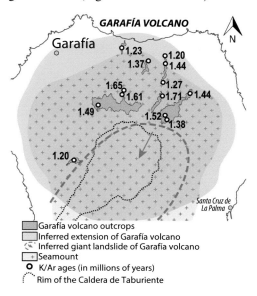

FIGURE 3.12

Inferred extent of the progressive seamount and Garafía volcanoes of La Palma. Modified after Carracedo et al., 2001.

FIGURE 3.13

The Garafía volcano (viewed from the north) can be inspected through a geological window on the northern flank of the Taburiente volcano that was excavated by intense erosion.

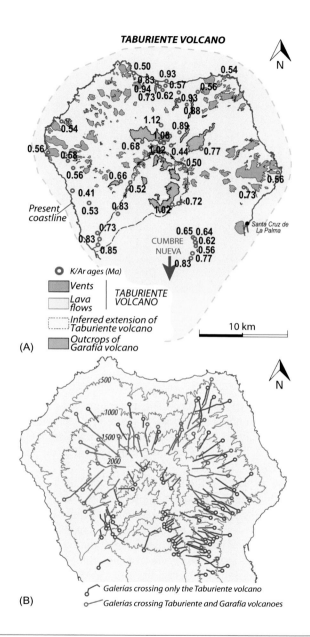

FIGURE 3.14

(A) Outcrops, ages, and estimated extent of the Taburiente volcano capping the Garafía volcano. Garafía sequences crop out in the deep barrancos of the northern flank (from Carracedo et al., 2001). (B) Galerías (water tunnels) that cross the northern shield of La Palma. Data from Plan Hidrológico Insular de La Palma, 2001; after Carracedo et al., 2001.

FIGURE 3.15

Taburiente volcano lava flows and pyroclasts overlying the Garafía deposits in the Caldera de Taburiente wall. The Taburiente sequence is approximately 1000 m thick, whereas the Garafia and submarine volcanoes crop out in the lower wall (see also Fig. 3.1).

inspected in the numerous *galerías* that were drilled through the northern shield, many of them sufficiently deep to reach the seamount basement. Galerías located to the north of the shield cross the Taburiente and Garafía volcanoes, whereas those at the south of the shield only penetrate the Taburiente volcano, supporting the aforementioned notion of a giant landslide from the Garafía volcano (Figs. 3.14B and 3.15).

When volcanism ceased in the northern shield approximately 500,000 Ma ago, a massive conical volcano, 25 km in diameter and maybe approximately 3000 m in height, had been built. Felsic explosive eruptions at the summit of Taburiente volcano indicate that basaltic magmas evolved toward more differentiated magmas during the final stages of its activity.

During the advanced stages of evolution of Taburiente, eruptions eventually migrated southward in a process that continues to the present day. As a result, the circular shape of the volcano was altered via a prolongation toward the south (see Fig. 3.14A). There, eruptive activity progressively concentrated to create the Cumbre Nueva "dorsal" rift while the northern shield gradually declined in activity and finally ceased approximately 400,000 years ago. A similar process may have occurred in the Miocene Central shield of Tenerife with the extension of this earlier volcano under the Anaga massif in the northeast.

The northern shield is now deeply eroded and a dense network of radial barrancos has been carved into it. Some barrancos are so deeply incised, particularly on the northern flank, and penetrate the entire sequence of subaerial volcanics. Sections through the Taburiente volcano are also visible in the walls of the Taburiente caldera (Fig. 3.15).

The Caldera de Taburiente

The Barranco de Las Angustias, which runs along the bottom of the Caldera de Taburiente, is the deepest and widest of all the barrancos on the island (see Figs. 3.1 and 3.8).

Leopold von Buch emphasized this circumstance clearly on the map that he drew during his visit to the island in 1815 (Fig. 3.16A). Indeed, the dimensions of the Barranco de Las Angustias and its basin, the Caldera de Taburiente, appears clearly larger on this map than all the other barrancos of the island. Moreover, von Buch may have exaggerated the size of the Caldera in accord with the importance he gave to this barranco as one of the most noteworthy examples of his "craters of elevation theory." On a similar map, Charles Lyell represented the basin of the Bco. de Las Angustias and the Caldera de Taburiente in a more proportionate manner in 1854, interpreting it as a typical erosive depression (Fig. 3.16B). However, it would be difficult to account for this enormous caldera on the basis of erosion only (Fig. 3.16C). The fact that this barranco is located in the part of the island with the lowest rainfall and is incised in a relatively recent formation (less than 500,000 years old) contrasts barrancos that are situated in areas with the heaviest rainfall and in considerably older volcanic formations (>700,000 years) (see also Fig. 3.31). The explanation for this incongruence lies in the fact that the Barranco de Las Angustias first began to form because of a tectonic process, a gravitational collapse of the western flank of the Cumbre Nueva rift, which subsequently widened and deepened by erosion to form the Caldera de Taburiente.

Von Buch believed the Caldera de Taburiente to represent an example of his "craters of elevation" theory, a process that explained the tilted lava flows due to deformation, because he believed lava flows to be always deposited horizontally on the ocean floor (Fig. 3.17). This theory was questioned by Charles Lyell and Charles Darwin only a few decades later.

No scientist of the time, including Lyell, could assume that lava flows, being fluid, could be arrested on slopes with a gradient of more than a few degrees. It is now known that lava is not a Newtonian fluid like water, which runs down-slope indefinitely, but rather a Bingham fluid, like honey or jam, with an internal cohesion that requires a thrusting force to advance. If this force wanes (eg, lava emission ceases), then the flow can be arrested on even the steepest slope (see, eg, Fig. 5.84A).

Thanks to his studies on Tenerife and La Palma and the excellent conditions for observation of such phenomena on these islands, Lyell finally claimed that lava flows are not formed by deposition of minerals precipitated from sea water, but rather are made from molten rock (magma), from intrusions and eruptions, consistent with newly arriving observations from Etna and the Andes at the time. Lyell concluded that flows could adapt to steep slopes, providing definitive proof to debunk von Buch's craters of elevation model and, by extension, Neptunism (the theory that basalt forms by mineral precipitation from sea water and subsequent deposition on the ocean floor to form horizontal basalt beds).

With the triumph of "Plutonism," the science of geology took a giant leap forward. Lyell ended by concluding that the Caldera de Taburiente was a product of erosion rather than deformation. Since then, this singular feature of La Palma has been considered the prototypical erosion caldera. As mentioned, the term caldera was introduced into the geological vocabulary by von Buch in his memoirs of his 1815 visit to the Caldera de Taburiente, and it simply stuck with the scientific community, especially in the Canaries, irrespective of the favored genetic model.

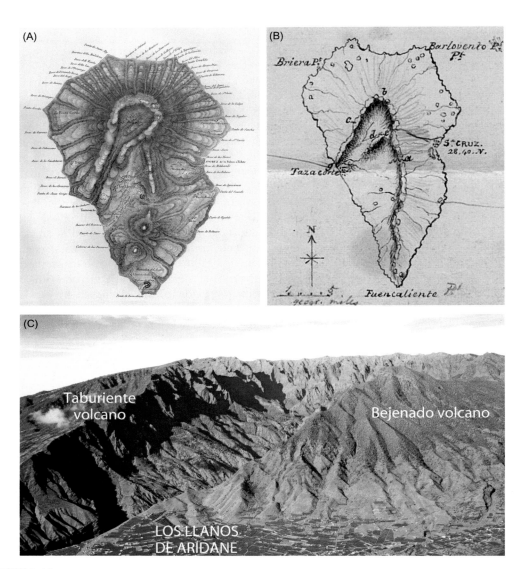

FIGURE 3.16

(A) Map of La Palma by L. von Buch (1815). The size of the Taburiente Caldera is exaggerated. (B) Map of La Palma by C. Lyell (1854). The size of the Caldera de Taburiente is more proportionate. (C) Caldera de Taburiente and Barranco de Las Angustias viewed from the southwest with the Bejenado edifice in the center of the image. Images (A and B) courtesy of Cabildo Insular de La Palma.

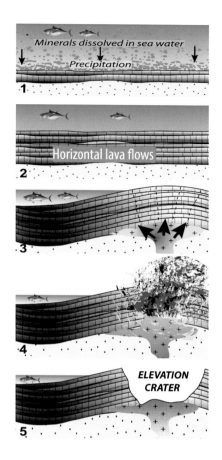

FIGURE 3.17

Cartoon explaining von Buch's "craters of elevation" model. Lava flows formed by precipitation of minerals dissolved in the ocean water and are originally horizontal. Tilted lava flows result from deformation as large volcanoes had been "pushed up" from underneath.

THE BEJENADO VOLCANO

At first sight, Bejenado volcano appears as a part of the east wall of the depression excavated by the Bco. de Las Angustias, similar to the western wall that forms a part of Taburiente volcano (see Fig. 3.16C). During his visit to La Palma in 1854, Lyell made a section of the Caldera de Taburiente, in which Bejenado is considered the eastern flank of Taburiente volcano (Fig. 3.18A). However, more recent radiometric dating has proven that Bejenado volcano is younger (Fig. 3.18B).

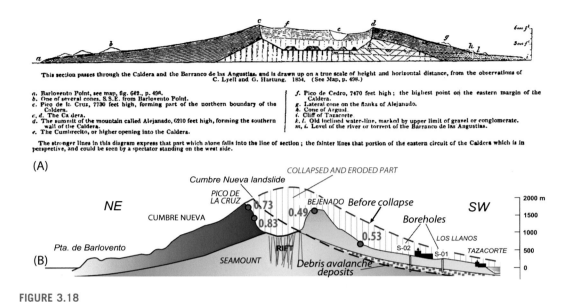

This section passes through the Caldera and the Barranco de las Angustias, and is drawn up on a true scale of height and horizontal distance, from the observations of C. Lyell and G. Hartung. 1854. (See Map, p. 498.)

a. Barlovento Point, see map, fig. 642., p. 498.
b. One of several cones. S.S.E. from Barlovento Point.
c. Pico de la Cruz, 7730 feet high, forming part of the northern boundary of the Caldera.
c, d. The Caldera.
d. The summit of the mountain called Alejanado, 6210 feet high, forming the southern wall of the Caldera.
e. The Cumbrecito, or higher opening into the Caldera.

f. Pico de Cedro, 7470 feet high; the highest point on the eastern margin of the Caldera.
g. Lateral cone on the flanks of Alejanado.
h. Cone of Argual.
i. Cliff of Tazacorte.
k, l. Old inclined water-line, marked by upper limit of gravel or conglomerate.
m, i. Level of the river or torrent of the Barranco de las Angustias.

The stronger lines in this diagram express that part which alone falls into the line of section; the fainter lines that portion of the eastern circuit of the Caldera which is in perspective, and could be seen by a spectator standing on the west side.

FIGURE 3.18

(A) Cross-section by C. Lyell (1854) of the Caldera de Taburiente (image courtesy of *Cabildo Insular de La Palma*). Both flanks were considered coeval. (B) Interpretation of Carracedo et al. (2001), where Bejenado volcano developed as a nested edifice inside the Taburiente depression that formed by the collapse of the flank of Taburiente volcano.

The Bejenado lavas overlie debris avalanche deposits interpreted as derived from the giant landslide of the western flank of the Cumbre Nueva volcano approximately 560 ka ago. Bejenado volcano formed nested in the collapse basin and is therefore younger than Taburiente volcano (Fig. 3.19). The two walls of Bco. de Las Angustias are therefore of different age and origin, a feature that can only reasonably be explained by tectonic processes, such as a flank failure and a subsequent basin refill (see, eg, Carracedo et al., 2011b).

CUMBRE VIEJA VOLCANO

The names given to 'Cumbre Nueva' and 'Cumbre Vieja' volcanoes are certainly contradictory because the former is geologically older. The naming relates to periods of reforestation, however, and not to the age of the volcanoes.

Eruptive activity in the past 150,000 years has been exclusively in the south of the island, forming the Dorsal or Cumbre Vieja volcano (Fig. 3.20A). This rift-type volcanic structure is 20 km long and 1950 m high, with a surface area of 220 km², and it is elongated in a north–south direction. The Cumbre Vieja rift represents the most recent geological feature of La Palma's evolution. The higher parts of the rift volcano are made up of an alignment of volcanic cones and fissures typical of volcanic rift zones (see Fig. 1.26). The greater part of the eruptive centers is concentrated along the rift axis (Fig. 3.20B), which projects southward below the sea, forming an alignment of recent submarine volcanoes (or seamounts), and seems to be similarly active relative to the emerged part.

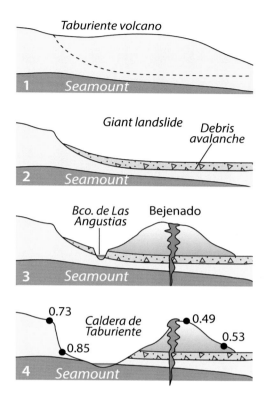

FIGURE 3.19

Cartoon explaining the geological and geochronological differences between the two sidewalls of the Caldera de Taburiente and the history of incision of the Bco. de Las Angustias (from Carracedo et al., 2001).

The Cumbre Vieja lava flows are predominantly basanitic in composition, but phonolitic rocks also occur, so they appear to have evolved much more noticeably than the preceding Garafía and Taburiente volcanoes (Klügel et al., 2000, 2005) (see Fig. 3.4). Indeed, phonolitic domes of different ages abound at the summit and on the flanks of the Cumbre Vieja rift. Most of the eruptions are strombolian and phreato-strombolian, with some recognizable episodes of dome collapse and associated small "*nuées ardentes.*"

CUMBRE VIEJA RIFT ERUPTIONS

The flanks of the Cumbre Vieja volcano, particularly the western one, have been intensely affected by marine erosion: the coasts have receded and vertical cliffs of varying heights have formed along the coast. Lava flows from recent eruptions have cascaded over the cliffs, often forming extensive coastal platforms.

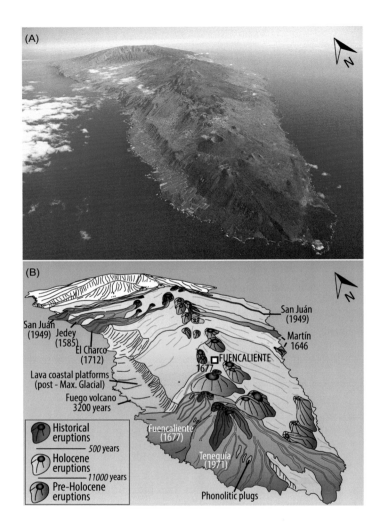

FIGURE 3.20

(A) Oblique aerial view of the Cumbre Vieja rift (photo courtesy of S. Socorro). (B) Ages and historic eruptions of the Cumbre Vieja ridge, La Palma. Modified after Carracedo et al., 2001.

Correlation of the radioisotopic ages obtained from cliff-forming lava flows with those that form a coastal platform has provided evidence that they define chronostratigraphic sequences. The ages of the cliff-forming lavas are consistently more than 20,000 years, which is the age of the last glacial maximum, whereas the ages of the platform-forming lavas are younger. It is thus evident that only the eruptions that took place after the 80- to 100-m sea level fall during and after the last glacial maximum could have formed coastal platforms at the present level. This

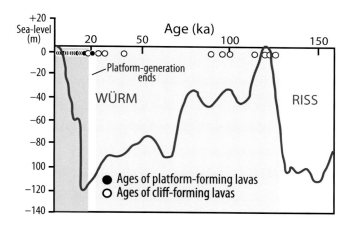

FIGURE 3.21

Correlation of ages of Cumbre Vieja lavas with sea level variations over the past 150 ky (after Carracedo et al., 2001).

circumstance can be used to separate Cumbre Vieja eruptions into two large units without resorting to radiometric dating: there are those eruptions that took place before and during the last glacial maximum with an age older than 20 ka and that form coastal cliffs, and there are those less than 20 ka in age that have fossilized the older cliffs and now form coastal platforms (Fig. 3.21).

The most recent lava flows can clearly be seen at the western coast of the Cumbre Vieja, where coastal platforms are well developed and protect the shoreline from advancing erosion (see Fig. 3.20). In contrast, where platforms are absent, marine erosion has caused the coast to recede and cliffs several hundred meters high form, such as those between *El Remo* and *Punta Llano del Banco*.

PREHISTORIC ERUPTIONS

Several Cumbre Vieja rift eruptions are very recent and have been dated by the ^{14}C radiocarbon method. These eruptions include the particularly noteworthy volcanoes of Birigoyo, Nambroque, Fuego, La Fajana-Las Indias, and Mña. Tacande (also known as Mña. Quemada, see Fig. 3.23A). This last eruption almost coincided with the colonization of La Palma (1493), as it was dated by ^{14}C to between 1470 and 1492 (Hernández Pacheco, 1982).

Similar to the historical eruptions, all these have their eruptive centers near the crest of the rift. We include San Antonio volcano in the list of prehistoric eruptions because San Antonio crater is not related to the 1677 eruption. It is actually more than 3000 years old but was covered by the 1677 scoria and lapilli, which led to some confusion (eg, Carracedo et al., 2001).

A frequent pattern in these eruptions is their relationship to phonolitic domes. The recent eruptions are usually located in the area immediately surrounding these types of domes or directly on top of them, probably taking advantage of the intensely fractured nature of these phonolite domes that likely offer an easy pathway for basaltic magma to reach the surface. The La Malforada—Nambroque cones, for example, are an interesting group of mafic and felsic eruptive vents (basanites and phonolites) that is situated on the crest of Cumbre Vieja and over an earlier highly fractured phonolitic dome (Fig. 3.22). Both La Malforada and Nambroque appear to be of a similar age, but only Nambroque has been dated and yields a ^{14}C age of 1045±95 BP.

A notable feature of the Cumbre Vieja is the curved alignment of eruptive centers in the crest of the rift zone (Fig. 3.23B), which might suggest stresses related to weakening of the western flank of the volcano (Day et al., 1999) or could be the surface expression of a larger dyke system at depth.

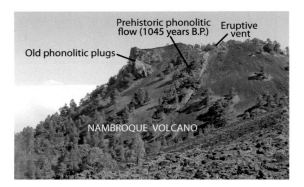

FIGURE 3.22

The Nambroque volcano is a cone with felsic eruptive vents (phonolites) situated over an earlier highly fractured phonolitic dome on the crest of the Cumbre Vieja. A phonolitic flow here yielded a ^{14}C age of approximately 1050 years BP.

SAN ANTONIO VOLCANO

Little information has been published on the historical eruptions, particularly the oldest ones, which led to frequent mistakes in the identification of the corresponding emission vents and lava flows. For example, until a few years ago, the Tacande eruption (also known as Mña. Quemada), dated by ^{14}C to between 1470 and 1492, was mistaken for the Jedey or Tahuya eruption that took place in 1585.

Another case of misrepresentation is that of San Antonio volcano, which is still widely considered to be the 1677 eruptive vent (Fig. 3.24) even though it seems to actually belong to an older eruption. Supporting evidence of this takes numerous forms. Lava flows from Fuego and La Fajana volcanoes, dated by ^{14}C at more than 3000 years, surround the San Antonio cone and thus must pre-date the 1677 eruption. Moreover, both San Antonio and the eruptive vent of La Caldereta (possibly contemporaneous), situated below *Las Indias*, are the products of phreatomagmatic eruptions of a very explosive nature. If these were indeed the vents to the 1677 eruption, then the eruptive magnitude would have erased the nearby town of *Fuencaliente*, yet nothing in this regard is recorded in the eye-witness accounts of that time. In turn, the materials related to the violent lateral explosions of San Antonio and La Caldereta volcanoes extend over a radius of several kilometers around their eruptive vents. However, the lava and lapilli of the 1677 eruption directly overlie the lava flows of the Fuego and La Fajana volcanoes and also Roque Teneguía, without any trace of interbedded explosive breccias.

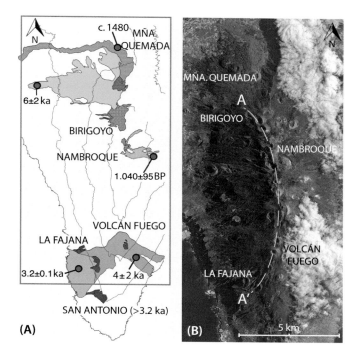

FIGURE 3.23

(A) Prehistoric (>500 years old) eruptions of La Palma (modified from Carracedo et al., 2001). (B) Satellite image of Cumbre Vieja (image courtesy of *NASA*). The curved alignment of eruptive centers in the crest of the rift zone (A-A′) may indicate stresses related to incipient weakening of the western flank of the volcano or, alternatively, mark the surface expression of a dyke system that intruded below.

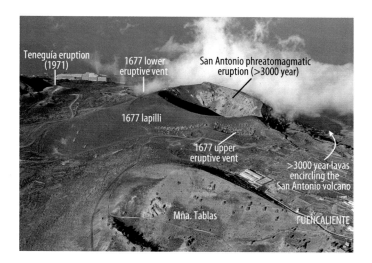

FIGURE 3.24

The San Antonio volcano is still widely considered to be the vent of the 1677 eruption, although it is actually the volcanic cone to an original event that occurred 3000 or more years ago (see text for details).

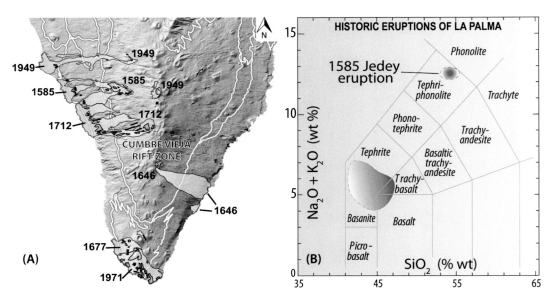

FIGURE 3.25

(A) Historical eruptions (<500 years) on La Palma (modified from Carracedo et al., 2001). (B) TAS diagram of lavas of the historical eruptions of La Palma (data from Hernández-Pacheco and Valls, 1982 and Klügel et al., 1999). Note the strongly bimodal distribution with basanite on the low silica side and phonolite on the high silica side of the series. The virtual absence of intermediate compositions is typical for ocean islands and is known as the Bunsen-Daly gap.

The confusion with the true 1677 eruption springs from the fact that San Antonio volcano is covered with lapilli from the 1667 eruption, giving it the false appearance of a recent volcano (see Fig. 3.24). This must have induced an error in the accounts of the historians who first documented the 1677 eruption and was aggravated by the loss of the original manuscript, which led to variations in transcribed reports.

Further evidence of the older age of San Antonio volcano is the presence of remains of shelters built by the primitive inhabitants (the *benahoritas*) on the flank of the San Antonio cone (Carracedo et al., 1996; Pais, 1997). The primitive inhabitants, however, disappeared from the island almost immediately after the Spanish conquest approximately two centuries before the 1667 eruption and, hence, the main San Antonio cone must predate the 1677 eruptive events.

In addition, phreatomagmatic deposits are totally absent among the pyroclastics of the 1677 eruption and its lava flows, even in the immediate area of San Antonio volcano. Considering the violence of the eruption that originally formed the San Antonio cone itself, this absence can only be explained if the cone is much older, probably predating the 3.2-ka-old encircling lava flows.

HISTORICAL ERUPTIONS

The fact that one of every two historical eruptions to take place in the Canaries was located in the Cumbre Vieja rift zone bears testimony to its intense eruptive activity (Fig. 3.25). Historic

Table 3.1 Historical Eruptions of La Palma

Eruption	Location[a]	Date and Duration	Years[b]	Composition	Type
Tahuya, Tajuya, or Jedey	Western flank	May 19–August 10, 1585 (84 days)		Basanites, tephrites	Strombolian and block and ash (small nuée ardente)
Martín or Tigalate	Eastern flank	October 2–December 21, 1646 (80 days)	61	Basanites	Strombolian
Fuencaliente volcano (previously confused with San Antonio volcano)	South end	November 17, 1677–January 21, 1678 (66 days)	31	Basanites	Strombolian
El Charco	Western flank	October 9–December 3, 1712 (56 days)	35	Basanites, tephrites	Strombolian, phreatomagmatic
San Juan (vents: Hoyo del, Banco, Duraznero, Hoyo Negro)	Summit and western flank	June 24–July 31, 1949 (38 days)	237	Basanites, tephrites	Strombolian, phreatomagmatic
Teneguía volcano	South end	October 26–November 19, 1971 (25 days)	22	Basanites, tephrites	Strombolian

[a]In relation to the Cumbre Vieja rift.
[b]From previous eruption.

eruptions include those that occurred in the past 500 years, because La Palma was conquered in 1493 and eye-witness accounts of the eruptions are only available since that date.

These eruptions have certain characteristics in common (Table 3.1), such as the aforementioned frequent associations with phonolitic domes and plugs where the magma possibly takes advantage of existing fractures and follows the least-effort route to the surface. Another common feature of these eruptions is the formation of multiple eruptive vents, which are generally aligned along a several-kilometer-long fissure that is often oblique with respect to the rift axis. The upper vents are predominantly explosive, whereas the lower ones are effusive and constitute the preferential exit route of lavas.

The last eruption to occur on La Palma was the 1971 Teneguía event. The analysis of inter-eruptive periods between successive eruptions seems to point to a mean of approximately 30 years. However, this is a crude estimate because between the 1712 El Charco eruption and the 1949 eruption, 237 years passed without any volcanic manifestation or event. One consistent pattern in the historical period, however, seems to be the continuous decrease in the duration of the eruptions (see Table 3.1).

ROCK COMPOSITION OF HISTORICAL LAVAS

The accurate separation of the past 500 years of eruptive activity of the island has little stratigraphical, petrological, or geochemical meaning. However, these historical eruptions have been

FIGURE 3.26

(A) Xeno-pumice fragment in 1971 Teneguía lava with bubbly texture and signs of disintegration that indicate syneruptive melting, gas loss, and partial assimilation. (B) Xeno-pumice fragment from the 1949 lavas that was caught in the initial stages of disintegration, displaying melt films and the first vesiculation in the open fractures, while still maintaining a degree of coherence.

analyzed separately from the remaining eruptions of the Cumbre Vieja volcano because they have detailed eye-witness accounts and are easily separated on geological maps (Figs. 3.3B and 3.25A).

The historical lavas are consistently basaltic (basanite), with the only mentioned exception of the juvenile hauyne tephriphonolite lavas and intrusives of the 1585 (Jedey) eruption (Fig. 3.25B). Notably, the geochemical characteristics of the historical eruptions confirm the general trends described for the Cumbre Vieja volcano (see Fig. 3.4).

An interesting aspect is the occurrence of xeno-pumice fragments in the Teneguía lavas (Fig. 3.26), similar to those founded in the 2011 eruption of La Restinga on El Hierro (Troll et al.,

2012) and in many other instances, such as the 1730−1736 eruption on Lanzarote (Aparicio et al., 2006) or in recent cinder cones on Gran Canaria (Hansteen and Troll, 2003).

These xeno-pumices were interpreted by Araña and Ibarrola (1973) as "fragments of acid pumice among the basic pyroclastics thrown out by the Teneguía volcano during its explosive phases." These authors propose that the pumice was produced by the fusion of the acid phase in a sub-volcanic complex located beneath the island of La Palma. However, because the La Palma xeno-pumice specimens frequently contain detrital quartz (whereas quartz is absent in the magmatic products of the Canaries), the xeno-pumice fragments on La Palma are most probably also sedimentary inclusions like their counterparts from Lanzarote, Gran Canaria, and El Hierro (see, eg, Carracedo et al., 2015a).

GEOLOGICAL ROUTES

Although the island of La Palma is relatively small (40 × 20 km), the abrupt topography and the bendy roads, particularly in the northern Taburiente shield, do increase driving time considerably.

The island is constructed of two main volcanoes with differing ages and morphology (eg, Galipp et al., 2006). In the northern, older, circular, and steep shield volcano, the population is scarce and disperse, and deep radial barrancos force roads to follow a continuously bendy course. Because of the intense marine erosion (eg, Carracedo et al., 2001; Klügel et al., 2005), the coasts are generally vertical cliffs in this area.

In the younger, elongated southern Cumbre Vieja rift, where volcanic construction prevails over marine erosion, relatively smooth coastal platforms allowed the construction of extensive banana plantations, and these areas concentrate the greater part of the island's population.

An introduction to the main geological features of La Palma requires at least 4 days (Fig. 3.27). The first day is dedicated to the oldest part, the seamount stage that crops out at the floor of the Taburiente caldera, whereas the second day focuses on the Quaternary Taburiente shield. Days 3 and 4 concentrate on the Plistocene-Holocene Cumbre Vieja rift.

The construction of an international astrophysical observatory at the summit of the Taburiente volcano now provides roads to access the younger stratigraphic sections of this edifice, whereas the crest of the Cumbre Vieja rift can be reached on foot only. However, the lava flows, particularly the historical ones (past 500 years), are crossed by the perimeter road from the capital *Santa Cruz de La Palma* to *Fuencaliente* and to *Los Llanos de Aridane* (Fig. 3.27) and are thus easily accessible.

For your comfort and safety, please be advised that the driver should be experienced in serpentine driving in mountains. Winding and narrow roads through mountainous terrain are common on La Palma and you will need horsepower (ie, do not take the smallest car available).

Take full equipment with you (eg., sun hat, gloves, rainwear, sun protection, swimwear, and towels) so that you do not have to miss any of the great experiences on the island.

Please be aware that some maps supplied by car rental companies are still using old road labels, which might lead to confusion. Please use the latest road maps when possible. The Canarian Government facilitates accurate maps of all the islands in the IDECanarias map viewer, which is available at http://visor.grafcan.es/visorweb/

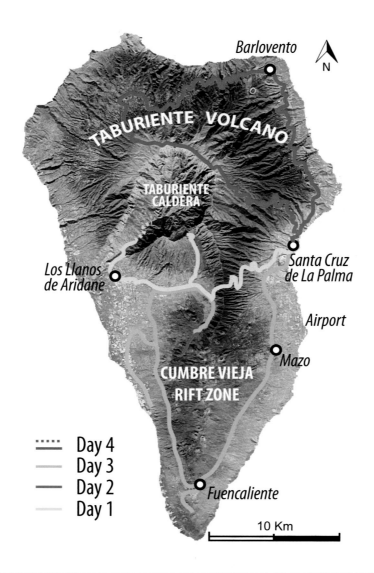

FIGURE 3.27

Google Earth map of La Palma with the Day routes. Each day is also shown with individual stops on more detailed maps below (eg, Fig. 3.26).

DAY 1: CALDERA DE TABURIENTE AND THE EMERGED SEAMOUNT

The tours are designed to start in St. Cruz de La Palma but can, of course, be started from anywhere else on the island. Make your way to *El Paso* (on LP-3, km 20 road sign) and stop at the visitor center (*Centro de Visitantes*) for an initial overview of the Caldera de Taburiente (open 9:00 to 18:00, free entrance).

FIGURE 3.28

Itineraries of Day 1 in and around the Caldera de Taburiente. Map modified from IDE Canarias, GRAFCAN.

Stop 1.1. Visitor Center (N 28.6538 W 17.8524)

Here, you can get an impression of geology, geomorphology, as well as plant and wild life, which will "sharpen your antenna" for the things to see inside the caldera. After you have digested the information provided at the visitor center, leave the car park and join LP-302 (green) right next to the visitor center and follow the brown signs for *Parque Nacional*.

Continue for a short stretch on LP-302 and follow the signs for *La Cumbrecita* (Fig. 3.28). Soon you will see lavas of the Bejenado volcano that crop out on your left and a quarry for crushing aggregates can be seen at the foot of the Bejenado massif. Continue on LP-302 toward *La Cumbrecita* and when you come to the road barrier at the entrance of the National Park; you may need to wait here for a short while. The barrier is designed to limit the number of cars at *La Cumbrecita* (it will be tight!). Once allowed to continue, drive up-hill on LP-302, and soon you will be driving through dense pine forests with a steep cliff to your right—the headwall of the 560-ka-old Cumbre Nueva giant landslide. The wall is several hundred meters high and is made up of lavas of the old Taburiente shield volcano, but it is riddled with dykes and other types of sheet intrusions. Upon reaching *La Cumbrecita*, park the car and have a look around.

Stop 1.2. Caldera de Taburiente, Mirador La Cumbrecita (N 28.6979 W 17.8566)

Here, the Bejenado massif touches the Taburiente caldera wall and it becomes apparent that the collapse of the caldera preceded the growth of the Bejenado volcano as a nested edifice inside the massive collapse

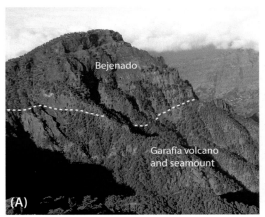 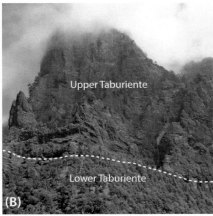

FIGURE 3.29

(A) View of Bejenado volcano from *Mirador de La Cumbrecita*. This escarpment is the hanging wall of the landslide that removed part of the Taburiente section and cut into the Garafía volcano, creating the floor of a collapse basin onto which Bejenado volcano grew. (B) View from *Mirador de La Cumbrecita* onto the Cumbre Nueva collapse scarp in a sector of the Barranco wall dominated by Taburiente lava sequences.

scar (see Figs. 3.19; 3.29A). Bejenado is possibly a direct response to large-scale decompression due to removal of the rock mass of the giant Cumbre Nueva collapse (see, eg, Carracedo et al., 2001, 2011b).

A trail starts at the parking site and goes down into the caldera for approximately 13 km. Walk a short distance down the track to observe the outcrops and to stretch your legs. After approximately 100 m, a small fork allows you to turn right to inspect the Cumbre Nueva collapse wall in more detail (Fig. 3.29B) instead of following the long trail down-hill into the Caldera de Taburiente (left). The trail to the right is only 5 km long, but it will take you up to a small "mirador" right at the headwall after approximately 800 m. Consider visiting the mirador before you return to your vehicle.

If you want to do the full trail you may not have the time to continue on our tour, so we suggest doing this trail on another day. Also, note for your safety that this path is not in the best condition, and you should come fully prepared.

Alternatively, a trail to the south into the Bejenado massif commences on the other side of the parking place, in case you find the younger rocks more appealing, but note that the lower part is geologically different, with Garafía volcano units underlying the Bejenado sequence (Fig. 3.29A).

Once you meet LP-3 again (at a T-junction with LP-302), turn right and drive down-hill into *El Paso* and follow signs for *Los Llanos de Aridane*. Note that LP-3 will turn into LP-2 (orange) a short while after *El Paso*.

Drive back on LP-302. Continue on LP-2 into *Los Llanos de Aridane*. At approximately the km 49 LP-2 road sign (in *Los Llanos de Aridane*), the first brown signs for *Caldera de Taburiente* appear, as well as green ones for LP-214. Follow these signs and leave *Los Llanos* and its outskirts on LP-214 up-hill.

The road will soon descend again and will become very bendy. Make your way down to the bottom of Barranco de Las Angustias and use the car park provided in the floor of the valley (not recommended in wet weather because of possible flooding). From here, a winding and steep path leads to the "Camping Area" on the banks of the Taburiente River after approximately 7 km of hiking along the bed of Bco. de las Angustias. This hike will take an entire day and you may want to consider camping inside the caldera for the night. Note, however, you can only stay up to two nights there and it is mandatory to book in advance (**phone number: +34 922 497 277**).

You can, of course, also inspect only the first few kilometers that are a relatively easy stretch of the barranco, which will be the goal for our guided tour. In any case, bring water and sun protection!

Stop 1.3. Caldera de Taburiente, Bco. de las Angustias (N 28.6860 W 17.9092)

Oceanic islands rise from the ocean floor by accumulation of volcanic materials to form a seamount and, eventually, an island. In the Canaries, seamounts rest on mature oceanic crust and form on top of marine sediments deposited prior to initiation of seamount formation (Fig. 3.30A). Interaction of intruding magma and extruding lava with sediments may produce frequent xeno-pumice fragments in the eruptives (see chapter: The Geology of El Hierro) and will produce local peperites at the interaction site, as seen here on La Palma.

The bulk of the deep-water stage of construction on La Palma is characterized by piled-up non-vesiculated pillow lavas (Fig. 3.30B). As the seamount summit rises to shallower depths, explosive fragmentation starts (because load pressure is lower) and clastic material (breccias) and vesiculated lavas are produced.

The final stage of the shallow/intermediate growth is reached when the first subaerial eruptions occur (Fig. 3.30C). These continue to construct the island, eventually above the direct influence of the waves. This usually allows stabilization of an emerging edifice (Fig. 3.30D), but will eventually initiate long-term decay once eruptions cease (Fig. 3.30E).

This is a general pattern for the construction of an oceanic island and is also seen on Fuerteventura (see chapter: The Geology of Fuerteventura) and in the Cape Verde Islands (eg, Ramalho et al., 2010), for example. However, the submarine stage is only rarely exposed because of subsidence (eg, on Hawaii), or because the later subaerial edifice usually covers the submarine formations completely and erosion has not yet exposed these deeper sequences.

In the Canary Islands, the seamount stage (preserved in the 'basal complexes') only crops out on Fuerteventura and La Palma. In Fuerteventura, very old formations, including oceanic crust, have been exhumed by large collapses and subsequent erosion. On La Palma, however, very young and extensive outcrops of the submarine stage are preserved in the Caldera de Taburiente floor, which requires some further explanation.

Underplating and endogenous growth may account for tilting and ascent of the submarine core of the islands, but erosion is not sufficient to cause the seamount to outcrop even in the deepest barrancos. The scenario on La Palma is quite particular. Of all the barrancos of La Palma, the Barranco de Las Angustias cuts deeper and has the widest headwall basin (Fig. 3.31), despite being situated in the leeward and driest area of the island. The explanation is likely that the Barranco de Las Angustias was initially caused by a gravitational landslide that deeply incised the Taburiente volcano and formed an extensive basin that was then the site for intense erosion to set in (Fig. 3.32A,B), whereas later renewed volcanism then produced the Bejenado volcano (Fig. 3.32C).

Bejenado volcano is nested in the collapse basin and forced the drainage to incise deeply and rapidly into both the older and younger volcanic sequences, and eventually into the underlying Seamount Series (Fig. 3.32D).

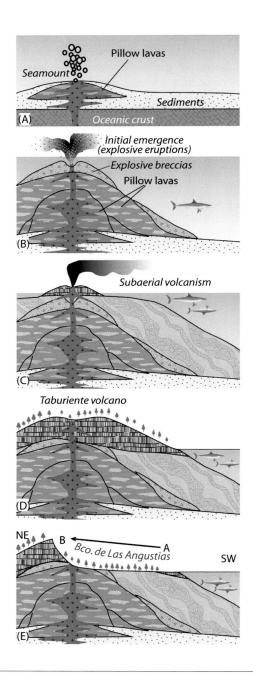

FIGURE 3.30

Cartoon depicting the evolution of La Palma during the seamount stage. The hike into Barranco de Las Angustias traverses the submarine volcano sequences, allowing us to look "beneath" the surface lavas that are commonly exposed in the Canaries and into a submarine oceanic volcano. Modified after Staudigel and Schmincke, 1981, 1984.

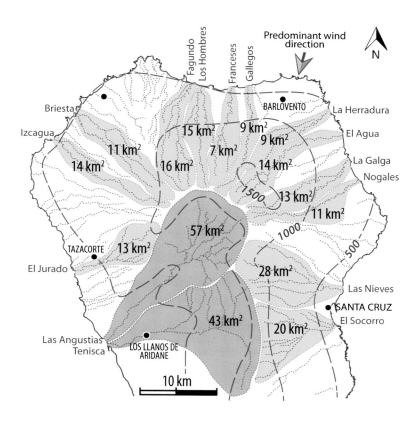

FIGURE 3.31

Barrancos of northern La Palma. Note that the Barranco de Las Angustias is the deepest on the island and has the widest headwall basin, despite being situated on the leeward and, hence, drier side of the island. Broken lines: Rainfall in l/m^2. Data from Plan Hidrológico Insular de La Palma; after Carracedo et al., 2001.

The bottom of the Caldera de Taburiente offers the extraordinary opportunity to observe the deep structure of the submerged stage of development of the island. Here, the submarine volcano has been uplifted by intrusive growth to 1500 m above sea level and tilts 50° to the southeast (toward the mouth of the caldera), implying that the route toward the interior of the caldera brings us deeper and deeper into the submarine volcanic edifice.

The up-river route through the caldera runs initially through breccias and shallow submarine lavas, with traces of oceanic sediments in the interstices, including warm-water fauna in the form of coral and foraminifera-bearing hyaloclastites that are interlayered with pillow lavas (Fig. 3.33A).

Further up-hill, shallow pillow lavas are readily identifiable due to the presence of vacuoles but are absent at depth because of the higher pressure (Fig. 3.33B−F).

Progression into the Caldera shows an overall increase in dykes and hydrothermal metamorphism that ranges from low-temperature alteration (<50°C) at the upper part of the section (in the lower course of the barranco), to mid-grade metamorphism (greenschist facies) in the interior of the barranco (see Fig. 3.34).

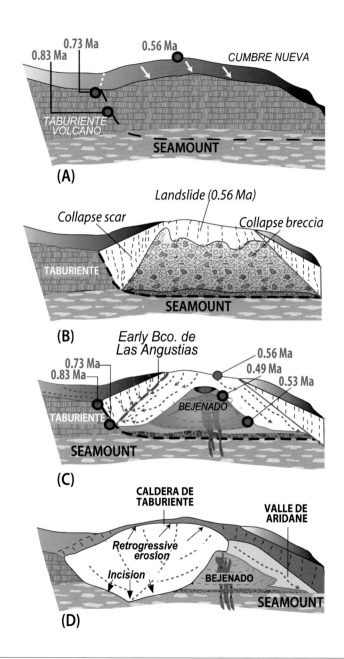

FIGURE 3.32

Cartoon sequence illustrating the sequential formation of the Bco. de Las Angustias. (A) Growth of Cumbre Nueva volcano to a point of instability. (B) Lateral collapse of its western flank. (C) Development of the Bejenado volcano nested inside the collapse embayment (Valle de Aridane), followed by initial incision of the Bco. de Las Angustias. (D) Enlargement of the current headwall of the Caldera de Taburiente by erosive retrogradation. Modified after Carracedo et al., 2001.

FIGURE 3.33

(A) The seamount exposed in Bco. Augustias has been dated to approximately Pliocene (4-3 Ma) through sediments in the interstices of the pillows that contain fossils of warm-water fauna such as foraminifera and corals (indicated with arrows). (B) Shallow water pillows. (C) Pillows with well-developed toes. (D) Deep-water pillows with incipient hydrothermal metamorphism. (E) Metamorphosed brecciated pillows of characteristic pistachio green color due to the presence of epidote. The metamorphic rocks here formed within a P-T range of ≥ 200 MPa (≥ 2 kbar) and at $250-450°$C. They can therefore be regarded as a low-grade metamorphic assemblage (greenschist facies). Mafic volcanic rocks metamorphosed under these conditions frequently develop green minerals such as chlorite and epidote (Fe^{2+}), the source of this characteristic color. (F) Interstitial jasper within pillow lava, a variety of silica that is red due to the presence of oxidized iron.

FIGURE 3.34

(A) Submarine lavas that formed in increasingly deeper water. Deeper pillow lavas are readily identifiable because of the absence of vesicles (former gas bubbles). Boreholes were locally drilled to sample for palaeomagnetic determinations. (B) Deeper inside the barranco, the path cuts into massive deep-water metamorphosed pillows. (C) Dykes in submarine breccias. (D) Dyke density increases as the barranco runs deeper into the Caldera de Taburiente. (E) Outcrop of plutonic rocks (gabbros) intruded by dykes at even deeper parts of the barranco. (F) Detail of a coarse-grain gabbroic plutonic rock inside the Caldera de Taburiente.

FIGURE 3.35

(A) *Google Earth* view of the Caldera de Taburiente indicating the main locations explained in the text. (B) The name '*Dos Aguas*' (Two waters) for this part of the barranco comes from the merging of Bco. Taburiente and Bco. Almendro Amargo (at 427 masl). The latter brings ochre-colored water due to high mineral content. (C) The warm and colored (orange to red) water from volcanic thermal springs is rich in gases and very oxidized (Bco. Almendro Amargo). (D) Cascada de Colores (the cascade of colors) located at 462 masl in Bco. Almendro Amargo.

The Barranco de Las Angustias is the only permanent stream in the Canaries due to the abundance of springs within the Caldera (Fig. 3.35). The water is channeled in the lower part of the barranco, but up-river it flows permanently. The extensive headwall area of the Caldera de Taburiente and the steep relief provide the conditions for spectacular flash floods in the rainy season, occasionally causing dangerous conditions. Return to the car at the end of the day. However, if you have decided to continue deeper into the Bco. de las Angustias, the spectacular but difficult 6- to 7-hour hike is best accomplished with an overnight stay in the camping facilities that are available in the heart of the Caldera.

If you have decided to proceed up-river to inspect these spectacular features, then climb through the Bco de las Angustias for approximately 5 km until you arrive at the confluence of two barrancos, the Bco. Taburiente to the left, with clear cold water, and Bco. Almedro Amargo to the right, with ochre-colored water (Fig. 3.35A,B). Some of this water is warm and strongly mineralized and gasified, turning intensely yellow or red when carrying high iron content. The name of Dos Aguas (two waters) comes from the merging of the two streams of different water composition, with the red colored water having higher mineral content (Fig. 3.35). Continue up Bco. Almendro Amargo for a short distance to the *Cascada de Colores*, a wall constructed to retain water from several springs, some of which emit highly mineralized water and magmatic gases (Fig. 3.35C), eventually forming a waterfall that displays a variegated combination of bright colors (Fig. 3.35D).

Return to Bco. Taburiente and follow the signposts to the camping site if you intend to stay overnight.

END OF DAY 1.

DAY 2: INSIDE A SHIELD VOLCANO

A shield volcano usually develops in a short time $(1-2\ Ma)$ and is geologically monotonous; it is formed dominantly by accumulation of lava flows with shallow dips of $10-15°$ away from the shield, with some interbedded pyroclasts that possibly occur locally. However, in the northern shield of La Palma, geochronological and structural observations show a more complex geological history, involving two successive shields that progressively covered the basal seamount: the Garafía and Taburiente volcanoes, which are separated by giant landslides.

Day 2 is dedicated to inspect this geological situation and its different internal structural components (Fig. 3.36).

Make your way to LP-1 by driving northward from *Santa Cruz*. After a few kilometers, turn left to join LP-4 in the direction of *Roque de los Muchachos* and *Observatorio Astrofísico* (36 km from this junction). LP-4 runs up-hill in a bendy fashion. Traverse Taburiente lavas and some intercalated volcano-clastic deposits for approximately 20 km before reaching the higher slopes of the old Taburiente shield.

Here, increasingly more scoria and reddish cinder deposits occur, indicating former vent systems in this part of the Taburiente shield. On your right you might see Teide in the distance from here, but this depends on the weather conditions.

At approximately the km 28 road sign on LP-4, stop in the lay-by on your right side.

Stop 2.1. Eruptive centers of the Taburiente volcano (N 28.7611 W 17.8675)

Here, black and reddish scoria beds are overlain by lavas in a spectacular road-side exposure, implying proximal deposition relative to an eruptive center or vent (Fig. 3.37). The red color of the lapilli, changed from its original black, is due to the oxidation of iron, a main component of basalt. When basalt is exposed to oxygen in a hot state for a certain length of time, the iron component is oxidized ($4\ Fe + 3\ O_2 = 2\ Fe_2O_3$), transforming the ferrous iron into red to reddish-brown ferric iron oxide (eg, hematite, from Greek *haimatites lithos*, blood-like stone). This process is also responsible for the red to brown coloration of many sedimentary rock and iron ores when oxidation is coupled with water uptake even at lower temperatures. Lunar basalts, in contrast, have higher iron contents than their terrestrial counterparts; however, because of the absence of a lunar atmosphere (and surface water), lunar basalts show a general lack of oxidation or significant secondary hydration. Instead, the reddish-brown color of Martian rocks (Mars is also known as the "Red Planet") is caused by iron-rich minerals, especially a reddish, fine-grained form of hematite, that

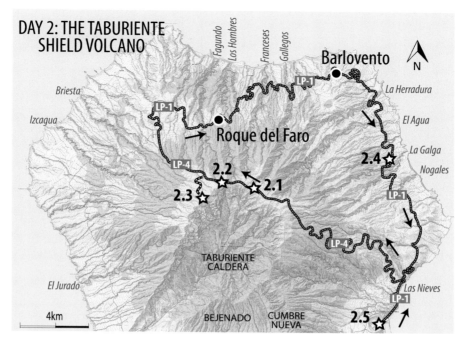

FIGURE 3.36

Map indicating the itinerary for Day 2, a tour around the northern shield of La Palma. Map modified from IDE Canarias visor 3.0, GRAFCAN.

FIGURE 3.37

Reddish scoria beds from a buried cinder cone overlain by lavas on a roadside exposure near the top of Taburiente volcano. The deposits indicate a nearby vent system of the old Taburiente edifice.

FIGURE 3.38

View from *Mirador de los Andenes* looking south onto the Taburiente caldera and onto the Bejenado and the Cumbre Vieja volcanoes. The volcanic edifices become progressively younger along this alignment (ie, toward the south).

occurs there as dust, likely implying that early Mars (billions of years ago) was wetter and warmer than today and that some water is likely still present in Mars's atmosphere.

Drive up-hill for a short while and just before the astrophysical observatory, park at *Mirador de los Andenes* (also known as *Degollada de Los Franceses*) at the headwall of this barranco. From the mirador there is a spectacular view of the main volcanoes forming the island of La Palma (if not covered by clouds) (Fig. 3.38).

Here, you can inspect the Caldera de Taburiente from above, looking at yesterday's stops 2 and 3 and the valleys and gorges inside the Caldera. The magnitude and scale of events affecting the old Taburiente shield become apparent at this site, for example, uplift and tilt of the submarine basement, construction of the subaerial shield, lateral collapse, and subsequent erosive incision to create the current Bco. de las Angustias.

Continue up-hill on LP-4, passing lavas with reddish to orange (palagonite-rich) horizons in places cut by thick dykes that likely fed eruptions of lavas in the area. Soon you will see the large telescopes and screens of the astrophysical observatory.

At the right side of the road there appears a sequence of sub-horizontal lavas (the Tamagantera section) several hundreds of meters thick. Intriguingly, it is not known how a sequence of horizontal lavas can accumulate at an altitude of more than 2200 m on the slopes of a shield volcano.

Stop 2.2. The Tamagantera section (N 28.7643 W 17.8801)

The plateau of horizontal lavas here is a significant geological feature of La Palma (Figs. 3.13; 3.39A,B), but we have seen similar features in other islands, such as La Gomera (see chapter: The Geology of La Gomera).

Radiometric dating and palaeomagnetic correlation using geomagnetic reversals place this sequence in the geo-magnetic Jaramillo period (see Fig. 3.2), a short normal polarity excursion from 1.07 to 1.0 ka within the reverse-polarity Matuyama epoch (Fig. 3.39C). The sequence

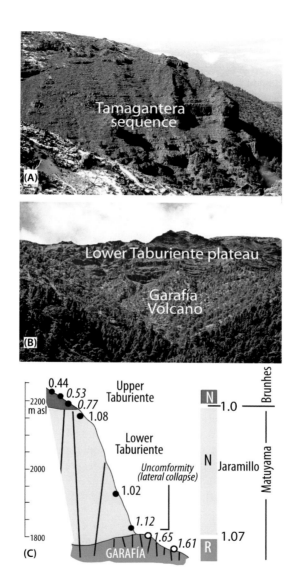

FIGURE 3.39

(A) At Tamagantera, a sequence of horizontal basaltic lavas several hundreds of meters thick is exposed that caps the Garafía volcano. (B) The Garafía volcano crops out through an erosional window excavated at the northern flank of Taburiente volcano. (C) Extremely fast emission of the Tagamantera lava plateau is indicated by paleomagnetic and radiometric determinations (in Ma) (after Carracedo et al., 2001).

overlies the Garafía volcano lavas unconformably (see Figs. 3.39B; 3.13), and similar correlations are observed in the galería *Los Hombres*, located in the northern flank of the Taburiente massif that crosses the same formations (see Fig. 3.3A).

The lava flows of the plateau have apparently been emitted at extremely fast rates (>6 mm per year), a common characteristic of postcollapse volcanism (Longpré et al., 2009; Manconi et al., 2009; Guillou et al., 2001; Carracedo and Troll, 2013; Kissel et al., 2014). A plausible explanation, consistent with the geochronological and stratigraphical data, suggests that the horizontal plateau sequence formed within the collapse basin of a lateral landslide from the south flank of the Garafía edifice (Fig. 3.40), possibly similar to the horizontal flows deposited by Teide volcano inside the Las Cañadas Caldera (see Fig. 5.34).

Continue up-hill and follow signs for *Observatorio Astrofísico* and *Roque de los Muchachos*. Here are some of the largest telescopes in the world. Star gazing is particularly easy in this spot due to the relatively high altitude and frequently clear skies.

At some point soon, a small road goes off to your left. Follow brown signs for the *Roque de los Muchachos* viewpoint and park when you arrive there.

Note that the wind can be quite chilly here (2400 masl), but if you are lucky the sun will be out and the Caldera will be clear of clouds. Here, you will have the opportunity to enjoy the very same impressions Leopold von Buch had when he coined the term "caldera" in 1815 (apparently at this exact spot here). Later, the term was modified and is now anchored in the geological vocabulary, mostly used in the context of vertical collapse calderas (ie, "cauldron-like" volcanic depressions).

Stop 2.3. Caldera de Taburiente; Roque de los Muchachos (N 28.7545 W 17.8852)

Walk up from the parking space and down the path that is several hundreds of meters long and brings you along the rim of the Caldera. On a good day, Teide on Tenerife is visible toward the southeast, and so might be El Hierro to the southwest and, of course, the inside wall of the Taburiente embayment. The path that starts here offers some amazing views "into the old volcano" and will take you approximately half an hour. There are more hiking trails here; two of them are spectacular and safe but long. These track the rim of the caldera to the east and to the west. Note that both need a complete day and some logistical preparations, so these are probably not for right now.

The lavas and scoria deposits of the caldera wall are cut by numerous dykes of variable orientation (Fig. 3.41). Note, however, that the back wall to the Caldera here is most probably not the collapsed wall of the Taburiente slide, but rather represents a former landslide scarp that experienced back erosion. The Caldera thus represents enlarged expression of the earlier landslide embayment, forming a set of radial gorges inside of the present caldera embayment. The Taburiente giant landslide occurred approximately 560 ka ago, leaving ample time for erosion to sculpture the landscape in this way.

Return to LP-4, and there enter LP-4 going west, following signs for *Santo Domingo de Garafía*. Soon the road will descend again and you will drive through extensive pine forests on basaltic peripheral flows of the Taburiente volcano.

After a good drive through the forest you will meet LP-1. Join LP-1 in the direction of *Garafía* and *Barlovento* to get to the leeward side of La Palma.

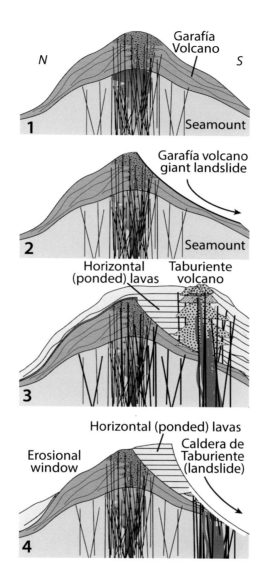

FIGURE 3.40

Cartoon depicting a probable origin of the plateau of horizontal lavas near the top of Taburiente volcano.
(1) Garafía volcano capping the seamount in an unstable, overgrown fashion. (2) Catastrophic landslide of the southern flank affecting the Garafía edifice and the submarine volcano. (3) The Taburiente volcano developed and nested inside the collapse depression, forming a sequence of horizontal lavas attached to the scarp. (4) Erosion now exhumed this fossilized Taburiente lava plateau (after Carracedo et al., 2001).

FIGURE 3.41

(A—H) Impressions of the upper wall of the Caldera de Taburiente, showing lava flows, interbedded pyroclastic units, and frequent cross-cutting intrusive sheets.

Soon you will pass the small village of *Roque de Faro*, which will make a nice spot for your lunch. From there, you will have a view of the plateau of the horizontal Taburiente lavas overlying the Garafía volcano (see Figs. 3.3B and 3.13).

Continue eastward on LP-1 now in the direction of *Santa Cruz*. The road cuts monotonous basaltic Taburiente flows while crossing the radial barrancos (see Fig. 3.42).

A little later, just before a tunnel, a "small" landslide scarp can be seen on your left. This is the source of the landslide that destroyed the old road a few years ago, which prompted the construction of the tunnel.

Continue driving through the tunnels. When reaching the north shore, you will find that the lavas dip mainly northward (ie, toward the sea), implying that you are now on the northern flank of the old Taburiente shield.

Continue on LP-1, following signs for *Santa Cruz*. After *Barlovento*, you will drive by several cinder cones that are part of the Taburiente volcano (0.7 Ma) and that form a rift zone (ie, a surface vent alignment on the north flank of the volcano). Banana production is common here, and often exploits the basaltic-derived soils and lapilli that are rich in this part of the island and the moist climate here on the northern slopes of the island.

Continue on LP-1 until *Mirador Jardín de las Hespérides* and park there. The mirador reminds us of the Greek writer and geographer Herodotus, who referred to the Canaries as "The Garden of Hesperides."

Stop 2.4. Mirador Jardín de Las Hespérides *(N 28.7722 W 17.7666)*

Here, you can view one of the deep barrancos of La Palma. The rocks at the bottom are more than 780 ka (Matuyama reverse-polarity epoch), whereas higher in the barranco the lavas are of positive magnetic polarity and thus less than 780 ka (Brunhes normal-polarity epoch) (see Fig. 3.42C). The presence of the Matuyama/Brunhes (M/B) limit is a consistent feature of the Taburiente lavas and is easy to define with a portable magnetometer (eg, Guillou et al., 2001; Carracedo et al., 2001; Singer et al., 2002). This technique was the main tool to establish the stratigraphy of this volcano that is mantled by recent (Brunhes) lavas. The giant landslide that formed the Taburiente caldera and the deeply incised erosional barrancos also exhumed these older (Matuyama) formations. These observations also imply extremely fast development of this shield volcano (see Figs. 3.2 and 3.42), with major growth cycles that operated on a 100-kyr time scale.

Return on LP-1 to *Santa Cruz*, and cross the town to join LP-2 to reach the spectacular tuff cone of La Caldereta, located at the south end of the town above LP-2. Leave LP-2 at the southern end of town. At the large roundabout that marks the southern entrance to the city, turn right and drive up-hill and directly into the Caldereta crater. Park on the little plateau inside the crater.

Stop 2.5. La Caldereta *(N 28.6771 W 17.7723)*

The impressive tuff cone of La Caldereta is composed of highly compacted yellow (palagonitic) tuffs and has a base measuring approximately 1.5 km and a crater diameter of 1 km. It is thus among the largest tuff cones in the Canary Islands (Fig. 3.43). Indeed, this hydrovolcanic cone falls in the Wcr/Wco diagram near the limit for tuff cones and close to the field of collapse calderas (see Fig. 5.20). In its interior, a magmatic eruptive center provides evidence that at the end of the eruption, the magma was isolated from the sea water, changing the eruptive style from hydrovolcanic to strombolian (see also Fig. 5.66 in chapter: The Geology of Tenerife).

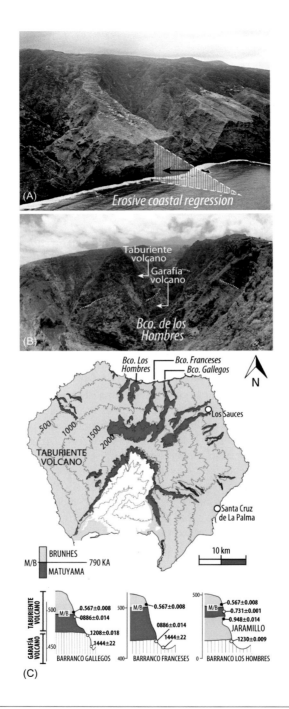

FIGURE 3.42

(A) Estimate of the scale of regression of the northern windward coast of the northern shield. (B) Barranco de los Hombres is incised into the Taburiente and Garafía volcanic sequences. (C) Distribution of the exposure of the older (Matuyama epoch) lava sequences in the northern shield (after Carracedo et al., 2001).

FIGURE 3.43

La Caldereta in *St. Cruz de La Palma*, one of the largest tuff cones in the Canaries. Note the lava flows of the Taburiente volcano that encircle the tuff cone. Image from Google Earth.

This cone, located in a rural area until recently and with its geological interior intact, has now been obstructed by growing housing complexes despite having been declared a protected Natural Monument some years ago.

As frequently happens in the Canaries, only the inaccessible parts have been legally preserved with a "monument status," leaving the areas possible for construction unprotected. The interior of La Caldereta is a dangerous housing area in the event of heavy rain because of the large basin and the highly impermeable components of the volcano, but it offers an attractive site for people who want to combine a life close to *Santa Cruz* with a touch of rural landscape and a spectacular ocean view.

END OF DAY 2.

DAY 3: TOUR OF THE CUMBRE VIEJA RIFT ZONE

One of the youngest and most active volcanic structures in the Canary Islands, the Cumbre Vieja rift, has developed over the past 125 ka. Half of the historical eruptions of the Canary archipelago occurred on this rift (see Table 3.1). Today's tour will offer the opportunity to inspect several of these historical lava flows closely (Fig. 3.44). Their vents, however, are usually located near the crest of the rift and require a separate 1-day expedition (Day 4).

From *Santa Cruz de La Palma*, make your way to LP-2 (red) going south. Pass Cumbre Nueva lavas in the cliffs to your right when leaving the town.

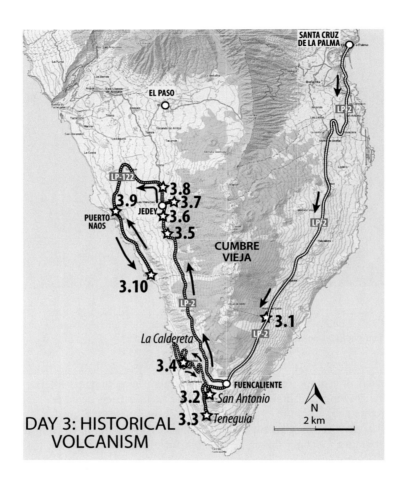

FIGURE 3.44.

Map indicating itinerary stops of Day 3 for inspection of the Cumbre Vieja rift and the associated historical volcanism of La Palma. Map modified from IDE Canarias visor 3.0, GRAFCAN.

Note that LP-2 goes off to the right after a short while; do not miss the turn! Follow LP-2 now for approximately 20 km in the direction of *Fuencaliente*. Pass increasingly younger lavas and some cinder cones on the way.

Approximately 2 km after the municipality of *Fuencaliente* sign, a lay-by on the left of the road opens up. Park there to inspect the 1646 AD Martin eruption products.

Stop 3.1. Martin eruption, 1646 AD (N 28.5213 W 17.8197)

Here, several massive lava beds are separated by top and bottom breccias, implying a stack of individual tongues make up this lava deposit. The occasional plutonic xenolith (gabbro, syenite, and periodite) can be observed. Note the fresh colonization of the lavas by vegetation (lychen,

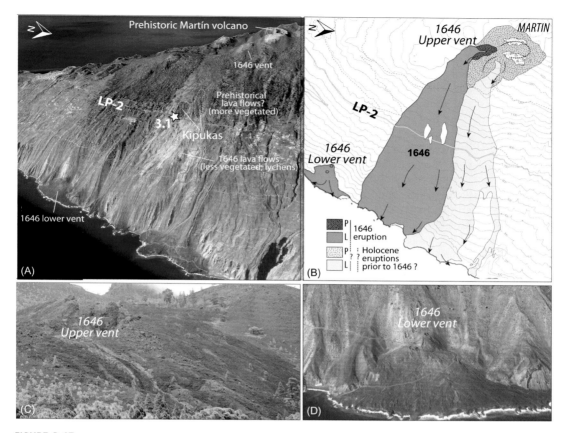

FIGURE 3.45

(A) Eruptive vents and lava flows of the 1646 Martín eruption (image from *Google Earth*). (B) Geological map of the 1646 eruptives (red color). A well-vegetated area to the right (pink color in the map) could be lavas of the 1646 eruption covered with airfall lapilli of an early phase of the eruption, or could correspond to a previous prehistorical eruption. (C) Upper vent of the 1646 eruption. (D) Lower vent of the 1646 eruption near the sea. Image (D) from Google Earth.

pines, and some endemic plants); it is scarcely vegetated in comparison with the adjacent older lavas and particularly with some older rocky islets here (kipukas) (Fig. 3.45A,B).

Unfortunately, there are limited original accounts of the 1646 eruption, known also as the Tigalate or Martín eruption. Two eruptive centers are described: an upper vent (1300 masl) at the base of the cone of the prehistoric Martín volcano, and a lower vent near the coast (Fig. 3.45). Both formed long lava channels and emitted very fluid lava flows that reached the sea in several places.

Continue on LP-2 for a while, and then turn left onto LP-209 (green) and follow brown signs for *Volcan San Antonio* and its visitor center. When you arrive at the visitor center, park your car.

Stop 3.2. San Antonio volcano (N 28.4871 W 17.8484)

The Volcan San Antonio visitor center shows original video footage of the 1971 eruption, and you will find a small souvenir and coffee shop as well as bathroom facilities here. The entrance fee is €5 for nonresidents, and for large groups it is €3 per person. Business hours are 10:00−19:00 during summer and 09:00−18:00 during winter.

After having taken in the information offered by the visitor center, walk up the well-laid hiking path to the top of the San Antonio crater (Fig. 3.46). Looking south, in the far distance you can see the 1971 (Teneguía) vent system, but you will also have a view inside the San Antonio crater when looking east. From here, when looking west and southwest, you can also see the phonolite plug that carries the famous aboriginal carvings that were almost buried by the 1971 lavas.

Note the gray-brown colors of the deposits of the San Antonio crater that are overlain, for example near the visitor center, by dark and fresh 1677 lapilli and scoria deposits (Fig. 3.46A). The main crater is considerably older than the 1677 events and very likely prehistoric. When walking back to the visitor center, note all the light-colored pumiceous inclusions in the lava blocks of the constructed walls. These are frothy sedimentary inclusions and have been termed xeno-pumice (Troll et al., 2012; Carracedo et al., 2015a). Xeno-pumice is abundant in the lava rocks here. The often glassy disintegration textures of these inclusions are notably similar to those in the Lanzarote's Timanfaya deposits or the recent "restingolites" of the El Hierro 2011 submarine eruption near the coastal village of *La Restinga*. The term "xeno-pumice" is, by now, applied to all of these examples (see Carracedo et al., 2015a).

Another important aspect is the confusion that still prevails regarding the age of the actual San Antonio cone. At the entrance of the road to the visitor center you might have seen a standing stone

FIGURE 3.46

(A) Close-up present-day view. (B) Vintage photograph of the San Antonio cone (circa 1930s) (image courtesy of *Cabildo de Insular de La Palma*). The small cinder mount to the left of the main San Antonio cone is the upper vent of the 1677 eruption, which was unfortunately removed during the construction of the visitor center. The 1677 lapilli that cover much of the *San Antonio* cone were erupted from here.

with the inscription *"Volcán San Antonio*, 1677"; this date is widely accepted for this volcano, reflecting that the 1677 eruption has traditionally been associated with the 150-m high and 1-km wide San Antonio cone. However, this concept has been revised in recent years, concluding that the San Antonio cone is an older explosive (phreatomagmatic) vent encircled by lavas that are 3000 years or older in age, whereas the dark and fresh 1677 lapilli and scoria deposits were ejected from a small vent next to the San Antonio cone that is still identifiable in old pictures (Fig. 3.46B) (Carracedo et al., 2001).

Eye-witness accounts describe the 1677 eruption as not particularly destructive and only a few houses were burned near the coast; damage in the nearby town of *Fuencaliente* was negligible. No human casualties were documented from this eruption, although a contemporary drawing depicts a shepherd being asphyxiated together with a flock of goats (Fig. 3.47).

This lack of significant damage to the nearby town of *Fuencaliente* is incompatible with the characteristics of the eruption of the large San Antonio cone, which included very violent episodes of phreatomagmatic activity (interaction of groundwater with magma), and even lateral explosions known as surges, which should have had devastating effects on the town of *Fuencaliente* if the construction of the entire San Antonio cone occurred in 1677 (Fig. 3.48A).

The materials related to the violent lateral explosions of the prehistoric San Antonio and La Caldereta volcanoes extended over a radius of several kilometers around their eruptive vents, whereas the lavas and lapilli of the 1677 eruption directly overlie the lava flows of the Fuego and La Fajana volcanoes and also the Roque Teneguía, but without any trace of interbedded explosive breccias (Fig. 3.48B,C). The absence of such phreatomagmatic deposits among the 1677 tephra precludes a highly explosive origin. It is thus evident that the San Antonio cone was formed several thousand years earlier. Moreover, the 1677 (Fuencaliente) eruption had two emission centers around this older San Antonio cone, one in the summit area of the cone and the other at the base of the San Antonio cone (Figs. 3.46 and 3.49). In contemporary accounts it is described as follows: *"the town's people would climb Mña. del Corral to observe the vents of the lower volcanoes and one day, after the curious had already disbanded, a great vent opened in the same mountain, which, had it caught them unaware, would have caused a most painful fatality."*

This description coincides surprisingly well with events as they occurred in the Teneguía eruption of 1971, in which the summit of the San Antonio volcano was the preferred viewpoint of local folks from which to observe the eruption. It would appear that the reference to the *"lower volcanoes"* points to the open vents at the base of San Antonio volcano (Figs. 3.49 and 3.50) that emitted large volumes of lava, whereas the *"large mouth in the same mountain"* appears to refer to the more explosive upper eruptive vent through which the eruptive system degassed (see Fig. 3.46), and indeed, the erupted 1677 pyroclastics contrast strongly with the older San Antonio phreatomagmatic deposits (Fig. 3.48).

Interestingly, the descriptions of this 1677 eruption in contemporary accounts is very precise for the first 10 days, likely because the general concern about the *Fuente Santa* (the Holy Spring) that was located at the southern tip of the island and was threatened by the lava flows. Once the Fuente Santa was destroyed (on November 26), the reports died down and the only additional information provided is in regard to the end of the eruption on January 21, 1678 (see Fig. 3.49).

After visiting Volcán San Antonio, join LP-209 (green) again for a short while and follow signs for *"Volcán Teneguía."* After a short drive, a sharp left turn will bring you onto LP-2091 (yellow road sign), which brings you closer to the 1971 Teneguía vents. A little further, however, the road becomes a dirt track; after a few minutes on the track, a sharp right bend brings you to the 1971 vent sites. There you will find a parking site. Walk for the last few hundreds of meters to the main Teneguía 1971 vents along the well-laid path.

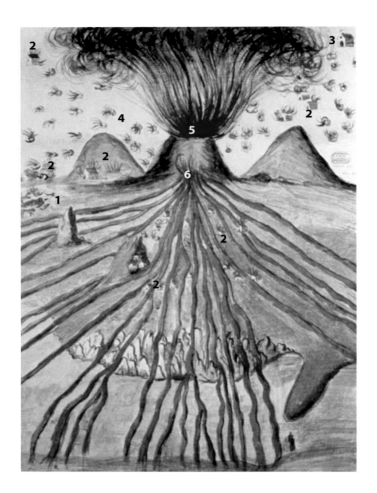

FIGURE 3.47

Contemporaneous illustration of the 1677 Fuencaliente eruption of La Palma. Eye-witness accounts (eg, see Miguel Santiago, 1960, for a compilation of contemporary reports) describe minor damages from this eruption (except for the destruction of the Fuente Santa) and no human casualties. The contemporaneous painting, however, shows a shepherd being asphyxiated together with a dozen goats (indicated by 1), and a number of houses on fire around the volcano especially to the south and west of the eruptive vents (2) (see also Fig. 7.4A). The painting thus seems exaggerated, but it indicates that the damage was mainly directed toward the south and southwest and not to the north toward the village of *Fuencaliente*, as probably implied by the undamaged church of the town (3). The painting also illustrates the explosive upper eruptive vent (5) and the lower vents that emitted large volumes of lava to the sea (6). Image courtesy of Archivo Histórico Nacional de España, Madrid; after Carracedo et al., 2001.

FIGURE 3.48

(A) Lapilli of the 1677 eruption cap the phreatomagmatic slabs of the older San Antonio cone. (B) Proximal cinder and bombs of the 1677 events are overlying lava dated at 3.2 ka. Note the absence of intercalated phreatomagmatic ash. (C) The 1677 lavas encircle the 56-ka Teneguía phonolite plug (see also Carracedo et al., 2001).

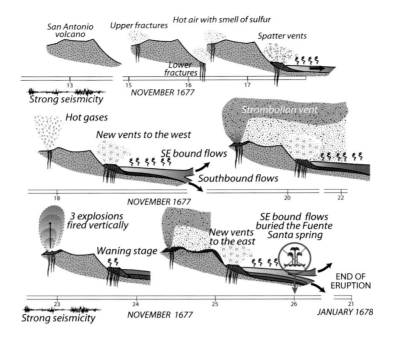

FIGURE 3.49

Cartoon of the 1677 Fuencaliente eruption reconstructed from eye-witness accounts and geological observations (after Carracedo et al., 2001).

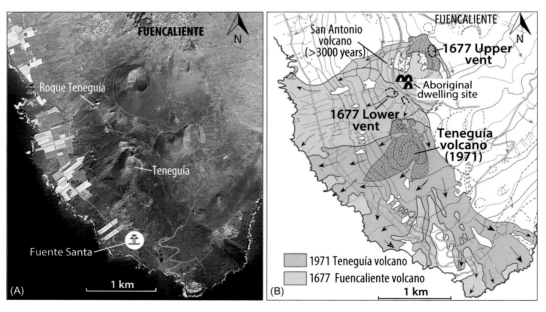

FIGURE 3.50

(A and B) Southern tip of La Palma, where the two recent eruptions of 1677 and 1971 took place. Archeological studies discovered remains of an aboriginal settlement located on the south flank of the San Antonio cone (Pais, 1997), and recent efforts claim to have uncovered the Fuente Santa, which was buried during the 1677 events. Image in (A) from Google Earth.

Stop 3.3. The 1971 Teneguía volcano (N 28.4789 W 17.8504)

Walk up the path to the crater rim. Once on top, you will get some very rewarding views. Looking south, you will see *Salinas* and *Faro de Fuencaliente*. To the north, you will see the lapilli of the 1971 eruption on the flank of the San Antonio cone. Also, the vents of the 1677 eruption at the base of the San Antonio cone are now visible. To the northwest, the Roque de Teneguía plug (light colored rock) is visible, which comprises older erosive remnants of phonolite (dated at 56 ± 2 ka, Carracedo et al., 2001) that stick out from the younger lava flows surrounding it. Looking west to the coast, the Teneguía lavas have been artificially covered with soil to grow cash crops, such as bananas, tomatoes, or pineapple.

The Teneguía eruption is the most recent to occur on La Palma and the second-youngest of the archipelago (after the 2011−2012 submarine eruption off El Hierro). The 1971 deposits are located at the extreme southern end of the island, near the phonolitic Teneguía dome, from which the eruption takes its name. They overlie the 1677 volcanics next to the town of *Fuencaliente* and also the 1646 and 1677 eruptive centers (see Figs. 3.50 and 3.51).

The 1971 eruption was preceded by strong seismicity in the Cumbre Vieja. A few days before, the onset of the eruption seismic activity concentrated around the village of *Fuencaliente*. Seismicity was

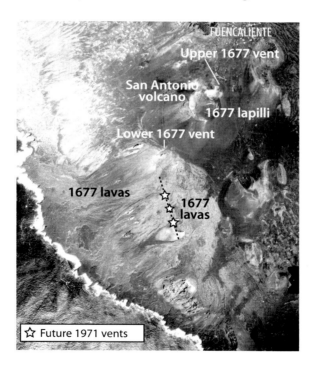

FIGURE 3.51

Aerial photography taken in 1968 that depicts the area where the Teneguía eruption occurred 3 years later (white asterisks). Note that the dark patch at the eastern side of the San Antonio cone are the lapilli from the upper vent of the 1677 eruption. Image modified from IDE Canarias visor 3.0, GRAFCAN.

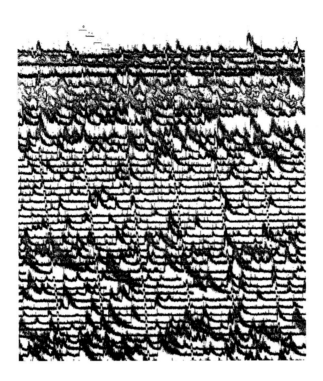

FIGURE 3.52

Thousands of unusual submarine "sounds" were recorded off La Palma in the days preceding the 1971 eruption by the US Sound Surveillance System (SOSUS), which was widely deployed during the Cold War. Hydrophones (two lines of 3 and 8.5 miles) were installed only a few years prior to the eruption at the coast near *Puerto Naos.* Courtesy of H. Castro, former staff of the surveillance base.

relatively strong (sufficient to produce some cracks in houses and the tower of the town church). No seismic monitoring instruments were deployed on La Palma at that time, but an array of hydrophones (two lines of 3 and 8.5 miles) were previously installed at the coast in *Puerto Naos* several years prior to the 1971 eruption and recorded thousands of microtremors in the days preceding the 1971 events (Fig. 3.52). The facility near *Puerto Naos,* then a branch of the US-based Columbia University in New York, used equipment that belonged to the US Navy Sound Surveillance System (SOSUS) that was originally developed and deployed for screening the passage of nuclear submarines through the Atlantic Ocean (http://minoslas.blogspot.com/2011/10/la-estacion-sosus-de-puerto-naos-en-la.html).

In 1971, this station provided an outstanding service to the people of La Palma and the study of volcanism on the island because several days before the first earthquakes of the imminent eruption were noticed, the Columbia University hydrophones captured "strange" underwater sounds that were rapidly identified as of volcanic origin. The epicenter was located in the south of the island, near the coast (see Fig. 3.52), and in late October the 1971 eruption started near the town of *Fuencaliente.*

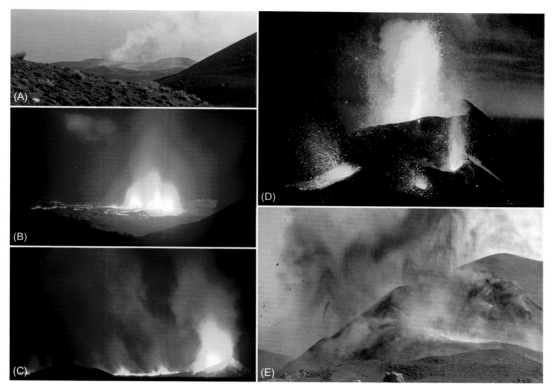

FIGURE 3.53

The 1971 Teneguía eruption. (A) Opening of the initial fracture occured at 17:00 on October 26, 1971. (B) The first hornitos develop at the upper part of the eruptive fissure. (C) At 23:00, the eruption had constructed a volcanic cone emitting lava flows down-slope. (D) Later, at mid-eruption, a main volcanic cone with several peripheral vents developed. (E) The Teneguía volcano close to the end of the eruption (November 18, 1971), showing only mild strombolian activity. Photographs courtesy of A.M. Diaz Rodríguez (A, B, and C).

The opening of a fissure several hundreds of meters long began in the afternoon of October 26, 1971, and gave rise to lava flows that quickly descended down-hill toward the south and southeast (Fig. 3.53A). Between October 26 and November 18, a group of volcanic cones then formed (Fig. 3.53B−E), emitting lava flows that in some instances reached the sea to form a typical but small coastal lava platform. The eruption ceased abruptly on November 18, leaving two victims who died of asphyxiation: a fisherman working in proximity of the lighthouse and a photographer taking pictures of the lava flows at the coast.

The descriptions of this eruption are rather detailed (see, eg, the Teneguía special issue of Estudios Geológicos in 1974); a set of continuous lava and gas samples was taken and the successive emission centers and the trajectory of the lava flows were mapped (eg, Fig. 3.54). The eruption evolved in two main stages. The first was characterized by the formation of two eruptive centers

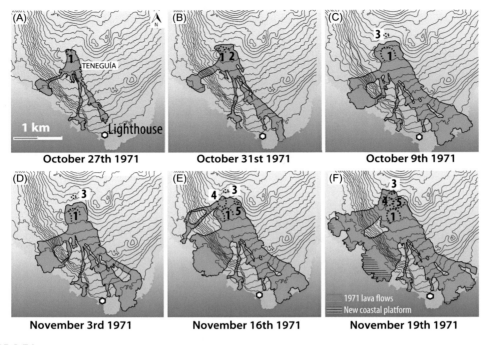

FIGURE 3.54

Evolution of the vent sites and lava cover during the 1971 Teneguía eruption. Note that vent 2 was destroyed as the eruption progressed. A new coastal platform formed late in the eruption and is presently used for banana cultivation. Modified after Hernández-Pacheco and Valls, 1982.

(1 and 2 in Fig. 3.54A,B), aligned along a fissure of 300 m. These eruptive vents emitted lava from the very start of the eruption and continued practically until the end of the events. The second stage was characterized by the formation of several new eruptive vents, starting on November 8, that likewise emitted lava flows, with the exception of the number 3 vent. Lava was never issued from vent number 3 because it was dominantly exhalative in nature (Fig. 3.54C–F).

The 1971 eruption resurfaced an area of approximately 135,000 m^2, whereas approximately 290,000 m^2 (29 hectares) were gained from the sea (Fig. 3.54F). The total volume of emitted materials was approximately 40 million m^3, the equivalent of filling a football field to a height of 5.7 km. The mean thickness of the stacked lava flows was approximately 12 m, whereas the main cone reached a height of approximately 100 m above the pre-eruptive land surface (Fig. 3.55).

Gases liberated into the atmosphere during the 1971 eruption became progressively simpler in composition until they became reduced to carbon monoxide (CO) and carbon dioxide (CO_2). The latter is much more abundant and is heavier than air, therefore displacing air (and consequently oxygen) from depressions. This likely caused the two fatalities by asphyxiation (see also Fig. 3.47).

Walk back down from the vents and explore the lavas of the Teneguía eruption along the well-laid paths that lead into the lava field. Once you have had sufficient time with the rocks, walk back to the car.

FIGURE 3.55

(A) Aerial view of the vents and flows of the 1971 Teneguía eruption. (B) 1971 lavas forming a coastal platform. (C) Lava flow cascading down a coastal cliff. The arrow indicates a developing acretionary ball. (D) Advancing front of the 1971 lava from *Teneguía* with one of the authors (J.C.C.) for scale.

Return to LP-209 the way you came and join it by making a left turn (down-hill). Park the car at the km 7 road sign to inspect the phreatomagmatic deposits of La Caldereta de Las Indias.

Stop 3.4. The Caldereta de Las Indias explosive (phreatomagmatic) volcano (N 28.5102 W 17.8686)

La Caldereta de Las Indias is a small cinder cone that was initially explosive (phreatomagmatic), expelling ash and breccia material several kilometers around the vent followed by emission of lavas. This is visible through interbedded tephra with basaltic lava flows in the road cut here. The thicker deposits, closer to the vent, have been quarried to construct a water reservoir (*Balsa de La Caldereta*). La Caldereta de Las Indias is likely related to the San Antonio phreatomagmatic eruption (Fig. 3.56A).

Here at the stop, phreatomagmatic beds are overlain by basanite lava and tephra rich in xeno-pumice (Fig. 3.56B). The xeno-pumice samples are probably the "empty shells" of degassing of formerly molten sedimentary fragments from the ocean crust (note, they contain quartz) and may

FIGURE 3.56

(A) View of the west coast (*Las Indias*), with the phreatomagmatic vent of La Caldereta. Red arrows indicate localities with abundant xeno-pumice inclusions (image from *Google Earth*). (B—E) Examples of xeno-pumice inclusions in basanite lava near *Las Indias*. (B—C) Note the spherical shapes. (D—F) Close-up of xeno-pumice samples in an advanced stage of disintegration. These samples are very similar to the occurrence of xeno-pumice near El Hierro in 2011.

have been a main cause for the phreatomagmatic behavior. Sedimentary rocks can be very water-rich (up to 25% even at several kilometers of burial depth) due to their mineral water, but especially due to pore water. In this respect, we note that, for example, the 1949 Hoyo Negro vent is also rich in xeno-pumice and was the most explosive vent of the 1949 eruption with distinct phreatomagmatic phenomena recorded (Klügel et al., 2000; Carracedo et al., 2001).

Turn around and return to LP-2 via LP-209. Continue northward on LP-2 until the km 33.5 road sign. Park the car at the entrance to a track on the side of the main road to inspect the 1712 eruptives.

Stop 3.5. The 1712 lavas (N 28.5498 W 17.8665)

Here, you can inspect a lobe of the 1712 eruption. We pass other lobes later (Fig. 3.57). This basanitic lava was emitted from Volcán del Charco (also known as Mña. Lajiones), which has whitish

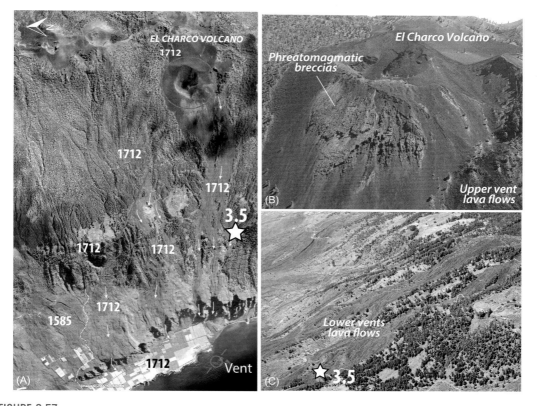

FIGURE 3.57

(A) Eruptive vents and flows of the 1712 El Charco eruption (image from *Google Earth*). (B) Oblique aerial view from the west of the main eruptive vent into the 1712 eruption. The Spanish name of this mountain, Mña. Lajiones or Lanchones, refers to the thick "slabs" of phreatomagmatic breccias that formed at the final stages of the eruption. (C) The lower vents of the 1712 eruption formed an alignment of hornitos that erupted several lava flows. These progressed independently down-slope and into the sea after crossing over the preexisting coastal platform.

slabs of phreatomagmatic deposits attached to the south flank and forms an alignment with several eruptive vents along a 2.5-km fissure that runs in a northwest—southeast direction (Fig. 3.57A). Lava flows were issued from all these vents along the fissure and cascaded over the cliff to form an extensive coastal platform.

The 1712 eruption is a clear example of two characteristics typical of the recent Cumbre Vieja eruptions. The general pattern of the feeder dykes is to align parallel to the rift axis, whereas the "flank effect" induced by the slope can cause feeder dykes or a fissure to adopt directions at angles with the rift axis (see Fig. 3.57A). This is because a propagating dyke that approaches a volcano flank will be diverted toward the flank in an effort to travel the shortest distance to the free surface; this occurred in the 1646 and 1712 eruptions and was repeated during the 1949 eruptive events.

The other characteristic feature of the 1712 vents is that the upper vents degassed the eruptive system by a "chimney effect." Gases migrated under pressure toward the higher parts of the rift, and the higher eruptive vents were exhalative and more explosive, including some phreatomagmatic pulses (Fig. 3.57B). The lower vents, instead, were predominantly effusive, emitting lava flows that progressed down to the sea (Fig. 3.57C).

This very poorly documented eruption of 1712 was followed by the 1949 event (ie, following a period of 237 years of eruptive repose).

Continue on LP-2 going north. Just a little further, we pass the 1712 lavas again (eg, near the municipality *El Paso* sign) and another time at the km 35 sign on LP-2. Just after the km 37 sign, the rock changes to a fresh dark lava that belongs to the 1585 basanite eruption (Fig. 3.58). Park at the side of the road for an inspection.

Stop 3.6. The 1585 lavas (N 28.5762 W 17.8766)

The lava here once more carries plutonic xenoliths (gabbro and pyroxenite), and large pyroxene crystals sometimes occur as isolated xenocrysts.

The location of the vents for the flows of the 1585 Jedey eruption was uncertain until a few years ago. Little was known apart from the date of the eruption (May 20, 1585), and an imprecise location description that read *"on the western flank of the Cumbre Vieja"* that was recorded in some contemporary accounts. There are several historic lavas, however, on the southwestern flank of the Cumbre Vieja (eg, San Juan or Nambroque, and El Charco) and the prehistoric Mña. Quemada or Tacande lavas, all of which share similar characteristics (see Table 3.1 and Fig. 3.3B).

These ambiguities were resolved when Mña. Quemada (or Tacande) was dated by ^{14}C as prehistoric in age (pre-1492 AD, Hernández-Pacheco and Valls, 1982) and when eventually the 1585 and 1712 flows were fully mapped (Carracedo et al., 2001). The eruptive center of the 1585 eruption was located at the Roques de Jedey, a set of phonolite spines that protruded during the eruption (Fig. 3.59).

Moreover, recently rediscovered contemporary description of the 1585 events was provided by a highly credible eye-witness, the Italian military engineer Torriani, who arrived on the island in 1584 and was there during the eruption. He reports that the initial vent opened in a small barranco above the village of *Jedey*. He also reports that during the first week of the events, this zone was uplifted to form a mountain topped with two large rock spines (Fig. 3.59B,C).

An interesting aspect of this eruption is the emission of different mafic to evolved melts. The most evolved (tephriphonolite) was highly viscous, and it may have pushed up the older rock "spines" to produce a phonolite cryptodome. This process was, according to Torriani, associated with strong seismicity and was soon followed by emission of basaltic pyroclasts and lava flows.

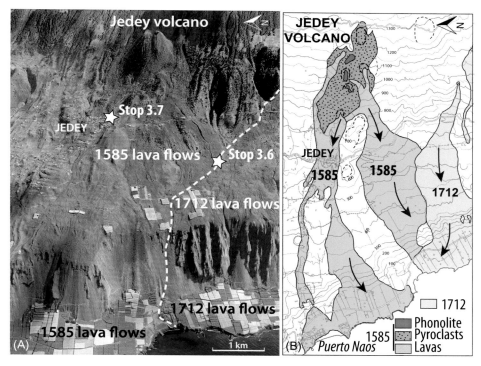

FIGURE 3.58

(A) Aerial view of the western flank of the Cumbre Vieja near the village of *Jedey*. Note how the 1585 and 1712 flows "merge" into each other, rendering them difficult to distinguish (image from *Google Earth*). (B) Geological sketch map of the 1585 and 1712 eruptions (after Carracedo et al., 2001).

The basaltic lavas frequently incorporated older phonolite blocks that can be seen to range from solid or partially molten to almost completely fused and reabsorbed (Fig. 3.60).

Field evidence for the presence of juvenile phonolites in the 1585 (cf. Johansen et al., 2005) eruption is ambiguous, because the majority of phonolites were extruded as solid plugs and only minor volumes erupted as melts, which therefore potentially originated as molten phonolitic blocks. However, juvenile phonolite flows occur on the Cumbre Vieja, for example, at Nambroque, Malforada, or Cabrito, (see Fig. 3.23A), but perhaps not in this eruption.

Continue your drive on LP-2 toward the north. After a short drive, a sharp turn to the right off LP-2 brings you onto a small road to another 1585 outcrop. Follow this road for a few hundred meters up-hill and stop before the road swings left in a wide bend.

Stop 3.7. Disused quarry with phonolite blocks dissolved in basanite, 1585 AD (N 28.5838 W 17.8786)

Here, in a small disused quarry on your right, you will find large blocks of partially digested phonolite clasts visible in the 1585 basanite lavas (Fig. 3.60).

FIGURE 3.59

(A) Eruptive center of the 1585 eruption (image from *Google Earth*). (B) Phonolite spines that protruded during the eruption. (C) Close-up view of the phonolite spines of the 1585 eruption.

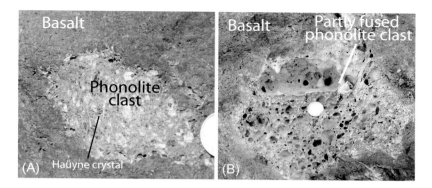

FIGURE 3.60

(A) Phonolite clast enclosed in basaltic lava of the 1585 Jedey eruption. (B) Another phonolite clast almost completely fused and was partly assimilated. This form of "xeno-pumice" is noteworthy and must not be confused with those specimen seen at *Stop 3.4* that likely derived from melted sediment and not from recycled lava.

Note that this bubbling and dissolution of phonolite clasts with all stages of the process are recognizable. In particular, a large volume of gas bubbles seems to form during the process. Here, the "xeno-pumice principle" (Troll et al., 2012) is superbly displayed in its variable evolutionary stages, but the protolith here is of a magmatic rather than a sedimentary derivation.

If you have time, drive up-hill for a closer view of the 1585 vent and the phonolite spines; otherwise, return to LP-2 and continue northward. Pass through the village of *Las Manchas* and right after the village, turn left onto LP-211 (green) in the direction of *Todoque-Puerto Naos* right after the crossroad. Park the car and walk back to inspect the road cuts. These show piled-up pahoehoe lavas from the 1949 eruption (please wear a high-visibility vest).

Stop 3.8. The 1949 lavas (N 28.6044 W 17.8820)

Here, you can investigate a cross-section of the 1949 lava flow in the road cut, which is a xenolith-bearing basanite with pronounced pahoehoe structures.

During the 1949 eruption, three vents opened at different stages and are distributed along a 2-km fissure (Fig. 3.61). The upper vents, Hoyo Negro and Duraznero (Figs. 3.62A−C and 3.63), opened on the Cumbre Vieja summit and are aligned with the rift axis. The lower vent, in contrast, was located on the western flank at an altitude of approximately 1300 m, inside a barranco at a site known as "Llano del Banco" (Figs. 3.62D and 3.63C).

Hoyo Negro was a degassing vent with very violent phreatomagmatic episodes but little juvenile magma emission overall. In contrast, the Duraznero and Llano del Banco centers were eminently effusive. The former gave rise to a lava lake that spilled toward the east coast (see Figs. 3.62 and 3.63) but was arrested soon before entering the sea (Fig. 3.61B). The latter vent, however,

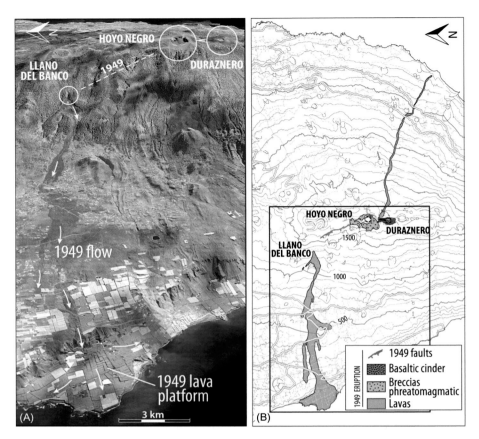

FIGURE 3.61

(A) Eruptive vents and flows of the 1949 San Juan volcano (image from *Google Earth*). (B) Map of the eruption indicates the three eruptive vents and the western and eastern lava flows. The former reached the coast and constructed a wide coastal lava platform, whereas the latter was arrested a few hundred meters from the coast (after Carracedo et al., 2001).

emitted large volumes of lava that ran down the western flank of La Palma and entered the sea, eventually forming an extensive coastal platform (Figs. 3.61, 3.62D−F, and 3.63C).

The initial stage of the 1949 eruption began with the opening of the Duraznero vent on June 24 that ended with a rather violent phreatomagmatic explosion on July 6 (Fig. 3.64; Klügel et al., 1999, 2000). On July 8, however, a new eruptive vent opened on the western flank of the Cumbre Vieja in the area of Llano del Banco, approximately 3 km to the north and approximately 500 m

FIGURE 3.62

Vintage photographs of scenes during the 1949 eruption (see Bonelli Rubio, 1950). Different views of the Hoyo Negro (A−C) and Llano del Banco (D) vents. In (C), a particularly bold person provides a scale for the height of the column. (E) The 1949 lava flow inundating a road. (F) Lava flow from Llano del Banco vent entering the sea. Images courtesy of Cabildo de Insular de La Palma.

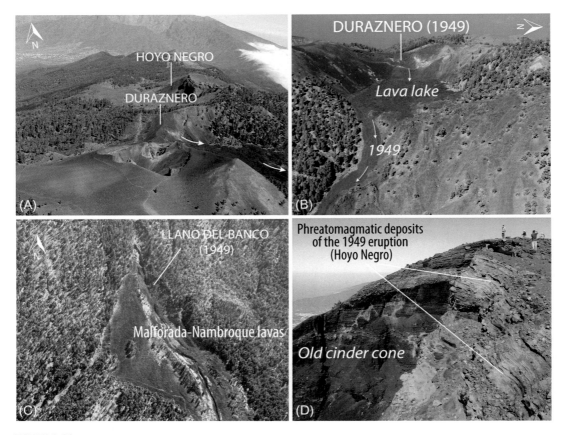

FIGURE 3.63

(A) The 1949 eruption. A lava lake formed east of the Duraznero vent. (B) Overflow of this lava lake caused a lava flow that almost reached the eastern coast. Llano del Banco vent (C) and Hoyo Negro (D) phreatomagmatic vent were connected by an underlying fissure.

lower in altitude. This vent continued to emit large volumes of channeled lava until July 26. The lava reached the coast and formed an extensive (6 × 3.5 km) coastal platform, which is now largely occupied by banana plantations (see Figs. 3.61A and 3.62F).

The third eruptive vent, Hoyo Negro (Figs. 3.62A−C, 3.63D), opened 700 m north of Duraznero on July 12. Activity began with strong phreatomagmatic explosions that opened a large funnel in an old volcanic cone (Figs. 3.62A, 3.63D, and 3.64). The very dark phreatomagmatic deposits that cover the surface of this depression account for the name given to it (*Hoyo Negro* means black hole). No lava was emitted here, only gases under great pressure. The eruption ended with the emission of a new lava flow from the Duraznero vent on July 30 that continued until the total cessation of events on August 4 (Fig. 3.64). This final lava flow progressed over the east flank of the Cumbre Vieja and became arrested approximately 300 m from the coast (see Fig. 3.61B).

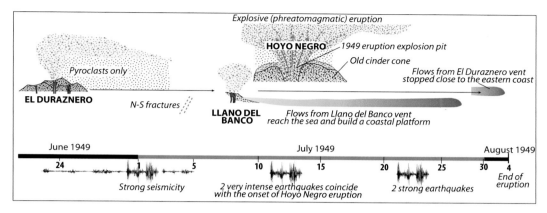

FIGURE 3.64

Timeline of the 1949 San Juan eruptive events reconstructed from eye-witness accounts and eruptive deposits. Modified after Klügel et al., 1999.

During the 1949 eruption, strong earthquakes were felt on the western flank, with their epicenters surrounding the village of *Jedey* (Fig. 3.65). The tremors were associated with developing fissures and faults along a 3-km portion at the crest of the rift (Fig. 3.66). The northern tip of the fault system outcrops in the mountains southwest of *La Barquita*. The interpretation of this fault system is still uncertain (Klügel et al., 1999). On the one side, it could be the surficial expression of a deep fault or, more probably, the result of shallow intrusions connecting the different vents of this eruption. Unfortunately, the Cumbre Vieja rift lacks galerías to observe its deep structure. This is because it is actively degassing, and the water yield in the few galerías excavated was too mineralized for human or agricultural purposes. Note, however, that the fault system did not move during the 1971 eruptive events, making it likely that it represents merely the surface expression of the feeder dyke that supplied the 1949 eruption from depth.

Contrasting faults on other oceanic islands (eg, the Hilina fault system in Hawaii), no indications of any further activity of the La Palma faults beyond the 1949 eruption have been observed. The Hilina fault system, on the south flank of Kilauea Volcano on Hawaii, experienced major flank earthquakes ($M > 7$) that involved slip along a sub-horizontal detachment fault and coastal subsidence of several meters. In contrast, seismicity at La Palma is lacking, with no major seismic activity recorded on or in close vicinity to the island in the past 20 years (*IGN Catálogo y boletines sísmicos*), which is consistent with an absence of major ground deformation on the island. The ground deformation network established to identify any displacement of the western flank of the Cumbre Vieja, by using infrared electronic distance measurement and the global positioning system (GPS), showed that apparent displacements recorded are within the error margins of the techniques used (Moss et al., 1999).

Therefore, there is no direct seismic or structural data to support that the 1949 fracture along the crest of the volcano is the surface expression of a major weakness zone along which a large-volume detachment and major failure can occur. Ward and Day (2001), using numerical studies,

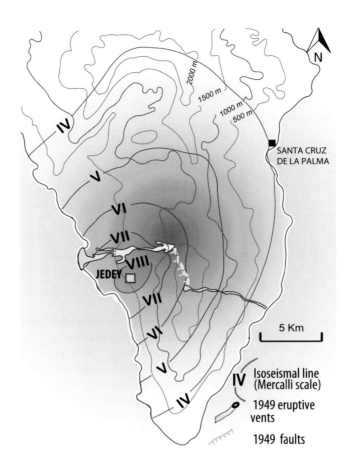

FIGURE 3.65

A major earthquake (magnitude VIII on the Mercalli scale) occurred on July 2, 1949, with its epicenter near the village of *Jedey*. This strong shock preceded the opening of a north−south trending fracture and the onset of eruptive activity at the more southern Duraznero vent. Data from Bonelli Rubio, 1950.

forecasted that during a future eruption of the Cumbre Vieja Volcano on La Palma a catastrophic failure of its western flank may occur, dropping 150 to 500 km^3 of rock into the sea. According to these authors, waves generated by the collapse could transit the entire Atlantic ocean basin and arrive at the coasts of the Americas with a height of 10 to 25 m (Fig. 3.67). Although no specific time frame for this postulated giant landslide was provided, it was inferred that it could be triggered by a future volcanic eruption of the Cumbre Vieja.

This notion created a widespread global media impact at the time (more than half a million entries on Google for "Tsunami La Palma"), with significant consequences for the island's economy, which is dominantly based on tourism.

FIGURE 3.66

(A and B) "En echelon" fissure of the 3-km fault system that opened at the crest of the Cumbre Vieja rift during the 1949 eruption (see Bonelli Rubio, 1950; photos (A and B) courtesy of *Cabildo de Insular de La Palma*). (C) The fault appears to connect the 1949 eruptive vents and likely represents the surface expression of a feeder dyke associated with the eruption, rather than a deep fault that would define a detached block on the western flank of the Cumbre Vieja.

This forecast is likely grounded on unrealistic assumptions regarding the present slope instability of the Cumbre Vieja and incorrect estimates of near and far field terminal effects, thus probably overstating the tsunami threat (Pararas-Carayannis, 2002). Although well documented in the Canaries, catastrophic flank collapses of island stratovolcanoes are extremely rare phenomena and none has occurred within recorded history (the last giant landslide on La Palma is approximately 560 ka old).

The widespread concern is partly caused by the publication of results from probabilistic numerical modeling studies without sound clarification of what constitutes a near-future risk versus what may be merely a geologically plausible but rare and far field event that occurs only once within hundreds of thousands of years.

Continue to follow LP-211 down-hill for a while until you reach a T-junction. There, join LP-213 toward *Puerto Naos* and follow this road down-hill until you reach *Mirador de Las Hoyas* on your right, where you may park the car for the next stop.

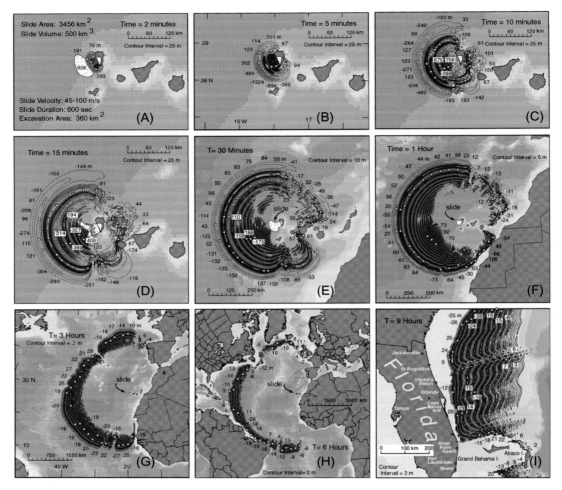

FIGURE 3.67

Conjectural model of the resulting tsunami wave from a possible collapse of the western flank of the Cumbre Vieja that might affect not only the Canary Islands but also the Atlantic coast of the Americas (from Ward and Day, 2001). According to Ward and Day (2001), this potentially catastrophic landslide could be triggered by another eruption on the Cumbre Vieja ridge.

Stop 3.9. Mirador de Las Hoyas; *historical coastal lava platform (N 28,5962 W 17,9122)*

From the *Mirador de Las Hoyas*, the lower coastal plains can be inspected when looking out onto the sea toward the west. The platform hosts enormous banana plantations today. Although these are all built on the 1949 lavas, a lot of soil has been shipped in to cover the lavas and create "mature new land" on the 1949 *malpaís* (Fig. 3.68A). Note that it takes at least 900 to 1200 years to degrade lava to soil if conditions are sufficiently warm and moist. Otherwise, it takes considerably longer.

FIGURE 3.68

(A) View of the coastal platform created by lavas of the historical 1585, 1712, and 1949 eruptions on the western flank of the Cumbre Vieja. (B) The coastal lava platform of the 1585 and 1712 eruptions, now viewed from the sea, is extensively used for banana cultivation. Some geologically pristine features remain nevertheless accessible (eg, Stop 3.10).

    Continue down-hill to have a spectacular view of the wide coastal platform created by the 1712 and 1585 lavas and that also host extensive banana plantations (Fig. 3.68B). Continue through *Puerto Naos*. Drive into the village and then through the long stretch of banana fields created on the historical lava fields. At some point near the coast, the road ends. However, it brings you directly to a lobe of the 1712 eruption (see Fig. 3.68B). Park the car in a nearby parking area just before the road ends.

Stop 3.10. El Remo; *Historical coastal lava platform (N 28.5521 W 17.8860)*

Walk for a few steps toward the dark 1712 lavas and you will come to a deep flow channel approximately 50 m wide that is bordered by enormous levees that are up to approximately 10 m high. In the outcrop, the rock is dark, with many pyroxene needles and fragments of gabbro and pyroxenite. Occasional syenite and large feldspar xenocrystals also occur.

A small path to the right from the car park crosses the lavas safely, so there is no need to climb down the levee, unless you must study levees in detail.

Once you have taken in the outcrop and the sea breeze, return to *Santa Cruz* along the road you came (LP-2), or cut across the mountainous center of La Palma (recommended) via LP-3. Before you head back to your base, you may want to consider having dinner in the fishing village of *El Remo* or in *Puerto Naos*, where you will find a selection of seafront restaurants to suit all tastes and budgets and where you can admire the sunset on the open western horizon.

END OF DAY 3.

DAY 4: CUMBRE VIEJA HIKE

This day includes a few hours of trekking to climb to the top of Cumbre Vieja to reach the area of the 1949 vents (Fig. 3.69). Bring clothes for all weather situations, plenty of drinking water, and food.

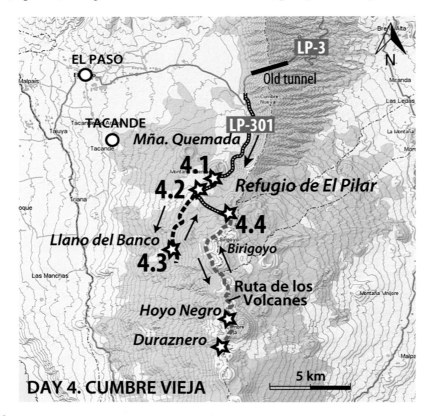

FIGURE 3.69

Map indicating itinerary for *Day 4*, the northern Cumbre Vieja rift, and the initial part of the *Ruta de los Volcanes* up to the 1949 vents. Map modified from IDE Canarias visor 3.0, GRAFCAN.

FIGURE 3.70

(A) Montaña Quemada volcano (also known as Mña. Tacande) was dated by radiocarbon to 1470–1492 AD (Hernández-Pacheco and Valls, 1982) (ie, to only a few years before the Spanish conquest of the island) (image from *Google Earth*). (B) *Cascada de nubes* (cloud waterfalls) across the Cumbre Nueva ridge. The air cools on ascent and later warms again when descending beyond the ridge. This leads to an adiabatic foehn-type effect.

Make your way to LP-3 (orange) and drive toward the center of the island. Cross the Cumbre Nueva ridge using the old tunnel (in a East to West direction), pass km 17 on LP-3, and after approximately 300 m join LP-301 (green) toward *El Pilar Zonas Recreativas* (left). At km 4 on LP-301, park on the open ground on the right to inspect the prehistoric Mt. Quemada flows (the Burned Mountain; Fig. 3.70A).

Stop 4.1. Montaña Quemada (N 28.6226 W 17.8426)

The Mña. Quemada eruption is also referred to at times as Mña. Tacande, because lava flows destroyed a village of the same name. The eruption has been dated by radiocarbon to just before the Spanish conquest of the island (Hernández-Pacheco and Valls, 1982). The dark lapilli beds are of olivine and pyroxene-bearing fresh basanite.

FIGURE 3.71

View from *Stop 4.1* (*Mirador Llano de Jable*), probably one of the best places to inspect the Taburiente volcano and its southern extension (the Cumbre Nueva). Highlights include the northern tip of the Cumbre Vieja above Mña. Quemada (foreground), the scarp of the 560-ka giant landslide in the middle ground (right), with the Bejenado volcano nested inside the collapse basin (middle ground left), and the wall of the Taburiente caldera in the far distance.

Continue up-hill, passing extensive lapilli fields of the Mña. Quemada eruption on your left. After a short drive you will reach *Mirador Llano del Jable* (*jable*, from the French *sable*, is a common name for lapilli in the Canaries).

Stop 4.2. Mirador Llano del Jable. *Panoramic view of Taburiente Caldera (N 28.6173 W 17.8485)*

The Cumbre Nueva landslide scarp is frequently capped by clouds that sometimes cascade down the cliff because of the Foehn effect (Figs. 3.70B and 3.71). Foehn (or *Föhn* in German) refers to a warm southerly wind coming over the Alps. However, the word is now used to describe similar meteorological effects on mountains all over the world. On the Cumbre Nueva, the cold and humid northeasterly tradewinds find the mountain range in their way and are forced to rise to avoid the obstacle and to get to the other side. As the temperature decreases with height, the moist air becomes saturated and condenses to form clouds and, hence, deposits dew on the mountain. After passing the ridge, the air descends along the lee side of the mountain and becomes warmer again, so the clouds disappear and the result is a dry and often hot wind.

The mirador is the start of the dirt track that travels across the western flank of Cumbre Vieja to reach *Fuencaliente* (approximately 30 km). Drive down the trail for approximately 2 km to inspect the lower vent of the 1949 eruption from above.

Stop 4.3. Llano del Banco 1949 vent (N 28.6026 W 17.8533)

At this location you will find yourself right above "Llano del Banco," the lowest of the three 1949 vents (see Figs. 3.61, 3.62D, and 3.63C). This particular vent gave rise to voluminous lava flows that reached all the way to the sea (Fig. 3.62E,F). Return on the track to LP-301 and continue the drive up-hill to *Refugio de El Pilar* and the Cumbre Vieja Visitor Center to start the *"Ruta de los Volcanes"* hike from here.

Stop 4.4. Refugio de El Pilar (N 28.6142 W 17.8360)

Park the car here and take all your valuables with you. Pack clothes for all weather situations and plenty of drinking water because you will be out for most of the rest of the day. You may wish to spend a little time in the Visitor Center to get some information on the geology, fauna, and flora of the *"Ruta de Los Volcanes"* before you start.

The *"Ruta de Los Volcanes"* is a spectacular 25-km-long trail, ending in the town of *Fuencaliente* after a 7 to 9-hour walk (Fig. 3.72A). However, we recommend you travel the volcanologically most interesting part of this trail (Fig. 3.72B) to the 1949 Hoyo Negro and El Duraznero vents. This will take 4 to 5 hours in total.

A hike along the "Ruta de los Volcanes"

It is difficult to imagine coming to La Palma without the plan to explore the Bco. de Las Angustias in the Taburiente Caldera or to hike the crest of Cumbre Vieja. However, in both cases you should take all precautions (eg, clothing, water, sun protection, good weather). Enjoy the hike and return to the parking site and eventually to your base whenever you have had sufficient exposure to sun and volcanic deposits. Alternatively, you may decide to continue all the way to Fuencaliente.

The full 25-km trail starts in the *Refugio del Pilar* and finishes at the *Faro de Fuencaliente* (Fuencaliente lighthouse) and takes approximately 7 to 9 hours to hike (see Fig. 3.72A). It is advisable to form a two-car group, leaving one car at the starting point and the other at the end of the walk, or you will need to use public transport or taxis after you arrive.

During the full hike you will be traveling on the crest of the most active rift zone in the Canaries, boasting three historical eruptions (1949, 1712, and 1646) and many other Holocene eruptive centers (Fig. 3.72C). This, no doubt, will be a fine finale for your visit to La Palma.

END OF DAY 4.

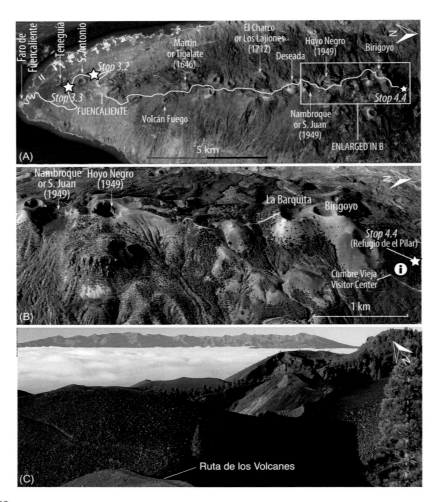

FIGURE 3.72

(A) The "*Ruta de los Volcanes*" trail allows one to tour the eruptive vents of the Cumbre Vieja rift volcano; it is a spectacular 25-km-long trail ending in the town of *Fuencaliente* after a 7 to 9-hour walk on the crest of La Palma. (B) We suggest a shorter tour (this will take 4–5 hours) that covers the most interesting part of this trail, including the historical (1949) vents of Hoyo Negro and El Duraznero. (C) Spectacular views may be enjoyed along the "*Ruta de los Volcanes*" trail.

THE GEOLOGY OF LA GOMERA

The Geology of the Canary Islands. DOI: http://dx.doi.org/10.1016/B978-0-12-809663-5.00004-9

THE ISLAND OF LA GOMERA

The island of La Gomera, which is $370\,km^2$, 25 km in diameter, and a maximum altitude of 1487 m, is situated approximately 40 km west of Tenerife. Although relatively small, La Gomera provides a wealth of geological features. This is because La Gomera's age and lack of recent volcanic activity allowed erosion to deeply incise the island and lay bare rock sequences that would remain buried in more active islands (Fig. 4.1).

La Gomera shares few geological similarities with La Palma and El Hierro. The simple explanation for this is that the latter islands are in an earlier stage of evolution (shield stage), whereas La Gomera appears more similar to the central Canary islands of Gran Canaria and Tenerife, which are in a postshield evolutionary stage. The central islands, in turn, contrast the oldest eastern islands of Fuerteventura and Lanzarote that form the termination of the archipelago. By classifying the islands in this way, their geological characteristics define geomorphologically and geologically different groups.

Many geological similarities (age, mineralogical, and isotopic composition of the lava, morphology, and structure, etc.) exist between La Gomera and the Teno and Anaga massifs on Tenerife, and also with the southwest of Gran Canaria (see chapter: The Geology of Tenerife and chapter:

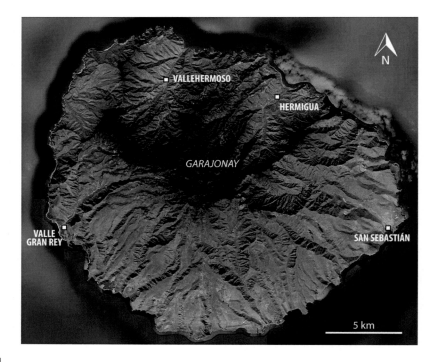

FIGURE 4.1

Satellite image of La Gomera (courtesy of *NASA*). Like Gran Canaria, the island of La Gomera shows a semi-circular outline, modified in the north by a Miocene giant landslide. The deep radial barrancos are related to the lack of voluminous volcanism in the past 4 Ma and a dominance of erosional processes.

The Geology of Gran Canaria), whereas the western and juvenile islands of La Palma and El Hierro are separated in age from La Gomera by a time interval of 7 million years (Ma) or more.

The main difference that sets La Gomera apart from the remaining Canary Islands, however, is a long, and still continuing, eruptive break. The island is currently in the postshield erosional stage (see chapter: The Canary Islands: An Introduction), in contrast to Lanzarote, Gran Canaria, Tenerife, La Palma, and El Hierro, which all had eruptions occurring up to the Holocene and also in historic times. Even Fuerteventura, which is far away from the postulated main region of magmatic activity underneath the western Canary Islands and widely considered extinct, had a significant Pleistocene basanite eruption approximately 136,000 years ago (eg, Malpais Grande). Compare this to La Gomera, which saw its last eruption, a monogenetic event, approximately 2 million years ago (Paris et al., 2005b). Apart from this single event, Gomera's last continuous phase of activity seems to have ceased even earlier, approximately 4 million years ago (Cantagrel et al., 1984). From this point of view, some volcanologists claim that the island is extinct. However, such projections are naturally of a probabilistic nature, and La Gomera's proximity to the active island of Tenerife also allows contemplation of other possibilities. Currently, however, there are no signs of recurring volcanic activity anywhere on the island of La Gomera.

In La Gomera's present geological stage, we are able to distinguish the main episodes of evolution, like the shield stage volcanism and postshield activity, and we can access the roots of an extinct and eroded felsic stratovolcano, which was once likely rather similar to Teide on Tenerife. We can also see the scar of a giant landslide, which has been refilled, and, most uniquely, the island shows us an inner core of countless dykes and plugs and other variably hydrothermally altered intrusive rocks. However, before we take a deeper look into La Gomera's main volcanic units and their significance, there is much more to learn from the island's age systematics.

Probably the most controversial aspect of the geology of La Gomera is the age of the oldest emerged formations and the debated existence of outcrops in the northwest of the island that may represent a much older and independent submarine edifice.

THE SUBMARINE EDIFICE OF LA GOMERA

The submarine part of La Gomera shows that the base of the island rests on a somewhat shallower ocean bed than El Hierro and La Palma, particularly to the east, where it overlaps with, and is structurally supported by, the island of Tenerife (Fig. 4.2).

Apart from radiometric ages, several lines of evidence support the view that La Gomera was the fifth of the Canary Islands to have emerged through the sea surface. First, the submarine slopes of La Gomera overlap the submarine portion of Tenerife (see chapter: The Canary Islands: An Introduction). Considering that these submarine aprons form by sedimentation of various materials from existing islands (see Fig. 4.2), the juxtaposition of La Gomera on top of submarine parts of Tenerife provides a strong argument for the younger age of La Gomera.

Another indicator for the age and evolution of La Gomera is its wide coastal platforms. These platforms are submarine remainders of the island where it has been laterally eroded by wave action. Because these areas are underneath the sea and therefore hidden from sight, research vessels were necessary to conduct so-called bathymetry, a form of sonar scanning of the seafloor to measure its morphology (see Fig. 4.2). The shelves of La Gomera are rather wide and imply that Gomera has lost a great part of its initial volume. Just as with the subaerial erosion and the resulting formation

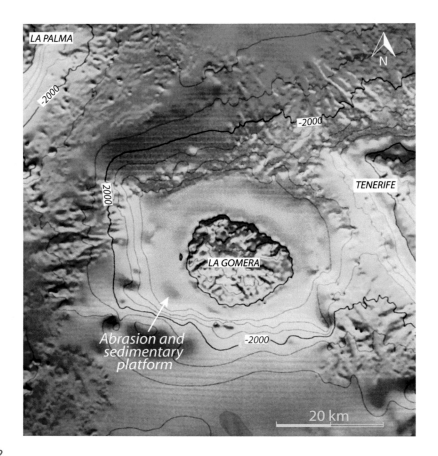

FIGURE 4.2

The submarine edifice of La Gomera is surrounded by a wide platform of erosional detritus. This large abrasion platform suggests that La Gomera has been eroded to half of its original size. Note the separation from Tenerife by a deep (≥2000 m) submarine canyon, indicating the development of La Gomera as an independent island edifice. The two islands, however, provide structural support for each other. Image from Masson et al., 2002.

of deep valleys, a necessary condition for the formation of coastal shelves is a long eruptive hiatus during which no new, fresh lava is deposited to consolidate the island, and therefore wind and wave action may work its way unhindered and uncompensated.

An additional important feature is the steeply sloping submarine edifice on the northern face of the island, which probably relates to a giant Miocene lateral landslide.

THE EMERGED EDIFICE OF LA GOMERA

The emerged edifice of La Gomera is semi-circular in shape, with no trace of rift zones, but these may have been removed by erosion. Instead, a radial network of deep, frequently amphitheater-headed

FIGURE 4.3

View of La Gomera from the south. Note the wide mouth of *Bco. de Benchijigua* in the center of the image, hosting the village of *Playa de Santiago*. The island shows the characteristic "shield" profile of an oceanic shield volcano and has lacked significant postshield eruptive activity over the past 4 Ma.

barrancos exists that is similar to those of the island of Oahu, Hawaii (see Fig. 4.1). This topography gives the appearance of an old island that has had very little eruptive activity for the past 4 Ma and none during the Quaternary.

The general profile of the island is reminiscent of a lying warrior's shield (see Fig. 4.3), which is precisely the name given to the initial stage of development of volcanic ocean islands (shield stage). This tells us that the posterosional or rejuvenated stage of this island probably did not yet commence, or that it did not involve a large volume of erupted volcanic materials and has already been removed by erosive events. Although the long period of erosion may have been sufficient to dismantle such structures, this option is less likely because distinct feeder systems should also exist if this were true.

AGE AND PALAEOMAGNETISM OF THE ROCKS ON LA GOMERA

A large number of radioisotopic dates have been collected from La Gomera. However, paradoxically, the geochronology of this island is not straightforward and gave rise to considerable debate. The radiometric ages were obtained by different groups with different methods and sampling criteria; therefore, the possibilities of some dates deviating from the true ages of the formations increase. The problem is most severe when it comes to altered rocks (eg, from the oldest volcanic or plutonic formations) that have experienced potassium (K) remobilization or contamination through foreign material (eg, xenocrystals or crustal assimilation).

The possibilities of error further increase in whole rock age determinations, as illustrated by the following example. La Gomera has been assigned a much greater age than Tenerife, forming the exception to the systematic progression of ages predicted by the hot spot model (Ancochea et al., 2006). This has come about by postulating ages of 15 to 20 Ma for an intrusive formation (Fig. 4.4), which is much older than the oldest known emerged volcanism on the island. Volcanism dates at approximately 11 Ma for La Gomera (Paris et al., 2005b), which is only marginally in contrast to the approximately 12 Ma established for Tenerife (Guillou et al., 2004a). However, when the reliability of ages was initially tested with various methods, it was found that the age of the oldest intrusions was actually younger than the age of the oldest emerged rocks. One of these two very old ages, 19 Ma, was obtained by dating a basaltic dyke (Abdel Monem et al., 1972). However, when a sample of the same dyke was re-dated with ^{40}Ar-^{39}Ar on separate feldspar crystals, 10 out of 12 crystals had a similar age of approximately 5 Ma, whereas 2 of them had an age of more than 130 Ma (see Fig. 4.4). Obviously, these last two ages represent xenocrystals dragged from the Jurassic ocean crust

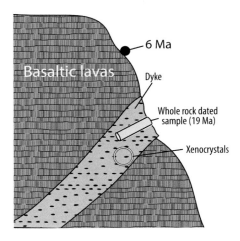

FIGURE 4.4

K/Ar dating may yield erroneous ages because of contamination with foreign (xeno-)crystals from the oceanic crust picked up by the ascending magma. This phenomenon is exemplified in a dyke intruding lava flows of the Miocene shield. The lavas were previously dated at approximately 10 to 8 Ma, whereas a sample of the dyke yielded a whole rock with K/Ar age of 19 Ma (Abdel Monem et al., 1972). When the dyke was recently re-dated with ^{40}Ar-^{39}Ar on separate feldspar crystals, 10 out of 12 crystals gave a similar age of 5.1 Ma, whereas 2 crystals gave an age of >130 Ma (K. Hoernle personal communication). It appears that 2 of the 12 crystals were xenocrystals from, for example, the Jurassic ocean crust. Averaging of the results of whole rock age dates may thus produce an artificial age of ≥15 Ma for the dyke, although the true age of the dyke intrusion is only ∼5.1 Ma, consistent with the older age of the Miocene lavas it intruded (∼10 to 8 Ma).

below the base of the island. Averaging of the results produces an artificial age of approximately 15 Ma, although the true age of the intrusion is probably only 5.1 Ma (K. Hoernle personal communication).

Most of the ages obtained by different groups are, however, consistent (Fig. 4.5), and the mentioned problems are specific to the oldest formations (the "Basal Complex" of several authors; Fig. 4.5). The more reliable age of the oldest activity on La Gomera would be approximately 10 or 11 Ma when discounting the three earlier determinations of between 19.8 and 14.6 Ma (Abdel Monem et al., 1971). Notably, Cantagrel et al. (1984) dated the same formation and confirmed the problem of conflicting ages by reporting a range of 15.5 to 9.1 Ma.

In our opinion, the oldest ages reflect the limitations of the K-Ar method for dating altered and probably contaminated rocks, and these three ages should be questioned. To establish criteria to ensure that the age determinations are as exact as possible, a fundamental approach is to sample stratigraphic series, which allow verification of a consistent age/stratigraphy correlation. Further help to reduce such errors comes by combining radioisotopic ages with, for example, magnetic polarity determinations that record the inversions of the Earth's magnetic field and by analysis of different mineral and matrix fractions of the same rock for radiometric ages.

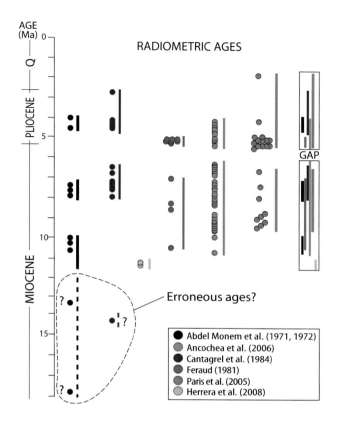

FIGURE 4.5

K/Ar and ^{40}Ar/^{39}Ar ages for the lavas of La Gomera. Note that the ages of different groups are fairly consistent, except the oldest ones shown with a question mark. Most likely, these ages are too old and erroneous (as discussed). These older ages are the main support for an alleged submarine edifice on the northwest of the island.

STAGES OF CONSTRUCTION AND MAIN VOLCANIC UNITS OF LA GOMERA
A SUBMARINE EDIFICE EXPOSED ON LA GOMERA?

Modern research on the island goes back approximately 50 years to Blumenthal (1961), Bravo (1964), Hausen (1971), Cendrero (1971), Cubas (1978), and, more recently, Ancochea et al. (2003). A comprehensive review of the geology of the island can be found in Paris et al. 2005b, whereas a review of the geochronology has been provided by Ancochea et al. (2006).

La Gomera's general stratigraphy comprises three main rock sequences (Fig. 4.6) that have been modified by various degrees of erosion or mass wasting: (1) a Miocene basaltic shield, including a basal plutonic complex; (2) a nested felsic stratovolcano; and (3) the youngest (Pliocene) volcanism.

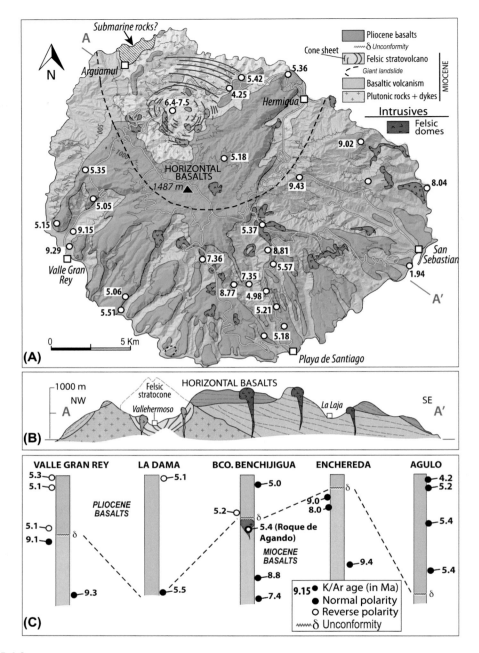

FIGURE 4.6

(A) Simplified geological map of La Gomera, with recent ages from Paris et al. (2005b). The felsic stratovolcano and cone sheets are simplified after Rodríguez-Losada and Martínez-Frías (2004). Note the proposed outcrop of the submarine edifice at the northwest part of the island. The most recent ages obtained from this formation (~ 11 Ma, Herrera et al., 2008) seem to exclude a 20- to 15-Ma submarine edifice separated from the La Gomera shield stage by an erosional gap of 4—5 Ma, as postulated by several earlier works (eg, Bravo, 1964; Cendrero, 1971; Ancochea et al., 2006). (B) Simplified geological cross-section of La Gomera. (C) Simplified stratigraphic columns and radiometric ages for the Miocene and Pliocene basaltic sequences of La Gomera.

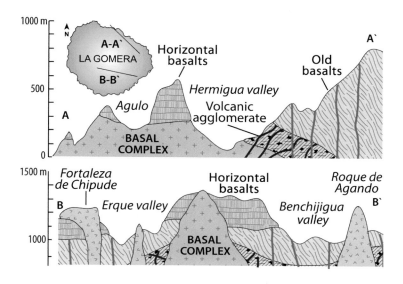

FIGURE 4.7

Geological cross-sections of La Gomera after Bravo (1964). This work was the first to define a "basal complex" on La Gomera, comprising submarine volcanics and plutonic rocks densely intruded by dykes, analogous to the basal complexes of La Palma and Fuerteventura. More recent data point to this basal complex as being the plutonic roots of the larger shield volcanoes, perhaps including some locally uplifted submarine volcanic rocks, but as yet without full confirmation.

It is important to note that, in our opinion, these three sequences are of subaerial origin, that is, they were formed after the island had already broken through the sea surface. This contrasts with the reports by some authors of outcrops of a submarine basal part exposed at the northwest tip of the island (Cendrero, 1967, 1971), and which Bravo (1964) named the "Basal Complex" (Fig. 4.7, but see also the chapter on Fuerteventura for a discussion on the origin of the "Basal Complex" concept).

We have been unable to locate conclusive submarine lavas or sediments that define a basal complex. Because of the mentioned lack of island subsidence and the deep erosion of the northwest sector of La Gomera, outcrops of submarine lavas can be found, but these represent lavas that entered the sea during shield volcano construction (eg, see chapter: The Geology of El Hierro, Fig. 2.24) and are not part of a "seamount series" as seen, for example, on La Palma.

Therefore, the question remains whether this submarine volcanism corresponds to an independent, much older submarine edifice, or to the basal parts of the main present-day emerged Miocene shield that, according to Ancochea et al. (2006), also contains submarine volcanics in its initial phase. The presence of a submarine formation separated from the shield stage by an erosive gap (Cendrero, 1971; Ancochea et al., 2006) would be relevant to the island's evolutionary history, but this would not be the case if the submarine products are related to the emerged Miocene shield volcano.

On the one side, there are two ages already mentioned (ages with a question mark in Fig. 4.5), 19.8 and 14.6 Ma (Abdel Monem et al., 1971), and 15.5 Ma (Cantagrel et al., 1984) that seem to support the old, independent submarine edifice. However, Cantagrel et al. (1984) also obtained a contrasting age of

9.1 Ma in rocks of the same formation. Recently, Herrera et al. (2008) dated the 'submarine' part of La Gomera using the $^{40}Ar/^{39}Ar$ method for the first time and obtaining ages of 11.49 ± 0.66 Ma and 11.4 ± 1.6 Ma. These radiometric ages reveal that the submarine volcanics are approximately 11.5 Ma, and thus would have formed very close to the beginning of the Miocene subaerial shield stage.

THE MIOCENE BASALTIC SHIELD
THE BASAL PLUTONIC COMPLEX

A Miocene shield volcano (~ 11 to 8 Ma) forms the bulk of the emerged part of La Gomera Island. The deeper levels to this Miocene episode are exposed as a sequence of dykes and plutonic intrusions that feature complex cross-cutting relationships and pervasive hydrothermal alteration. This plutonic complex crops out in the north of the island, between *Arguamul* in the west and *Hermigua* in the east (see Figs. 4.6A and 4.8). The main types of rocks are gabbros and pyroxenites, which are densely intruded by predominantly basaltic dykes that form 80% or more of the rock mass in places (Fig. 4.9, see also Stop 1.6).

THE MIOCENE BASALTS

This Miocene volcanic phase (see Fig. 4.8A) mainly consisted of basaltic eruptions, which are preserved as a thick succession of lava flows that appear parallel to each other. The characteristic attribute of the Miocene shield is the inclination of its lavas toward the sea (Fig. 4.8B). Deeply incised by erosion, these inclined lavas form the sidewalls of the majority of La Gomera's steep valleys where they are visible as a pile of layers (ie, in a layer cake fashion). As a result, their dip is visible in many of the valleys and can be readily distinguished from the much shallower, dipping, younger (Pliocene) lavas, which rest unconformably on top of the Miocene sequence (Fig. 4.8A).

The second main stage of Gomera's evolution is related to an approximately 8-Ma sector collapse of the Miocene shield toward the north. Inside the 10- to 15-km-wide collapse scarp of this landslide, which mass-wasted approximately 25% of the surface of the Miocene shield, a new subsequent volcanic edifice was constructed in the form of a large stratovolcano (Fig. 4.9).

Judging by the wide erosion platform surrounding La Gomera (see Fig. 4.2), the diameter of the basaltic shield that formed during the Late Miocene must have been perhaps twice the diameter of the present-day island. Moreover, the Miocene shield basalts are crossed by a dense network of dykes, many of which are fractured and even tilted or folded (Fig. 4.8C). In the north–northwest part (from *Alojera* to *Agulo*), the density of both dykes and plutonic intrusions (gabbros) increases, forming a plutonic complex (see Fig. 4.6A,B). The age of these intrusions is very similar to those of the shield volcanism they fed, and all of those are dated between 9.5 and 8 Ma (see Fig. 4.6C).

The shield eventually grew to a level of instability and underwent at least one large lateral collapse. This occurred approximately 8 million years ago, and a large volume of material was shed from the northern flank of the shield. Remnants of the collapse are observed in sonar images of the northern part of the submarine edifice (see Fig. 4.2). Onshore, the collapse basin is marked by a thick formation of breccia generated by the event (eg, the breccia crops out in the road from *Vallehermoso* to *Parque Marítimo*, at the coast).

FIGURE 4.8

(A) View of La Gomera from the south (image from *Google Earth*) showing the main volcano-stratigraphic units of the island. (B) Sequence of basaltic lava flows (dipping toward the sea) and dykes of the Miocene shield volcano. Playa de Arguamul. (C) Dense dyke swarm of the initial phases of construction of the Miocene shield. Playa de Arguamul. (D) Volcanic breccia intruded by a dyke, forming the base of the Miocene basalts at the eastern wall of Bco. de Hermigua. (E) Seaward-dipping flows of the Miocene basalts at *El Pescante*, the eastern part of Playa de Hermigua. (F) Thick sequence of Miocene basaltic flows at the eastern wall of Bco. de Hermigua. The topographic expression of this formation, after exposure to several Ma of erosion, bears resemblance to the "pali"-type valleys on the island of Oahu on Hawaii.

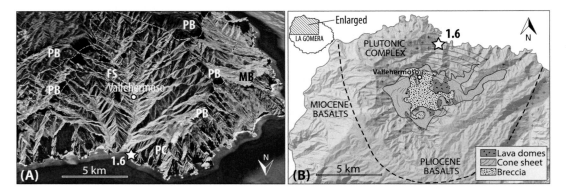

FIGURE 4.9

(A) La Gomera viewed from the north (image from *Google Earth*) showing the wide valley of Vallehermoso, which was the site of a Miocene felsic stratovolcano (FS) that is now completely eroded, however. PC: Plutonic complex; MB: Miocene basalts; PB: Pliocene basalts. (B) Present-day extent of the remains of the Vallehermoso felsic volcanic complex (simplified after Rodríguez-Losada and Martínez-Frías, 2004).

The best outcrops of the Miocene basaltic shield formations are found in the northwest of the island (eg, the coast of *Alojera* and *Tazo*) and in the barrancos of Valle Gran Rey (Fig. 4.6A) and Hermigua (Fig. 4.8D–F).

THE VALLEHERMOSO STRATOVOLCANO

The second main stage of Gomera's evolution is related to this approximately 8-Ma sector collapse of the Miocene shield toward the north. Inside the 10- to 15-km-wide collapse scarp of this landslide, post-collapse volcanic activity constructed a large stratovolcano (Fig. 4.9), probably very similar to the Teide volcanic complex on Tenerife. This volcano, named the Vallehermoso Trachyphonolitic Complex by Cendrero (1971), likely produced abundant phonolite domes and eruptions, judging from the many phonolitic dykes that can be observed today in this area. However, the edifice of Vallehermoso volcano has been completely eroded by now, leaving behind only the shallow roots of a conical dyke system as well as breccias and domes (Ancochea et al., 2003). The geometry of this root zone hints at the former size of this volcano. It must have been comparable to today's Teide volcano on Tenerife. Notably, both volcanoes formed in a collapse basin, became progressively differentiated, and gave rise to terminal peripheral felsic domes (Carracedo et al., 2007, 2011). Although the La Gomera stratovolcano was active between approximately 7.5 and 6.5 million years ago and has been continuously eroded since, the Teide volcano has not yet reached the end of its activity, but it will likely share the fate of the Vallehermoso volcano on La Gomera in only a few Ma from now.

THE PLIOCENE BASALTS

The activity of the differentiated Vallehermoso volcanic edifice ceased at the end of the Miocene and the island entered a period of eruptive repose and erosion lasting at least 1 Ma. Eruptive

FIGURE 4.10

(A) Horizontal Pliocene basaltic flows near *Arure*. (B) View of the unconformity between the Miocene basalts and the horizontal Pliocene basalts (near *Arure*). Note that the Miocene—Pliocene unconformity is erosive, whereas the Pliocene pyroclasts and horizontal basalts reflect fillings of a giant landslide scarp (the Garajonay landslide). (C) View from the sea of dipping Pliocene basalts. Here these are similar to the Miocene basalts in their inclination, but they can be clearly separated by fewer dyke intrusions and younger radioisotope ages (~5.5−4.2 Ma).

activity resumed in the Pliocene (~5.5 to 4.2 Ma), and eruptions were again basaltic, with the eruptive centers preferentially located in the higher altitude parts of the island.

The presence of the collapse basin and the nested felsic stratovolcano favored ponding of Pliocene lavas between the southern flank of the stratocone and the collapse scar, and some lavas were consequently emplaced with a horizontal orientation (the horizontal Pliocene basalts, Fig. 4.10A). Upon filling of the depression, lava flows spilled over the southern flank of the island (ie, over the Miocene basalts), a scenario very similar to what is discussed for the Teno shield on Tenerife (see chapter: The Geology of Tenerife) and for the Taburiente volcano, La Palma (see chapter: The Geology of La Palma).

The Pliocene basalts overlie the Miocene shield basalts in an angular unconformity that is clearly observed at many sites (Fig. 4.10B). The Pliocene basalts (5.5 to 4.2 Ma) surround the former stratovolcano and extend in an outcrop toward *Hermigua* in the north and beyond the southern flank of the island (ie, from *Valle Gran Rey* to *San Sebastian*) (Fig. 4.10C). At the final stages of the Pliocene activity, eruptions sharply declined. Some younger (~3 to 2 Ma) disperse basalts occur and may represent either a very poorly developed posterosive rejuvenation or the latest manifestation of the postshield stage. In the latter case, the island of La Gomera would be in the

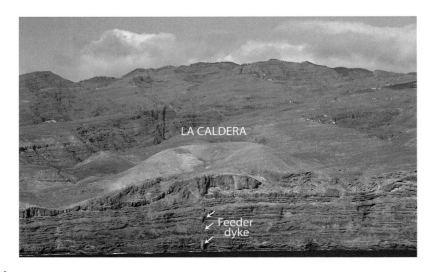

FIGURE 4.11

View of *La Caldera* from the sea south of La Gomera. This 4.2-Ma volcanic cone is the only well-preserved volcanic cone on the island.

postshield eruptive repose period, and the posterosive rejuvenation stage has not yet begun. Only one recognizable volcanic cone remains (La Caldera), although it is highly altered and partly eroded. This 135-m high, approximately 4.2-Ma-old cone (Cantagrel et al., 1984) is located west of the airport and is covered with a layer of calcrete (Fig. 4.11). Furthermore, a lava flow at the mouth of Barranco del Machal, south of *San Sebastián*, has an age of 1.94 Ma (Paris et al., 2005b). Although this suggests some scattered late Pliocene and post-Pliocene eruptive activity, La Gomera is the only island of the Canarian archipelago that shows no documented Holocene volcanism.

GEOLOGY AND GROUND WATER RESOURCES

Groundwater reserves are the main source on La Gomera for the population supply and irrigation due to the lack of significant surface water resources. A large number of traditional wells, drilled for supply and agriculture during the past 500 years, have been complemented with several modern boreholes drilled to supply water for the increasing water demands of our modern society.

The volcano's stratigraphic architecture defines the island's main aquifers: a basal aquifer in the impermeable Miocene shield that is intruded by the numerous impervious vertical dykes and an upper aquifer mainly linked to the Pliocene basalts. There, particularly in the horizontal basalts, water collects at the unconformity with the Miocene formations in the central parts of the island, forming a high-altitude "perched" aquifer (Leal et al. 2014; Fig. 4.12). At present, La Gomera has sufficient groundwater resources because of the abundant hydro-dams, wells, and boreholes, whereas only a few decades ago the geologically controlled natural springs served as the main water supply to the islanders.

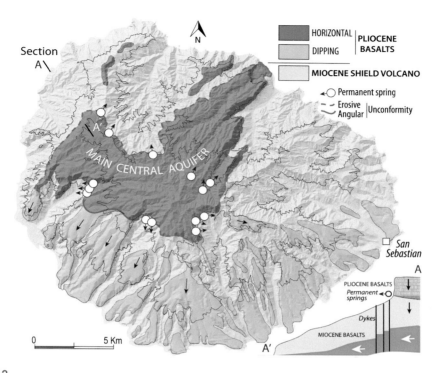

FIGURE 4.12

Simplified hydrostratigraphy of La Gomera, which is related to the island's volcanostratigraphy. During the greater part of the year, springs are located at the unconformity between the Miocene and Pliocene basaltic sequences. Apart from this upper (perched) aquifer, there is a main basal aquifer with water stored in the Miocene lava flows between impervious vertical dykes (see cross-section in the inset). This ancient groundwater reserve is exploited by traditional and modern wells and boreholes. Modified from Izquierdo, 2014.

DYKES AND DOMES

On La Gomera, where erosive degradation has been prevalent over the past millions of years due to an absence of significant volcanism, a wide range of eroded volcanic landforms is naturally dominant. These landforms include spectacular examples of relief inversion that leave the most resistant formations as the main topographic elements of the landscape, which are principally the feeder conduits of previous eruptions. When eruptions are from fissures, the feeders take the form of dykes (blade-like intrusions), and when they are centrally fed, the feeder conduits are later preserved as domes or plugs (such as Agando).

FIGURE 4.13

(A−D) Examples of basaltic dykes in the walls of Bco. de la Villa, above *San Sebastián*. These frequently kilometer-long sheet intrusions cross the Miocene basalts but stand proud during subsequent weathering and erosion. On La Gomera, this type of partly eroded dyke is known as "*taparucha*."

DYKES (TAPARUCHAS)

The most spectacular examples of dyke intrusions in the Canaries can be seen on La Gomera (Fig. 4.13). Many examples of eroded out dykes that resemble walls can be traced for hundreds of meters through the Gomeran landscape. This eroded form of dyke is known on the island as "*taparuchas*", with the most spectacular examples in Barranco de la Villa, above *San Sebastián* and at *Degollada de Peraza* (see Stop 3.1).

On occasion, volcanic cones can be seen to have a conduit through which the lava passed to reach the surface to feed an eruption. They are generally hidden, however, but erosion may uncover them (eg, in the coastal cliffs) (see Fig. 4.11).

LAVA DOMES

La Gomera is characterized by a large number of trachytic and phonolitic domes and coulées distributed over most of the island, although the highest concentration is in the center. There, the felsic intrusions form a semi-circular arrangement below and within the Pliocene basalts (Fig. 4.14).

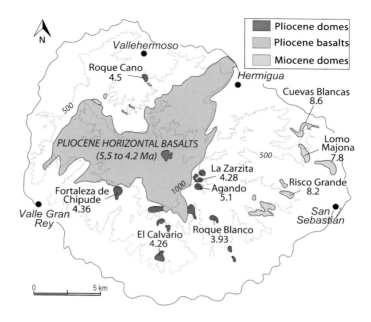

FIGURE 4.14

Simplified map of La Gomera showing the main felsic lava domes of the island. Ages are from Cantagrel et al. (1984), Ancochea et al. (2006), and Paris et al. (2005b). Note that most of the Pliocene lava domes are of an age similar to the Pliocene basalts, whereas Miocene domes group mainly in the east of the island and are considerably older.

If the domes form on a slope, as occurs most frequently, then they usually advance in thick flows (coulées) that later give rise to erosion remnants (buttes or stacks) due to relief inversion (Fig. 4.15). However, on flat terrain they produce round domes or necks of various types that form eventually prominent features in the landscape of the island by erosion-induced relief inversion (Fig. 4.16A–D).

Felsic domes are frequently associated with explosive eruptions in the form of projections of pumice or lateral collapses that produced block and ash deposits. On La Gomera, most of these deposits have been eroded, but local sediment traps do still record these airfall pumice beds (eg, in prominent road cuts in the central part of the island) (Fig. 4.16E,F). Note that pyroclastic fall deposits adapt to the topography. If they fill a depression (eg, a small valley), then the layers will form a carpet that will have the shape of a syncline, although no folding has actually occurred. The materials simply adapted to the preexisting relief (Fig. 4.16E).

GEOLOGICAL ROUTES

La Gomera is the second-smallest island of the Canaries, with a population of only 21,150 inhabitants as of 2013. It is easily reached from Tenerife by daily ferry boats that take approximately 1 hour to travel from *Los Cristianos* to *San Sebastian* (the administrative capital of La

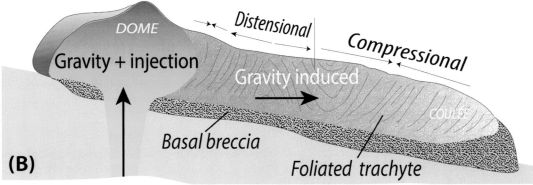

FIGURE 4.15

Roque de Aluce lava dome has experienced flow from the original position of ascent, thus resembling a combined lava dome and lava flow (coulée) situation (compare Blake, 1990).

Gomera). The most visited tourist areas are, however, *Valle Gran Rey* at the southwest coast and *Playa de Santiago* at the south coast of La Gomera (Fig. 4.17).

Therefore, our tours start in *Valle Gran Rey* (Fig. 4.18). It is one of the key areas with tourist accommodations available and, hence, is a likely starting point for many visitors. In addition, it also serves as a harbor from which daily boat trips to the northwest coast and to Los Órganos can be arranged (Fig. 4.17).

We have planned 3 days of geological excursions, including a half-day boat trip. We highly recommend the boat trip because it is the only way to inspect the spectacular Los Órganos intrusion (equivalent to Fingall's cave in Scotland or the Giant's causeway in County Antrim, Northern Ireland). Also, it will help you to form an impression of the oldest part of the island that will be visible from the boat in the coastal cliff sections.

FIGURE 4.16

(A) Upright appearance of the 4.5-Ma phonolitic dome of Roque del Cano, near *Vallehermoso* (Stop 1.5).
(B) Cluster of felsic lava domes at the center of La Gomera (Stop 2.3). (C) View from the sea looking north onto
the trachytic lava dome of El Calvario, near *Alajeró* (Stop 3.4). (D) Trachyte ring dyke of Bco. de Benchijigua
(Stop 3.3). (E) Dyke intruding felsic explosive products of Roque de Agando (5.1 Ma). The hot sheet intrusion
backed the pyroclasts forming a characteristic zone (aureole) of metamorphosed rock surrounding the sheet
intrusion. (F) Air fall deposits from explosive eruptions related the felsic lava domes of the center of the island
(area of 'Los Roques'). The layers appear to form a syncline, although no folding has actually taken place.
The materials simply filled the preexisting negative relief.

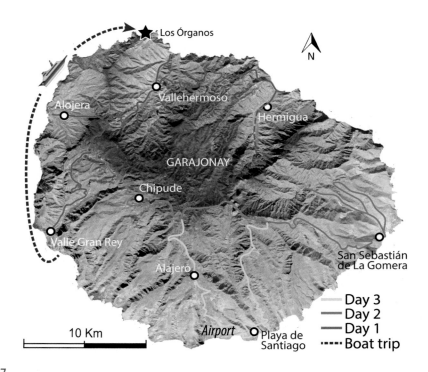

FIGURE 4.17

Geological itineraries on La Gomera (image from *Google Earth*). See also Figs. 4.18, 4.26, and 4.34 for further details on the geological excursions of the individual days (eg, stops, names of the roads, etc.).

DAY 1: THE EARLIER STAGES OF SUBAERIAL GROWTH

Stop 1.1. Valle Gran Rey

We start the day in *Valle Gran Rey* (Fig. 4.18). The village grew on the flat alluvial plain at the mouth of the barranco and is framed on both sides by vertical cliffs with thick scree deposits at their base (Fig. 4.19).

The approximately 600-m-high walls of the barranco expose a lava sequence of sub-horizontal Pliocene basalts that are unconformably overlying Miocene seaward-dipping basaltic flows.

The valley gets its name from the legendary King Hupalupa of the pre-Spanish era. The fertile valley was his home and that of his successor Guanche kings (gran = great, rey = king). Sheltered from the cool and humid trade winds, the valley benefits from a warm climate, plenty of fresh water from natural springs at the valley head, and numerous palm trees and, more recently, banana plantations among the coastal gardens and plantation fields. This setting and one of the best beaches on the island explain the spectacular touristic development of this location over the past three decades.

Leave *Valle Gran Rey* on GM-1 in a northerly direction toward *Arure*, climbing along the western wall of the valley. After km 56 on the GM-1 road sign, you will see a long parking strip on the

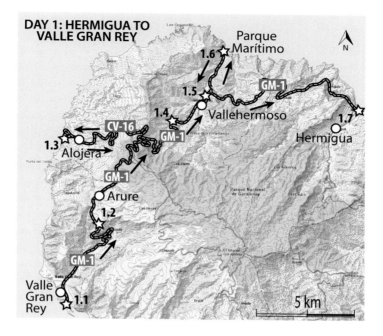

FIGURE 4.18

Detailed itinerary map for *Day 1*, with the stops and other relevant main roads in and around the Miocene basalts, the Vallehermosos felsic stratovolcano, and the north–northwest plutonic complex. Note that the main roads on La Gomera were renamed in 2010 from the prefix TF in previous road maps to new signposts that use GM as the road number prefix. Map from IDE Canarias visor 3.0, GRAFCAN.

right side of the road. Park there to visit the adjacent mirador. It offers one of the best panoramic views of Barranco Valle Gran Rey (Fig. 4.20A).

Stop 1.2. Mirador de Cesar Manrique *(N 28.1196 W 17.3159)*

You can visit either the restaurant on the left end of the mirador or one of the balconies to the right of the restaurant, which are free of charge. On the opposite side of the valley, a major unconformity is usually well visible. Seaward-dipping Miocene basaltic units in the lower valley are seen to be overlain by sub-horizontal Miocene–Pliocene basalt sequences, implying a major period of erosion between these two large magmatic episodes (Fig. 4.20B). It is this erosion event that also removed the now lost Vallehermoso volcano, and we must speculate that this period must have lasted several millions of years (perhaps 2 or 3).

Stop 1.3. Alojera *old pier (N 28.1646 W 17.3343)*

Pass through *Arure* and continue on GM-1 in the direction of *Vallehermoso*. Follow GM-1 for a good while up-hill until a left turn for *Alojera* appears. Turn left here and continue on the road to *Alojera* (which had no road number at the time of writing). Drive through *Alojera* and down to the sea. Park your car above the little old harbor and walk toward the exposure (Fig. 4.21A).

FIGURE 4.19

The village and harbor of *Valle Gran Rey* have developed on the alluvial delta at the mouth of the barranco at the foot of 600-m high cliffs made of Miocene and Pliocene basaltic sequences.

Here, Miocene lava flows with zeolite fillings are exposed. These are among the oldest rocks of the island ($\sim$9 to 10 Ma). The flows dip gently toward the sea and are made of many thin individual flow units, and thus are of dominantly pahoehoe origin. The Miocene basalts are cut by abundant dykes and sills (black-colored units). On the right valley side, a light-colored felsic (trachytic) intrusion is visible (Fig. 4.21A,B). Vintage ages of almost 20 Ma have been derived from the Miocene basalts, dated by K-Ar (Abdel Monem et al., 1971), but are likely compromised by zeolite formations because zeolites require vast volumes of warm water to pass through the rock to mobilize alkali elements (like K and Na) and to precipitate them in the form of silica-bound zeolite crystals. The K-Ar method is thus not strictly suitable for this rock formation here. Therefore, the absolute age of the rocks exposed here remains uncertain.

Enjoy the breeze before returning on the road on which you came to join GM-1, but now in the direction toward *Vallehermoso*, which is approximately 8 km ahead (once you are back on GM-1). At approximately the km 43 road sign, you will start to see many dyke intrusions in adjacent road cuts, the first signal of the cone sheets of *Vallehermoso* (see, eg, Ancochea et al., 2003, 2008). Approximately 1 km further and just before the km 42 road sign, park the car in a small viewpoint

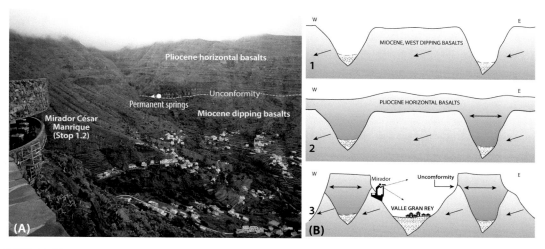

FIGURE 4.20

(A) View from the balconies next to *Mirador César Manrique*. This is a good viewpoint to observe the unconformity between the horizontal Pliocene basalts that overlie a dipping Miocene basaltic sequence. A group of permanent springs has developed at the contact (see Fig. 4.12). (B) Sketch showing the evolution of the Baranco Valle Gran Rey by erosion and relief inversion (after Carracedo and Day, 2002).

with a lay-by on your right. Here, you can enjoy the view into the Barranco and also inspect the first cone sheets in the road cut in more detail.

Stop 1.4. Barranco de Macayo *cone sheets (N 28.1663 W 17.2777)*

The cone sheets here dip roughly into the valley (ie, toward the northeast) (Fig. 4.22A,B), but some stray dips do occur. Different states of hydrothermal alteration are apparent, with older sheets usually showing brownish colors and a fragile appearance, whereas younger sheets tend to be darker and more massive. Naturally, each new sheet within such a swarm will thermally overprint the previous sheet(s) in its vicinity, thus explaining highly variable degrees of thermal metamorphism in a cone-sheet swarm (Donoghue et al., 2010). The cone sheets are arranged in concentric fashion around a cone-sheet center, which on La Gomera describes a ring of sheets approximately 10 km in diameter at sea level and with an estimated depth of focus approximately 2100 m below the present day sea level (Fig. 4.22C; Ancochea et al., 2003).

A short hiking path (∼2 km) commences a few meters further from the parking site if you want to stretch your legs a little more or inspect the barranco and the cone-sheet swarm in further detail.

Otherwise, continue down-hill on GM-1 toward *Vallehermoso*. Just before reaching the town, a large rock mass sticks out in the skyline straight ahead to your right. This is Roque Cano, a trachyte plug. It postdates the Miocene basalts and the cone sheets, and it is more resistant to erosion, thus forming a towering feature in the landscape (Fig. 4.23). Soon after leaving *Vallehermoso* and soon before the first tunnel after the town, a vantage point/mirador appears on your left. Park the car here to inspect the Roque Cano trachytic dome from below and to get further views of the valley.

FIGURE 4.21

(A) View of the western coast of La Gomera. The local sequences of Miocene basalts is deeply eroded and weathered and is intruded by numerous dykes, both basaltic and felsic in composition. (B) Close-up view of Miocene basalt rocks that are cut by a thick and now altered trachyte dyke, at *Playa de Alojera*.

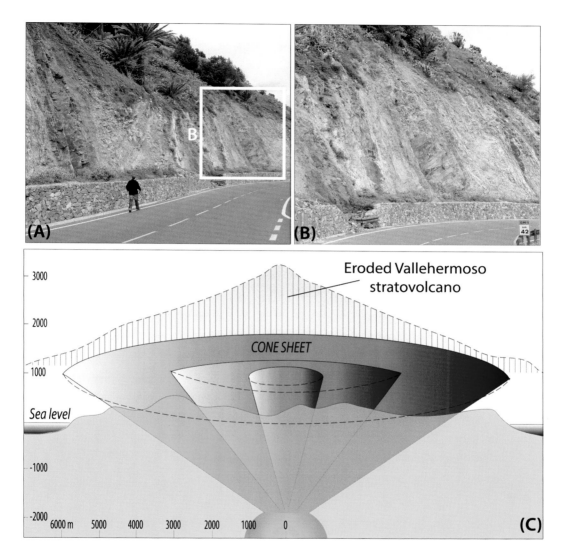

FIGURE 4.22

(A) Trachytic dykes of the Vallahermoso conical intrusion swarm (cone sheets). (B) Close-up of the white square in (A) showing the dense swarm of inclined parallel felsic dykes. (C) Sketch of the cone-sheet swarm with dykes in concentric arrangement around the cone-sheet source describing a ring approximately 10 km in diameter at sea level and with an estimated depth of focus of ∼2100 m below the present sea level (after Ancochea et al., 2003).

FIGURE 4.23

Roque Cano, a trachyte intrusion of Pliocene age ($\sim$4.5 Ma), with pronounced columnar jointing. Towering over *Vallehermoso*. This dome is the centerpiece of the local municipality's emblem. Inset courtesy of Ayuntamiento de Vallehermoso.

Stop 1.5. Roque Cano *(N 28.1827 W 17.2620)*

Roque Cano rises almost 250 m above its surrounding and forms a widely visible landmark just outside *Vallehermoso* (Fig. 4.23). In fact, it is the centerpiece of the local municipality's logo emblem (inset in Fig. 4.23). Roque Cano is a trachyte intrusion of the Pliocene age ($\sim$4.5 Ma; Paris et al., 2005b) and displays thick columnar jointing.

Return to *Vallehermoso* and follow signs for *Parque Maritimo*. Continue down-hill after the town and pass some interesting sheet exposures on your left in road cuts before you reach the coast. Park your car in front of the *Parque Maritimo* building near the coast and take the paved road on the left for a short walk. Avoid driving there because this stretch of road is extremely prone to rockfalls on good and bad weather days.

FIGURE 4.24

The plutonic complex at *Parque Marítimo* near *Vallehermoso*. A general view at the end of the road. Note the high density of dykes intruding the plutonic host rock.

Stop 1.6. Parque Maritimo (N 28.2028 W 17.2530)

Walk along the coastal road and inspect a dense sheet swarm dipping dominantly to the northwest at this site. The intrusive density is very high in this location and the rock mass consists almost exclusively of dykes intruded into sheets or dykes (Fig. 4.24). This dense, intrusive complex marks the heart of the main Miocene shield volcano, which must have once existed above our current position (eg, Ancochea et al., 2008).

Return to *Vallehermoso* and GM-1. Make your way through the town and leave *Vallehermoso* on GM-1 in the direction of *Agulo*.

Continue up-hill on GM-1. At approximately the km 30 road sign on GM-1, the road becomes flat and you will find yourself next to semi-horizontal Miocene−Pliocene basalts with strong columnar jointing. Continue toward *Agulo*. Soon the road will go down-hill again and you will drive once more through hydrothermally altered Miocene rocks cut by many former sheet intrusions. Continue down-hill, and just when approaching the major tunnel before *Agulo*, the semi-horizontal Miocene−Pliocene basalts are visible again, right above the tunnel entrance. However, we have been driving down-hill for some time now, implying a pronounced topography that must have existed prior to deposition of the Miocene−Pliocene lavas and that was filled completely by the Miocene−Pliocene volcanism.

Drive through *Agulo* in the direction of *Hermigua*. A short while after *Agulo*, the semi-horizontal Miocene−Pliocene basalts give way to the underlying rocks, which are now basaltic lavas of the old Miocene shield volcano. In *Hermigua*, look for signs to *El Pescante* (the old pier) and follow them to the coast. Just before the coast, directly in front of the last row of houses prior to the shoreline, take a right turn and drive the last few hundred meters toward the old and badly damaged pier on the right side of the bay (looking seawards). Park the car near the coast in a larger car park and walk to the pier (approximately 400 m).

Stop 1.7. El Pescante: *The old pier (N 28.1773 W 17.1772)*

At the old pier (Fig. 4.25A), pahoehoe lavas with frequent zeolite and calcite vesicle fills are cut by darker and much fresher looking dykes (Fig. 4.25B,C). Some of the dykes are ankaramites (high proportion of large pyroxene and olivine) and are worth a closer inspection. Notably, the chilled outer margins of these dykes are often poorer in crystals, implying segregation of crystals to the interior of the dyke during magmatic transport (flow differentiation).

Looking inland, one can see the semi-horizontal Miocene−Pliocene basalt succession resting on less regularly arranged Miocene basalts and scoria. Note that the lower boundary of the Miocene−Pliocene group does occur at variable horizontal levels, emphasizing the notion of a rugged and wild topography prior to their deposition.

This marks the end of Day 1. Return to *Valley Gran Rey* via the route you came or drive to *Hermigua* and leave *Hermigua* village in the direction toward *San Sebastián* (GM-1), and then follow signs for GM-2 and *Valle Gran Rey*.

Consider stopping again at the first stop of the morning (the *Mirador de Cesar Manrique*) on your return journey because the "photo light" is best there in the late afternoon.

END OF DAY 1.

DAY 2: DYKES, "PLUGS," AND BAKED CINDER CONE DEPOSITS

Leave *Valle Gran Rey* on GM-1 toward *Arure*. Once you arrive in *Arure*, look for a turn to *Chipude* on your right. Follow this small road up-hill (un-numbered at the time of writing) for a few kilometers and continue to follow signs for *Chipude*, which requires another turn after a while. Note, this is a sharp right turn. A few minutes after this turn, you will arrive in *El Cercado*. Keep left at the fork and drive through the village in the direction of *Chipude*. Pass through *Chipude* and turn left toward *Igualero* at the fork in the village (Fig. 4.26). Only a few seconds after *Chipude*, the upstanding rock '*La Fortaleza*' (the Fortress) will appear on your right. Stop at the next lay-bay on your right to take in the view.

Stop 2.1. La Fortaleza: *The fortress (also known as La Torta) (N 28.1083 W 17.2796)*

'*La Fortaleza*' is a 4.36-Ma-old trachytic intrusion that displays a flat top (Fig. 4.27A). This unusual roof shape may perhaps be due to near-surface ballooning and lateral expansion of the intrusion in a highly viscous state (eg, like toothpaste), lacking the energy to violently break through to the surface. The 300-m-long Fortaleza hill is also one of the most important archaeological sites of the island, because it acted as a refuge for the aboriginal population during the Spanish conquest of La Gomera (hence the name "Fortaleza").

Continue up-hill and just before the hamlet of *Igualero*, a larger mirador appears on your right. Park the car and look at '*La Fortaleza*' from the "back side".

Stop 2.2. Mirador de Igualero: *La Fortaleza feeder system (N 28.0994 W 17.2550)*

From here, the overall mushroom shape of the Fortaleza intrusion becomes apparent by exposing the feeder system to this high-level intrusion (Fig. 4.27B). Perhaps a viscous surface dome formed during its active period (eg, like at Mt. Rajada on Tenerife), but a fully intrusive origin is also conceivable. In any case, the trachyte magma likely behaved like cold honey or stiff toothpaste when it was squeezed through the conduit and into cracks and opening cavities to form this intrusion (Fig. 4.27C,D).

FIGURE 4.25

(A) Due to an increasing number of banana plantations in Bco. de Hermigua in the beginning of the 20th century, the trading company *La Unión* began the construction of an ocean pier to be able to ship and sell fruit. The construction of a road connecting *Hermigua* with *San Sebastián*, and the associated access to the larger harbor in *San Sebastian*, rendered the pier obsolete in the early 1920s. Today, only the towering concrete pillars of the former loading station remain (image courtesy of *Cabildo de Insular de La Gomera*). (B) Basaltic, mainly pahoehoe, sequence of the Miocene basalts at the "*Pescante*" pier. (C) Close-up view of a Miocene basaltic rock in this area, with zeolites and carbonates filling the former vesicles (coin for scale).

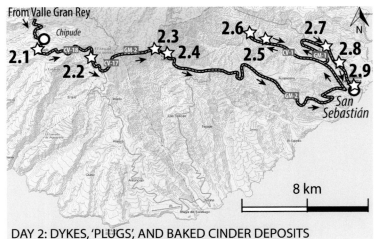

DAY 2: DYKES, 'PLUGS', AND BAKED CINDER DEPOSITS

FIGURE 4.26

Detailed map of *Day 2*, with the individual stops marked. Map from IDE Canarias visor 3.0, GRAFCAN.

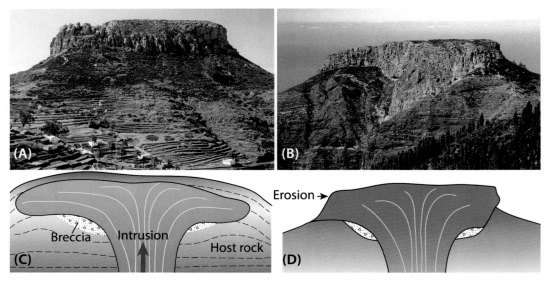

FIGURE 4.27

The morphology of lava domes is variable. On La Gomera they are typically thick, steep extrusions, but their shapes can vary from circular, low-profile domes (*tortas*) to cylindral spines with thick talus slopes (Peléean domes). (A) View from the north onto the '*Fortaleza de Chipude*', a 350-m-long, 80-m-thick trachytic lava dome. (B) View of the Fortaleza de Chipude intrusion from *Mirador de Igualero* to the east. From here, the feeder system to this high-level intrusion becomes visible, and the thick short lava or sill that was emplaced toward the south is apparent. Sketch illustrating the possible intrusion (C) and erosion (D) of the dome.

FIGURE 4.28

(A) Cluster of felsic domes in central La Gomera (image from *Google Earth*). (B) Closer view of Roque de Agando trachyte dome, 5.1 Ma. Note the prolate spheroidal "onion skin" joints implying that this dome was significantly larger before erosion (or collapse) of its southern flank. (C) View of Roque de Ojila dome from the southeast (in the foreground) and Roque de Agando in the background. (D) Roque de la Zarcita phonolite dome (4.25 Ma).

Continue through *Igualero*. After a short while, you will meet a fork. Turn left in the direction of *San Sebastián* and, after a short while again, you will meet another turn. There, join GM-2 in the direction of *San Sebastián* and, after only a few kilometers on GM-2, a mirador opens up on your right, which offers some impressive views onto the cluster of spectacular felsic domes exposed here (Fig. 4.28A).

Stop 2.3. Mirador de los Roques *(N 28.1091 W 17.2146)*

Looking south from the Mirador, the large spine of Roque Agando stands proud, rising for approximately 100 m above the road, whereas the south face reaches 220 m as it drops into the valley (Fig. 4.28B). Roque de Agando was a place of worship for the aboriginal population. Remains of a

Guanche sacrificial shrine have been found on the summit. These remains were in reasonable condition up to the 1980s, but they were looted by a film crew that shot a documentary at the site, leading to public outrage but unfortunately little else. During the 1980s and 1990s there were several climbing routes, mainly on the south face of Roque Agando, but climbing is not allowed anymore and these central trachyte domes have now been declared "natural monuments."

Looking north, the Roque de Ojila dome (Fig. 4.28C) and Roque de la Zarcita (Fig. 4.28D) form the steep upstanding hills in the foreground. The "roques" are all trachytic or phonolitic in composition and, like La Fortaleza, they represent plugs and dome complexes, which are sites of former felsic magma injections close to the surface (ie, up to several hundreds of meters below the former land surface).

Continue on GM-2, and right around the next bend you will find a larger mirador on your right that is closer to the foot of Roque Agando. Park there.

Stop 2.4. Roque de Agando *(N 28.1074 W 17.2136)*

From here, you can walk toward Roque Agando along the small path on the left side of the road. In addition, a small path starts right behind you at the mirador and leads you with only a few steps onto one of the smaller trachyte intrusions in the area. Note the light-colored versus dark-colored rocks on the ground and in the hillside, representing a mixture of mafic and felsic intrusives in this part of the island.

At the foot of Roque Agando, you can inspect columnar jointed trachyte, frequently with visible high-temperature potassium feldspar (sanidine).

Continue on GM-2 in the direction of *San Sebastián*. Just after passing Roque Agando, you will find yourself driving through breccias made largely of trachyte. This is Agando's autobreccia. Sometimes domes are produced by outpourings of extremely viscous lava pushed up from the vent like toothpaste from a squeezed tube. As the dome grows, the expanding crust breaks and forms a heap of angular rock fragments (autobreccia) around the base (see Fig. 4.29).

Lava domes can take many forms, including circular and flat-topped ones "tortas," circular and spiney "peleean" ones, or piston-shaped ones (upheaved plugs), and sometimes they can even take on a hybrid form between a lava flow and a lava dome (ie, a coulée, see Fig. 4.15) (Blake, 1990). The form that a dome takes is a function of many factors but, primarily, it is controlled by the viscosity of the lava and the topography (slope) of its surroundings.

If lava domes grow rapidly and become unstable, then they can collapse and give rise to deadly pyroclastic density currents. Lava dome collapses have been the cause of some of the large volcanic disasters (eg, in *St. Pierre* on the island of Martinique in 1902), but the phenomenon is also known from Merapi or Mount St. Helens in Indonesia and the United States, respectively.

Continue all the way to *San Sebastián* and just when you arrive in the outskirts of town, turn left after the river to drive into Barranco de la Villa (toward *Chejelipes*). Naturally, a stop in *San Sebastián* is possible and may even be advisable to escape the usually intense mid-day sun for a short refreshment break.

Once you drive up the barranco, you will note that it has a flat bottom, defining a U-shape, which indicates a mature stage of valley evolution (eg, chapter: The Geology of Lanzarote, Fig. 7.34). After a few kilometers, the road crosses the river and follows the southern wall of the barranco for a while. After another few kilometers, a steep "wall-like" feature with a lay-by directly in front of it appears on your left. Park the car in the lay-by to your left.

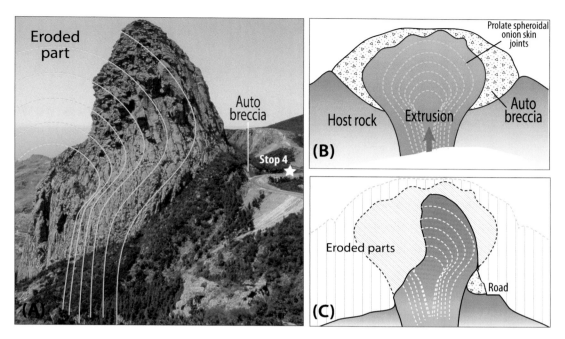

FIGURE 4.29

(A) Roque de Agando's breccia crops out in the road cut. (B) Note the internal structure of the joints. Upon cooling and contraction, concentric joints form that now resemble onion skin layers. (C) Unstable domes can collapse and produce dangerous (even pyroclastic) density currents. Eventually they cool and erosion reduces their size by fracturing along the dome's joints.

Stop 2.5. Massive Taparucha (N 28.1136 W 17.1589)

Here, the side wall of a large dyke intrusion is exposed and can be traced for hundreds of meters into the hillside (Fig. 4.30). These parallel dykes are the feeding conduits of fissure eruptions, characteristic of oceanic island basaltic volcanism. Similar dykes have been observed in Miocene basalts in Fuerteventura, Gran Canaria, and Tenerife, and will be exposed as "*taparuchas*" after erosive dismantling. These dykes feed the most recent rift zones of the Canaries, such as the northwest rift zone of Tenerife, the Cumbre Vieja on La Palma, or the 1730−1736 eruption on Lanzarote.

The polygons on the sidewall of the dyke are cooling joints that usually form perpendicular to the cooling surface. Because dykes are very long relative to their thickness, the columns will form preferentially across the thickness of a dyke.

Continue up-hill for a short while until the road gets a little bendy. Soon a major water dam will appear, first on your right side and then a little later in front of you. Drive toward the dam and park on the open space on the right side of the water reservoir. Turn left when reaching the dam.

FIGURE 4.30

Basaltic dyke (*taparucha*), up to 5 m thick and several kilometers long, that intruded the Miocene basalts near *Los Chejelipes*, in Bco. de la Villa, above *San Sebastián*.

Stop 2.6. Hydro-dam, Barranco de la Villa *(N 28.1164 W 17.1678)*

Here, several long and weathered dykes (*taparuchas*) are visible in the backdrop of the lake. These strike east−west to southeast−northwest and form a semi-parallel swarm of intrusive sheets (Fig. 4.31). The sheets are offset in "*en enchelon*" style and reflect the direction of stress that acted during their emplacement (ie, σ3 is perpendicular to the swarm's long axis).

The highly compacted and altered Miocene lavas and pyroclasts here are crossed by numerous massive dykes that form impermeable "compartments" in the regular rocks that are suitable for the construction of water reservoirs. Remarkably, and partly for this very reason, La Gomera is the island with the best water per capita reserves in the Canaries.

FIGURE 4.31

Several parallel dykes (taparuchas) in Miocene basalts, Bco. de la Villa. The combination of compacted and altered basaltic lavas and the more massive dykes create basins that retain water, allowing the construction of highly efficient dams. These water reservoirs allow La Gomera to have the largest water reserves per capita in all of the Canaries.

Once you have sampled the views, return to *San Sebastián* on the road from which you came. In *San Sebastián*, look for signs that guide you to GM-1. Leave the town on GM-1 in the direction of *Hermigua*. After a few kilometers only, and soon after *El Molinito*, a turn toward the right with signs for *La Lomada* takes you to a large and rather new viewpoint (mirador) right next to a reddish, strongly bedded, and fossilized scoria sequence in the Miocene basalts. Park the car on the upper one of the two lay-bys.

Stop 2.7. Abandoned scoria quarry along new road (N 28.1110 W 17.1262)

This site is an old quarry that is now used to accommodate the new road on which you came. The rock here is scoria and lapilli beds that were buried under Miocene basalt flows and experienced subsequent "burial metamorphism" (Fig. 4.32A,B). This is witnessed by abundant vesicle fills of calcite and zeolites, as well as the calcite-lined fracture planes that record intense secondary mineralization. This "second heating" event baked the lapilli together, allowing to cut blocks from the formerly semi-loose pyroclastic material. This rock type is one of the most popular building stones on La Gomera, and it has been used since the colonization of the island approximately 500 years ago and was widely used in houses, churches, and public buildings (Fig. 4.32C). Notably, it is still widely used in modern constructions all over the island (Fig. 4.32D).

FIGURE 4.32

(A) Abandoned quarry now used to accommodate the new road. The rock here is a crystal-rich scoria and lapilli sequence that was buried under Miocene basalt flows and experienced subsequent "burial metamorphism" (image courtesy of *C. Karsten*). (B) The rock is sufficiently compact to allow cutting blocks for edifice construction. Many sacral and municipal buildings on La Gomera use this red agglomerate (*tosca* or *toba roja*), like on the front of the *Iglesia de San Sebastián* (C), built at the end of the 15th century, or at the front door of the new airport (D) that opened in 1999.

Continue up-hill for a few hundred meters and park your car at a large mirador in the next major road bend.

Stop 2.8. Location: Viewpoint above old quarry (N 28.1044 W 17.1208)

The majority of sedimentary formations on the Canary Islands are derived from the mechanical disintegration and accumulation of products from earlier volcanic eruptions. On an older island such as La Gomera, accumulations of fragmented volcanic materials abound. Most have been deposited in the bottom of the barrancos, forming alluvial beds of rounded *boulders*, *cobbles*, and *sand*, which are also typical for many of the barrancos of La Gomera. Although V-shaped barrancos are the most frequent on the younger islands, on La Gomera sediment filling frequently gives rise to U-shaped valley profiles at the mouths of some barrancos (eg, Bco. de la Villa) in which the capital *San Sebastián* is located. This feature is also frequently observed in the older eastern islands, such as in the Famara shield on Lanzarote, for example (see chapter: The Geology of Lanzarote, Fig. 7.34).

From here, you will have a beautiful overview of *San Sebastián* (Fig. 4.33A) and, once more, of Barranco de la Villa and its valley head (Fig. 4.33B), where we have inspected the dyke intrusions earlier. Enjoy the overview and continue up-hill to turn at the roundabout on the hilltop to return to *San Sebastián* the way you came.

If you have time left, make your way to the old church in the town center of *San Sebastián*, which is constructed from the "baked scoria" that we just inspected.

Stop 2.9. Old church in San Sebastián (N 28.0926 W 17.1112)

Partial "burial metamorphism" by subsequent volcanism and resultant compaction of scoria and lapilli by the load of overlying formations are combined with secondary mineralization to transform the original loose scoria into a hard rock (locally called *tosca*). This "baked lapillistone" can be cut in blocks and has historically been a very popular building stone on La Gomera that is still widely used in modern constructions all over the island (eg, the main entrance of the island's airport; Fig. 4.32D). The old church (*Iglesia de San Sebastián*) was built in the 17th century on

FIGURE 4.33

(A) View of *San Sebastián*, the island's capital that developed in the coastal opening of *Bco. de la Villa*. (B) Upstream, the alluvial filling of the barranco provides a flat bed that is well suited for farming and dwellings.

the site of a late-15th-century monastery and is a testament to the functional and decorative use of the "*tosca*" lapillistones.

From *San Sebastián*, return to *Valle Gran Rey* via GM-2 and the subsequently signposted route.

On your return journey through barranco *Valle Gran Rey*, please note that the cemetery and the funeral hall in *Valle Gran Rey* as well as the *Ermita de San Antón*, the little chapel along the road, are partially made from this rock type also, reflecting its use not only through several centuries but also as an element of style all over the island.

END OF DAY 2.

DAY 3: TRACHYTE INTRUSIONS

This day is dedicated to the southern part of the island (Fig. 4.34), where felsic intrusions abound, and to Los Órganos, for which you will need to take a boat trip (see Fig. 4.17). The boat trip can be taken either in the morning before the field stops or in the afternoon following the field itinerary (or on any other day if you wish). From *Valle Gran Rey*, boat tours usually leave twice per day. A range of options exists, from regular boat cruises of 4 hours to fast-boat trips of 2 hours, and prices vary between €30 and €45 per person, depending on how fast you like to travel. We recommend doing the boat trip in the morning (usual start is at 10:00 AM or later) because more boats

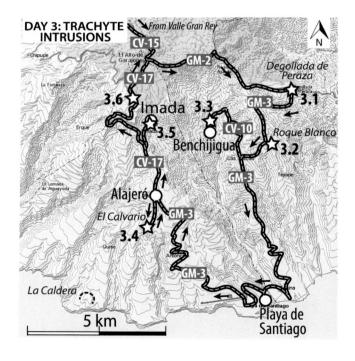

FIGURE 4.34

Detailed map of *Day 3*. Map modified from IDE Canarias visor 3.0, GRAFCAN.

will do the tour at that time than in the afternoon, and you also do not want to have to rush back from the outcrops to catch the boat.

If you make your way to the harbor at approximately 9:00 am, you should have plenty of time to shop around for the type of vessel you like to board. The geological text for Los Órganos can be found at the end of the day. Prebooking is possible via the Internet and several tour operators also offer a range of options online.

Field stops: Leave *Valle Gran Rey* on GM-1 and join GM-2 after passing through *Arure* in the direction of *San Sebastián*. GM-2 takes you over the central highlands of La Gomera again. Pass the Roque de Agando trachyte spine once more and continue in the direction of *San Sebastián*.

At km 15, pass the crossroad with GM-3, and after another 70 m park at the mirador.

Stop 3.1. Degollada de Peraza *(N 28.0996 W 17.1848)*

The mirador offers a wide panoramic view to the north of the Bco. de la Laja, which is carved into the seaward-dipping Miocene basalts. Abundant dyke-like walls several kilometers long crop out because they are more resistant to erosion than the lava flows (the *taparuchas* of Stop 2.5).

Return to the cross road with GM-3 and turn off toward *Playa de Santiago*. Past the km 4 road sign you will see the trachyte dome of Roque Blanco in front of you. Just before the first tunnel on GM-3, park on a stretch of old disused road to the left of the tunnel entrance.

Stop 3.2. Roque Blanco *trachyte lava dome (N 28.0804 W 17.1991)*

The semi-annular intrusion may represent the remaining part of a 3.9-Ma ring-dyke structure (Cubas et al., 2002). Here, you can inspect a trachyte intrusion close up. Strong hydrothermally altered domains alternate with fresher ones, with the latter being characterized by pristine and transparent alkali-feldspar (sanidine). Those domains that are characterized by "Liesengang structures" (fluid flow indicators), show white (hydrated) or brownish stained feldspar (see also Tindaya intrusion on Fuerteventura, Fig. 8.43).

Return to GM-3 and continue down-hill through the tunnel. Soon you will drive through another tunnel, and after another approximately 220 m take a deviation by following signs to *Benchijigua* (on CV-10). The bendy road now descends into Bco. de Benchijigua, which is the largest valley on the island. Continue until you reach the bottom of the barranco. Drive to the end of the road and park the car.

Stop 3.3. Trachyte ring dyke of Barranco de Benchijigua *(N 28.0845 W 17.2183)*

The Bco. de Benchijigua is a characteristic "Oahu-type" horseshoe-head valley carved into Pliocene and Miocene basalt formations. A spectacular trachytic ring dyke crops out at the central part of the barranco (Fig. 4.35A).

On closer inspection along the road, it can be observed that the annular dyke intrudes the Miocene basalts (Fig. 4.35B,C). We recommend this stop not only to examine an interesting geological feature but also because of the beauty of this spectacular caldera-like depression, similar to those in the Hawaiian valleys, such as in the island of Oahu, which gave the name to this type of canyons.

Return to GM-3 and continue down to the coast. Pass *Playa de Santiago* and the airport. The road then takes you up-hill again to *Alajeró*. Make your way into the village center of *Alajeró*, and

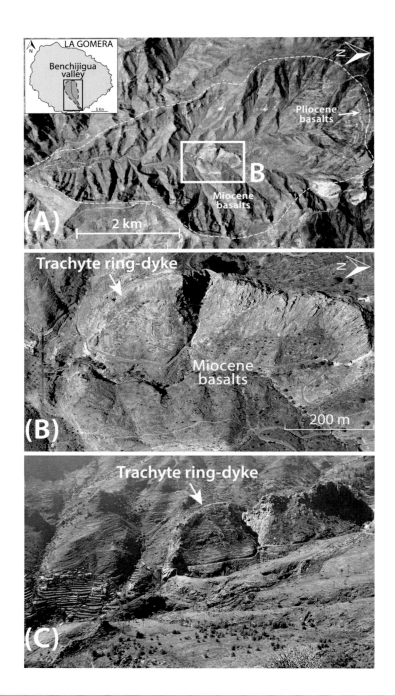

FIGURE 4.35.

(A) Bco. de Benchijigua with a ring dyke type of structure in the river bed (image from *Google Earth*). (B) Detail of the ring dyke. (C) View of the ring dyke from the top of the barranco wall.

from there follow signs for *Ermita de San Isidro*, which is built on a trachyte plug. Park in front of the trachyte hill and take a walk up to the *Ermita*.

Stop 3.4. Ermita de San Isidro; Mña. Calvario *(N* 28.0557 *W* 17.2413*)*

Note the rock here is strongly altered. Even the freshest pieces show the shiny surfaces of mica growth during alteration, implying that the rock was kept at elevated temperatures for quite some time and likely interacted with ground or meteoric waters to allow abundant white mica and related sheet silicates to form.

Viewed from the west, the dome shows several flow units that were emplaced toward the north, whereas the southern flank has been eroded or collapsed (Fig. 4.36A,B). The flat top of the dome, once a Guanche necropolis, is now topped by the *Ermita*, which is the focus of an annual religious pilgrimage on foot to the top of the mountain (*Romería de San Isidro*) to make offerings to St. Isidro before gathering for a picnic on the hill and enjoying the views.

The top of the dome offers a view of the south of the island, including La Caldera (looking southwest), the only preserved volcanic cone in the island (Fig. 4.36C). La Caldera (N 28.0278 W 17.2608), 280 m in height, is a trachyte scoria cone dated at 4.2 Ma (Cantagrel et al., 1984). To inspect the cone (optional for another day), take the road in *Alajeró* signposted *Quise-Las Canteras* and drive for a few kilometers (always down-hill) until the end of the road at the village of *Quise*. From there, you have to walk another 4.3 km to La Caldera.

Return to GM-3 and continue westward (ie, turn left onto GM-3). Soon after leaving *Alajeró* village, you will pass another "baked scoria" deposit that was quarried in its former days but is now home to storage tunnels and hosts the local water reservoir.

A few hundred meters up-hill, a sharp turn to the right takes you to *Imada*. Soon you will enter a new barranco, and the pass into this barranco is flanked by two more trachyte plugs. The left one is Roque de Imada. In the distance, the spine of Agando will now become visible. Make your way to the village center of *Imada* and park on the terrace next to the village bar.

Stop 3.5. Roque Imada *and Imada village (N* 28.0859 *W* 17.2411*)*

Here, you can gaze upon a beautiful view of Roque de Imada (Fig. 4.37). The dome is cut by the road and can be inspected and sampled. The little village usually allows for a short refreshment stop. Return to GM-3 and continue westward (ie, turn right onto GM-3 when back at the cross-roads). Soon the road passes under the Agalang trachyte plug before the road climbs up-hill again, and a quick stop here is optional.

A few kilometers further the road passes onto the higher plateau of La Gomera and right in a sharp bend before the treeline, a very thick trachyte sheet comes up for inspection on your left. Park the car on the open area on the right of the road.

Stop 3.6. Lava dome near Igualero *(N* 28.0929 *W* 17.2479*)*

A deeply eroded remnant of one of the many trachyte sheet-like feeders of the southern flank of La Gomera crops out here. Walk up to the contact between the intrusion and the underlying Miocene scoria, which is strongly baked adjacent to the intrusion. The contact is inclined and dips southward. Walk along the intrusion to inspect the rocks in detail and (importantly) safely, because you are off the road now (Fig. 4.38).

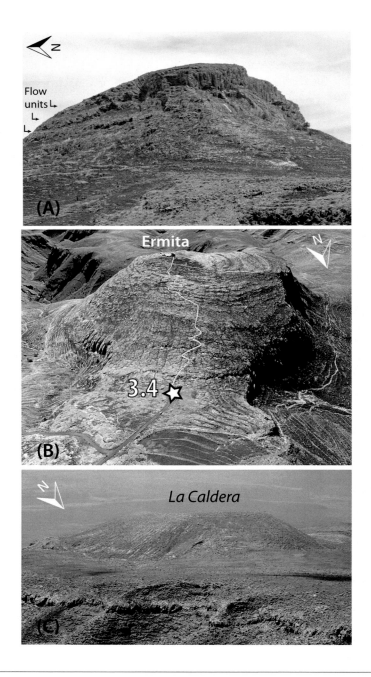

FIGURE 4.36

(A) View from the west side of Mña. Calvario near *Alajeró*, a trachytic dome with several thick, very short flow lobes toward the north. (B) View of Mña. Calvario showing the pressure ridges (ogives) formed by the flow of viscous lava (image from *Google Earth*). (C) La Caldera seen from Mña. Calvario, the only cinder cone type of vent preserved on the island.

FIGURE 4.37

Roque de Imada, with the village of *Imada* in the backdrop.

FIGURE 4.38

A characteristic deeply eroded remnant of one of the many trachyte intrusions (plugs and sheets) that crop out at the southern flank of La Gomera. This one is near *Igualero*.

Continue the drive up-hill, and right after the bend you will see more felsic rocks on the continuation of the ridge to your right (on top of the hill that extends away from the road). Continue for a short while up-hill and then turn left to *Chipude*, from where you can go back to *Valle Gran Rey* via *Arure* and GM-1 or, alternatively, continue on GM-3 until you meet GM-2, which also brings you to *Valle Gran Rey*. The former route is a little slower and bendier, but it offers more rock exposure than the higher route (strongly forested) and, on a good day, the lower route affords some spectacular views of La Palma in the far distance (we recommend it).

Boat trip to Los Órganos

If you have not done the boat trip in the morning, we suggest continuing the day with the boat trip to Los Órganos.

The 12 nautical miles from *Valle Gran Rey* to Los Órganos are accompanied by spectacular views of imposing valleys and coastal regions, small towns, and bays. There is also a good chance of seeing whales and dolphins *en route*.

Los Órganos (The Organ Pipes) is an imposing, partially eroded trachyte dome that crops out in a cliff at the northwest tip of the island (Fig. 4.39A). The phonolitic dome of *Punta de las Salinas* (Los Órganos) has been partially removed by marine erosion and its inside comprises perfect and

FIGURE 4.39

(A) Large trachytic dome north of *Playa de Arguamul*. (B) Erosion has removed half of the dome, thus exposing an internal structure of vertical columnar joints know widely as '*Los Órganos*' (the organ pipes). (C) Detail of the columnar jointing showing the geometry of the retraction joints. (D) Sketch explaining the generation of regular joints in case of undisturbed cooling.

spectacular rock columns, like in giant organ pipes. The internal structure of this dome is character-ized by the presence of thousands of concentric and radial fractures (Fig. 4.39B). Concentric frac-tures reflect cooling isotherms, progressing toward the interior of the domes, like layers of an onion, losing definition as they approach the center. The radial fractures originate by retraction caused by loss of volume on dome cooling (eg, shrinkage).

When lava continues to flow during cooling and the plasticity limit is exceeded, fracturing is irregular or curved, whereas if the lava cools undisturbed, as occurs in many of the Gomera domes, cooling fractures are regular, indicating high-viscosity magma.

Least-effort fracturing gives rise to triple junctions with regular 120° angles (Fig. 4.39C), simi-lar on a smaller scale to that postulated for the origin of the triple-armed rift structures in ocean islands such as the Canaries. The association of these fractures breaks the rock into hexagonal prisms, producing perfectly shaped columns that centuries ago were believed to have been man-made (Fig. 4.39D). Rock columns have always attracted the attention of humans, with the most famous rock columns arguably being *Fingal's Cave* on the Isle of Staffa in Scotland (Mendelssohn was inspired by this awesome Scottish landscape and wrote "Fingal's Cave" for his Hebrides Overture) and the *Giant's Causeway* (in County Antrim, Northern Ireland), both of which are com-posed of basaltic columns.

END OF DAY 3.

THE GEOLOGY OF TENERIFE

The Geology of the Canary Islands. DOI: http://dx.doi.org/10.1016/B978-0-12-809663-5.00005-0

THE ISLAND OF TENERIFE

Tenerife is located in the central part of the Canary archipelago. The archipelago is a plume-related, approximately 800-km-long and 450-km-wide belt of seamounts and islands (see Fig. 1.11). Tenerife not only occupies a central position within the archipelago but also represents an intermediate evolutionary stage relative to the eastern and the western islands of the island chain. In contrast to the Hawaiian and most oceanic islands elsewhere, the Canaries show remarkable long-term island stability, with little overall island subsidence (see Fig. 1.15). Mass-wasting and gradual erosion eventually outpace volcanic growth and reduce the size of the islands until they are eroded to sea level (after ∼20 Myr, eg, Fuerteventura). The age-dependent ratio of subaerial to submarine volume in the Canary Islands increases from the youngest western to the oldest eastern islands. However, the increase is not linear, but rather shows a maximum in the central island of Tenerife, indicating that the western islands have not yet attained a mature stage of island growth, whereas the eastern islands are already in an advanced state of erosive decay. Teide, in turn, is currently at its peak of growth, at more than 3700 m in height (Fig. 5.1).

Tenerife is the largest (2034 km^2), highest (3718 meters above sea level (masl)), and most populated (0,88 million inhabitants, 5 million annual visitors in 2015) island of the Canaries. The island played a crucial role in the development of modern geology and volcanology, and the height of Mt. Teide was already a major scientific topic in the 18th century. In fact, Teide was considered to be the highest mountain on Earth until Mont Blanc and the Andean volcanoes were measured and observed to be higher. Notably, modern volcanology has recently reinstated Teide among the highest "volcanic" structures on the planet (only surpassed by Mauna Loa and Mauna Kea, on the island of Hawaii). If the base level of the edifice is taken to be the ocean floor and not sea level, then Mt. Teide rises more than 7000 m in height (3718 masl).

The role played by the Canaries and Mt. Teide in the historical development of volcanology changed with the arrival of the "era of naturalists" that included names such as Leopold von Buch, Charles Lyell, Alexander von Humboldt, and Georg Hartung, among others. During the 18th century, geology was at the center of a long-standing controversy between those who held the view that all rocks, including what we now see as volcanic rocks, were marine deposits formed by chemical

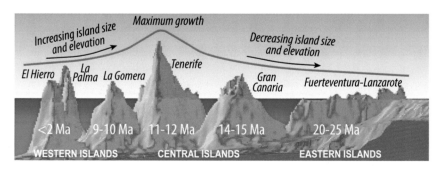

FIGURE 5.1

Computer-generated cross-section of the Canary Islands showing age versus height. At present, Tenerife represents the peak of evolutionary development in the Canaries because the western islands have not yet attained a mature stage of island growth, whereas the eastern islands are already in an advanced phase of erosional decay (from Carracedo et al., 1998).

precipitation in the oceans (Neptunists, after the god of the sea in Roman mythology) and those who believed that volcanic rocks resulted from the solidification of molten rock from the Earth's interior (Plutonists, after Pluto, Greek god of the underworld). The controversy was greatly advanced with the observations made in the Canaries, which favored Plutonism. To von Buch, for example, we owe the basic concept that minerals in lava form by magmatic crystallization and, in this respect, the islands have contributed decisively to the development of geology and volcanology as modern sciences.

MAIN GEOMORPHOLOGICAL FEATURES OF TENERIFE

The island's overall shape is that of a truncated tetrahedron, with its flat top formed by a partially filled caldera (Caldera de Las Cañadas) and a nested stratovolcano (Teide) forming the apex of the island. The three rift zones extend toward the northwest, northeast, and south (Fig. 5.2A) and further conspicuous features are two pronounced horseshoe-shaped embayments, the lateral collapse valleys of La Orotava and Güímar. In addition, two deeply eroded massifs outcrop at the northwest and northeast edges of the island, and a less apparent third one outcrops in the southwest, corresponding to the old (Miocene-Early Pliocene) shields of Teno, Anaga, and Roque del Conde. The latter is also widely referred to as the Central Shield.

The stratigraphic relationships between these main geological features are shown in Fig. 5.2B. The extensive and thick felsic (phonolitic) pyroclastic deposits covering the south and southeast flanks of the island are an interesting geological and geomorphological phenomenon. White pumice and ignimbrite outcrops form a characteristic geological landscape related to large explosive (plinian) postshield eruptions of the Las Cañadas Volcano (LCV). The color and morphology of these deposits, for example, in the Bandas del Sur, are in strong contrast with the mafic (dark-colored) volcanic landscape that otherwise dominates the island.

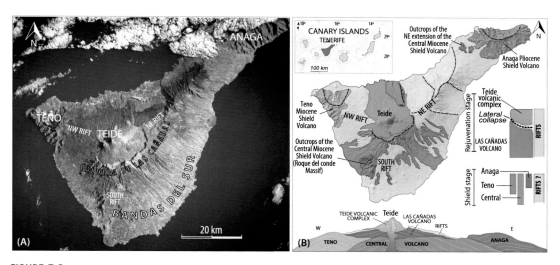

FIGURE 5.2

(A) Satellite image (courtesy of *NASA*) and (B) simplified geological map and cross-section of Tenerife showing the main structural and geological units (from Carracedo et al., 2007b).

GENERAL STRATIGRAPHY OF TENERIFE

The overall stratigraphy of Tenerife is based on field, geochronological, and palaeomagnetic data from outcropping formations and galerías (Fig. 5.2). The main volcano-stratigraphic units were initially defined as the Old Basaltic series, the Cañadas series, the trachytic-trachybasaltic series, the Basaltic series II, and the recent series (Fúster et al., 1968d). Later, Carracedo et al. (1998) proposed a simpler stratigraphy in accordance with the stages of development of ocean islands adopted for the Hawaiian Islands, such as the submarine, the shield, and the rejuvenation stages (Stearns, 1946; Walker, 1990). This simple division is complemented with the definition of the different volcanic complexes or episodes (eg, the Teide-Pico Viejo Volcanic Complex, TPV), the Anaga shield volcano, the Las Cañadas Volcano, etc.), allowing a relatively precise differentiation of the sequence of geological events that formed the island as we know it today.

THE MIO-PLIOCENE SHIELD VOLCANOES

The remains of the apparently disconnected Miocene—Pliocene shields crop out at the vertices of the island, which prompted an interpretation that Tenerife formed by several independent island-volcanoes (Fig. 5.3A) that later unified into a single edifice by postshield volcanism (Ancochea et al., 1990). Recent radiometric dating, geological mapping, isotope geochemistry, and observations of the deep structure of the central part of the island through galerías (Carracedo et al., 2007b, 2011b; Delcamp et al., 2010, 2012; Deegan et al., 2012) suggest that the island is composed of a main Miocene central shield with the Teno and Anaga shields superimposed on its flanks

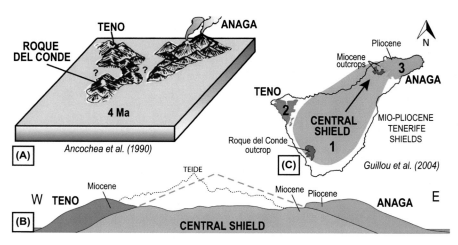

FIGURE 5.3

(A) Early evolutionary models for the island of Tenerife considered three Miocene shield volcanoes (after Ancochea et al., 1990). (B) and (C) More recent evolutionary models for Tenerife. The island was more likely dominated by the Miocene central shield, and the peripheral massifs of Teno and Anaga are later edifices that grew onto the flanks of the central shield (after Guillou et al., 2004; Carracedo et al., 2007b, 2011a).

(similar to the growth of Kilauea on the southern flank of Mauna Loa). The peripheral edifices thus probably grew after the central shield attained a critical altitude (Fig. 5.3B). This assumption is supported by age, isotopic and palaeomagnetic data, and, in particular, by outcrops adjacent to and underlying the Anaga massif. There, rocks similar to that of the central shield occur, suggesting an extension of the central shield volcano under the Anaga massif and further to the northeast (Guillou et al., 2004a; Carracedo et al., 2011b; Fig. 5.3C).

CENTRAL EXPLOSIVE VOLCANISM: THE LAS CAÑADAS VOLCANO

Eruptive activity of the central shield probably lasted from approximately 8 to 12 million years (Ma), after which a period of quiescence ensued until the onset of the rejuvenation stage that is marked by Las Cañadas Volcano (LCV) at approximately 3.5 Ma. The period of repose of the central shield coincided with the growth of the peripheral Teno and Anaga volcanoes (Carracedo, 1979; Ancochea et al., 1999). The corresponding erosive unconformity was likely removed by northbound massive landslides, explaining the absence of central Miocene formations in the northern flank of the island.

Long residence times of magmas during this period favored magmatic differentiation processes to become more significant in the center of Tenerife, which may explain the felsic and explosive nature of LCV volcanism. The Las Cañadas Volcano grew perched on top of the remnants of the central shield, perhaps attaining an altitude similar to that of the present Teide volcano (Fig. 5.4). The LCV then developed a summit caldera complex, which gave rise to sustained pyroclastic activity (Booth, 1973; Bryan et al., 1998; Brown et al., 2003; Edgar, 2003; Pittari et al., 2005). This prominent location and the prevailing winds that blow from the northeast (the trade winds) directed

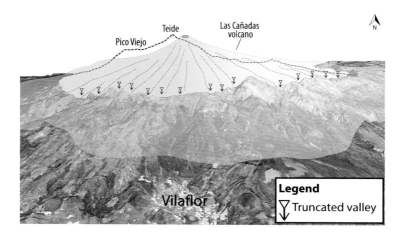

FIGURE 5.4

Schematic reconstruction of the pre-Teide Las Cañadas Volcano on a *Google Earth* image. Explosive (Plinian) volcanism mantled the southern flank of the island (the Bandas del Sur) with air-fall pumice and ignimbrites. The Teide and Pico Viejo volcanoes that developed during a later stage are shown for reference. During the progressive evolution of the Las Cañadas Volcano (LCV), the Las Cañadas Caldera (LCC) likely started to form as a vertical collapse caldera but was later accentuated by a large northbound landslide (the lateral Icod collapse).

ash and pumice fallout as well as ignimbrites from the successive plinian eruptions dominantly to the south and southeast slopes of the island, where they accumulated to form the spectacular Bandas del Sur region.

Eruptions in that epoch were very different from the characteristic basaltic volcanism of oceanic islands, which is predominant in the volcanic history of Tenerife. The largest of the LCV eruptions probably ejected up to 30 km^3 of tephra, likely in a style similar to observed caldera-forming eruptions, such as at Katmai in 1912 or Pinatubo in 1991 (Edgar et al., 2007), or like many of the Miocene ignimbrites of neighboring Gran Canaria (eg, Troll and Schmincke, 2002).

The LCV on Tenerife is one of a few intraplate ocean-island volcanoes with an extended eruptive history including major Plinian explosive events. A detailed study of the thick Bandas del Sur deposits in the past decades (eg, Booth, 1973; Bryan et al., 1998; Brown et al., 2003; Edgar, 2003; Pittari et al., 2005; Edgar et al., 2007; Dávila-Harris, 2009) documented the eruptive record and pyroclastic stratigraphy of the past 2 Ma by defining 18 to 20 soil-separated eruption units (Fig. 5.5). This succession indicates that the LCV underwent a series of major explosive eruptions over four main activity cycles, each of which involved successive ignimbrite-related caldera collapse and/or large sector collapse events, interspersed with basaltic volcanism from the coeval but noncentral rift zones.

Eruptions during each explosive cycle increased in volume, with the largest eruption occurring at the end of the cycle (Brown et al., 2003). In general, these eruptions started with a plinian phase followed by emplacement of voluminous nonwelded to highly welded ignimbrite deposits from pyroclastic density currents.

More than 10 major explosive eruptions vented moderately large volumes of phonolite magma during the past two cycles alone. Culminating each explosive cycle was the emplacement of a relatively large-volume (10 km^3) ignimbrite with coarse, vent-derived lithic breccias, interpreted to represent a major phase of caldera collapse in each case (Booth, 1973). In the extracaldera record, the main explosive cycles are separated by periods of nonexplosive activity characterized by erosion and remobilization of pyroclastic deposits and by local soil formation processes.

The current period of relatively low explosive activity is characterized by the construction of the Teide−Pico Viejo volcanic complex, which is nested within the Las Cañadas Caldera. It could therefore be that eruptive hiatuses in the older extracaldera record of the LCV may reflect effusive activity and stratovolcano building phases confined to within the Las Cañadas Caldera.

Several of these Quarternary explosive episodes are particularly well-preserved and studied, such as the Diego Hernández Formation (DHF) that erupted between 180 and 600 kiloannum (ka) ago (see Fig. 5.5), and represent the ultimate cycle of phonolitic explosive volcanism on Tenerife (eg, Bryan et al., 1998; Wolff et al., 2000; Pittari et al., 2006; Edgar et al., 2007). It includes the 273-ka caldera-forming Porís ignimbrite (Edgar et al., 2002; Brown and Branney, 2004) and the late Pleistocene (188 ka) Abrigo eruption, the last major caldera-forming eruption of the LCV and, in fact, on Tenerife overall (Pittari et al., 2006). The period of the LCV also includes the 735-ka Abona giant landslide associated with the Los Helechos eruption of the Lower Bandas del Sur Volcanic Group (Cycle 2), which has recently been identified (Dávila-Harris, 2009). Finally, a spectacular example of a felsic phreatomagmatic maar, the Caldera del Rey, formed near the southeast coast of Tenerife during this episode, showing superbly preserved base surge deposits, stratified lapilli tuff beds, and ballistic bombs with deep impact craters (Paradas and Fernández Santín, 1984).

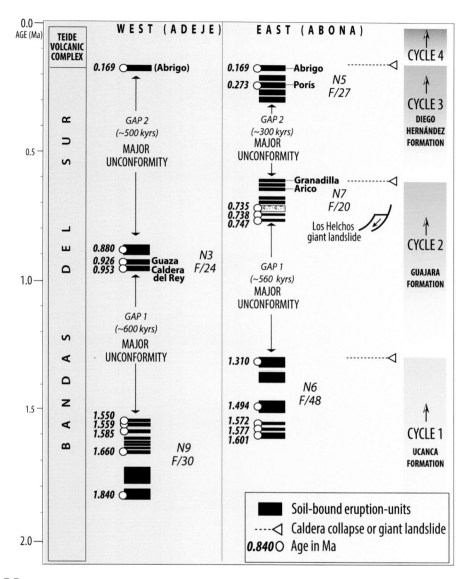

FIGURE 5.5

Generalized eruptive sequence diagram for pyroclastic units in the western (Adeje) and eastern (Abona) sectors of the Bandas del Sur. All ages (Ar/Ar) are from Brown et al. (2003) and Dávila-Harris (2009). Unit thicknesses are not to scale. N refers to number of eruptions in each cycle, and F refers to intervals between eruptions in kyrs. Individual eruptions mentioned in this figure are explained in more detail in the main text.

In contrast, explosivity has been very low in the present TPV complex when compared to the last (fourth) eruptive cycle of the LCV. Only one sub-plinian eruption of relatively low magnitude (Mña. Blanca, 2000 ka) has so far been identified in Teide's recent record (Ablay et al., 1995).

Although the explosive volcanic processes and deposits are very complex in detail, the most frequent type of explosive eruptions in the LCV were plinian events. These generally erupted as a high-speed jet of a hot pyroclast—gas mixture that formed a several-kilometer-high eruptive column that rose buoyantly into the atmosphere. There, it reached a neutral buoyancy level where the gas and particle plume spread out to form an umbrella region (Fig. 5.6A). The pyroclasts (ash and pumice) were then dispersed by high winds and fell to form tephra layers. As the eruption progressed, the initial momentum eventually wore out and the eruptive column collapsed. This spread more and more of the mixture of pyroclastic particles downward and sideways in the form of density currents, forming thick ignimbrite deposits when they came to rest at the surface (Fig. 5.6B,C).

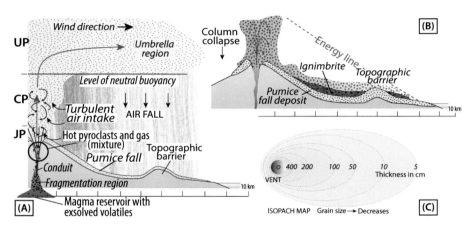

FIGURE 5.6

(A) Plinian eruptions usually generate large eruptive columns that are powered upward by the thrust of expanding hot gases, as dissolved gas in the magma exsolves and expands during ascent. Bulk density is greater than that of the atmosphere at the base of the plume, and the rise of the material is due to momentum obtained from gas exsolution and expansion (jet phase, JP in figure). When the bulk density becomes less than that of the surrounding atmosphere, rise of the plume is then promoted by buoyancy, where ascent is analogous to that of a hot air balloon (convective phase, CP), until a neutral level is reached from where the plume expands laterally (umbrella region, UP) and becomes dispersed by high winds (eg, the northeast—southwest trade winds) to produce ash and pumice-fall deposits. (B) The eruptive column will finally collapse as the eruption progresses and loses momentum, and the unsupported particle—gas suspension falls back and flows down-slope, forming pyroclastic density currents and, after deposition, initial ignimbrites. Therefore, the typical plinian eruption deposit will consist of pumice-fall layers overlain by ignimbrite and, finally, by a fine-grained layer of ash that has settled slowly after deposition of the ignimbrite (see, eg, Sparks, 1986). (C) Isopach maps, together with information on particle size, are widely used for analyzing past eruptions (ie, magnitude, height of the eruption column, and mass discharge rate).

THE SOUTHWEST SECTOR OF THE BANDAS DEL SUR (ADEJE)

A detailed study of the pyroclastic succession of the southwest flank of the LCV, around the town of *Adeje*, has been conducted by Huertas et al. (2002), and more recently by Dávila-Harris (2009). This latter author has been able to separate 11 eruption units, equivalent to one explosive event every 25 to 30 ka during the height of LCV activity.

This period of explosive activity of the LCV lasted from approximately 0.8 to 1.8 Ma and comprises large volumes of predominantly phonolitic ash, pumice, and ignimbrite deposits with a cumulative thickness that exceeds 300 m. The deeply eroded basaltic Miocene relief then allowed accumulation of a pyroclastic apron that extended around most of the southern flank of the island. This felsic, highly explosive volcanism, which is rather exceptional on oceanic volcanoes, represents the most violent and catastrophic period of the eruptive history of Tenerife, probably recurrently devastating animal and plant life in large parts of the island and, particularly, on its southwest and southeast flanks (see also chapter: The Geology of Gran Canaria).

THE ADEJE FORMATION (1.56 MA)

The "Adeje red ignimbrite" (Wolff, 1983) or Adeje-type *siena tostada* (burnt sienna ignimbrites) (Fúster et al., 1994) is a nonwelded to partly welded unit outcropping southwest of the town of *Adeje*. The dark orange-red and more intensely welded part of the ignimbrite displays fiamme (first named "eutaxites" by Fritsch and Reiss in 1868, from the Greek term for banded appearance). For these reasons, the rock has been traditionally quarried as an ornamental building stone in this part of Tenerife (Fig. 5.7).

FIGURE 5.7

The "*tosca colorada*" (orange-red tuff) near *Adeje*. The outcrop shows two cooling units of ignimbrite topped by a basaltic flow. Note that the rock is a popular building stone in the area due to its light porous character and attractive natural color.

The phonolitic Adeje formation formed during the oldest recorded ignimbrite-forming explosive episode of the LCV, at approximately 1.56 Ma (Dávila-Harris, 2009). The eruption began with an initial plinian phase that deposited ash and pumice across southwest Tenerife (Fig. 5.8A). The eruptive column then collapsed and generated pyroclastic flows, depositing an ignimbrite sheet that extended perhaps approximately 15 km into the ocean (Fig. 5.8B). As observed in similar recent explosive eruptions (eg, the Montserrat's Soufrière Hills volcano eruption in 1995), pyroclastic

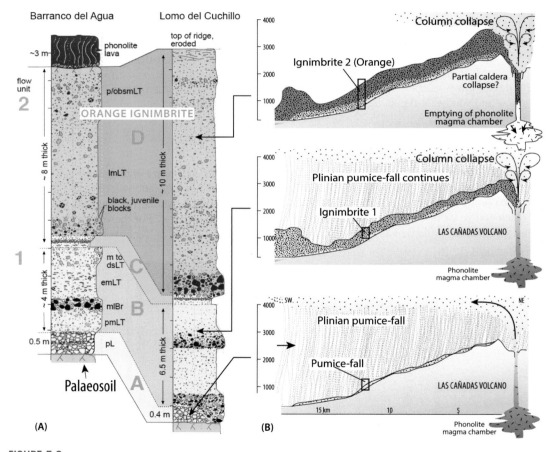

FIGURE 5.8

(A) The Adeje ignimbrite at Bco. del Agua and Lomo del Cuchillo in southwest Tenerife. The Adeje Formation comprises a Plinian pumice-fall deposit (layer A), followed by an ignimbrite flow unit (layer B). A 25-cm-thick layer of ash aggregates formed during a pause between ignimbrites B and D (layer C), with layer D being the characteristic orange to red massive lapilli tuff. (B) Schematic reconstruction of the main phases of the Adeje eruption. A plinian eruptive column deposited ash and phonolite pumice lapilli to approximately 1 m in thickness over southwest Tenerife at up to 10 km from the source. Collapse of the eruptive column deposited ignimbrite B. Finally, partial caldera collapse deposited the orange-red ignimbrite D (after Dávila-Harris, 2009).

flows on Tenerife may have traveled long distances over water after reaching the coast (eg, Freundt, 2003). A short pause, indicated by a thin, accretionary, lapilli-bearing pellet fallout layer, was followed by the climactic phase of the Adeje eruption and is recorded by a 10-m-thick ignimbrite deposit.

THE SE SECTOR OF BANDAS DEL SUR (ABONA)

The excellent record of explosive volcanism of the southwest sector of the Bandas del Sur is only surpassed in the southeast, with abundant, well-preserved sequences of cycles 2 and 3 of the LCV, including the most recent large plinian eruptions on the island.

THE ARICO IGNIMBRITE (0.66 MA)

The Arico ignimbrite (0.66 Ma, Brown et al., 2003) comprises a distinctive pumice fall overlain by an extensive orange to pale cream ignimbrite that contains a partly welded ash-rich section that is best exposed around the town of *Arico* (Fig. 5.9). The pumice fall layer comes from a rising plinian

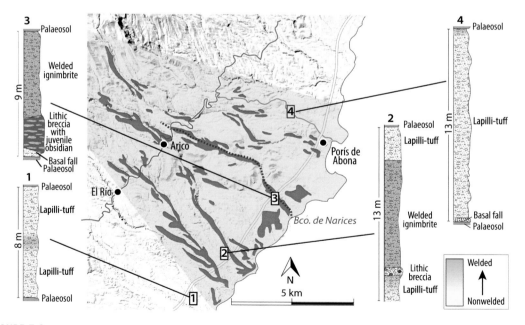

FIGURE 5.9

Extent (orange) and outcrops (red) of the Arico ignimbrite after Alonso et al. (1988). 1—4: Representative logs of the Arico formation showing lateral variations in depositional lithofacies and in welding intensity (after Brown et al., 2003).

column, represented here by a thin (5−10 cm) basal ash and pumice fall deposit (basal fall in the logs of Fig. 5.9). This fall layer is overlain by massive ignimbrite (ash flow) deposits from the collapse of the eruptive column (logs 3 and 4 in Fig. 5.9), which grade into lapilli-tuff at the edges of the ignimbrite deposit (logs 1 and 4) but form welded ignimbrite in the central part of the area (logs 2 and 3).

The apparent small volume of this pumice fall was considered to potentially represent a *nuée ardent* (glowing avalanche-type eruption) caused by the gravitational collapse of a phonolitic dome (Alonso et al., 1988). However, Brown et al. (2003) view the eruption as having a similar magnitude and style as the other major explosive events recorded in the Bandas del Sur.

Most of the outcrops of the Arico ignimbrite consist of nonwelded lapilli tuffs. Outcrops of intensely welded ignimbrite are located in Bco. de Narices (Fig. 5.10A), near the position of stratigraphic column 3 in Fig. 5.9. Here, the ignimbrite shows a bright orange color due to oxidation, whereas the eroded lower part of the unit, incised by the barranco, shows the original pale cream color (Fig. 5.10A).

Another characteristic features of the Arico ignimbrite is the pentagonal columnar jointed outcrops of eutaxitic ignimbrite locally observed (Fig. 5.10B), such as the parts of welded ignimbrite with fiamme (eutaxites). These are very hard and so strongly welded that the rock breaks with transgranular fractures (eg, through the lithics; Fig. 5.10C).

Several quarries (eg, at *Cantera Guama*) commercialize this ignimbrite (locally referred to as *piedra chasnera*) for use as building and ornament stone (Fig. 5.10D). The Arico ignimbrite is and has been a popular building stone on Tenerife for many centuries and is frequently used for walls and paving (Fig. 5.10E). This can be seen in many historical edifices, such as, for example, at the 17th-century main church of *Arico* village (Fig. 5.10F).

THE GRANADILLA PLINIAN ERUPTION (0.57 TO 0.6 MA)

The second cycle of eruptive activity of the LCV culminated with the large plinian Granadilla eruption, dated at 0.57 Ma (Bryan et al., 1998) to 0.60 Ma (Brown et al., 2003). This eruption produced one of the most extensive and archetypical phonolitic pumice deposits on Tenerife (Fig. 5.11), covering more than 800 km^2 (Booth, 1973).

The Granadilla member comprises a plinian fall deposit at the base (Fig. 5.12), the main Granadilla pumice (approximately 35 km^3), and the overlying ash-flow deposits of up to 30 m in thickness (approximately 5.5 km^3; Booth, 1973). The pumice spread more than 500 km^2 by prevailing northwesterly winds (Bryan et al., 2000), with a maximum thickness in the area of *Granadilla de Abona* (see Fig. 5.11). The dispersal area is consistent with the particles being transported and released from a plinian eruption column up to 25 km high and with the umbrella cloud elongated toward the southeast of the island (see Fig. 5.11).

Bedding and grain size stratification of fall units (U_1 and U_3 in Fig. 5.12A) seem consistent with an unsteady, probably pulsatory, plinian column, whereas the thin unit U_2, the "cement-like" band of Booth (1973), likely represents a short pulse of phreatic activity. Fall unit U_4, 10 m thick in places, is the main component of the Granadilla pumice (Fig. 5.12B). An estimate of the energy of this eruption is provided by the presence of very large (up to 20 cm) pumice fragments dispersed at distances of more than 15 km from the vent.

FIGURE 5.10

Different aspects of the Arico ignimbrite. (A) Thick ignimbrite (ash flow) deposit in Bco. de Narices. The red color at the top is due to oxidation of the late fallen ash (top of ignimbrite deposit). The pale cream color in the eroded lower part of the outcrop marks the original color of the materials that have been deposited from a hot and fast-moving ash cloud. (B) Part of the welded ignimbrite shows regular prismatic columns that formed on cooling, like sometimes seen in basaltic lava. Inset: Detail of the columns. (C) The Arico ignimbrite is in places so intensely welded that it breaks with transgranular fractures (ie, through the former lithics). (D) Quarry in the Arico ignimbrite at *Guama*. (E) Details of Arico welded ignimbrites with fiamme texture seen on the polished slabs. (F) The church of *Arico* village (17th century) was built from Arico ignimbrite.

Finally, collapse of the plinian column (probably following progressive widening of the vent and decreasing volatile content from the evacuation during U$_3$) led to the generation of pyroclastic flows that emplaced the nonwelded Granadilla ignimbrite on top of the successive fall units.

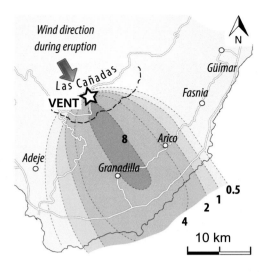

FIGURE 5.11

Isopach maps of the Granadilla pumice. Isopach contours are given in meters (after Booth, 1973).

FIGURE 5.12

(A) The air-fall pumice of the phonolitic 0.57 Ma Granadilla plinian eruption is divided into several stratigraphic units comprising four main fall units (U1 to U4) and a thin ash layer (U2) described as cement-like by Booth (1973). The latter forms a widespread marker horizon in the region. Plots at the left indicate the frequency curves (weight percent retained by each sieve grade against sieve aperture) for the ignimbrite and the air-fall pumice for the Granadilla deposit, showing a higher ratio of pumice to lithics in the former. (B) The Granadilla pumice (modified after Booth, 1973).

A minimum volume estimate of the preserved Granadilla ignimbrite is 5 km^3, but according to calculations of Bryan et al. (2000), the total erupted volume may have been up to five-times greater, especially when accounting for the part of the eruption that possibly went into the sea.

THE DIEGO HERNÁNDEZ FORMATION

The third complete cycle of volcanism of the LCV (0.37−0.18 Ma) is represented by the Diego Hernández Formation (DHF) (Edgar et al., 2007). The fourth cycle is the latest cycle and is represented by Teide volcano and the active rifts.

The DHF (yellow in Fig. 5.13) shows a maximum thickness of approximately 250 m in the southeast of the island. The formation is composed of plinian fall and ignimbrite deposits erupted from a vent located south of La Fortaleza, the northeast rim of the Las Cañadas Caldera, and is interbedded with basaltic volcanics from the northeast rift. Although most of the proximal sequences were removed by the subsequent northbound Las Cañadas giant landslide, the succession crops out in the eastern rim of the Las Cañadas Caldera. The more distal pyroclastic sequences accumulated on the flank of the island (eg, Bandas del Sur, see also above), where they appear in discrete packages separated by palaeosoils and erosional unconformities. This observation indicates that alternations of intense eruptions and periods of repose and erosion were common at that time.

FIGURE 5.13

Map of Tenerife showing the present-day distribution of the Diego Hernández Formation (sand color) and the Porís eruption (red). Las Cañadas volcanism prior to the Diego Hernández Formation is shown in gray (Guajara and Ucanca cycles) (modified after Edgar et al., 2002).

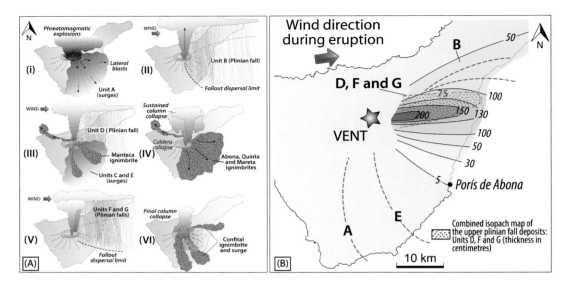

FIGURE 5.14

(A) Schematic reconstruction of the main phases and likely vent site of the Porís eruption. (B) Isopach map of the successive Plinian fall units of the Porís eruption (thickness in cm). Dashed green lines indicate the approximate extent of surges A and E (modified after Edgar et al., 2002).

THE PORÍS ERUPTION 273 KA

The best preserved, most voluminous, and most widespread sequence of this cycle corresponds to the 273-ka caldera-forming Porís eruption, occupying the southeast apron of Tenerife (red in Fig. 5.13). Its total tephra volume was estimated by Edgar et al. (2002) to between 13 and 14 km^3 (~ 4 km^3 DRE). The authors also defined two surge deposits related to phreatomagmatic events, as well as five plinian fall units and five major ignimbrites.

Successive lateral explosions that swept the southeast slopes of the island left pyroclastic surge deposits (Fig. 5.14A), with a maximum combined thickness of 20 cm (Edgar et al., 2002). Interbedded with these surges appears a 2-m bed of plinian fall pumice (Fig. 5.14B), dispersed mainly toward the Güímar valley. Subsequent plinian falls then also reached the Güímar valley.

The five ignimbrite "packages" of the Porís eruption appear interbedded with plinian fall pumice layers, except for the last one (the Confital ignimbrite) that overlies the entire sequence (Fig. 5.15). The first ignimbrite on the southeast and north aprons (Manteca) reached the northwest coast, near *Los Silos* (Fig. 5.15A). The subsequent ignimbrites (eg, the Porís eruption) are only seen on the southeast flank and coast, except the Abona and Mareta ignimbrites. These are also recorded along the north flanks, reaching the north coast west of the Orotava Valley (eg, near the town of *Los Realejos*) (Fig. 5.15B—E).

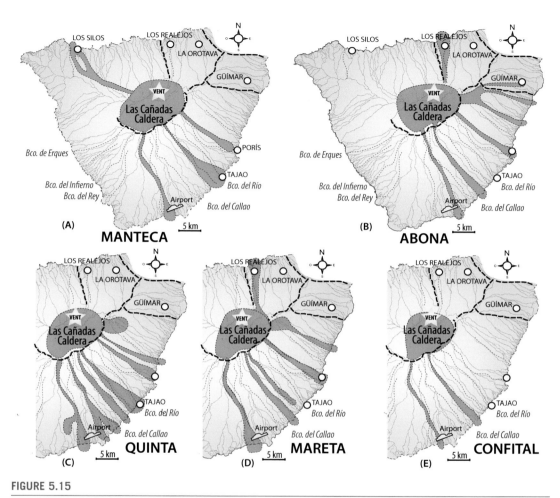

FIGURE 5.15

Distribution maps of the five ignimbrite pulses of the Porís eruption: (A) Manteca, (B) Abona, (C) Quinta, (D) Mareta, and (E) Confital (thickness in meters) (from Edgar et al., 2002).

THE ABRIGO ERUPTION (188 KA)

The last major caldera-forming event on Tenerife is the El Abrigo eruption, dated at 130 ± 20 ka (Ancochea et al., 1990), 168 ± 2 ka (Brown et al., 2003), 179 ± 90 ka (Mitjavila and Villa, 1993), or 188 ± 19 ka (Pittari et al., 2006). This eruption, the uppermost unit of the DHF (see Fig. 5.5), likely covered the entire island of Tenerife except Anaga (Fig. 5.16A). The Abrigo event immediately preceded the growth of Teide within the Las Cañadas Caldera.

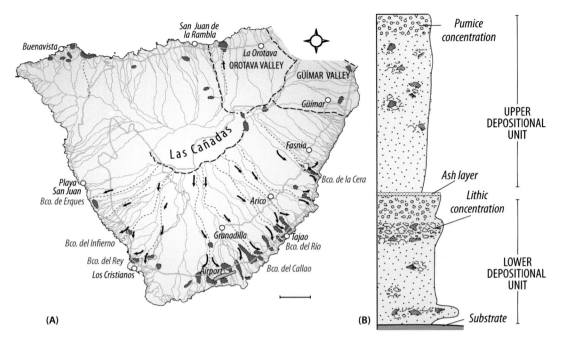

FIGURE 5.16

(A) Map of central Tenerife showing outcrops of the Abrigo ignimbrite, which are generally associated with major radial barrancos. Note that this eruption affected the entire central part of the island. (B) Simplified composite stratigraphy of the El Abrigo ignimbrite (after Pittari et al., 2006).

This eruption has been studied in detail by Nichols (2001), Pittari and Cas (2004), and Martí et al. (2008). Opinions are divided among workers whether the initial phase was a plinian fall or whether the entire eruption was purely ignimbritic. Although the Abrigo eruption may have begun with a short-lived plinian event, the bulk of the eruption is certainly formed by ignimbrites. Pyroclastic flows erupted from a vent system, likely inside the area later occupied by the present-day Las Cañadas Caldera, and spread radially onto the flanks of the Las Cañadas Volcano. Notably, these flows were mainly channeled along major barranco systems but fanned out at lower altitudes (Fig. 5.16A).

On the Bandas del Sur, the Abrigo ignimbrite shows two main massive lithic-rich depositional units (Fig. 5.16B). The lower of the two contains discontinuous cobble- to boulder-sized lithics, whereas the upper unit is lithic-poor and is frequently eroded from outcrops.

At the time of the eruption, the magma was undersaturated in water and, therefore, volatile overpressure was unlikely to have been the trigger mechanism for the eruption (Martí et al., 2008). In turn, intrusion of fresh basaltic magma at the bottom of the felsic magma chamber probably triggered the event (Fig. 5.17) to produce the massive El Abrigo ignimbrite that is probably the most widespread deposit of this type on Tenerife. Its minimum onshore volume is 1.8 km^3 (with up to 0.74 km^3 of lithics), and a maximum thickness of at least 25 m is recorded (Pittari et al., 2006).

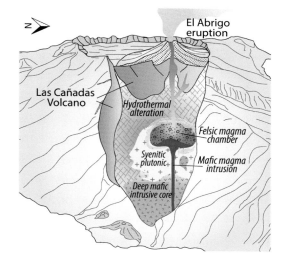

FIGURE 5.17

Interpretation of the active volcanic system at the time of the El Abrigo eruption. The eruption was likely triggered by an intrusion of fresh basaltic magma at the bottom of the felsic magma chamber (after Pittari et al., 2008).

EXTRACALDERA FELSIC ERUPTIONS

The vents of plinian explosive felsic eruptions on Tenerife were located at the summit area of the LCV, in the region of the present Las Cañadas Caldera, with the exceptions of Montaña de Taco at the northwest coast and the Caldera del Rey and Guaza eruptions near the south coast. These off-center events occurred during the second cycle of LCV activity (Mña de Taco, 706 ka, Carracedo et al., 2007b; Mña. de Guaza, 926 ka, Carracedo et al., 2007b; and Caldera del Rey, 953 ka, Dávila-Harris, 2009).

CALDERA DEL REY (953 KA)

This spectacular double-cratered trachy-phonolite tuff ring was investigated by Paradas Herrero and Fernández Santín (1984) and by Huertas et al. (2002), who dated this eruption at 1.13 Ma. Dávila-Harris (2009) conducted a detailed study of this monogenetic volcano, including $^{40}Ar/^{39}Ar$ dates from sanidine-bearing juvenile pumices, which yielded an isochron age of 953 ka. The result is consistent with the age of the neighboring and overlying Montaña Guaza phonolite dome, which was dated by K/Ar to 926 ka (Carracedo et al., 2007b).

The Caldera del Rey eruption comprised magmatic and phreatomagmatic events, forming a succession of stratified ignimbrites and thick pumice-lapilli fall layers exposed around *Los Cristianos* and *Las Americas* in southwest Tenerife (Fig. 5.18A). Thin distal deposits of the Caldera del Rey eruptives have been found to occur several kilometers from the source (eg, in the southeast, near Montaña de Guaza).

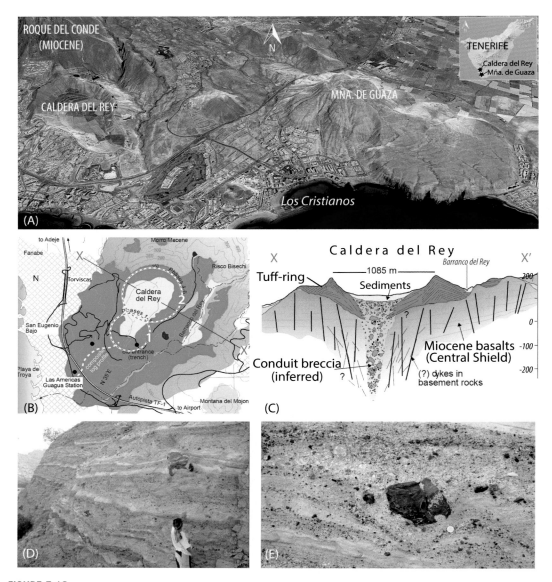

FIGURE 5.18

(A) Oblique *Google Earth* view of the south of Tenerife. There, extracaldera felsic volcanoes that were active during the second cycle of Las Cañadas activity are exposed, including the phreatomagmatic tuff ring of Caldera del Rey and the lava dome of Guaza, west and east of *Los Cristianos*. (B) Geological sketch of the 953-ka double-cratered trachy-phonolite tuff ring of Caldera del Rey on the southwest coast of Tenerife. (C) Northwest—southeast geological cross-section of Caldera del Rey (after Dávila-Harris, 2009). (D and E) High-energy bedding structures and block-impact sags from balistic ejecta in the wall of the Caldera del Rey tuff ring.

Notably, the Caldera del Rey edifice rests unconformably on basalt lavas of the central (Roque del Conde) Miocene shield (Guillou et al., 2004a) and consists of a shallow tuff ring formed by two overlapping explosion craters (Fig. 5.18B,C). The craters show diameters of approximately 675 m and 1100 m, respectively, producing a structure with the plan view of an "8" (Paradas Herrero and Fernández Santín, 1984). At the walls and rim of the craters, explosive products such as lithic-rich, stratified, and cross-stratified tuffs with abundant accretionary lapilli are well exposed (Fig. 5.18D,E), except in the northern rim, where the Miocene basalts dominate.

The generalized eruptive sequence of the LCV (see Fig. 5.5) shows a long period without ignimbrite eruptions, approximately 550 ka in southwest Tenerife, in which activity was reduced to effusive and scoria cone eruptions, mainly related to the south rift zone of the island. Explosive pyroclastic activity also resumed in the southwest coast of Tenerife at that time with the Caldera del Rey phreatomagmatic eruption (eg, Bryan et al., 1998).

According to Dávila-Harris (2009), the Caldera del Rey eruption likely started with a trachy-phonolite intrusion causing small initial phreatomagmatic explosions that were followed by an intense and sustained eruption that produced thick proximal pumice-fall deposits. Increasing magma−groundwater interaction gave rise to rhythmic phreatomagmatic explosivity generating block-impact structures from balistic ejecta and high-energy bedding features (Fig. 5.18E). These phases correspond to the southern tuff ring. The next three phases relate to the north shift of the eruption to form the northern tuff ring that followed a pattern similar to the previous phases but involved more water. It is likely that a small lake might have existed, or a fractured basement aquifer (Paradas Herrero and Fernández Santín, 1984). Increasing water-to-magma ratio reduced the conversion efficiency of thermal energy to explosive mechanical energy and thus changed the eruption to a "wet and cool" surtseyan style, with coarse lithic ejecta frequently impacting into moist ash. The eruption ended with a highly explosive but "dry" event that generated the most widely dispersed deposit with a thin ignimbrite and ash-fall sequence preserved more than 10 km from the vent.

MONTAÑA DE GUAZA (926 KA)

This spectacular lava dome and the Caldera del Rey tuff ring developed in the area of the south rift zone of Tenerife. Mña. de Guaza is a 425-m-high cone made of trachyte tuffs and breccia from an initial explosive stage that is seen to overlay the Caldera del Rey deposits. The construction of the cone ended with an effusive period in which a 100-m-thick coulée extended to the south, forming the prominent cliffs east of *Los Cristianos* (Fig. 5.19). The flow yielded a K-Ar age of 0.926 ± 0.02 Ma (Carracedo et al., 2007b).

Another felsic eruption coeval with the second cycle of activity of the LCV is seen at Montaña de Taco at the northwest coast of the island (near the town of *Buenavista*). That eruption has been dated by K/Ar to 706 ± 15 Ma (Carracedo et al., 2007b). It comprises a 315-m-high cone of trachyte tuffs with some features indicative of water−magma interaction (eg, palagonite and accretionary lapilli), and a thick trachyte lava flow can be seen to extend from there to the coast.

Interestingly, the topography profiles of Caldera del Rey and Mña. de Guaza show considerable differences in morphometric parameters (Fig. 5.20A). They plot in the tuff cone and cinder cone fields, respectively, as defined in a cone height relative to crater diameter diagram (Hco/Wcr), which is especially clear when the characteristic cones and calderas of the Canary Islands are plotted also (Fig. 5.20B, after Carracedo, 1984).

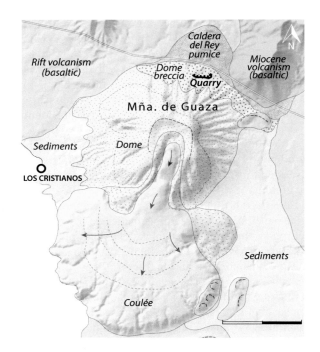

FIGURE 5.19

Simplified geological map of the trachyte lava dome of Mña. de Guaza dated at 926 ka (after Carracedo et al., 2007b). The dome displays a thick (100 m) coulée east of *Los Cristianos*.

RIFT ZONES ON TENERIFE

Felsic explosive volcanism, as described, is rather exceptional in ocean islands, and basaltic volcanism, usually in the form of fissure eruptions, is commonly predominant. A frequent architectural pattern derived from the grouping of basaltic eruptive centers is the 120° spaced three-armed rift zone arrangement (Fig. 5.21). There, numerous surface eruptions are fed via extensive dyke systems at depth (eg, Carracedo, 1994; Delcamp et al., 2010, 2012; Carracedo et al., 2011b).

Following detailed studies on Hawaii and Tenerife, rift zones have been shown to be one of the most pronounced and persistent structures in the development of oceanic volcanic islands, and may even be an invariable characteristic of ocean volcanoes. They control the construction of the insular edifices, possibly from the initial stage, that form the main relief features (shape and topography), concentrate eruptive activity, and frequently play a key role in the generation of flank collapses and the catastrophic disruption of established volcano plumbing systems.

The rift zones of Tenerife were first deduced by MacFarlane and Ridley (1968) from conspicuous gravity ridges in the Bouguer anomaly map of Tenerife, explained as a high concentration of dykes (see Fig. 1.24). The subsequent ground study of rift zones took advantage of the numerous

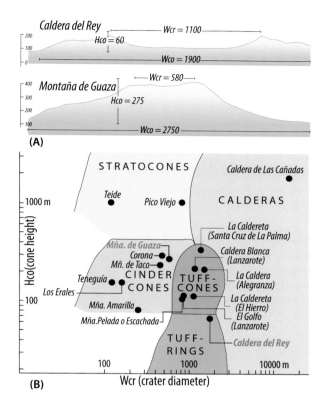

FIGURE 5.20

(A) Morphometric parameters (in cross-section) for Caldera del Rey and Mña. de Guaza. Wco: cone base diameter; Wcr: crater diameter; Hco: cone height. (B) Cone height versus crater diameter for characteristic cones and calderas of the Canary Islands (after Carracedo, 1984 and Urgelés et al., 2001).

water tunnels on Tenerife used for groundwater mining (locally called *galerías*). These are 2×2 m tunnels, usually several kilometers long, with a combined length for the entire island that exceeds 1600 km. These galerías facilitate access to the deep structure of the rift zones (Fig. 5.22A), providing a unique opportunity for direct observations and sampling, which facilitate improved understanding and quantitative assessment (Carracedo, 1994; Delcamp et al., 2010, 2012; Carracedo et al., 2011b; Deegan et al., 2012).

Rifts in ocean island settings can represent the surface expression of initial plume-related fracturing in response to vertical upward loading. Up-doming magma pressure causes a swell (Fig. 5.23A), eventual rupture (Fig. 5.23B), and consequent injection of blade-like dykes (Fig. 5.23C). The result is a narrow, dense swarm of parallel or sub-parallel dykes and related collinear clustering of emission centers at the surface that pile up along narrow dorsal ridges (Fig. 5.23D) (Carracedo, 1994; Carracedo et al., 2011b).

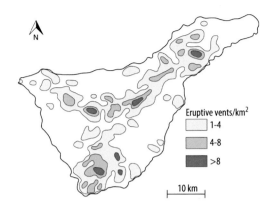

FIGURE 5.21

The clustering of Quaternary eruptive vents and centers in Tenerife form a 120° three-armed rift system, the classical "Mercedes-star" pattern also seen on many other oceanic islands (eg, El Hierro and Mauna Kea, Hawaii) (from Carracedo, 1994). Rift zones have been interpreted as regional fractures associated with the Atlas tectonics or the oceanic plate (Anguita and Hernán, 1975; Geyer and Martí, 2010), but no systematic regional or oceanic fractures have yet been demonstrated that can explain the phenomenon (see Carracedo, 1994, 1996a; Krastel and Schmincke, 2002a,b; Carracedo et al., 2011a, 2011b).

Although endogenously driven mechanisms are thought to play a major role in establishing these axial volcano architectures, an alternative model relates these rifts to extensional fissures due to volcano instability and spreading, which develop once a volcano has grown to a certain height and instability (Walter and Troll, 2003; Walter et al., 2005; Delcamp et al., 2010, 2012).

At present, there is no definitive evidence in favor of either of these models (ie, endogenously driven least effort fracturing or rifting by spreading and creeping of volcano flanks). Both mechanisms, although very different at the start, provide similar results. A plausible assumption is that large, deep, triple-armed rift zones develop at the early stages of island construction by plume-related up-doming and fracturing, with later modifications due to volcano edifice stability issues, whereas smaller rift systems (not necessarily multiple) might form entirely from gravitational spreading and associated structural rearrangements in large unstable volcanoes (Carracedo and Troll, 2013).

Rift zones are exceptionally well represented on Tenerife, particularly in two examples: the northeast and northwest rift zones. The northeast rift zone (NERZ) has been inactive for more than 100 kyrs along most of its length. The resulting erosion, combined with the effects of two giant landslide scarps that show only scant subsequent fill-in, allows an in-depth study of the rift's internal structure, including the complex network of dykes exposed (Delcamp et al., 2010, 2012; Carracedo et al., 2011b; Deegan et al., 2012). In contrast, the northwest rift zone has been highly active during the Holocene (eg, Carracedo et al., 2007b; Wiesmaier et al., 2011, 2013). Consequently, it serves as a record of the spatial and temporal distribution of Tenerife's activity up to the recent volcanism and allows monitoring of geochemical and petrological variations in that time window (Deegan et al., 2012; Carracedo and Troll, 2013; Wiesmaier et al., 2013).

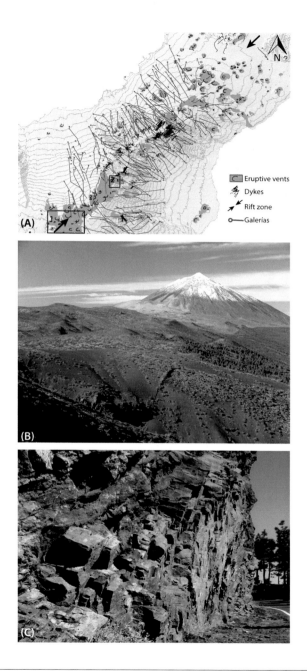

FIGURE 5.22

Northeast rift zone of Tenerife (NERZ). (A) Map of the NERZ indicating the concentration of eruptive vents, the orientation of dyke swarms along the axis of the rift, and the numerous existing galerías that cross the rift zone. (B) Cluster of volcanic cones in the recent part of the NERZ (box in A). (C) The central and intensely eroded part of the NERZ shows dykes that are approximately in parallel with the axis of the rift (small box in A).

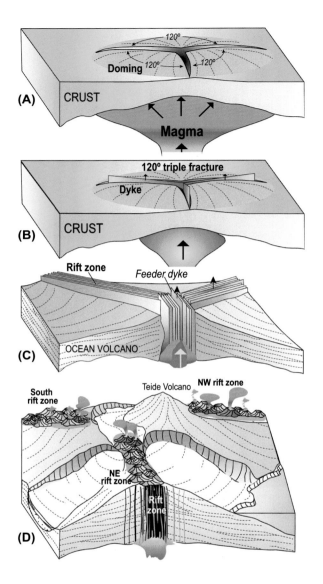

FIGURE 5.23

Three-armed rifts, spaced at 120°, seem to be the naturally preferred configuration as a response to least-effort fracturing associated with magmatic up-doming (after Carracedo, 1994; Carracedo and Troll, 2013). The genesis of Canarian rifts has been related to crustal tectonic structures that were present in the basement on which the islands have grown. Therefore, this model is based on structures that cut through the lithosphere to cause and control the location of the volcanism. These "tectonic" models were largely influenced by the geophysical (seismic and gravity) investigations of Bosshard and Macfarlane (1970), explaining the origin of the Canaries in association with fracture zones connected with tectonic movements that took place on the neighboring African continent. However, no evidence has been found to prove the existence of any fault connecting the Atlas with the Canaries in any detailed geophysical studies of the area (Martínez and Buitrago, 2002). Moreover, the regional fracture model has severe difficulties accounting for the large volumes of magma required to develop the Canary Volcanic province (McKenzie and Bickle, 1988; White and McKenzie, 1989). See text for further details.

THE NORTHEAST RIFT ZONE

The northeast rift zone of Tenerife (NERZ) represents a superb example of the entire cycle of activity of an oceanic rift system (Fig. 5.24). The evolution of the NERZ includes a period of very fast growth toward instability (between approximately 0.83 and 1.1 Ma), followed by three successive large landslides: the Micheque and Güímar collapses, which both occurred approximately at 830 ka

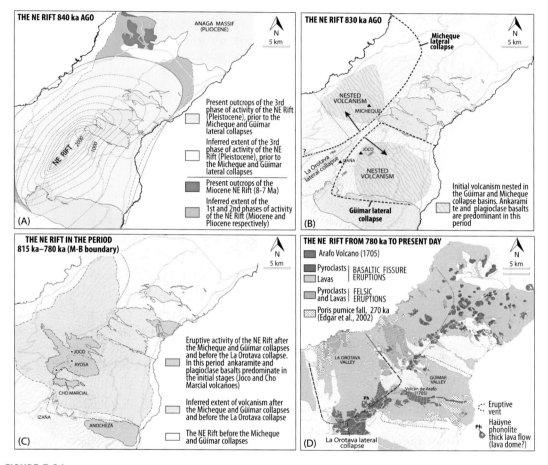

FIGURE 5.24

Different stages of development of the northeast rift zone (NERZ). (A) The initial NERZ edifice grew discordantly onto a southwest extension of the Miocene central shield (Guillou et al., 2004a). This early NERZ developed into an overly steepened and unstable ridge that went through successive lateral collapses and subsequent refilling events of the resultant collapse embayments (B and C). In the final stage (D), rift activity resumed with basaltic fissure eruptions, with the latest one in 1705 AD from Volcán de Arafo in the Güímar valley (from Carracedo et al., 2011 and Carracedo and Troll, 2013).

but on either side of the rift, and the La Orotava landslide, which occurred between 566 ± 13 and 690 ± 10 ka.

The eruptive and structural history of the NERZ comprises a complex succession of constructive stages. A Miocene basement to the rift, probably formed by a northeast extension of the central shield, crops out at the northeast edge of the NERZ and was dated to Upper Miocene (8.05 Ma, Thirlwall et al., 2000; 7.27 Ma, Carracedo et al., 2011b). This Miocene formation is overlain by the Pliocene western flank of the Anaga massif (Fig. 5.24A). The NERZ then had a second constructive phase in the Upper Pleistocene at approximately 2.7 Ma (Carracedo et al., 2011b) that can be accessed through galerías (see Fig. 5.25). The third and latest volcanic phase of the NERZ started approximately 840 ka ago with the construction of a collinear rift zone that may have reached an altitude of 2000 masl. This stage only crops out on the northeast flank of the ridge (Fig. 5.25).

The northeast flank of the ridge rapidly progressed toward critical instability and collapsed approximately 830 ka ago (the Micheque landslide, ~60 km^3), which probably immediately followed as a response to the earlier collapse by the massive Güímar slide on the southeast flank (~47 km^3; Carracedo et al., 2011b; Delcamp et al., 2012) (see Fig. 5.24A,B). Both collapse basins were immediately refilled by subsequent volcanism. The former was refilled almost completely, and the fill extends beyond the coastline, thus concealing the scarp and the avalanche breccia of the Micheque slide. The latter Güímar slide scarp was only partially filled, however, and much of the collapse wall is still preserved today (Figs. 5.24C,D and 5.25A,B). Rift activity resumed and the NERZ underwent another (the latest) lateral gravitational collapse that formed the "Valle de La Orotava," with an estimated volume of 57 km^3 (Fig. 5.24D). Precise dating of the Micheque and Güímar collapses has not been feasible, and the time of occurrence of the Orotava landslide has only been constrained between 566 ± 13 and 690 ± 10 ka (Carracedo et al., 2011b), placing it after the Micheque and Güímar landslides (see Figs. 5.24D and 5.25C). Following these three collapses, the rift entered into a stage of stabilization with progressively decreasing eruptive activity (Delcamp et al., 2010). Simultaneously, the eruptive centers, previously grouped preferentially on the crest of the rift, increasingly dispersed, particularly at the distal northeast end of the rift zone (see Fig. 5.24D).

The northeast rift zone currently represents a volume of approximately 510 km^3, but may have reached a maximum volume of approximately 615 km^3 before the collapses (between 840 and 550 ka). Growth rates have been assessed in the north scarp of the Güímar Valley. This section of 500 m of lava flows, dated in detail using geomagnetic reversals and radiometric ages, erupted between 806 ± 18 ka and 929 ± 20 ka, yielding a growth rate of more than 4 m/kyr (Carracedo et al., 2011b; Kissel et al., 2014). The highest eruptive rates, however, are those determined for the initial fill of the Micheque landslide basin of the first lateral collapse. This was estimated in Los Dornajos galería, which extends approximately 2800 m into the fill-in sequence until terminating at the actual avalanche breccia (Fig. 5.25B). The ages obtained at the beginning and end of the initial fill-in sequence of the landslide basin in this galería yielded a growth rate of approximately 12 m/kyr. This is the highest observed rate in the Canaries so far, but it is actually comparable in magnitude to the rates reported for other nested volcanoes, such as Bejenado edifice on La Palma, which grew up to 600 m in approximately 50 Thousand years (kyr) (Carracedo et al., 2001), or eruptions during major shield stages, like those of Gran Canaria (Schmincke, 1993).

The reconstruction of the volcanic history of the NERZ and observations on rift zones of other islands in the Canaries outline notable common characteristics. Rifts are recurrent features

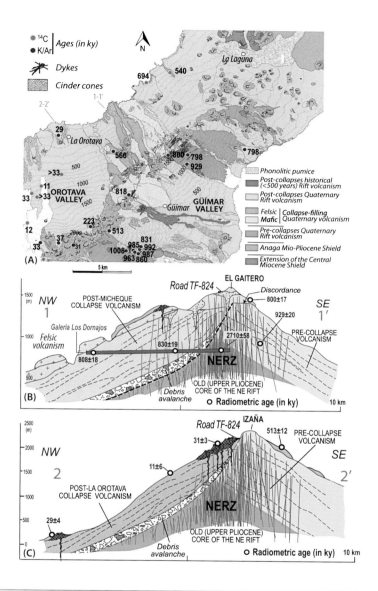

FIGURE 5.25

(A) Geological map of the NERZ of Tenerife (simplified from Carracedo et al., 2011a, GSA data repository item 2011011). (B) Geological cross-sections of the northeast rift zone (after Carracedo et al., 2011a). The Pliocene core of the rift has been dated using samples from the galería *Los Dornajos*. Note that the northwest flanks in both sections show a partially filled lateral collapse. These represent successive landslides rather than a single event (the Micheque and the La Orotava giant landslides at >800 and at ~560 kyrs).

that show cyclic behavior with a broad sequence of events: (1) growth; (2) instability; (3) flank collapse; (4) nested volcanism; and (5) eruptive decline and dispersion. The duration of these cycles is typically approximately 500 kyr, whereas nested volcanism generally lasts for 150 to 300 kyr (Carracedo et al., 2011b). Moreover, although felsic volcanic complexes in the Canaries may originate from a variety of processes, a considerable volume of differentiated volcanism in the Canaries appears to be associated with rift flank collapses. Landslides are usually followed by abundant and prolonged nested volcanism that evolves from initially mafic post-collapse magmatism to terminally felsic compositions in response to lateral collapses. A collapse implies disruption of the established feeding system of a rift, which initially allows for dense mafic magmas to ascend to the surface by edifice unloading and is often followed by magmatic differentiation (eg, Longpré et al., 2009; Manconi et al., 2009).

THE NORTHWEST DORSAL RIDGE

Most of the Holocene basaltic volcanism on Tenerife is located in the northwest side of the island, between the west flank of Pico Viejo and the Miocene Teno massif. The northwest rift zone (NWRZ) was already active prior to the 200-ka lateral collapse that removed part of the Las Cañadas Volcano's superstructure and during the past 30 ka comprised eruptions of Pico Viejo lavas and the most recent eruptive products on the island, the 1909 eruption. Precollapse volcanics from this rift zone are, however, only exposed at the most distal westernmost edge near the Teno massif and can be inspected inside a small number of galerías.

The recent volcanism along the NWRZ, in turn, forms a characteristic ocean island rift zone alignment, with the eruptive centers clustering along the crest in a northwest−southeast direction. The lava flows of the NWRZ drape both flanks of the ridge (Fig. 5.26). Relatively frequent eruptions occurred along this rift during the entire Holocene, including four historical eruptions (1−4 in Fig. 5.27A): Boca Cangrejo (1492 AD); Garachico (1706 AD); Chahorra (1798 AD); and Chinyero (1909 AD). The eruption of Montaña Reventada (see *Stop 3.5*) dates back to approximately 895 ka AD (5 in Fig. 5.27A) and shows an extraordinary example of magma mixing within a single lava unit in the transition zone between the dominantly basaltic NWRZ and the evolved TPV central edifice (eg, Wiesmaier et al., 2011).

The eruptions of the NWRZ actually form two distinct volcanic chains: the Chío chain and the Garachico chain, along with a saddle between them (Fig. 5.27). This feature is crucial for controlling flow directions. The lava flows from vents in the Chío chain consistently advance down on the southern flank of the rift, whereas those in the Garachico chain flow persistently down the northern flank. Those located between both structures flowed into the saddle area, where the lavas concentrate in a southeast−northwest trending corridor (eg, Chinyero and Montaña Reventada, labeled 1 and 5 in Fig. 5.27A).

One of the most interesting features of the NWRZ is the compositional arrangement of eruptions, displaying a distinct bimodal distribution. Here, basanite and phonolite dominate at the distal and proximal parts, respectively, resulting in mixed-mode eruptions in the area where the two regimes overlap (Carracedo et al., 2007b; Wiesmaier et al., 2011, 2013).

FIGURE 5.26

Panorama view of the northwest rift zone (NWRZ) of Tenerife from the summit of Pico Viejo. In the foreground, Mña. Reventada (895 ka AD), including the short phonolitic coulées of Las Lenguas (the tongues) that also come from the Mña. Reventada eruption. The NWRZ extends all the way to the Miocene Teno massif in the far distance.

GIANT COLLAPSES ON TENERIFE

Giant landslides play an integral role in the geological evolution of Canarian volcanoes. In contrast to most other intraplate oceanic volcanoes, subsidence is not significant in the Canary Islands (Carracedo, 1999), and the islands remain emerged until they eventually are worn down by erosion. Giant collapses, most common when the island volcanoes are at their peak developmental stages, can remove large volumes of these islands in geologically short times (Carracedo et al., 1998; Delcamp et al., 2012). Eleven giant landslides have been identified in the islands of La Palma, El Hierro, Tenerife, Fuerteventura, and Gran Canaria by means of on-shore and off-shore data (Fig. 5.28). Similar collapses, but with less well-preserved landslide structures, may have occurred on the other islands.

Repetitive injection of blade-like dykes progressively increases the anisotropy of a rift zone, forcing new dykes to wedge their path parallel to the existing intrusions (like a knife between the pages of a

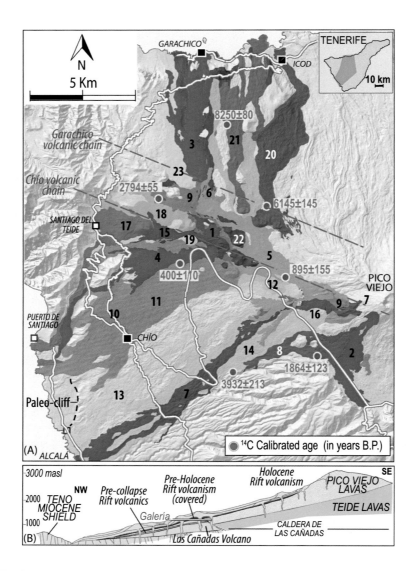

FIGURE 5.27

(A) Geological map of the Holocene volcanism of the northwest rift zone of Tenerife. Eruption sites are: (1) Chinyero, (2) Chahorra, (3) Garachico, (4) Boca Cangrejo, (5) Mña. Reventada, (6) Volcán Negro, (7) Mña. Cuevas Negras, (8) Los Hornitos, (9) El Ciego, (10) El Espárrago, (11) Mña. Cascajo, (12) Mña. Samara, (13) Mña. la Botija, (14) Mña. de Chío, (15) Montañetas Negras, (16) Mña. Cruz de Tea, (17) Mña. Bilma, (18) Mña. Cruz, (19) Mña. del Estrecho, (20) Cuevas del Ratón, (21) Mña. Liferfe, (22) Mña. de Abeque, and (23) Mñas. Negras. (B) Geological cross-section along the axis of the NWRZ. Ages are radiocarbon dates expressed in years BP (from Carracedo et al., 2007b).

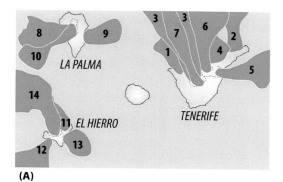

(A)

N°	Giant Landslide	km²	km³	Age (Ma)
	TENERIFE			
1	Teno	-	-	-
2	Taganana	-	-	~4.5
3	Roques de García	5500	1000	0.6-1.3
4	Micheque	-	-	0.83
5	Güímar	1600	>120	0.83
6	La Orotava	5500	1000	0.56-0.69
7	Icod	5500	1000	0.18-0.20
	LA PALMA			
8	Garafía	1200	650	1.0-0.08
9	Santa Cruz	1000?	-	-
10	Cumbre Nueva	780	95	0.56
	EL HIERRO			
11	Tiñor	-	-	0.88
12	El Julan	>1600	60-120	>0.15
13	Las Playas	700	25-35	>0.17
14	El Golfo	2600	150-180	0.80-0.13

(B)

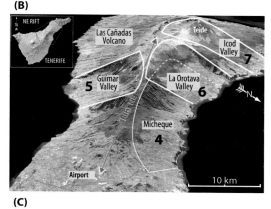

(C)

FIGURE 5.28

(A) Giant landslides identified on Tenerife and in the western Canaries. (B) Characteristics of the main collapses (from Watts and Masson 1995; Carracedo et al., 1999, 2001; Urgelés et al., 1999, 2001; Krastel et al., 2001) (the numbers refer to A). (C) Giant landslides of Tenerife in an oblique Google Earth view (from the northeast). Several large collapse basins are still visible today, except the Micheque collapse that has been completely filled by subsequent nested volcanism. In this case, the identification of the landslide was based on geological studies (geochronology and stratigraphy) (see also Figs. 5.24 and 5.25).

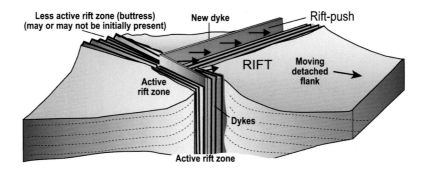

FIGURE 5.29

In triaxial rift zones, two of the three arms are usually more active, whereas the third acts as a passive buttress. Repetitive injections into the active rifts forces one of the enclosed blocks between the rift arms outward and, eventually, cause collapse by "rift-push." Post-landslide volcanism is likely to concentrate within and around the collapse scar and along the opposing third rift arm (eg, Walter and Troll., 2003; Walter et al., 2005; Carracedo et al., 2015a).

book), and therefore preferentially guide the path of successive intrusions (Fig. 5.29). However, a volcano can only accommodate a number of dykes if the overall structure of an edifice is volumetrically confining. Because repetitive dyke intrusions will progressively increase internal overpressure, new injections will incrementally push the flanks of the rift zone apart, a phenomenon known as "rift-push" or "flank-push." This means that in growing rift zones the flanks must be able to allow displacement; if not, then injections will arrest because the dyke swarm in the rift will encounter an increasingly compressional setting, preventing subsequent intrusions and thus terminating rift activity. Once flanks can move, rift-push will cause them to gradually oversteepen and creeping will eventually set in, then accelerate, and finally induce a basal detatchment (decollement). When a critical threshold is then reached, a massive landslide is initiated (eg, Delcamp et al., 2012).

LAS CAÑADAS CALDERA: VERTICAL COLLAPSE VERSUS GIANT LANDSLIDE

The most spectacular example of a collapse embayment on Tenerife is the Las Cañadas Caldera (LCC), one of the best exposed calderas in the world, although the detailed collapse history is not yet fully resolved (ie, vertical vs lateral collapse). Vertical collapse is believed to have dominated the early development and formed several classic (vertical) collapse calderas on the Las Cañadas Volcano at a time between 1.2 and 0.18 Ma (eg, Ridley, 1971; Booth, 1973; Martí et al., 1994; Bryan et al., 1998). However, a range of workers have considered the present day Las Cañadas Caldera (LCC) as a primarily lateral landslide scar (Navarro Latorre and Coello, 1989; Ancochea et al., 1990, 1999; Carracedo, 1994; Watts and Masson, 1995; Urgelés et al., 1997, 1999; Masson et al., 2002). In fact, evidence exists for a lateral collapse at approximately 200 ka (the Icod lateral collapse), as evidenced by radiometric ages of lava flows overlying the basal debris avalanche to this collapse in galerías (Carracedo et al., 2007b). This lateral collapse is clearly linked to a large

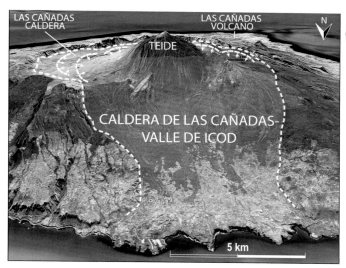

FIGURE 5.30

Oblique *Google Earth* view of the northern flank of Tenerife, showing the former Las Cañadas edifice (approximately 0.2 to 1.2 Ma), the northern breakout that formed the Icod Valley at approximately 200 ky, and the Teide–Pico Viejo edifice that grew after the Icod giant landslide. In the small drawing to the right, the Caldera de Taburiente on La Palma is shown in orange for comparison.

submarine debris avalanche deposit on the northern flank of the island (Watts and Masson, 1995). Moreover, a continuous layer of debris avalanche deposits extends inside the LCC and below the present Teide stratocone (Márquez et al., 2008), providing strong support for a landslide origin of the currently visible depression of the LCC.

Likely, a series of vertical collapses preceded the formation of the present-day LCC, before a giant lateral landslide occurred. Ocean islands represent unbuttressed free surfaces, and if pre-weakened by, for example, repeated vertical inflation and subsequent caldera subsidence, then entire "cake slices" may break out of an island's edifice by lateral instability (see, eg, Troll et al., 2002). Therefore, the combined effects of vertical and lateral collapses may have given rise to the present-day Las Cañadas Caldera (Fig. 5.30). The most recent modification, however, is the Teide and Pico Viejo complex that is currently growing inside the embayment of the approximately 200-ka (Icod) lateral collapse.

THE 735-KA ABONA GIANT LANDSLIDE

Another type of landslide, probably related to explosive volcanism, has recently been discovered on the south flank of the LCV. The avalanche deposit of debris from the Abona landslide (Dávila-Harris et al., 2011) has been dated at 733 ± 3 ka and covers approximately $90 \, km^3$ of the south–southeast slope of the island (Fig. 5.31). Several lines of evidence relate this avalanche to an ignimbrite-forming explosive eruption (the Los Helechos event), with successive eruptive

FIGURE 5.31

Map of the Abona lateral landslide and the associated debris-avalanche deposit (dark red) in south Tenerife. After Dávila-Harris et al. (2011).

phases. The avalanche deposit is interbedded within ignimbrites and plinian fallout pumice, all from a single eruption, because pumice blocks that were hot during landslide emplacement yield the same $^{40}Ar/^{39}Ar$ age as the associated ignimbrite and fall deposits (Dávila-Harris et al., 2011).

This lateral collapse is noteworthy because it is the first well-preserved large landslide deposit recognized on the LCV. It has been directly dated with a high degree of accuracy and preserves evidence of being associated with, if not triggered by, a large explosive eruption (Dávila-Harris et al., 2011).

GIANT LANDSLIDES AND MAGMATIC VARIABILITY

A comparative analysis of the evolution of different Canary Islands rift zones, including those of Tenerife, outlines common characteristics (Fig. 5.32). A particularly interesting one of these is the apparent association of giant landslides with magmatic variation in post-collapse nested volcanism that often tend toward differentiated magmas, such as trachytes and phonolites (Fig. 5.32).

As mentioned, rifts are recurrent features that show cyclic patterns of growth, instability, flank collapse, nested volcanism, and eventually eruptive decline and dispersion (Carracedo et al., 2011b; Carracedo and Troll, 2013). Variations in magma composition appear to occur in response to lateral collapses, because a collapse effects disruption of an established feeding system to a rift. This allows initially dense mafic magmas to ascend to the surface (eg, by edifice unloading) (Fig. 5.33A), resulting in the concentration of progressively centralized eruptions in the interior of the landslide scarp, thus gradually filling the collapse embayment (Carracedo et al., 2007b, 2011b; Longpré et al., 2009; Manconi et al., 2009; Carracedo and Troll, 2013). The likely density barrier of early erupted crystal-rich (ankaramitic) magmas will cause arrest of new magma at shallow depths within this growing nested volcanic edifice. This allows for potentially extensive

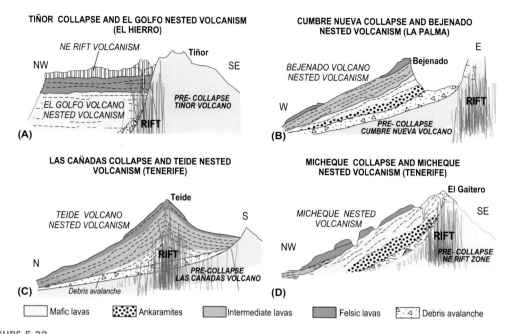

FIGURE 5.32

Examples of rift zones and associated landslides in the Canary Islands. Subsequent nested and differentiated volcanism and progressive magmatic differentiation seems to be common in the sequences filling the landslide scarps. After Carracedo et al. (2007b, 2011a).

modification of magma through progressive magmatic differentiation that can produce felsic compositions (trachytes, phonolites). These compositions become more and more dominant due to a gradual increase in height of the nested volcano inside the landslide embayment (Fig. 5.33B) where within-edifice storage will increasingly promote magmatic differentiation.

Although felsic volcanic complexes on Tenerife may originate from a variety of processes, a considerable volume of differentiated volcanism on the island thus appears to be linked to rift flank collapses that are followed by abundant and prolonged nested volcanism. Regularly, these eruptions evolve from initially mafic to terminally felsic compositions (Fig. 5.33C). Lateral collapses therefore represent a major cause for structural and petrological variability in ocean islands (Carracedo et al., 2007b, 2011b; Longpré et al., 2009; Manconi et al., 2009), with Teide volcano probably being the prime example for this phenomenon (Carracedo et al., 2007b).

THE TEIDE VOLCANIC COMPLEX

Upon arrival of the great naturalists on Tenerife in the 18th and 19th centuries, such as Leopold von Buch, Charles Lyell, and Alexander von Humboldt (among others), Teide became a natural

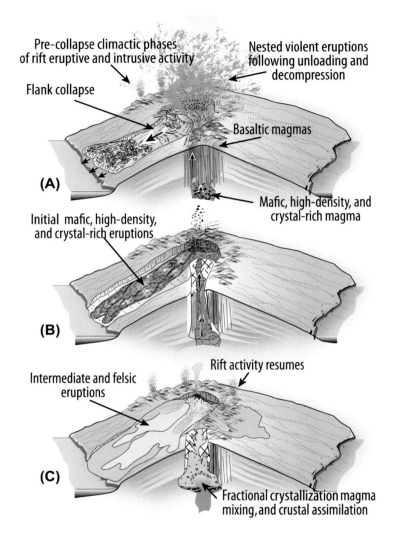

FIGURE 5.33

Schematic model of rift evolution in Tenerife. Following catastrophic collapse (A), the plumbing system readjusts, leading to structural and petrological modifications within the rift. (B) Initial post-landslide mafic eruptions evolve to (C) felsic eruptions during extended post-collapse nested volcanism (after Carracedo et al., 2011a; and Carracedo and Troll, 2013).

laboratory for testing long-lasting controversies between Neptunists and Plutonists. Teide is among the highest volcanic structures on the planet (only surpassed by Mauna Loa and Mauna Kea on the island of Hawaii). If the base level is taken to be the ocean floor and not sea level, then Mt. Teide rises to more than 7000 m in height.

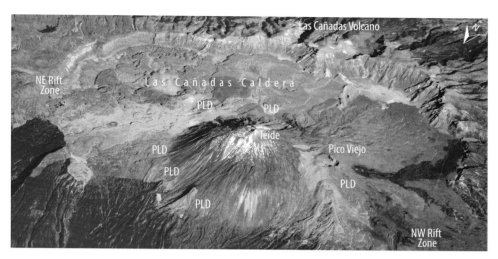

FIGURE 5.34

Oblique satellite view from the northwest of Teide volcano into the Las Cañadas Caldera. PLD, Peripheral lava domes. Photograph (*ISSO13-E-23272*) taken on June 8, 2006 from the International Space Station. Courtesy of NASA.

The Teide Volcanic Complex (TVC) is the most emblematic volcano in the Canaries. However, the eruptive history of the TVC was insufficiently understood until very recently, mainly because of the lack of geochronological information (Carracedo et al., 2007b). Dating the TVC was initially considered unfeasible due to the supposed impossibility of applying K/Ar and ^{40}Ar/^{39}Ar techniques to this young period of volcanic activity and due to the alleged absence of suitable organic matter for radiocarbon dating. In 2007, 54 new K/Ar and ^{14}C radiometric ages became available and provided, for the first time, precise age constraints on the recent eruptive history of Teide volcano (Carracedo et al., 2007b). These new data have helped greatly to establish a geochronological framework of the structural and volcanic evolution of the TPV complex and allowed a more precise and realistic assessment of potential eruptive hazards (eg, Carracedo and Troll, 2013).

Although Teide and Pico Viejo volcanoes are geographically well-defined stratovolcanoes built on the floor of the Las Cañadas Caldera (Fig. 5.34), the simultaneously active rift zones are also considered part of the TPV volcanic system. The boundaries between these parts are not always clear in spatial, temporal, or compositional terms, because the rift zones and the Teide—Pico Viejo volcanoes function as a coeval and also an interactive volcanic system (Fig. 5.35A). To facilitate the study of this volcanic complex, Carracedo et al. (2007b) defined five main components (Fig. 5.35B): Teide Volcano; Pico Viejo Volcano; the youngest peripheral lava domes that cluster around the base of Teide; the northwest rift; and the northeast rift. Because the northwest and northeast rifts have already been described, we now focus on the differentiated units of the active Teide and Pico Viejo complex.

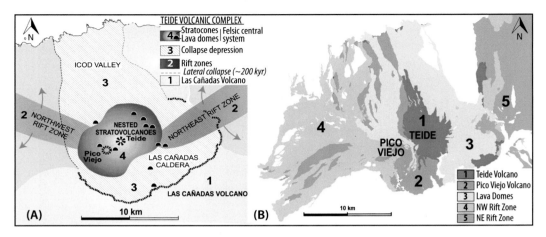

FIGURE 5.35

(A) Geological features of the TPV composite volcanoes (see text for details). (B) Simplified geological map with the main volcano stratigraphic units: (1) Teide Volcano, (2) Pico Viejo Volcano, (3) Peripheral domes of the main stratocones, (4) NWRZ volcanism, and (5) NERZ volcanism. Modified from Carracedo et al., 2007b.

THE CENTRAL STRATOCONES: TEIDE AND PICO VIEJO

Teide and Pico Viejo composite volcanoes tower over the island of Tenerife and, in fact, all of the Canary Islands, with 3718 m and 3100 masl, respectively (Fig. 5.36A). The geometry and stratigraphy of these two overlapping stratocones is difficult to define, particularly without radiometric data (the first radiometric ages of the stratocones were published by Carracedo et al., 2007b). Before then, the only age data available were the 2-ka age for Mña. Blanca (Ablay et al., 1995). This lack of geochronological control led to a number of conflicting interpretations of how the volcano system operates and developed. Generally, Pico Teide and Pico Viejo are considered to be twin volcanoes with the Teide edifice constructed on top of Pico Viejo (Fig. 5.36B) (Fúster et al., 1968d; Ablay and Marti, 2000). However, this perception is deceiving. Based on 54 new ^{14}C and K/Ar ages on the TPV volcano, the volcanic stratigraphy shows Pico Viejo to be a later "parasitic" stratocone, which apparently developed after the Teide stratocone was already in its terminal phase. The youngest age obtained for Teide, except for the 1150 ± 140 years Before Present (BP) Lavas Negras summit eruption, is approximately 32 ka, whereas the oldest Pico Viejo age is only 27 ka. Therefore, the 3100-m altitude of Pico Viejo is reached because its base is located at 2500 m, sitting on top of the western flank of Teide (Fig. 5.36C). The altitude reached by Teide (3718 masl) may have favored this later growth of Pico Viejo because observational evidence suggests that lithostatic load begins to limit the altitude of composite volcanoes from approximately 3000 m of edifice height (Davidson and de Silva, 2000), thus probably forcing vent migration from Teide toward Pico Viejo in its later stages of evolution (Fig. 5.37).

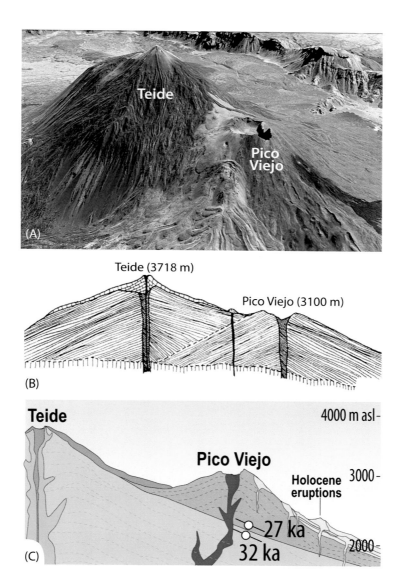

FIGURE 5.36

(A) Teide and Pico Viejo (TPV) composite volcanoes top the island of Tenerife with altitudes of 3718 m and 3100 masl, respectively. In the foreground, the Roques Blancos peripheral lava dome and thick phonolitic coulees are visible. (B) The central stratocones have been considered to be twin volcanoes, with the Teide edifice developing on top of the Pico Viejo stratocone (figure from Fúster et al., 1968d). (C) Volcanic stratigraphy based on new ^{14}C and K/Ar ages show the Pico Viejo edifice to be younger than the Teide stratocone (after Carracedo et al., 2007b).

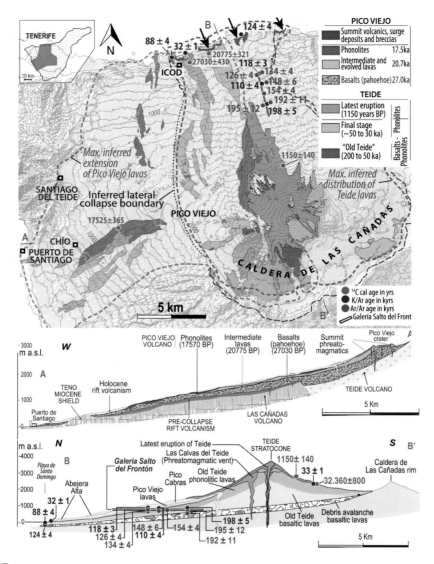

FIGURE 5.37

Geological map and cross-sections of the Teide and Pico Viejo volcanoes (modified from Carracedo et al., 2007b). While some Pico Viejo lavas flowed unrestricted toward the north and southwest coasts, Teide lavas were dominantly confined by the Las Cañadas Caldera (LCC). This is supported by coastal outcrops of "Old Teide," only visible at the mouth of this giant collapse basin (indicated with arrows in the upper center of the map). The cross-section A-A' shows Pico Viejo as a peripheral stratocone on the flank of Teide volcano. Hot and fluid Pico Viejo lavas, particularly the initial pahoehoe plagioclase basalts, mantled most of the western flank of the former Las Cañadas Volcano, flowing toward the northwest and southwest but confined in the west by the Teno massif and in the east by the LCC. Ages from samples within the deep structure of Teide (eg, the lavas filling the lateral collapse basin from the Icod debris avalanche) (cross-section B-B') bracket the development of the Teide edifice to the past 200,000 years. An evolution of magmas from initial mafic compositions to the later differentiated felsic (phonolitic) magmas is observed (after Carracedo et al., 2007b; Wiesmaier et al., 2012).

TEIDE'S PERIPHERAL PHONOLITIC LAVA DOMES

Lava domes and coulées that encircle Teide and Pico Viejo volcanoes are related to the TPV system. Basaltic magmas are denser than their differentiated equivalents, and the excess pressure required for magma to ascend to the surface must exceed the lithostatic pressure on the magma, which increases with the altitude of the cone (Fig. 5.38-1). The lateral pressure of the magma increases at the base of the volcano (Fig. 5.38-2); thus, a "density filter" is established as the height of the nested stratovolcano increases (Davidson and De Silva, 2000; Pinel and Jaupart, 2000; Carracedo et al., 2015b). Therefore, on reaching a critical altitude, estimated empirically at approximately 3000 m, the vertical pressure forces the formation of radial fractures and, eventually, the intrusion of lateral cryptodomes and associated eruption of lavas (Fig. 5.38-3). Summit eruptions then turn more sporadic and flank eruptions at the basal perimeter of the stratocones are favored, forming peripheral parasitic lava domes in Teide's circumference (Fig. 5.38-4).

Although silicic, generally andesitic and trachytic lava domes are common, and although the literature pertaining to these features is abundant, phonolitic lava domes and coulées of this volume and flow length are rare in oceanic islands. These domes were emplaced into sub-horizontal strata (eg, in the interior of the Las Cañadas Caldera) to form thick, short, typically flat-topped lava flows

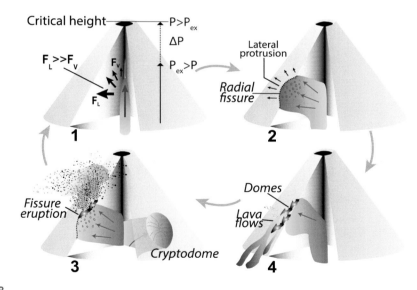

FIGURE 5.38

Schematic diagram explaining the relationship between increasing lithostatic pressure of a growing stratocone (eg, Teide volcano) and magmatic overpressure required to sustain summit eruptions. (1) When a critical height is approached (eg, 3000 m), the vertical push of magma changes to a lateral one. (2) Bulging may result in radial dykes and related fissures. (3) Eventually, evolved fissure eruptions from radial fractures commence, sometimes with an initial explosive phase. (4) Finally, effusive phases produce thick evolved (phonolitic) lava flows and coulées (after Carracedo and Troll, 2013; Carracedo et al., 2015b).

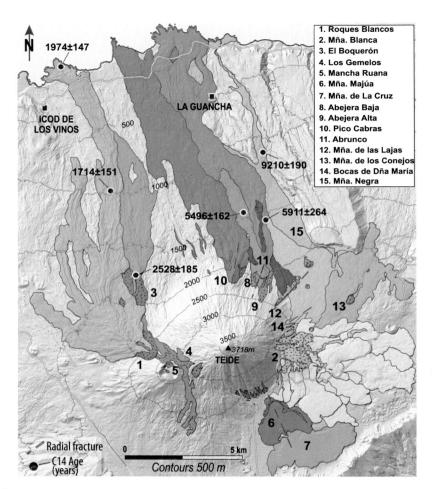

FIGURE 5.39

Phonolitic lava domes and associated lava flows of the TPV volcanic complex. Ages (^{14}C) in years BP (after Carracedo et al., 2007b). Note that the domes are mainly located at the basal circumference of Teide, particularly at the eastern flank and away from the direct influence of the Pico Viejo volcano (see also Fig. 5.38).

that accumulate in a radial fashion around the emission center (eg, from Montaña Rajada, Montaña Majúa, Montaña de la Cruz, Tabonal Negro). Domes located on a slope, in turn, gave rise to lava flows that ran over very long distances (coulées) and that frequently reached the coast approximately 15 km away (eg, Abejera Alta, Abejera Baja, Abrunco, Pico Cabras, see Fig. 5.39). Solidification of an outer lava crust appears to be the mechanism contributing to keep the lava thermally isolated, which may explain the fact that these phonolitic lavas can travel over distances usually only permitted to hotter, and thus lower-viscosity, basaltic lavas.

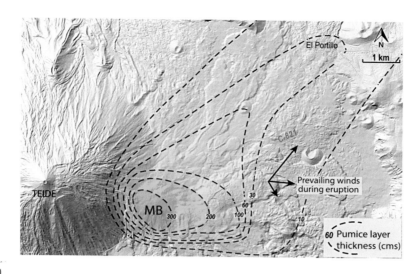

FIGURE 5.40

Isopach map of the Montaña Blanca pumice deposit (contours in centimeters). Montaña Blanca, the site of the most recent explosive eruption of the TPV stratovolcanoes. The southwest–northeast dispersal axis is consistent with the prevailing northeast trade winds. Note that the distal effects of this event, considered the only substantial postcaldera explosive eruption, are relatively modest (eg, Ablay et al., 1995).

Most of the exposed lava domes and coulées have been dated by Carracedo et al. (2007b) and range from 1714 ± 151 to 9210 ± 190 years BP (Fig. 5.39). In general, lava dome eruptions can show associated explosive episodes with extensive air-fall deposits of pumice and collapse of the front of the coulées to give rise to hot avalanches and block-and-ash flows or surges (eg, Newhall and Melson, 1983; see also chapter: The Geology of La Gomera).

However, no evidence of pyroclastic flows associated with these young lava domes has been observed on Tenerife, although block-and-ash deposits related to this type of small hot avalanches have recently also been found on the islands of La Palma and El Hierro (Carracedo et al., 2001). Air-fall pumice deposits, in turn, are ubiquitous around Teide volcano and are generally composed of thin (centimeter scale) layers. A thicker (>1 m) layer of pumice is related to the approximately 2-ka Montaña Blanca eruption, the only event in the TPV complex with significant (sub-plinian) explosivity (Ablay et al., 1995). Pumice from this event, with a vent site located at the foot of Teide, was dispersed over much of the east and northeast of the Las Cañadas Caldera (Fig. 5.40) and is still widespread today.

Comparison of the eruptive behavior of the TVC with the previous cycles of activity of the LCV reveals that although all of them follow a similar petrological pattern, there is a significant difference in their explosivity. Pre-Teide central activity is mostly characterized by large-volume (up to >20 km^3, DRE) eruptions of phonolitic magmas, whereas the TPV complex is dominated by effusive eruptions, with large explosions being rather exceptional (Martí et al., 2008; Carracedo and Troll, 2013).

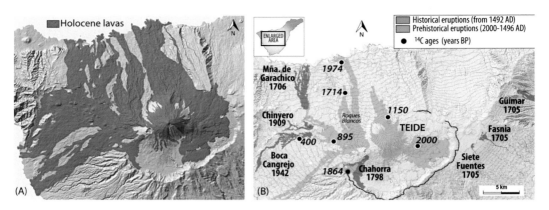

FIGURE 5.41

(A) Extent of volcanic resurfacing of the northwest of Tenerife by Holocene eruptions (blue). Note that Holocene volcanism focused in the central volcanic TPV complex and on the northwest rift zone. An exception is the 1705 eruption that occurred on the northeast rift zone. (B) Prehistorical eruptions (blue) and historical eruptions (red) of the TPV stratocones. Note that lava flows from most of the recent eruptions (9 out of 12) terminated on the flanks of the rift zones without actually reaching the sea.

RECENT VOLCANISM ON TENERIFE

Volcanic activity, involving both mafic and felsic eruptions, has been relatively frequent in the TPV complex over the past 2000 years. Significant parts of the central and western regions of Tenerife were resurfaced by eruptions during the Holocene (Fig. 5.41A).

The pre-Hispanic population of the Canaries, most probably of North African origin and known as "the Guanches," colonized Tenerife in the middle of the first millennium BP, coinciding with an epoch of intense volcanism on the island (Fig. 5.41B). The Guanches endured several moderately explosive phonolitic events (probably explaining the name Echeide "Hell" for Teide in the Guanche language). The most relevant are the Montaña Blanca sub-plinian event of approximately 2000 years ago (~ 0 AD) and the last summit eruption of Teide approximately 1240 years ago (ie, approximately the 8th century in our calendar system).

HISTORICAL ERUPTIONS ON TENERIFE

References to alleged eruptions of the TPV complex, mainly associated with Teide volcano, are numerous but, for the most part, they are likely erroneous (Fig. 5.42). The frequent presence of orographic clouds (the well-known *toca del Teide* or Teide's headdress), periods of intense fumarolic activity within Teide's crater, or forest fires (either natural or caused by the Guanche inhabitants) may explain the recurrent references to possible eruptions of Teide by 15th- and 16th-century seafarers who mostly sailed past the island at a considerable distance. Mapping and dating (using ^{14}C) now permits correlation of these possible eruptions of Teide to actual events. This approach

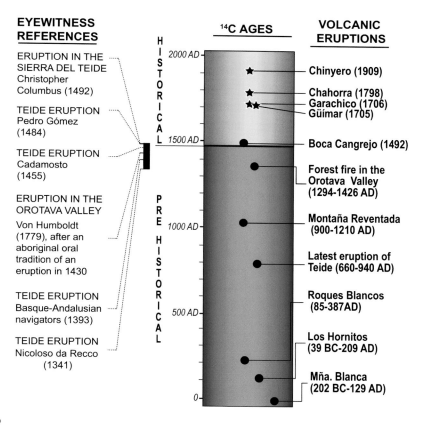

FIGURE 5.42

Eye-witness references to possible volcanic eruptions of Teide volcano (on the left) versus the established stratigraphy of the past 2000 years on the basis of geological relationships and ^{14}C dating. After Carracedo and Troll (2013).

reduced the number of presumed summit eruptions of Teide volcano in the past 2 kyr to a single event dated by ^{14}C at 1240 ± 120 years BP (calibrated age 660–940 AD).

This medieval age (Early Middle Ages) excludes a previously accepted Teide summit eruption in the year 1492, which was based on a report by Christopher Columbus. He describes "a fire on the sierra of Teide" when sailing past the south coast of Tenerife in August 1492. Only one eruption complied with Columbus's description, the Boca Cangrejo Volcano, that gave a radiocarbon calibrated age of 400 ± 110 years BP, thus overlapping with the epoch of the report by Christopher Columbus in 1492. Therefore, the 1492 Boca Cangrejo eruption can be considered to be the first eruption in the historical record of Tenerife (Carracedo et al., 2007a), increasing the number of witnessed and documented eruptions on Tenerife to five: 1492, 1705, 1706, 1798, and 1909 (see Figs. 5.41B and 5.42).

ROCK COMPOSITION AND VARIATIONS

Relationships and distribution of lava types in the TPV complex show two main regimes at play: the basanite-fed rift zones and the phonolite-dominated central complex. Consequently, this volcanic complex is a rare example of an oceanic volcano where an entire Ocean Island Basalts (OIB) evolutionary series is represented (Fig. 5.43).

The compositional evolution of volcanism nested inside the LCC shows initially mafic lavas (200 ka–30 ka) and, later, highly differentiated phonolite from 30 ka to now (see also cross-section B-B' in Fig. 5.37).

The effect of spatial distribution is well illustrated by the compositional bimodality between the felsic Teide–Pico Viejo complex on the one side, and the exclusively mafic lavas of the western edge of the northwest rift zone on the other (Fig. 5.44). Intermediate composition magmas have erupted only in that part of the rift zone that borders the central complex (Ablay et al., 1998; Carracedo et al., 2007b), and intermediate magmas on Tenerife may thus form through the interaction of two end member–type magmas: basanite from the rift zones and phonolite from evolved shallow magma chambers of the central complex (eg, Wiesmaier et al., 2011, 2012). This is manifested, for example, in the lava flow of the prehistoric Montaña Reventada volcano, where basaltic and phonolitic magmas have erupted simultaneously, or at least in very rapid succession.

This model implies that a closed system evolution is likely oversimplified because it does not consider the mixing of magmas to produce intermediate compositions or crustal recycling of preexisting rocks, which is also recorded in the Teide lavas (Wiesmaier et al., 2012, 2013).

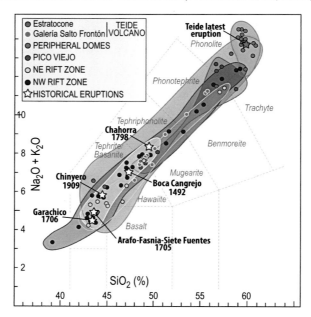

FIGURE 5.43

Total alkali versus silica diagram (TAS) for lavas of the TPV, which encompass the entire OIB series from primitive basanites to highly evolved phonolites throughout its development. Data from Rodríguez-Badiola et al., 2006.

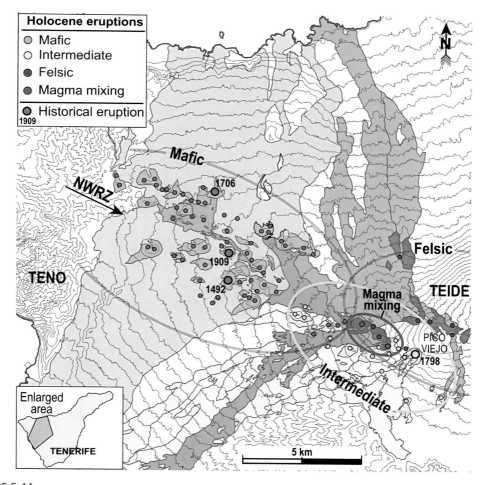

FIGURE 5.44

Distribution of vents along the Northwest Rift Zone on Tenerife. Note the clustering of felsic eruptions within the central complex, whereas mafic lavas characterize the rift zone. Transitional lavas fall into the geographical and chemical transition between the mafic and felsic regimes. Magma mixing occurs close to the boundary of the felsic central complex, likely through the intersection of shallow felsic differentiated magma pockets by mafic rift magmas from depth (see text for details).

Indeed, Sr–Nd–Pb–O isotope data demonstrate that open system behavior for the petrogenesis of Teide–Pico Viejo phonolites and isotope mixing hyperbolae require an assimilant of predominantly felsic composition. Variations in isotope composition across the evolved Teide–Pico Viejo succession imply older nepheline syenites to be a likely contaminant. These intrusive remnants from pre-Teide activity occur as fresh and altered xenoliths in pre-Teide ignimbrite deposits. Concentration of mafic activity into the central part of the island at approximately 30 ka ago has therefore likely remobilized parts of Tenerife's older plutonic core, thus defining a much wider variability of magmatic differentiation processes at Teide—Pico Viejo than previously thought.

On the basis of trace element composition, the mafic lavas show typical OIB signatures, whereas transitional to evolved trachytes and phonolites are enriched in incompatible trace elements but often depleted in Ba and Sr. Linking the spatio-chronological distribution of eruptions with these compositional groups, it appears that mafic activity migrated from the outskirts of the rift zones toward the central complex until approximately 30 ka ago, when the arrival of this migrating activity beneath the central complex gave rise to the onset of trachyte and phonolite eruptions (Fig. 5.44).

The spatial distribution of lava compositions of dated eruptions along the NWRZ therefore defines the geometry and evolution of the felsic central magmatic system. According to this pattern, the chamber system has experienced a progressive contraction in the past 5 kyr. The phonolitic eruptions in Fig. 5.45 seem to group in a linear pattern, apparently following the trend of both the

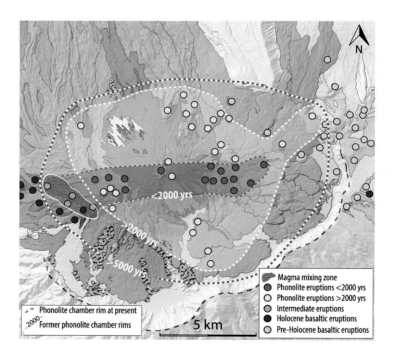

FIGURE 5.45

Bimodal distribution of the composition of lavas in the eruption products of the active northwest and northeast rifts. The greatest concentration of highly evolved eruptions (phonolites) occurs in the area where the rifts converge and where the Teide—Pico Viejo stratovolcanoes and their peripheral domes have formed (eg, Wiesmaier et al., 2012). In contrast, less evolved magmas (basalt, basanite) erupt at the distal ends of the rifts. Intermediate composition lavas frequently show signs of magma mixing (basalt—phonolite) and occur in the parts of the rifts closest to the central TPV complex. Note the reduction in size of the phonolitic system over the past 5000 years (after Carracedo and Troll, 2013).

northwest and the northeast rifts. This arrangement likely reflects a progressive decline of Teide's activity, which may thus be in a terminal stage of evolution (Wiesmaier et al., 2012).

Several geophysical and gas geochemistry studies have investigated the size, depth, and conditions of the potential magmatic chamber of Teide, the presence or absence of one or more active magma pockets, their precise location, and their potential eruptible volume. The results suggest that at present, Teide's magma reservoir seems to be approximately 4 km in diameter, at approximately sea level or just below, and is in a quasi-solid state with temperatures less than 400°C (Albert-Beltran et al., 1990). This reservoir likely contains isolated pockets of magma, but seems dominated by semi-solidified crystal mushes.

Conversely, eruptive activity has increased in the northwest and northeast rift zones in the Holocene, perhaps indicating the waning activity of the central Teide and Pico Viejo stratocones. The contraction of the felsic magma system associated with the decline of felsic eruptions is perhaps replaced by increasing rift activity. If correct, then future phonolite eruptions will be related mainly to the deep injection of basalt into isolated pockets of relatively cool felsic magma to produce mixed eruptions, as exemplified in the Montaña Reventada event (Wiesmaier et al., 2011, 2012) (see *Stop 3.3*).

THE CLIMATE OF TENERIFE

Tenerife, at the same latitude as the Sahara desert, enjoys a year-round warm climate because of the cooling effect of the cold sea currents of the Canary Islands. Major climatic contrasts on the island are due to variable altitude and because of the trade winds. In winter it is possible to enjoy warm sunshine at the coast and snow on Teide, only approximately 20 km away. The windward north side of the island receives more than 70% of all precipitation on the island, whereas the southern flank is distinctly arid.

A frequent condition on the island is the presence of clouds on the windward side at an average altitude of between 1000 and 1200 m, locally known as *mar de nubes* (sea of clouds). The sea of clouds forms on Tenerife as a result of two superimposed layers of the trade wind system separated by a thermal inversion layer (Fig. 5.46). The lower layer of cool and humid air that is forced to ascend is intercepted by the northern slopes of the island and is trapped by the thermal inversion boundary, a fact that conditions the climate of the islands, which are generally humid and rainy in the north and hot and sunny in the south. These circumstances explain why present-day holiday resorts are dominantly located in the South of Tenerife. However, tourism on Tenerife actually started in the Orotava Valley, in *Puerto de la Cruz*, in 1886, and became only predominant on the warmer south coast during the second half of the 20th century.

The trade winds that influence the northern flank of Tenerife also provide a very favorable environment for rain forests, including the remnants of a formerly widely distributed Neogene laurel plant community, which developed in the circum-Mediterranean region approximately 20 Ma ago (Sperling et al., 2004).

The arrival of these plants on the island, at present separated from their source area by a distance that makes direct colonization unfeasible, has recently been explained through the hot spot model for the Canary Volcanic Province (see Fig. 5.47). In this context, large and high islands may have been continuously available in the region for very much longer than the oldest current island (~27 Ma),

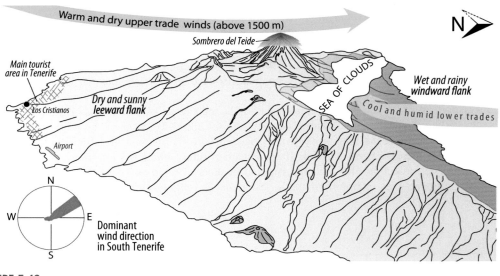

FIGURE 5.46

Altitude and the trade wind regime determine the climate of Tenerife. Dry and sunny in the south and humid and cloudy on the windward northern side. The *mar de nubes* (sea of clouds) is formed by two superimposed layers of the tradewinds, separated by a thermal inversion layer.

possibly for as long as 60 Ma (eg, Geldmacher et al., 2001), thus acting as stepping stones for radiations of plant and migratory animal groups (Fernández Palacios et al., 2010).

GEOLOGICAL ROUTES

Distances on Tenerife are relatively important because the island is 45 km wide and 80 km long and displays some severe topography. Most visitors arrive and stay in the south of the island, where the main airport and tourist resorts are located (eg, *Los Cristianos* and *Playa de Las Américas*). The "virtual" starting point for the day trips will therefore be the main roundabout near the entrance to the south motorway (TF-1) at *Los Cristianos* (Fig. 5.48).

The extent and geological complexity of this island is explained in 5 field days that aim to provide a general overview of the most relevant geological and volcanological features of the island (Fig. 5.49). Each day will focus on a specific topic. Day 1 investigates the oldest (Miocene) volcanism of Tenerife, with a focus on the Miocene shield volcanoes (Central and Teno shield volcanoes), representing the basaltic volcanism characteristic of young and emerging oceanic islands (eg, sequences of lava flows, dykes). Day 2 explores the explosive volcanism related to the posterosional Plio-Quaternary central felsic Las Cañadas Volcano, with spectacular outcrops of pumice-falls and ignimbrites that resulted from

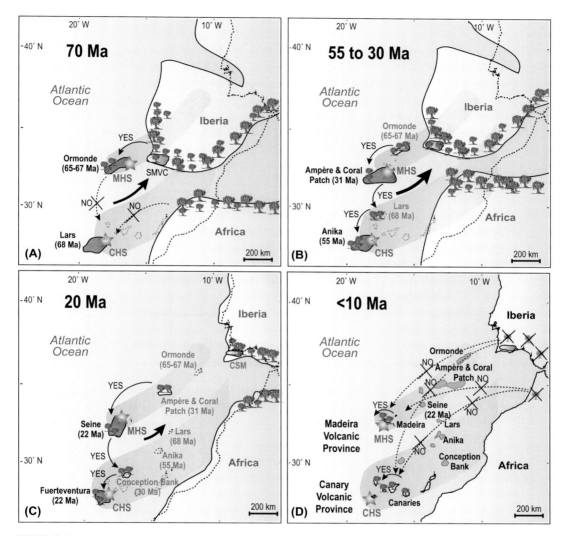

FIGURE 5.47

The presence of palaeo-endemic floral elements in the laurel forest of the Canaries (eg, Tenerife) is difficult to explain because of the age differences and excessive distances from the paleotropical sources to the present Canarian and Madeira archipelagos (due to limited ocean-crossing dispersal abilities of species). A new approach linking radiation of paleotropical flora to the Macaronesian archipelagos and the hot spot model has been proposed by Fernandez-Palacios et al. (2010). This concept suggests that large and high islands may have been continuously available in the region for as long as 60 Ma (Geldmacher et al., 2005), thus functioning both as stepping stones and as repositories of paleoendemic forms, as well as providing crucibles of neo-endemic radiations of plant and animal groups. MHS, Madeira Hot Spot; CHS, Canary Hot Spot; SMVC, Sierra Monchique volcanic Complex. Modified from Fernández-Palacios et al. (2010).

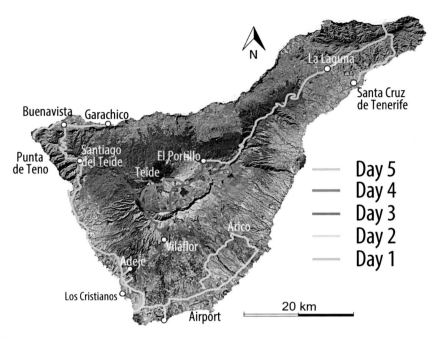

FIGURE 5.48

Map of the planned geological daytrips on Tenerife. Image from Google Earth.

highly explosive plinian eruptions. Days 3 and 4 offer a tour along the recent (mostly Holocene) basaltic northwest rift zone and the central Teide and Pico Viejo stratocones, respectively. Day 5 will then close the exploration of Tenerife with a different view of a rift, in this case the older (Plio-Quaternary) northeast rift zone (NERZ), which is deeply dismantled by erosion and allows insight into the spectacular giant landslide embayments of Güímar and La Orotava. This day also includes the option of touring the Miocene-Pliocene Anaga shield volcano after having completed the NERZ outcrops.

DAY 1: MIOCENE SHIELD VOLCANOES

We spend the day touring the two oldest Miocene shield volcanoes; the central and Teno shield massifs (see Fig. 5.2). In contrast to the earlier ideas of Tenerife having formed from three independent island volcanoes (see Fig. 5.3A), radiometric dating and geological superposition now suggest that the island is composed of a main Miocene central shield, with the Teno and Anaga shields constructed essentially on its flanks (see Fig. 5.3B). The sequential development of shield massifs results from the eventual growth of the central shield beyond a critical altitude, favoring a shift of the eruptive focus into the northeast and northwest flanks (eg, Guillou et al., 2004a). A similar process may have constructed the five separate shield volcanoes Kohala, Mauna Kea, Hualalai, Mauna Loa, and Kilauea that form the islands of Hawaii, where one edifice is effectively overlying the other.

A detailed map of the stops on Day 1 is provided in Fig. 5.49.

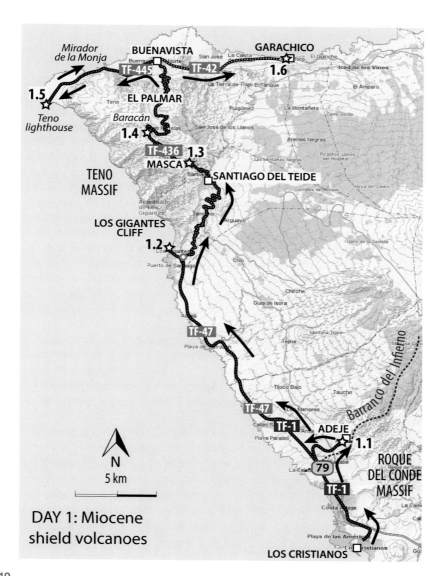

FIGURE 5.49

Detailed map of west Tenerife with stops for Day 1 indicated. The day is dedicated to the Miocene shield volcanoes of the island: The Roque del Conde and Teno massifs. Detailed road maps of Tenerife are recommended because of the dense traffic network, particularly in the south of the island. Map from IDE Canarias visor 3.0, GRAFCAN.

Stop 1.1. Roque del Conde (N 28.1229 W 16.7255)

From the starting point in *Los Cristianos*, take the motorway (TF-1) at exit 69 and drive west in the direction of *Adeje*. Drive for approximately 6.5 km and take exit 79 to join TF-5121 toward *Adeje*. On arriving in the center of *Adeje*, drive past the old church and park the car near the church on the big new parking area in the center of town. Walk back to the church and into a wide plaza and mirador next to the church that offers superb views over the Bco. del Infierno (N 28.1231 W 16.7237) and onto a spectacular 1000-m-high mountain that stands proud above the surrounding lava field (Fig. 5.50). This massif here is the only large outcrop of the Miocene central shield on the island. The central shield developed between 8.5 and 12 Ma and formed the first emerged island of what was to become Tenerife (Guillou et al., 2004a). Volcanism initially focused on the center of the present-day island and then propagated toward the northeast to form the Anaga shield and the northeast rift zone (see cross-section in Fig. 5.2 and Fig. 5.3) (Guillou et al., 2004a; Deegan et al., 2012).

The oldest age obtained from subaerial volcanics on Tenerife (11.86 Ma; Guillou et al., 2004a) is from a sample from the upper part of the Bco. del Infierno that is right in front of you. The Roque del Conde mountain, in the background, is topped by lava flows dated at 8−10 Ma and that form a flat plateau, probably a remnant palaeorelief of the central shield (see cross-section in Fig. 5.50). To hike there is possible from *Vento*, a small village west of *Arona*, where a trail crosses Bco. del Rey at N 28.1008 W 16.6901 and climbs 367 m to the top of Roque del Conde. This will take approximately 4 to 5 hours and, if you decide to go, you should consider an additional day for it, but the views are worth the extra effort.

An interesting aspect is the relative constancy of isotope ratios for the central shield through time (Deegan et al., 2012), suggesting that the central part of Tenerife had very similar sources of magmatism from Roque del Conde throughout the Las Cañadas episode and now in a part of Teide. However, Teide shows a mildly different signal that has been interpreted by these workers to indicate a progressive decline of productivity of the Teide complex relative to the earlier episodes mentioned.

Return to TF-1 and join the motorway going west; stay on TF-1 until it terminates as a motorway. There, join TF-47 toward *Playa de San Juan* and *Los Gigantes* (see Fig. 5.49).

Topographic contrast is evident between the abrupt relief of the old, deeply eroded Teno massif (to the east in the distance) and the gentle seaward-dipping slopes of the Las Cañadas Volcano, mantled with recent basaltic flows from the northwest rift zone (in the foreground on the right side of the road). Pass the village of *Alcalá*, after which the road runs on an old marine-abrasion platform (dated at 0.9 Ma; Carracedo et al., 2007b). A palaeo-cliff on the right side of the road likely represents an ancient coastline.

Continue on TF-47 and, after a while, the road crosses pahoehoe lavas from the early eruptions of Pico Viejo. Spectacular ropy structures and pressure ridges can be observed at the km 19 road sign on TF-47 (see map and cross section A-A' of Fig. 5.37).

Continue on TF-47 until it meets TF-454. Go left for a few hundred meters and park the car at the upcoming mirador to view the Los Gigantes cliffs.

Stop 1.2. Los Gigantes (N 28.2401 W 16.8367)

The spectacular 600-m-high cliff of Los Gigantes (Fig. 5.51A) is the result of coastal regression, which has been dramatically reducing the originally significantly larger Miocene shield volcano

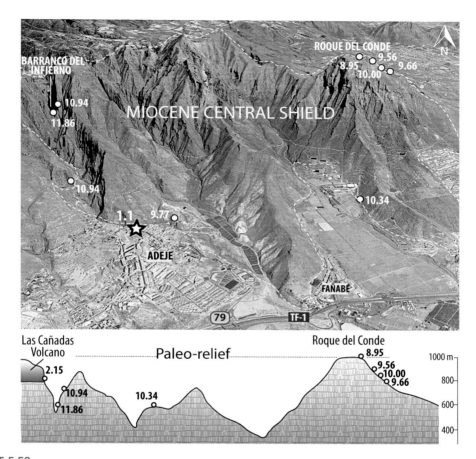

FIGURE 5.50

(A) Oblique *Google Earth* image from the west of the *Roque del Conde* massif. The massif represents the main outcrop of the Miocene central shield on Tenerife. This 1000-m-high massif also represents the oldest emerged part of Tenerife and is dated at approximately 12 Ma in its older part. Notably, the isotope composition of this early central volcanism on Tenerife appears consistent with the later Las Cañadas episode, implying a similar magmatic source for this part of Tenerife (eg, Deegan et al., 2012). Ages in Ma (Guillou et al., 2004a).
(B) Schematic cross-section.

over the past 5 Myr. The lower and upper parts of the Teno lava pile, composed of southwest-dipping lava flows crossed by numerous dykes and several interbedded cinder cones, have been dated at 5.5 and 5.0 Ma, respectively.

Following Guillou et al. (2004a), note that the cliff section corresponds to the south—southwest overflow of basaltic lavas from the collapse basin of the northern flank of the Teno massif, following the complete fill of a Miocene collapse embayment (Fig. 5.51B).

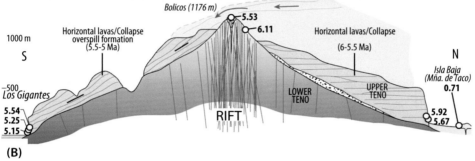

FIGURE 5.51

(A) Los Gigantes is an imposing cliff made up of a sequence of basaltic lavas of the Miocene Teno Shield volcano. (B) Note that the ages of the lava flows at the base of the southern cliff are the youngest in the Teno massif, and are younger than those forming the northern scarp. Radiometric dating implies that the youngest lavas correspond to the south−southwest overflow of the collapse basin in the northern flank of the Teno massif. Note the fast (∼1 Ma) development of the Teno volcano, which is similar to the island of El Hierro, the Taburiente volcano on La Palma, and the northwest Miocene shield of Gran Canaria.

A short walk along the base of the cliff shows remains of palaeodunes attached to the basaltic sequence, likely remnants of more extensive accumulations of aeolian sands onto the shallow coastal abrasion platform during periods of low sea level and which are now gradually eroded.

Return on TF-454 and continue north toward *Santiago del Teide*. Travel through the "Valle de Tamaimo" (see Fig. 5.49) that is cut into the Miocene Teno sequences that form the valley walls. This steep and bendy road is particularly scenic. At the village of *Tamaimo*, join TF-82 and in *Santiago del Teide*, right before the church, take a left onto TF-436 to *Masca* and *Buenavista*. The road winds over a steep pass now; just before going down into Barranco de Masca, park the

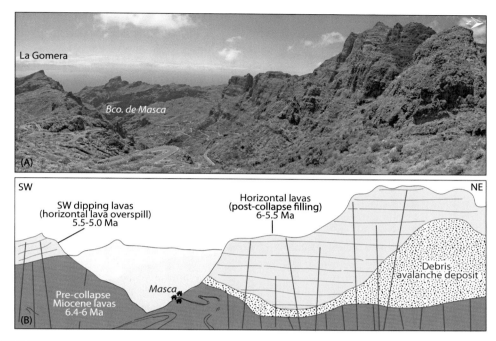

FIGURE 5.52

(A) View of Bco. de Masca from *Stop 1.3*. The interior of the Teno massif records different episodes of lateral collapses and post-collapse volcanism. Some lavas were partially filling the collapse depression (flat lavas), whereas others were spilling over from the filled embayment (dipping lavas). Image courtesy of M.-A. Longpré). (B) The basalt formations are separated by major debris-avalanche deposits.

car at the mirador on the highest point of the pass. This location provides a particularly scenic view over the impressive landscape in this part of Tenerife (Fig. 5.52A).

Stop 1.3. Mirador de Cherfe, view over Bco. de Masca (N 28.2998 W 16.8239)

The mirador offers a spectacular view into the interior of the Teno massif. Walk up on the path on the right side of the road for approximately 50 to 100 m to get the best views over the Bco. de Masca in one direction and onto Teide in the other. The deep seaward-trending Barranco de Masca shows a vertical escarpment at the right side that allows a detailed structural analysis. Two stratigraphic units are unconformably separated: an upper sequence of sub-horizontal lavas and a lower unit of deeply altered basaltic lava flows, predominantly pahoehoe lavas that are cut by numerous dykes. A debris avalanche deposit that is densely intruded by dykes marks the unconformity. This debris avalanche deposit is likely associated with a northbound giant landslide (Walter and Schmincke, 2002; Longpré et al., 2008a, 2009) (Fig. 5.52B).

The stratigraphy of the Teno sequence at the Los Gigantes cliffs can be deceiving. The flows at the foot of the cliff appear to be the oldest in the stratigraphic sequence, with the lava pile over

Masca at the top of the section. However, radiometric dating and palaeomagnetic determinations reveal the opposite (see cross-sections in Fig. 5.51B and Fig. 5.52B). This apparent inversion is explained when considering that both the sub-horizontal flows at the top of the Masca section and those forming Los Gigantes cliffs are actually contemporaneous, corresponding to the filling and overspill of a northbound collapse embayment (Guillou et al., 2004a).

To observe the older, precollapse sequence, continue on TF-436 past *Masca*. At mirador *Cruz de Hilda* (N 28.3129 W 16.8459), a dyke intrudes alternating plagioclase-rich and ankaramite lavas opposite the mirador, and you may consider an optional stop here. Otherwise, continue on TF-436 and note frequently alternating crystal-rich ankaramites and plagioclase-rich lava deposits. A few hundred meters after the turn to *Los Carrizales*, park the car in a small lay-by on the right, in front of a columnar jointed ankaramite lava (N 28.3209 W 16.8553). A few meters above the car you can now inspect ankaramite lavas with large pyroxene crystals (Fig. 5.53C,D). When walking back toward the turn to *Los Carrizales*, a third breccia package can be inspected within the Teno basal sequence, likely representing volcano instability and collapse (eg, Walter and Schmincke, 2002; Longpré et al., 2009). Notably, ankaramites seem frequently associated with large landslides in the Canaries (see Fig. 5.32), and Teno is likely a key example of this phenomenon (eg, Guillou et al., 2004a; Longpré et al., 2009; Carracedo et al., 2011b).

Ankaramites are not unusual on ocean islands, yet they appear to be particularly abundant in situations where large landslides did occur. In fact, the pressure release from these landslides is sufficient to cause a distinct pressure drop down to a depth equivalent to the upper mantle (eg, Manconi et al., 2009), hence allowing mafic crystal-rich magmas to ascend in and around former landslide areas (eg, Longpré et al., 2008a, 2009; Carracedo et al., 2011b).

The road now climbs to the crest of the collapse scarp and toward the Pico Baracán. The *Mirador de Baracán*, at a sharp bend at km 2, provides a panoramic view of the entire massif (Fig. 5.54).

Stop 1.4. Mirador de Baracán (N 28.3279 W 16.8564)

The Lower Teno unit is consistently formed by basaltic flows and interbedded layers of basaltic pyroclast deposits that are intensely eroded and altered. The most distinctive feature is the abundance of dykes, with some protruding like walls from the ground (Fig. 5.53A). This type of protruding dyke is particularly abundant on La Gomera, where the dykes are known as *taparuchas* (see chapter: The Geology of La Gomera, Figs. 4.13 and 4.30). They form by differential erosion, as they tend to be harder than the surrounding lavas and pyroclasts, and thus stand proud relative to the weaker and more easily eroded units they intrude.

There is a path that allows hiking up to Baracán, but note that this will take several hours and is recommended for another day.

Continue on TF-436 toward *El Palmar* and *Buenavista*. The bendy road passes the twin basaltic cones of El Palmar, dated at 153 ka (Carracedo et al., 2007b). One of the cones shows the deep scars of lapilli mining. Lapilli ("picón") is widely used on the islands for road construction and agriculture (Fig. 5.55). The road now follows the lava flows of these young vents and offers a panoramic view of the *Isla Baja* (the lower island) coastal plain and the felsic vent of Mña. de Taco (with a water reservoir built into the crater). The latter is dated at 0.76 Ma (Carracedo et al., 2007b).

Continue on TF-436 until you reach *Buenavista*. In the village center, take a narrow road to the left (TF-445) to *Punta de Teno*. The scenic road continues westward all the way to *Punta de Teno* at the westernmost tip of the island and brings you along some spectacular vertical

FIGURE 5.53

(A) Differential erosion exhumed this dyke in Miocene basaltic lavas and pyroclasts of the Teno massif.
(B) Massive ankaramite lavas of the Teno massif with pronounced columnar jointing. (C and D) The ankaramites,
a mafic rock type with abundant large crystals and aggregates of pyroxene and olivine, are common in Teno and
tend to occur frequently in the initial stages of collapse basin fillings (eg, Longpré et al., 2008, 2009).

cliffs and screes (Fig. 5.56A). Note that this road may be closed after heavy rainfall and during
stormy weather.

You may want to stop at *Mirador de la Monja* along this stretch for a panoramic view of the
Isla Baja, *Buenavista*, and the Teno vertical cliffs (see Fig. 5.54A).

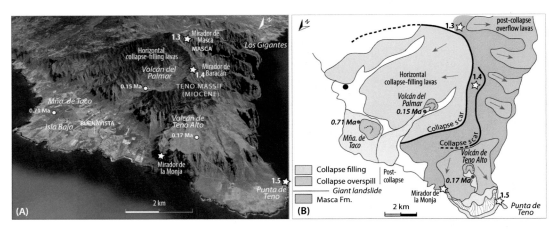

FIGURE 5.54

(A) Oblique *Google Earth* view of the Teno massif viewed from the northwest. (B) Schematic geological map showing the two northward-directed landslides that are marked by breccia horizons in the field.

FIGURE 5.55

Mña. de El Palmar, a 153-ka-old volcanic cone in northwest Tenerife that has been quarried for lapilli, which help to improve farm land and are used as aggregates for road construction.

Continue on TF-445 toward *Punta de Teno*. Park the car at the end of the road and walk to the small viewpoint at the top of the cinder cone for a spectacular view of the Teno massif and the Los Gigantes cliffs (Fig. 5.56).

FIGURE 5.56

(A) A recent coastal plain composed of northwest rift zone lava flows leads up to Teno cinder cone at *Punta de Teno*. Scree deposits locally preserve the Miocene Teno cliffs from rapid erosion. (B) Teno sea cliffs near *Punta de Teno*.

Stop 1.5. Punta de Teno cinder cone and lighthouse (N 28.3433 W 16.9198)

Several Quaternary basaltic vents are located at the top of the Teno massif (indicated in Fig. 5.54). Their lavas cascaded down the cliffs and formed the coastal platform in the foreground (Fig. 5.56A). These vents, dated at 178 ka (Carracedo et al., 2007b), and the Punta de Teno cinder cone belong to the prolongation of the northwest rift zone, suggesting that the rift zones act on the entire island and irrespective of preexisting lithology.

From the top of the cinder cone, the view to the west shows a wide coastal plain made of basaltic flows from the *Teno Alto* area topped by scree deposits that formed at the foot of this part of the Teno cliffs, because these deposits are protected from marine erosion by the coastal lava platform. To the east there is a full view of the Miocene vertical sequence of shield-stage basaltic lavas that are densely intruded by dykes (Fig. 5.56B).

Return to *Buenavista* and join TF-42 toward *Garachico* to tour the Quaternary lava platform of the Isla Baja and the northern dead cliff of Teno, fossilized by lava flows from Mña. de Taco (706 ka) and other volcanoes located beyond the cliff, such as Volcán de El Palmar (153 ka), Mña. Los Silos (194 ka), and Volcán Tierra del Trigo (261 ka) (Carracedo et al., 2007b).

Just before *Garachico*, on arriving at a mirador on your left in a sharp rightward bend, park the car at the mirador and walk to the nearby "Monument for the Canary Emigrant" (watch the traffic because you will need to cut across to the left side). This mirador was the traditional viewpoint to welcome ships from the Americas and to wave goodbye to those ships laden with emigrants departing to the New World. The mirador provides the best view of the lava flows that plunged down the cliff and partially destroyed the town and harbor of *Garachico* in 1706 AD.

Stop 1.6. The Garachico eruption of 1706 AD (N 28.3723 W 16.7697)

Although the historical volcanism of Tenerife will be addressed in *Day 3*, a tour to the northwest part of the island would be incomplete without visiting *Garachico*, a very prosperous harbor town in the 16th century that was, back then, the main port in the trade with the Americas. Unfortunately, its harbor

FIGURE 5.57

(A) Painting of the 1706 volcanic eruption reaching *Garachico* (anonymous contemporary painting copied by Ubaldo Bordanova in 1898) (courtesy of the Marqués de Vilaflor, private collection). (B) The lava flows cascaded down the northern cliffs and partially filled the harbor (red in the figure), which was the most important trading port with South America in the Canaries at that time (modified from anonymous contemporary drawing courtesy of Archivo General de Simancas, Valladolid).

was destroyed by the eruption of 1706 (Fig. 5.57). This caused a shift of trading activities from *Garachico* to *Santa Cruz*, the present-day capital of the island and also the island's largest port.

The Garachico eruption began on May 5, 1706, after a series of strong earthquakes the night before. The eruptive vent opened 1300 m above the town on a steep slope and approximately 6 km away as the crow flies. In only a few hours, lava flows were perched atop the cliff above *Garachico*, where they formed seven branches (see Fig. 5.57A). The eastern lobes then partially filled the harbor basin with lava, making it impracticable for heavy naval traffic (see Fig. 5.57B).

The eruption caused moderate damage to the town (Fig. 5.58) and, fortunately, there were no casualties, yet this eruption is nevertheless considered to be one of the most disruptive of all historical eruptions in the Canary Islands. The destruction of the port put a temporary pause to trade with the Americas and enforced a complete socioeconomic reorganization of the islands. Eventually, the

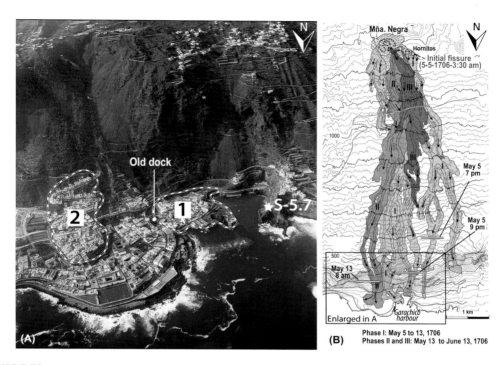

FIGURE 5.58

Oblique aerial view of the cliff above the town of *Garachico* showing remains of the 1706 lava flows. One lobe in particular partially filled the harbor basin (1), whereas another lobe destroyed the upper parts of the town (2). (B) Geological map of the 1706 eruption.

main port and capital were established in *Santa Cruz de Tenerife*, previously a small fishing village. This also marked the beginning of a shift in political power to the northeast of the island.

Drive to *Garachico* town and park the car in the large parking area at the village's entrance. Walk toward the old harbor and the town center for an inspection of the 1706 lavas and to get a first hand impression of the societal and infrastructural consequences for the town and the island at the time.

Eyewitness accounts primarily emphasize the damage to the port; for instance, Viera y Clavijo (1776) described the devastation: "Vineyards, springs, birds, port, trade and neighbors disappeared as a lava branch filled the harbor and forced the sea to retreat, leaving a cove impracticable even for small boats."

Consider a brief refreshment stop in *Garachico* before you return to your base on the same route that you came, or, alternatively, via the motorway around the island (TF- 42 to *Icod*, TF-5 to *La Laguna*, and TF-2 to join TF-1 to the south airport and *Los Cristianos*). Both routes are equally long, but using motorways might be a welcome change after today's bendy drives.

END OF DAY 1.

DAY 2: EXPLOSIVE VOLCANISM; THE BANDAS DEL SUR

The geographical distribution of the best outcrops to observe the different deposits of the explosive volcanism of the Las Cañadas Volcano (LCV) makes it so that the successive stops today do not follow

FIGURE 5.59

Itineraries of *Day 2*. The Bandas del Sur. Map from IDE Canarias visor 3.0, GRAFCAN.

stratigraphical order. The observations can be linked, however, with the general volcanic history of the LCV that is provided in the geological summary and with the stratigraphic column in Fig. 5.5.

WESTERN BANDAS DEL SUR

From *Los Cristianos*, take the motorway TF-1 in a westerly direction (ie, toward *Adeje*) for approximately 5 km and then take exit 78. Cross under TF-1 in the direction of *La Caleta* (Fig. 5.59) and follow signs for the local police station. After approximately another kilometer, park the car at the end of a pronounced bend in the road near a large arched gate that is on the wayside to your left.

Stop 2.1. The Adeje ignimbrite (N 28.1082 W 16.7351)

In the road section, the Adeje ignimbrite is overlain by a pale, creamy white lapilli tuff (Fig. 5.60A) (the San Juan Formation in the stratigraphic column in Fig. 5.60), whereas at the northern side of the road it is topped by a LCV phonolite lava flow (Fig. 5.60B). The orange ignimbrite can be seen to rest on a layer of plinian pumice-fall from the initial stages of the eruption.

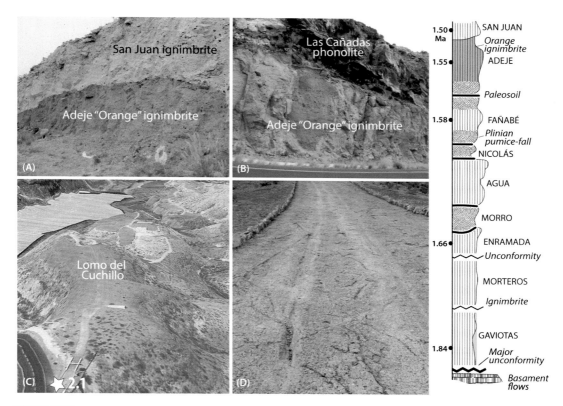

FIGURE 5.60

(A) The Adeje ignimbrite (orange-red) is unconformably overlain by the San Juan pale cream ignimbrite. (B) The Adeje ignimbrite is itself topped by a Las Cañadas volcano phonolite flow. Note the obsidian fragments preserved in the San Juan ignimbrite, likely representing chilled magma from the eruptive conduit walls. (C) "Lomo del Cuchillo" (Google Earth view looking south) is formed by the orange Adeje ignimbrite. (D) Track on Lomo del Cuchillo connecting *Adeje* with the port of *La Enramada*. The track was used to transport heavy loads of wine and sugar cane in steel-wheeled wagons that carved still visible ruts into the *tosca colorada* (orange ignimbrite). On the right, stratigraphic column from Dávila-Harris et al. (2009).

Walk through the gate and walk down-hill along the *Lomo del Cuchillo* (Fig. 5.60C,D), which is an excellent site to observe one of the oldest outcropping deposits of the LCV. This unit here corresponds to the latest phases of cycle I of activity of the LCV volcano (see Fig. 5.5). At the bottom of Bco. del Agua, the section described by Dávila-Harris et al. (2013) for this Adeje plinian eruption includes a basal pumice fall and two ignimbrites; the upper one has a distinct dark orange color (see also Fig. 5.7).

The Guanche aborigines used this trail on the orange-colored *tosca colorada* to bring their goats up to the grazing highlands south of the Las Cañadas Caldera. After the conquest of the island, this trail was improved to connect *Adeje* with the port of *La Enramada*, which was crucial for the economy of the area because it allowed shipping of the products from the rich farms of *Adeje* to *Garachico* and *La Orotava*

(Fig. 5.60D). Since the 16th century, this track, today called *Camino de la Virgen* (Footpath of the Virgin), is also used in religious processions, such as for the transfer of the image of the "Virgin of the Incarnation" from the church of St. Ursula in *Adeje* to the hermitage of *La Enramada* at the coast, which is traditionally associated with petitionary prayer to her to protect the islanders from pirate attacks.

Continue up-hill for approximately 1 km and join TF-1 in the direction toward *Santa Cruz*. Leave again at exit 28 for *Los Gigantes*. Cross the bridge over TF-1 and drive toward *Siam Park*. At the bus stop just before *Siam Park*, head up-hill (north) for approximately half a kilometer and take a track entrance at the head of a sharp bend to your right (just before a green house). Right after the turn, the track crosses a small bridge over a deep cut in the flank of Caldera del Rey's tuff ring. The ravine below shows a spectacular cut through a sequence of depositional features, but the place is not suited for a car stop. Continue up-hill and along the rim of Caldera del Rey for approximately 2 km. Park the car at the entrance to the water treatment plant and inspect the area and its spectacular phreatomagmatic eruption features.

Stop 2.2. Caldera del Rey (N 28.0851 W 16.7073)

Although basaltic phreatomagmatism is frequent in the Canaries, felsic examples, like the Caldera del Rey event, are rare and offer a unique opportunity to study this type of volcanic phenomenon.

Felsic phreatomagmatism can be very explosive, like the 12,900-year BP Laacher See eruption in Germany, where a Plinian eruption produced a 35-km-high eruptive column and covered an area of 1300 km^2 with pumice and ash (Schmincke et al., 1999), in addition to forming the 2-km-wide lake at *Laach*.

The sub-plinian Caldera del Rey eruption (Fig. 5.61) is of a lesser magnitude, however, and vented from two explosion craters with diameters of approximately 675 and 1100 m that produced a two-stage eruption (see also Fig. 5.18).

Walk around, especially above and below the road along the eastern rim of the caldera, which facilitates the observation of many characteristic features of this type of volcanism. This includes thinly bedded, frequently rhythmic, individual layers with bomb sags, wet and dry surges, and accretionary lapilli (Fig. 5.61).

From here, we return to the motorway (TF-1) and proceed toward the south Airport and *Santa Cruz de Tenerife* (see Fig. 5.59). Follow TF-1 for a short while and after the km 71 signpost, you will see the entrance to a petrol station. Park at the petrol station to catch a glimpse of the Mña. de Guaza trachyte dome complex.

EASTERN BANDAS DEL SUR

Stop 2.3. Mña. de Guaza dome (N 28.0592 W 16.6916)

This stop is to observe a section of the trachytic pyroclastic flow and ash deposits of the Mña. de Guaza dome complex (Fig. 5.62). This magnificent 435-m-high, 2-km-diameter trachytic dome erupted a steep-sided and thick (~100 m) trachyte coulée that expanded to the coast from the 600-m-wide crater and formed the coastal cliffs east of *Los Cristianos* (Fig. 5.63).

The aerial view shows this flow with prominent pressure ridges (ogives), formed as a result of compressional forces parallel to the flow of the coulée. Mña. de Guaza is a peripheral dome of the pre-Teide LCV, and it is dated at 0.93 Ma (Carracedo et al., 2007b).

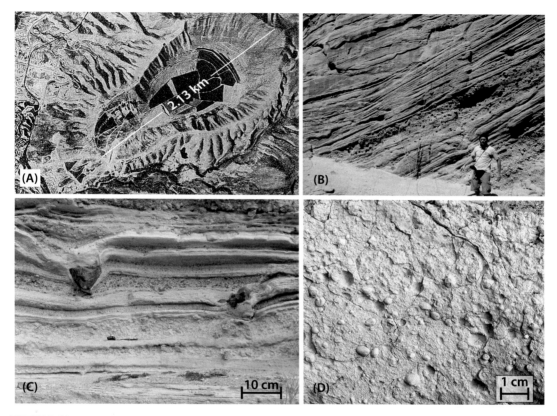

FIGURE 5.61

(A) Caldera del Rey, a 953-ka-old, double-cratered phonolitic tuff ring (maar-type). Examples of felsic phreatomagmatic explosive eruptions, like the Caldera del Rey event, are rare on Tenerife and offer a unique opportunity to study this type of volcanic phenomenon (vintage aerial image courtesy of *Cabildo Insular de Tenerife*). (B) Sequence of thinly bedded, frequently rhythmic layers characteristic of high-energy events. (C) Impact structures from balistic ejecta (bomb sags). (D) Accretionary lapilli are small spherical balls of volcanic ash that form from a wet nucleus falling through a volcanic ash cloud and are locally seen in the Caldera del Rey deposits.

Join TF-1 again and take the first exit to reach TF-66 toward *Las Galletas*. The road passes basaltic cones of the South rift zone and their associated lava fields. Continue on TF-66 until you reach *Las Galletas*. Stay on the left after the marina and use the periphery road around town until you see a sign for *Costa del Silencio*. Follow the sign and turn right. Stay on a straight stretch until, after a small roundabout, a brown sign for Mña. Amarilla becomes visible. Follow the sign to Mña. Amarilla, seen ahead, and park at the coast in front of the cone soon after a sharp left turn.

Bring your swimming gear if you would like to enjoy a refreshing dip in the colorful and rather sheltered bay.

FIGURE 5.62

An erosional channel near the base of Mña. de Guaza is filled with pyroclastic flow deposits (likely from Guaza) and with fall deposits from Caldera del Rey. Note the thickening of the upper deposit inside the channel in the center of the figure, thus moderating topography, a characteristic feature of pyroclastic flow deposits.

FIGURE 5.63

Aerial view (from the south) of Mña. de Guaza lava dome and coulée. Note the thickening of the viscous coulée was caused by the arrival of a 100-m-thick and 2.5-km-wide lava flow at the coast. The visible pressure ridges (indicated by arrows) formed when the emission of lava continued after the lava front reached the sea, where it solidified and became arrested. The pressure of the continued eruption then led to shortening of the coulées through arced folds now visible as ridges on the top of the lava flow.

Stop 2.4. Mña. Amarilla phreatomagmatic vent (N 28.0094 W 16.6397)

As one of the littoral phreatomagmatic vents located at the southern coast of Tenerife, Mña. Amarilla is morphologically different from most other tuff cones on Tenerife (Fig. 5.64A), because the eruption was initially rich in sea water but later evolved from phreatomagmatic to strombolian, and the ratio of Hco/Wcr is anomalously low (see Fig. 5.20). The names of phreatomagmatic volcanoes on the southeast coast of Tenerife related to this eruptive mechanism are Mña. Amarilla

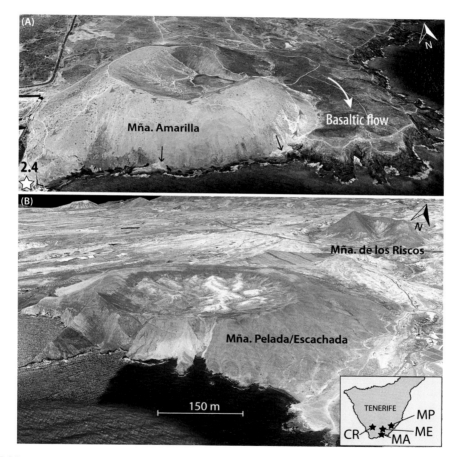

FIGURE 5.64

(A) Oblique *Google Earth* image of Mña. Amarilla viewed from the south. The black arrows indicate fossil dunes that are currently being removed by erosive coastal regression. (B) *Google Earth* image of Mña. Pelada (bare), also known as Escachada (squashed). Note the higher aspect ratio of Mña. de los Riscos (in the background), a characteristic strombolian cone without any magma/water interaction. The inset shows phreatomagmatic eruptions in the southern part of Tenerife. CR, Caldera del Rey; MA, Mña. Amarilla; ME, Mña. Los Erales; MP, Mña. Pelada.

(Yellow mountain, because of the bright color of the palagonite), Mña. Pelada (Bare mountain, because the hard, indurated tuff prevents vegetation growth), and Mña. Escachada (Squashed mountain, because of the low aspect ratio of the volcano) (see Table 5.1).

The tuff layers here at Mña. Amarilla show the characteristic yellowish color caused by palagonitization (interaction of glass with aqueous solutions) (Fig. 5.65). You can take a short hike (20−30 minutes) around the Amarilla crater (indicated with red arrows in Fig. 5.64A), where marine dunes are interbedded with scoria. Also, note the frequent small impact structures in the phreatomagmatic tuffs.

The fossil dunes overlying the tuff cone in the coastal part consist of weakly cemented, middle-to-coarse sands that show horizontal and planar cross-bedding (Fig. 5.65B,C). The beach sands are made of biogenic components (red algae, marine gastropods, and sea urchins), lithoclasts (scoria and mafic rock fragments), and fragments of minerals (olivine, plagioclase, pyroxene). The sands are cemented with a thin circumgranular cement of microcrystalline calcite and aragonite needles, comparable to a lower beach sequence (eg, Kröchert and Buchner, 2009). Locally, pumices are seen in the sand deposits, implying that "floating pumices" were also deposited here. These pumices likely come from eruptions further afield (eg, Caldera del Rey) or from offshore events similar to the floating rocks of El Hierro in 2011−2012 (eg, Troll et al., 2012; Carracedo et al., 2015a).

From here, you can also hike up to the crater. In the former crater, crater-fill deposits are well exposed and there is a late pulse of strombolian activity on one side of the crater (dark basaltic material). Take the path around the crater for some good local views.

Examples of phreatomagmatic activity that occur in typical rift-related basaltic vent alignments are relatively frequent in the Canary Islands (Table 5.1) and are not exclusive to coastal areas, like here at Mña. Amarilla. These inland vent alignments usually erupt in strombolian fashion, but aligned feeders may intersect different strata, variable structural features, or different hydrological situations, thus giving rise to possible phreatomagmatic activity that is not easily predictable in intensity and duration (Clarke et al., 2009).

Return to your car and to the main road and continue north to the next village (*Guargacho*). Once you arrive in *Guargacho*, keep right at the T-junction at the village entrance and continue on TF-652. Approximately 1 km after *Guargacho* village, you will pass a ravine on the right side, and a number of cinder cones become visible to your right in the distance. Park the car in a small open space on the right side of the road (close to Mña. de Los Erales) or at the upcoming restaurant on your left (300 m further).

Stop 2.5. Los Erales transitional phreatomagmatism (N 28.0432 W 16.6290)

Montaña de los Erales, a 70-m-high Quaternary cinder cone, belongs to a rift-related chain of vents (the South rift zone) in the Bandas del Sur region (Fig. 5.66A). These vents emitted basaltic magma, whereas the central Las Cañadas Caldera produced predominantly felsic magmas at this time. Mña. de los Erales shows an initial phase of activity that was largely driven by magma−water interaction. The eruption style of this vent changed progressively from an initial phreatomagmatic phase (Unit 1 in Fig. 5.66B), through a transitional stage (Unit 2), to one that was entirely strombolian (Unit 3). Palagonitization, extensive in the lower phreatomagmatic deposits, becomes less noticeable in the later strombolian deposits.

FIGURE 5.65

(A) Mña. Amarilla. The most striking macroscopic feature is the color of the palagonitized rocks showing the typical bright orange to yellow of phreatomagmatic eruptions. (B) Details of the strongly consolidated tuff and lapilli layers with some ballistic blocks. All the tuff deposits here show intensive palagonitization. (C) Fossil sand dunes interlayered with tuff deposits at the foot of Mña. Amarilla. These consist of weakly cemented, biogenic sands that show horizontal and planar cross-bedding and frequent pumice beds that accumulated from drift pumice.

Table 5.1 Recent Strombolian and Phreatomagmatic Eruptions in the Canary Islands

Eruption	Island	Date/Age	Duration (Days)	Composition	Eruptive Style	Proximal to Shoreline
Historic Strombolian Eruptions						
Boca Cangrejo	Tenerife	1492	Not known	Basanites, tephnites	Strombolian	No
Jedey/Tajuya	La Palma	1585	84	Basanites, phonolites	Strombolian-block and ash	No
Martin/Tigalate	La Palma	1646	80	Basanites	Strombolian	No
Fasnia-Siete Fuentes-Arafo/ Volcán de Güímar)	Tenerife	1704—1705	13	Basaltic	Strombolian fissure-related	No
Montaña Negra-Garachico	Tenerife	1706	9	Basalts-basanites	Strombolian fissure-related	No
Timanfaya/Mñas. del Fuego	Lanzarote	1730—1736	2067	Basaltic-tholeiitic	Strombolian	No
Chahorra	Tenerife	1798	92	Tephriphonolite	Strombolian fissure-related	No
Chinyero	Tenerife	1909	10	Basaltic	Strombolian	No
Teneguía	La Palma	1971	22	Basanites, tephrites	Strombolian	No
Phreatomagmatic Eruptions						
Mña. Goteras	La Palma	Holocene	Not known	Basaltic	Phreatomagmatic	Yes
Pico Viejo crater	Tenerife	Holocene	Not known	Phonolitic	Phreatomagmatic	No
Caldera de Bandama	Gran Canaria	Pleistocene	Not known	Basaltic	Phreatomagmatic	Yes
La Isleta	Gran Canaria	Pleistocene	Not known	Basaltic/basanitic	Phreatomagmatic	Yes
Mña. Escachada	Tenerife	Pleistocene	Not known	Basaltic	Phreatomagmatic	Yes
Mña. Amarilla	Tenerife	Pleistocene	Not known	Basaltic	Phreatomagmatic	Yes
Caldera del Rey	Tenerife	Pleistocene	Not known	Phonolitic	Phreatomagmatic	Yes
Teide (Calvas del Teide)	Tenerife	Pleistocene	Not known	Phonolitic	Phreatomagmatic	No
La Caldereta	La Palma	Pleistocene	Not known	Basaltic	Phreatomagmatic	Yes
Mña. Amarilla	Lanzarote-La Graciosa	Pleistocene	Not known	Basaltic	Phreatomagmatic	Yes
Caldera Blanca	Lanzarote	Pleistocene	Not known	Basaltic	Phreatomagmatic	Yes
Caldera de los Marteles	Gran Canaria	Pleistocene	Not known	Basaltic	Phreatomagmatic	No
Mixed Eruptions						
San Juan (Hoyo del Banco-) Durazneru — Hoyo Negro)	La Palma	1949	38	Basanites, tephrites	Phreatomagmatic opening phase-Strombolian	No
Nuevo del Fuego-Tinguatón-Tao	Lanzarote	1824	90	Basaltic	Phreatomagmatic with final phreatomagmatic phase	No
El Charco	La Palma	1712	56	Basanites, tephrites	Strombolian-phreatomagmatic opening phase	No
Fuencaliente volcano (previously) confused with San Antonio Volcano	La Palma	1677—1678	66	Basanites	Strombolian with Phreatomagmatic opening phase	No
Los Erales	Tenerife	Quaternary	Not known	Basaltic	Phreatomagmatic opening	No
El Golfo	Lanzarote	Quaternary	Not known	Basaltic	Phreatomagmatic opening	Yes

Data sources: Carracedo and Day (2002), Carracedo and Troll (2006), Carracedo et al. (2007), Klügel et al. (1999), Clarke et al. (2005, 2009).

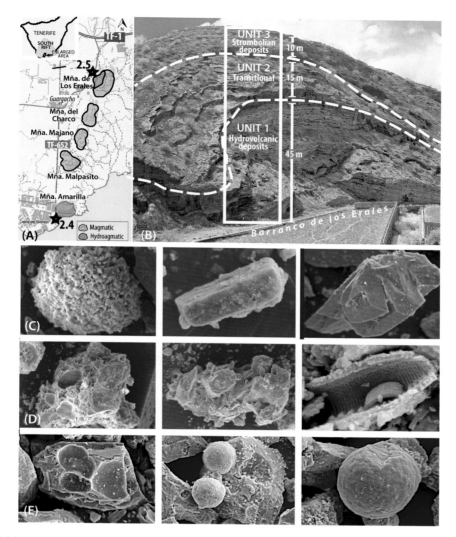

FIGURE 5.66

(A) Alignment of cinder cones with evidence of magma–water interaction indicated (green). Most of these vents may have been initiated as phreatomagmatic cones, but this is only visible at Mña. Amarilla, on the coast, and at Mña. de los Erales, a partially eroded cone a little up the barranco. (B) View of Mña. de los Erales. The eroded western flank of the volcanic cone shows three distinct successive stages: an early phreatomagmatic, an intermediate transitional, and a terminal magmatic (Strombolian) stage are recorded. (C) Globular and blocky tephra of the early phase (unit 1). (D) Transitional tephra (unit 2) with a diatom in the deposit (right hand image). (E) Spheroidal vesicles imprints and droplets (unit 3) indicate strombolian activity (after Clarke et al., 2009).

Grain size and morphology of tephra particles change up-section (Fig. 5.66B), indicating a more fluidal magmatic regime during the final stages of Los Erales eruptive history, reflecting an essentially strombolian end phase to this particular cone here. Notably, water-living diatoms, one of the most common types of freshwater and marine environment phytoplankton, were found in the lower tephra unit (Fig. 5.66D), confirming external water input as a key factor here.

Continue north on TF-652 and, if you have not already parked here, note that there is a larger restaurant located just a few meters up the road from Los Erales (look for a big purple house). This place maybe a pleasant refreshment stop if you feel like having a little break. Otherwise, continue for approximately 600 m until you meet a roundabout close to the motorway TF-1. Turn right (east) onto TF-655, which is parallel to the motorway, for approximately 1.7 km. Then, follow the signs for *Los Abrigos* and turn right onto TF- 65. Once there, pass through this town and follow TF-643 toward *El Médano*. Between km 5 and 4 on TF-643, you will be crossing sections of the Abrigo ignimbrite. Outcrops of this ignimbrite frequently appear overlying older ignimbrites or palaeosoils in erosive discordance (Fig. 5.67A). Here, the Abrigo ignimbrite is formed by two depositional units, an upper lighter one and a lower lithic-rich one, that rest unconformably over older substrate (Fig. 5.67).

The 188 ± 19-ka-old Abrigo ignimbrite (Pittari et al., 2006), the last major explosive caldera-forming eruption on Tenerife (see Fig. 5.5), is probably also the most widespread and voluminous deposit of this type on Tenerife. Main outcrops of this ignimbrite are found at the northeast sector Diego Hernandez Formatin (DHF) of the Las Cañadas Caldera wall and on the southern coastal plain, particularly between *Los Abrigos* and *El Médano*, as well as around *San Miguel de Tajao* (see Fig. 5.16A).

Continue for a short while and then turn left, following signs for *Cueva Santo Hermano Pedro* to inspect the depositional facies of the Abrigo ignimbrite in more detail. Stop at *Cueva Santo Hermano Pedro* and park the car next to the cave complex in the parking spaces provided. Note that there are bathroom facilities. Also note that this is a religious site, and we advise being careful of what you wear.

Stop 2.6. Cueva del Santo Hermano Pedro. *The Abrigo ignimbrite (N 28.0505 W 16.5534)*

Caves were often used as dwelling sites by the Guanches; many of them, particularly in the south of Tenerife, were excavated from softer layers such as paleosoils or air-fall pumice deposits that underlie harder units or beds. This cave here is particularly spacious, with a compact 2- to 3-m-thick ignimbrite unit forming the roof of the cave (Fig. 5.67B).

Walk around the little walkway for a closer inspection of the light gray phonolitic, lithic-rich ignimbrite deposit and its internal structure and features.

After the inspection, continue north on the small road until you can join TF-64 again (with direction TF-1 left). Cross TF-1 and enter *San Isidro*. Stay on TF-64 until a roundabout takes you onto TF-636 in the direction of *Chimiche* (to the right). Stay on TF-636 for a while until you reach TF-28 (the "old South road"), which you join, again in the direction of *Chimiche* (to the right). Pass through *Chimiche* on TF-28 and approximately 300 m after the village, and just after a sharp bend, there is a parking place on the right side, opposite to the entrance to the Chimiche quarry. Park there.

FIGURE 5.67

(A) Section of the Abrigo ignimbrite. The upper lighter (UU) and lower lithic-rich (LU) depositional units rest on an erosive unconformity with older ignimbrites and palaeosoils. (B) The cave of Santo Hermano Pedro, which was partly excavated by him. The site offers beautiful sections of the Abrigo ignimbrite.

Stop 2.7. The Granadilla pumice (Chimiche quarry) (N 28.1386 W 16.5303)

Walk up-hill for a few steps on the other side of the road and, after a short walk, the quarry face reveals a 12-m-thick cooling unit of light, creamy-white unwelded ignimbrite overlying an 8-m-thick plinian air-fall pumice (Fig. 5.68). This section is one of the thicker parts of the Granadilla ignimbrite deposit that covers more than 150 km^2 of the Granadilla-Arico sector. The deposit is attributed to the collapse of a plinian eruption column (see Figs. 5.11 and 5.12), which was most likely erupted from the area of the upcoming Las Cañadas Caldera.

The Granadilla pumice shows the dispersal pattern of an ellipse that may have covered more than 2400 km^2 with a 1-m-thick pumice deposit (see Fig. 5.11). A thin pumice layer associated with this eruption is believed to have spread over the entire island (Booth, 1973).

Pumice and ignimbrite from this eruption are extensively used in the south of Tenerife for paving (Fig. 5.69A), wall construction (Fig. 5.69B), and in artificial terrace arrangements used in

FIGURE 5.68

The Chimiche quarry. (A) Phonolitic pumice air-fall and ignimbrite from the plinian 0.6 Ma Granadilla eruption. (B) Close-up view of the pale cream, compacted ignimbrite. (C) Details of the air-fall pumice.

agriculture (Fig. 5.69C). Pumice rock offers excellent insulation due to its porosity (often >50% by volume), and therefore makes for a light but attractive building material. Moreover, it is easy to carve into the Granadilla deposits or to dig caves into the pumice beds, which were traditionally used as living quarters for humans or animals and continue to serve as cellars and storerooms (Fig. 5.69D).

Continue up-hill and pass through *El Río* village. Note the increasingly frequent terracing on the ignimbrite slopes in this region. Drive through to the next village and, once you have passed through *Las Cisneras*, approximately 1 km after the village, you will meet a sign for a "cul-de-sac" with a green label "Vera Herrera." Drive there and park the car at a convenient spot.

Stop 2.8. The Arico ignimbrite (N 28.1675 W 16.5068)

On your right, a small disused tarmac road allows inspection along the full thickness of the Arico ignimbrite deposit here at its locality (see Booth, 1973). The Arico ignimbrite was first described as late as the early 1970s, and this outcrop here is where Booth made his most critical observations. Along the section, you will pass the lithic and fiamme-rich base of the deposit, which grades up-section into a more ash-dominated and generally finer-grained depositional regime.

Once you have inspected the Arico deposit, return to the car and continue on TF-28 toward *Arico* village. Once you arrive in *Arico*, a small signposted road takes you to the village church (N 28.1661 W 16.5013). Note that the church is partly built from Arico ignimbrite (see Fig. 5.10F), as is the imposing colonial public building opposing the church.

FIGURE 5.69

The Granadilla ignimbrite is commonly used in the south of Tenerife for paving (A) and walls (B). The pumice is also frequently utilized to correct farm soil and build terraces for irrigation (C). Pumice beds were often carved out by the aborigines and early Castilian settlers to make dwelling caves for people and animals and cellars for food and crops (D).

As a distinctive characteristic feature of the Arico ignimbrite, this is the only ignimbrite on Tenerife that is intensely welded and very hard. For this reason, it has been traditionally used as building and ornament stones (see Fig. 5.10).

This is a good place to end the day, but for those still willing, another option to end the day is to inspect the actively working Arico ignimbrite quarry of *Guama*, but you may want to contact the quarry owners in advance (www.guamarico.com/piedra.html; info@guamarico.com). Note that the quarry is only open during working days. To get to the quarry (N 28.1319 W 16.5007), make your way back to TF-28 and continue toward the end of *Arico* village. There, join TF-629 in the direction of the motorway (TF-1) at a rather newly created roundabout. This road (TF-629) will bring you to TF-1. After 6.5 km on TF-1, take exit 49 and follow signs for *El Rio*. A narrow road (Carretera general de El Río) first runs parallel to TF-1 (but in the opposite direction) and then makes its way up-hill toward *El Río*. After approximately 4 km on this road, you will arrive at the entrance to a quarry signposted *Guama-Arico* (N 28.1287 W 16.5058). Take this narrow tarmac road and when you reach a junction, turn right. After a few minutes you will reach the quarry. After inspection of the quarry, return to TF-1 and to your base.

END OF DAY 2.

DAY 3: LAS CAÑADAS CALDERA AND THE TEIDE—PICO VIEJO VOLCANIC COMPLEX

This day aims to collect observations of three of the most spectacular geological features of Tenerife: the northwest rift zone, Las Cañadas Caldera, and the Teide and Pico Viejo stratovolcanoes. Teide—Pico Viejo are of particular global geological interest and deserve a closer inspection, including ascent to their summit (*Day 4*). For today, there will be possible refreshment stops, but there is no fuel station inside the Las Cañadas Caldera (LCC). Make sure you have plenty of fuel!

Make your way to the *Boca de Tauce* crossroads as the starting point of *Day 3* (Fig. 5.70). If coming from *Los Cristianos*, take the route to Las Cañadas and Teide via *Vilaflor* on TF-51. Pass *Vilaflor*, the highest village on Tenerife and in all of Spain, and take TF-21 to the entrance of the Las Cañadas Caldera. From *Vilaflor* to *Boca de Tauce* crossroads (~15 km), the road crosses thick phonolite lava flows and coulees. These erupted from vents and lava domes of the Las Cañadas Volcano (LCV) that are located at the top of the present LCC rim. Past km 64 on the road signs of TF-21, the entrance to a track provides a safe place to park the car. Walk up-hill on the track for approximately 350 m to inspect some of the phonolitic coulées, including the "Sombrero the Chasna."

Stop 3.1. Sombrero de Chasna (N 28.1771 W 16.6450)

You will see in front of you (looking north) the most spectacular lava dome in this part of the island, the "Sombrero (hat) de Chasna" (Fig. 5.71A—D). Differential erosion left the harder lava forming a flat-topped and approximately circular feature known as a "sombrero" on Tenerife because it has a resemblance to a hat. In other instances, they are known as "fortaleza" (fortress) (eg, *Fortaleza de Chipude* on La Gomera).

Continue on TF-21. Upon passing the km 62 road sign you will see, looking east, a characteristic welded ignimbrite flow cut by the road. It shows flow banding or sub-parallel platy joints presumably in response to ductile flow and cooling stresses (Fig. 5.71E).

Just past km 59 on TF-21, park the car in front of the entrance of a restaurant and walk 350 m up-hill on the road to inspect large accretionary balls.

Stop 3.2. Accretionary phonolite lava balls (N 28.1919 W 16.6674)

Phonolite flows on Tenerife do form accretionary lava balls, especially when deposited on a slope. A superb example crops out on a flow cut by the road here (Fig. 5.71F). Accretionary lava balls form as small fragments of solidified lava that detach from the rear of a flow and roll down-hill on the surface of the hot flow, thus accreting lava layers to the rolling mass, much like snow sticks to a rolling snowball (Fig. 5.71G).

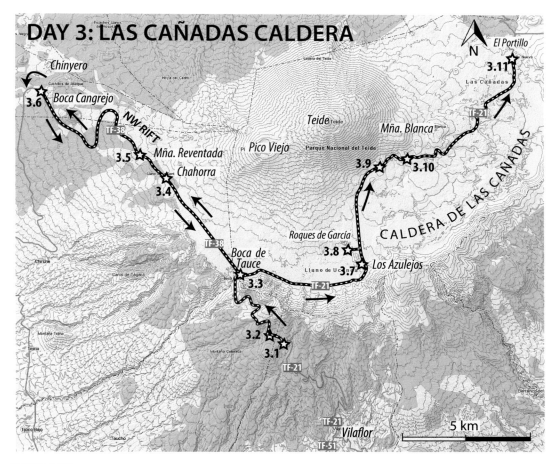

FIGURE 5.70

Map with the stops for *Day 3*. The Las Cañadas Caldera and Teide volcano. Map from IDE Canarias visor 3.0, GRAFCAN.

Continue on TF-21. On arriving at *Boca de Tauce* (the crossroads of TF- 21 and TF-38), park the car for a first view of the Las Cañadas Caldera (LCC) and some fascinating dyke structures that are exposed here at its margin due to preferential erosion.

Stop 3.3. Boca de Tauce cone sheets and ring dykes (N 28.2142 W 16.6783)

Several prominent dykes crop out on the side of the road, and phonolitic radial dykes, cone sheets, and semi-vertical concentric dykes can be inspected (Fig. 5.72A). These intrusions (part of the former Las Cañadas Volcano) can be traced for considerable distances around the caldera and reflect repeated cycles of inflation (cone sheets and radial dykes) and deflation (ring dykes) of the emerging Las Cañadas Volcano (Walter and Troll, 2001; Troll et al., 2002). Cone-sheet dykes commonly

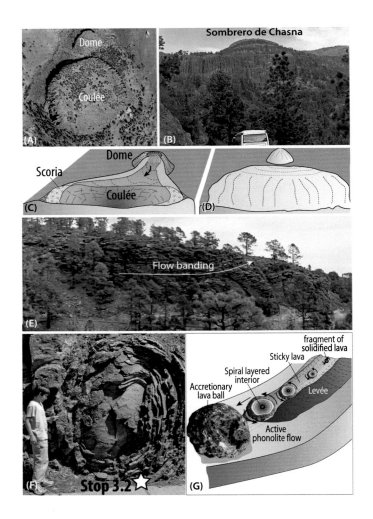

FIGURE 5.71

Some volcanic features on the road from *Vilaflor* to *Boca de Tauce*. (A) Vertical *Google Earth* image of Sombrero de Chasna. (B) Sombrero de Chasna dome complex viewed from *Stop 4.1*. (C) Sketch of the formation of Sombrero de Chasna lava dome and coulée system. (D) The same after erosion. (E) Flow foliation in a 30-m-thick welded phonolite rheomorphic flow (km 62 on TF-21). (F) Accretionary lava ball (km 59.3 on TF-21). (G) Cartoon showing the formation of accretionary lava balls. A small fragment of solidified lava rolls along the surface of an a'a flow. More and more lava adheres to the rolling mass, much like snow sticks to a rolling snowball, and individual balls can grow to considerable size.

converge at a common apex at depth, describing an inverted conical geometry, which gave rise to the term "cone sheet" (Fig. 5.72B; see also Figs. 6.20−6.22). Magma pressure can superimpose a tensional regime on the host rock and form inverted conical fractures, the intrusion of which produces cone sheets and radial dykes (Walter and Troll, 2001). Decrease in magma pressure, in turn,

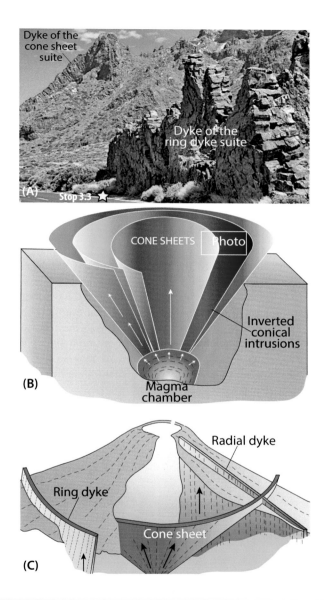

FIGURE 5.72

(A) Dyke intrusions (ring-type dykes, radial dykes, and cone sheets) exposed in the wall of the Caldera de Las Cañadas at *Boca de Tauce*. (B) Sketch illustrating these structures. (C) Formation of a cone-sheet intrusion by excess pressure from an underlying magma chamber.

will cause subsidence of the roof rock along outward-dipping fractures, which when intruded by magma will form ring dykes (Fig. 5.72C).

Forming and maintaining a crustal magma chamber depends strongly on the dyke-injection frequency (eg, Gudmundsson, 1990). The intersection of rift zones at the center of Tenerife (see Fig. 5.35A) is the region where the maximum volume of injections of magma through dykes from deep sources is expected. Phases of intense magma intrusion likely favored transition from vertical dykes to sills and, ultimately, led to the development of a shallow magma chamber system in the Las Cañadas and Teide situations (eg, Mathieu et al., 2008).

Take TF-38 to the west and drive for approximately 3.3 km to where the road crosses black Chahorra lavas and ochre Pico Viejo phonolites. The western Las Cañadas Caldera wall should now be on the left side in some distance. A "mirador" to the right then provides parking facilities and a closer view of Pico Viejo and the 1798 Chahorra vents and lavas on its flank.

Stop 3.4. Mirador de Chahorra (N 28.2384 W 16.6983)

The 1798 Chahorra eruption was described in detail and illustrated in contemporary accounts. Bory de Saint-Vincent arrived on Tenerife in 1801, 2 years after the Chahorra eruption, and obtained detailed eye-witness information and copied a color drawing of the active eruption by Bernardo Cólogan (Fig. 5.73A). The 1798 eruptive vents formed along a 1.2-km-long radial fracture on the southwest flank of the Pico Viejo stratocone (Fig. 5.73B). The eruption was mainly strombolian, but violent explosive events (probably phreatomagmatic pulses) were reported. Although located on the flank of the felsic Pico Viejo stratocone, the Chahorra eruption shows an unusual intermediate (tephriphonolitic) composition, most probably a function of magma mixing where deep NWRZ mafic magmas blended with phonolite magma from the shallow and differentiated Teide and Pico Viejo magma reservoirs of the central felsic complex.

FIGURE 5.73

(A) Copy (by *Bory de Saint-Vincent*) from *1801* of an original color drawing of the 1798 Chahorra eruption by **Bernardo Cologan** (image courtesy of *Cabildo Insular de Tenerife*). (B) View from *Stop 3.2* onto the 1798 Montaña Chahorra vents and lava flows. Note the aligned vents, indicating a radial eruptive fissure on the flank of Pico Viejo. Small tree for scale (circled).

FIGURE 5.74

Vent and lavas of the 1798 Chahorra eruption (black) that flowed over reddish and altered Pico Viejo lava flows. The Chahorra lavas were halted by the western wall of the Las Cañadas Caldera.

The main lava flow of the 1798 eruption, the longest (3 months and 6 days) of the historical eruptions on Tenerife and the second-longest in the Canary Islands (after the Timanfaya eruption on Lanzarote, 1730−1736), ponded to a considerable thickness in places (15−20 m), as was confined by the western caldera wall that formed a natural barrier to down-hill flow on that occasion (Fig. 5.74).

Continue on TF-38 in a westerly direction. Past the km 8 road sign, a long and steep bend takes you through the thick lava flow of Mña. Reventada. Park the car on the open ground before the outcrops and walk along the road section. You are near a tight bend, so please use the necessary safety measures (eg, high visibility vest, safety triangle, etc.).

Stop 3.5. Mña. Reventada composite flow (N 28.2724 W 16.7287)

Mña. Reventada, located on the northwest rift at the foot of Pico Viejo (Fig. 5.75), provides an example of mixed basanitic and phonolitic magma eruption within a single lava flow (see Figs. 5.44 and 5.45). The road cuts the 15-m-thick flow and exposes a massive phonolite/basanite flow resting on a basement of baked lapilli. Dense and partly columnar lava at the lower and central parts of the outcrop grade into a scoriaceous breccia at the top (Fig. 5.76A). A close inspection shows that the lower, columnar bottom is formed by massive basanite, whereas the central and top parts are made of porphyritic phonolite that carries basanite enclaves (Fig. 5.76B).

Mña. Reventada is an excellent example of petrogenetic bimodality in recent Teide activity and reflects interaction of a deep, basanitic magma feeding system to the northwest rift zone (Fig. 5.76C) that interacted with differentiated (phonolitic) magma from a shallow chamber or pocket of the central Teide−Pico Viejo complex (Wiesmaier et al., 2011, 2012).

FIGURE 5.75

Montaña Reventada eruptive vent and lavas. Pico Viejo and the Roques Blancos in the background, and Teide volcano in the far distance. Montaña Reventada is a complex eruption that vented in the zone of interaction between the northwest rift and the central felsic volcanoes Teide and Pico Viejo (TPV). This eruption therefore involved mixing of mafic and felsic magmas.

At Reventada, a basanite lava flow erupted and was soon followed by the eruption of phonolite lava from the same vent, thus forming a composite flow or cooling unit. The phonolite part of the flow contains abundant dark inclusions (enclaves) that appear to be related to the basanite part. Syneruptive fractures within the basanite base have been filled with phonolite (Fig. 5.76D), indicating a considerable temperature contrast and short-term interaction between basanite and phonolite only, which was probably insufficient to achieve homogenization. At Montaña Reventada, hybridization remained incomplete because mixing was apparently interrupted by eruption, and the inclusions reflect only short-term interaction between basanite and phonolite. However, prolonged interaction of basanite and phonolite would likely lead to homogenization of the liquid magma portions and may be one of the processes responsible for producing the rare intermediate magmas in the Canary archipelago and other ocean islands elsewhere.

Return to the car and continue down-hill for approximately 4 km. Past km 14 on the road sign, the road crosses the vent and lavas of the 1492 Boca Cangrejo eruption. After only a short distance (1.3 km), park the car at the entrance to a hiking trail to the 1909 Chinyero volcano. Note, accessing this track with a vehicle requires a valid permit.

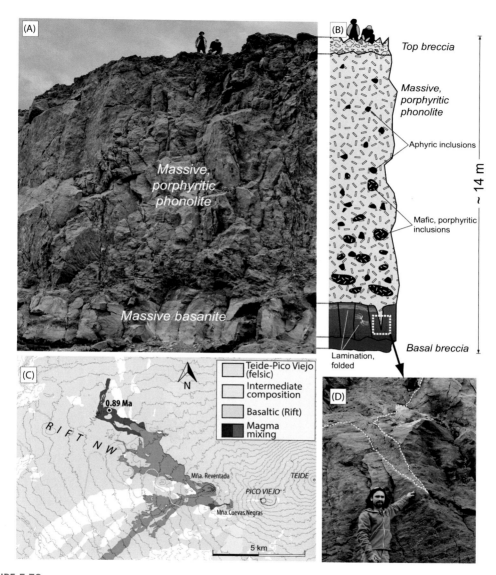

FIGURE 5.76

(A) Photograph of the main outcrop of the Montaña Reventada composite flow. Note people for scale. (B) A simplified stratigraphic column of this main outcrop. (C) Location map from Carracedo et al. (2007b). (D) An opened fracture within the basanite that has been filled with phonolite (images from Wiesmaier et al., 2011 and Carracedo and Troll, 2013).

Stop 3.6. Chinyero and Boca Cangrejo volcanoes (N 28.2846 W 16.7626)

Lava flows east of the entrance to the Chinyero trail are very fresh and completely devoid of vegetation. These come from a relatively small basaltic cone, the Boca Cangrejo volcano (Fig. 5.77A), whose radiocarbon age (400 ± 110 years BP; Carracedo et al., 2007a,b) is compatible with a report by Columbus in August 1492 AD. Columbus reported: "*A big fire in the Sierra del Teide*" (see Fig. 5.42) during his voyage from La Gomera to Gran Canaria to get one of his ships repaired before his journey to America (Fig. 5.77B). He records further: "*I explained to the sailors the cause of that fire with the example of Mount Etna, on Sicily, and many others, where the same thing had been observed*" (**C. Columbus, cited in** Carracedo et al., 2007a). The 1492 flows are of tephritic composition and spread into numerous branches or lobes once they detached from the main vent. Finally, the branches merged into a single flow that cascaded into Barranco Santiago del Teide and stopped approximately 600 m from the coast (see Fig. 5.27A).

FIGURE 5.77

(A) *Google Earth* image of *Stop 3.6*, with the Boca Cangrejo (1492) and Chinyero (1909) volcanoes at the upper right and left, respectively. (B) Columbus reported an eruption on the northwest rift of Tenerife during his voyage from La Gomera to Gran Canaria on August 24, 1492. Columbus arrived first on La Gomera from Spain on August 11, 1942, and returned to Gran Canaria to repair the rudder of one of his caravel ships (the La Pinta) on August 24, 1942. (C) The Chinyero eruption of 1909 (contemporary photograph courtesy of *Cabildo Insular de Tenerife, Centro de Fotografía Isla de Tenerife*). (D) Present-day view of the Chinyero vent and lava flows.

Looking north, the Chinyero vent is also visible (the 1909 eruption). Numerous eye-witness accounts, scientific articles, and newspaper reports describe this sole 20th-century eruptive event on Tenerife, which was also the first eruption to have been photographed in the Canaries (Fig. 5.77C).

Actually, both volcanoes are very similar in dimension and in their state of preservation, although Chinyero is approximately 500 years younger than Bco. Cangrejo (Fig. 5.77D). Walk up to the Chinyero vent (approximately 800 m) for closer inspection. Following the inspection return to your car, and from there return to *Boca de Tauce (Stop 3.3)*, from where we continue the tour of the LCC and of Teide volcano.

From *Boca de Tauce*, continue on TF-21 (ie, along the base of the caldera wall). On the left you will pass pahoehoe lavas of Pico Viejo, dated at 20.7 ka, and phonolite flows, dated at approximately 17.5 ka (see Fig. 5.37; Carracedo et al., 2007b). To the left extends the western Las Cañadas Caldera floor (*Llano de Ucanca*), forming a closed or endorheic watershed where the topography prevents drainage to the ocean. Water from rain and melted snow leaves the drainage system by evaporation and seepage, depositing sediments washed from the caldera wall. Very fine evaporite sediment (white colors in Fig. 5.78A) occurs frequently and forms from short-lived (ephemeral) lakes (Fig. 5.78B).

Continue on TF-21. Just after the km 48 road sign and a signpost for *Los Azulejos*, park the car for a close view of the colorful Los Azulejos formation (Fig. 5.78C). These natural colors are used in the traditional carpets prepared every year in the town plaza of *La Orotava* (Fig. 5.78D) using finely ground material from *Los Azulejos* and other sands and soils from the Teide and LCC areas.

Stop 3.7. Los Azulejos (N 28.2188 W 16.6279)

The color of volcanic rocks is of quite some interest and also of scientific relevance. Primary colors often provide very relevant bits of information. Dark colors are typical of mafic rocks, whereas light colors are characteristic of felsic ones. However, obsidian at Teide is an exception, because although its composition is phonolitic, its color is a deep black or brown. This is due to the fact that obsidian is an amorphous glass that blocks light through opaque impurities and microscopically small crystals. Teide's typical jet-black varieties of obsidian are likely derived from the abundant microscopic crystals of magnetite, whereas portions of the obsidian have mixed layers of black glass and light-colored pumice. The latter is also glassy, but bubble-rich and lightweight, and develops when gas escapes rapidly from the molten glass, resulting in obsidian that is "streaked" in texture and color (Fig. 5.78E). This concept will become much clearer the next time you have the opportunity to investigate frothy and liquid portions of, for example, a pint of ale, where color differences within a glass of the same liquid can be striking.

Please also note that the edges of conchoidal fracture surfaces of obsidian can be sharper than a razor. This had obvious advantages for the primitive Guanches, who used obsidian extensively as a tool and for tool making (Fig. 5.78F).

Regarding rock colors, secondary colors are generally due to weathering or hydrothermal overprint and geologists usually have developed an eye for primary and secondary rock colors. Los Azulejos is a good example of secondary rock colors. These formed by fault-controlled hydrothermal alteration that occurred during the lifetime of the large Las Cañadas Caldera volcano and prior to the giant landslide that formed the present-day Las Cañadas Caldera approximately 200 kyrs ago. Circulation of hot and mineralized sub-volcanic fluids through these fractures along the caldera wall and floor most likely did not exceed 350°C, because the precipitated minerals that give

FIGURE 5.78

(A) Sediment partially filling the Las Cañadas caldera at "Llano de Ucanca." (B) The finer evaporite sediment formed from ephemeral lakes. (C) Los Azulejos hydrothermal formation at the margin of the Las Cañadas Caldera, indicating hot fluid circulation in the faulted marginal domains of the caldera basin. (D) Traditional pavement carpets are painted annually on the town hall plaza of *La Orotava* from finely grinded material from Los Azulejos and other sand and soils from Teide (courtesy of *Ayuntamiento de La Orotava*). (E) Obsidian from the latest medieval eruption of Teide. (F) The obsidian was used by the guanches to produce working tools like knives and spearheads. Image courtesy of Cabildo Insular de Tenerife.

rise to the color of the Azulejos suite are characteristic of "hydrothermal" fluid—rock interaction (clays, chlorites, micas, zeolites, adularia, epidote), giving the rocks bright blue-greenish colors in some places, but pink to orange colors in others (see Fig. 5.78C). A similar case of hydrothermal alteration is seen on Gran Canaria along the margin of the Miocene Tejeda caldera (see Fig. 6.69) (eg, Donoghue et al., 2008).

Continue from the Los Azulejos stop for approximately 0.5 km and take the first turn left to park at the *Mirador Roques de García*.

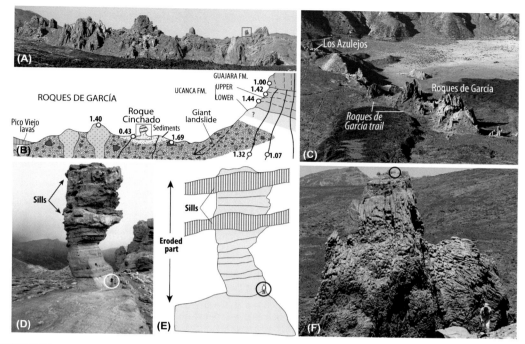

FIGURE 5.79

(A) View of the Roques de García alignment from the south (B) Geological cross-section. Ages (K/Ar) in Ma (after Huertas et al., 2002). Note that all the ages, including the youngest of 0.43 Ma from a dyke, indicate formations prior to the formation of the present Las Cañadas Caldera that formed from a large lateral landslide (the Icod collapse) at approximately 200 kyr BP. (C) *Google Earth* view onto the Roques de Garcia trail. (D and E) Roque Cinchado, an erosion remnant that resisted weathering and denudation. The hard rock of the younger sills have slowed erosion, whereas surrounding rock has been more efficiently eroded away. The Roque Cinchado (red squares in A and B) experienced greater erosion at its softer base (a person is circled for scale). (F) The cathedral, one of the imposing pinnacles of the Roques de García ridge, is an intensely eroded phonolitic plug (a person on top is circled to show scale).

Stop 3.8. Mirador Roques de García (N 28.2231 W 16.6307)

This mirador provides an exceptional view of the Roques de García alignment, the perimeter of the Las Cañadas western collapse basin (Fig. 5.79A).

The Roques de García comprises a complex lithological assemblage, including debris avalanche deposits, derived from various Las Cañadas Volcano landslides (dated at ∼1.69 Ma; Huertas et al., 2002). Moreover, sediments intruded by felsic volcanic necks (1.4 Ma) and mafic to felsic dykes (0.43 Ma) are preserved (see cross-section in Fig. 5.79B).

The Roques de García crest divides the floor of the present Las Cañadas Caldera into two collapse basins that differ in elevation by 100 to 120 m (Figs. 5.79C and 5.80A). This observation was interpreted by some workers (eg, Martí and Gudmundsson, 2000; Coppo et al., 2008) through the

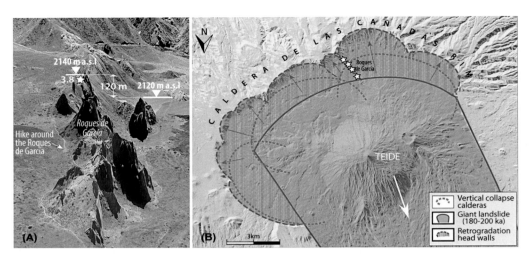

FIGURE 5.80

(A) The Roques de García alignment divides the present Las Cañadas Caldera into two basins that differ in elevation by 100−120 m (image from *Google Earth*). (B) Different interpretations of the multiarcuate configuration of the caldera rim exist: three vertical collapse calderas are thought to have built the current Las Cañadas caldera complex (dotted red ellipses) (eg, Martí and Gudmundsson, 2000), whereas, alternatively, a giant landslide scarp that was enlarged by retrograde erosion of the headwalls into abutting barrancos has also been considered (eg, Carracedo et al., 2007) (dotted blue ellipses). Recent and partly unpublished age data now seem to indicate that the vertical collapse regime of the Las Cañadas volcano was succeeded by a large lateral collapse at approximately 200 kyr (see also Huertas et al., 2002; Carracedo et al., 2007).

differential collapse of two separate vertical calderas. However, the observation can also be explained with a giant landslide depression that was created during several events, thus potentially preserving a series of former wall remnants. The giant landslide scar was then likely also enlarged by retrogradation of the head walls of abutting barrancos that developed along the walls of the unstable landslide depression (Fig. 5.80B). The uneven floors of the different collapse basins are then perhaps merely caused by their different rate of filling with pyroclastic deposits, reworked sediment, and lava flows. Notably, the main supply of lava to the western collapse basin ceased approximately 17 ka ago (see Figs. 5.37 and 5.39), whereas the filling of the eastern collapse basin continued during the Holocene (see Fig. 5.41A), which might be a sufficient reason to explain the different base levels across the LCC.

So, in summary, the scalloped rim of the Las Cañadas Caldera has been used as support of the existence of up to three overlapping vertical collapse calderas (red broken lines in Fig. 5.80B; Martí and Gudmundsson, 2000) or the multiarcuate configuration of the backward eroded headwall(s) of the giant Icod landslide (see, eg, Carracedo and Troll, 2013; Fig. 5.80B).

In a hazard respect, both of the two main basins of the LCC are confined and lava can only escape at the northern and western edges. The wall of LCC acts as an effective topographic barrier that protects the southern flank of Tenerife from resurfacing (Fig. 5.81A). This is well illustrated

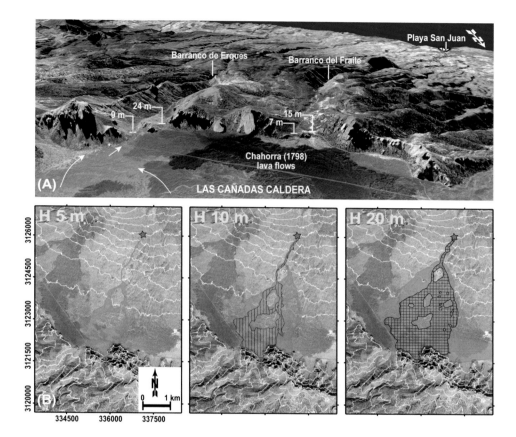

FIGURE 5.81

(A) Oblique view from the northwest onto the southern wall of the Las Cañadas Caldera (image from *Google Earth*) showing the breaches ("*portillos*") carved by the backward erosion of the main barrancos. These gaps would be the pathways for a potential overflow of lava during future eruptions inside the Las Cañadas Caldera, threatening villages on the southern coast. (B) GIS simulations show that for overflow to commence, a 20-m-thick lava flow or fill is needed. From Carracedo and Troll, 2013.

by the 1798 Chahorra lavas. A GIS simulation (Fig. 5.81B) shows that a 20-m-thick layer of lava is required to allow the lava to overflow the lowest gap (Barranco de Erques) and to spill toward the southern coast.

You may hike around the Roques de García ridge with its various spines, which will take you approximately 1.5 hours, starting at *Stop 3.8* (see Figs. 5.70 and 5.80A). Alternatively, continue directly to our next stop.

For that, return to TF-21 and continue for approximately 10 minutes in the direction toward Teide. Then, turn left into the entrance of the road up to the Teide cable car (at km 43). Here, you

FIGURE 5.82

Summit cone and black obsidian phonolite lavas of the last (medieval) eruption of the Teide stratocone that yielded a ^{14}C-calibrated age of 660 to 940 AD. Image courtesy of Teide National Park, Cabildo Insular de Tenerife.

have the option of a comfortable climb to the summit of Teide by cable car. We address this option on *Day 4* (Option B). The cable car costs a fee of €20, and it brings you to the summit in approximately 10 minutes.

Stop 3.9. Teide medieval eruption (Lavas Negras) (N 28.2544 W 16.6227)

The latest volcanic activity of Teide stratocone occurred in medieval times, between 660 and 940 AD (Carracedo et al., 2007b). This eruption built the 220-m summit cone, increasing Teide's elevation from approximately 3500 m to the present 3718 m, and erupted a number of lava tongues (Fig. 5.82). Approximately 0.66 km^3 of lava spread radially from the summit crater, partially covering the former stratocone with thick flows of glassy phonolites that are intensely black in color, hence their name, Lavas Negras (black lavas) (Fig. 5.83).

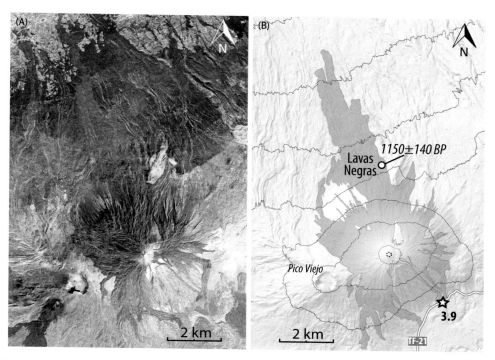

FIGURE 5.83

(A) Lavas Negras, the latest (medieval) summit eruption of Mt. Teide (image from *Google Earth*). (B) Geological map outlining the Lavas Negras and the extent to which the lavas have traveled.

The Early Middle Age occurrence of this eruption implies that local aboriginal populations would have witnessed this likely spectacular event, towering over the entire island and widely visible beyond.

Return to your car after your brief visit to Teide and continue on TF-21 for approximately 1.5 km, crossing thick coulées and pumice fields of the Mña. Blanca eruption (~2 ka).

Once you pass km 41 on TF-21, a "mirador" on the right side of the road offers you a panoramic view of the Las Cañadas Caldera, with the thick, dark red coulée (known as El Tabonal Negro) in the foreground. A brief stop here is optional. After another approximately 0.8 km, a road sign indicating Mña. Blanca and the track to *Refugio de Altavista* and Teide's summit will appear, which is the starting point for the potential hike on *Day 4* (Option A).

Continue on TF-21 to the *Minas de San José* area (another ~2 km) and park there.

Stop 3.10. Minas de San José (N 28.2649 W 16.5895)

The top of a thick coulée from the Mña. Rajada vent has been almost completely buried here by a deep layer of pumice from the Mña. Blanca's sub-plinian eruption in approximately 0 AD (~2000

FIGURE 5.84

(A) A viscous phonolite flow suspended mid-slope. Lava, in contrast to water (a Newtonian fluid), is a non-Newtonian (Bingham) fluid that requires overcoming a yield stress to flow. If this stress decreases (eg, if the lava stops flowing from the vent or if cooling is efficient), then a lava flow or tongue can become arrested even on steep slopes, as seen in the photograph. (B) Chaotic aspect of a phonolitic block lava from Mña. Rajada. Note the obsidian (arrow) at the crest of several blocks. Circled person shown for scale.

years BP). The pumice here has been mined and was traditionally used to produce lightweight concrete, abrasive toothpaste as well as in heavy-duty hand cleansers and body scrubs. The heavy environmental impact of the mining on the area forced the suspension of the activity in the late 1970s, and restoration of the area commenced after the declaration of Teide as a Spanish National Park in 1981. Have a stroll around here to take in the views.

An interesting feature here is the lava tongue to the left of the road, which is a lava flow that solidified mid-slope. This behavior is characteristic of non-Newtonian (Bingham) fluids (as opposed to Newtonian fluids such as water). The flow may have halted when the lava supply and the associated internal forward push ceased, or perhaps because it had been sufficiently cooled during flow on the slope due to thinning (or both) (Fig. 5.84A).

Walk back on TF-21 for a short distance to inspect "block lavas with obsidian." There, a spectacular, very viscous coulée from Mña. Rajada can be seen that is broken into large blocks. This happens when lava is forced to flow in a semi-solid state, and the glittering intensely black obsidian glass still shines on these chilled and fractured surfaces (Fig. 5.84B).

Return to the car and continue northward on TF-21 for approximately 6.5 km and park at the first restaurant you encounter (*El Portillo*).

Stop 3.11. El Portillo *(NE rift zone) (N 28.2937 W 16.5652)*

This stop is along the transition between the northeast rift zone and the rim of the Las Cañadas Caldera (Fig. 5.85). Walk back for a short distance to have a better view of the spectacular landscape, such as the Mña. Mostaza cone (meaning mustard, because of the color) of the northeast rift zone and, of course, of the Las Cañadas Caldera scarp. The caldera wall (here the Diego Hernández sector) shows a felsic succession that is interbedded with basaltic flows and pyroclasts that date back to the pre-Teide Las Cañadas Volcano (Fig. 5.86A).

You can take a break for refreshments in any of the restaurants in the area or drive for another 1.5 km to the Teide National Park Visitors Center, where you have a permanent display with information on the biology and geology of the Canary Islands, Tenerife and the Teide National Park, and a botanical garden area that preserves and explains local plant life in some detail (http://www. hellocanaryislands.com/nature-areas/tenerife/teide-national-park). The entrance is free of charge (opening hours 9:00 to 16:00).

From here, you may return to your base by retracing the route we used during the day.
END OF DAY 3.

DAY 4: HIKING EXCURSIONS TO TEIDE'S SUMMIT

The commented geological excursions have been planned to inspect relevant geologic formations of the island in a relatively short time and, therefore, act as a road guide. However, there are several geological features of importance that can only be reached by foot, either because there are no

FIGURE 5.85

Northeast rift vents located on the perimeter of the Las Cañadas Caldera (Volcán Arenas Negras). This eruption belongs to a cluster of basaltic cinder cones in this area with ages of approximately 30 ky.

FIGURE 5.86

(A) View of the northeast wall of the Las Cañadas Caldera (Pared de Diego Hernández). The stratigraphy (modified after Martí et al., 1994) is defined by four felsic sequences separated by basaltic rift eruptions. (B) View of Teide volcano and the Las Cañadas Caldera from the eastern caldera rim (photo courtesy of S. Socorro).

appropriate tracks or roads or because a special permit is required. Here on Tenerife, the obvious outstanding geological features that require a longer hike are the Teide and Pico Viejo stratovolcanoes.

These walking expeditions will be in high-mountain conditions and come with restrictions of a World Heritage and National Park site (Teide National Park is the most frequently visited national park in Europe and the second-most frequented one worldwide, after the Hawaii Volcanoes National Park). The hiking excursions require some careful planning; for visiting the peak, a special permit is mandatory. Sampling rocks or plants en route during the hike is not allowed because many plants here are endemic and because Teide National Park aims to preserve its botanical variety. For some of the excursions it would be advisable to form a two-car group, leaving one car at the starting point and the other at the end of the hike.

The Pico (the one-time name of Teide) was considered the highest mountain on Earth, with up to 15 miles of elevation being postulated (eg, in 1668 AD). To attempt to climb to its summit was once regarded as highly dangerous because the summit was believed to be so high that the sun's heat would be harmful. Later, Mont Blanc and the Andean volcanoes were measured and Teide was re-measured, and other mountains and volcanoes are now observed to be considerably taller (Fig. 5.87A).

Although for early scientists during the 18th and 19th centuries, the interest in Teide was primarily to measure its height (Fig. 5.87B,C) with the visits of Leopold von Buch (Fig. 5.87D), Alexander von Humboldt, Charles Lyell, and the other great naturalists, Teide played an increasingly crucial role in the development of modern geology and volcanology (see Carracedo and Troll, 2013).

The climb to the summit of Teide (3718 m) is only suitable for people in strong physical conditions and without heart problems. You will need appropriate clothing, sun protection, sufficient water, and, ideally, experience with regular mountaineering practices (see, eg, www.theuiaa.org).

Edward Barlow	1668	27 miles	
Allain manesson Mallet	1683	15 miles	
Robert Challe	1690	2730 toises	5320 m
Louis Feuillée	1724	2193 toises	4274 m
Manuel Hernández	1742	2658 toises	5180 m
John Cross	1742	2408 toises	4693 m
Thomas Astley	1744	2,25 miles	4162 m
Michel Adanson	1794	2052 toises	3999 m
Jean Charles Borda	1776	1095 toises	3713 m
Alexander von Humboldt	1799		3736 m
Charles Phillipe de Kerhallet	1851		3715 m
Charles Piazzi Smyth	1856		3717 m
Parque Nacional del Teide	1954		3718 m

(C)

FIGURE 5.87

(A) The island of Tenerife and a towering Mount Teide (Pico de Tenerife) in an engraving by Olfert Dapper from 1686. (B) Painting from the epoch of the measurement of Teide's elevation. Here, a view from the summit of Montaña Taoro, one of the stations used in the triangulation. (C) Recorded elevation of Teide volcano through time. (D) The "*Pic de Tenerife*" (Mt. Teide) and the encircling "uplifted crater" (the Caldera de Las Cañadas) viewed from the east in a drawing by Leopold von Buch. The images A, B, and D are courtesy of Cabildo Insular de Tenerife.

OPTION A. MONTAÑA BLANCA-REFUGIO DE ALTAVISTA-PICO DEL TEIDE HIKE

The ascent to the summit of Teide is possible via several routes (Fig. 5.88A), each of which differs in length and difficulty (Fig. 5.88B). The summit is restricted and requires a permit that can be obtained from the National Park office in *Santa Cruz de Tenerife* or online in advance (http://www. reservasparquesnareserv.es). If you consider staying overnight, the *Refugio de Altavista* (3270 masl, bed cost €20) also requires booking in advance (www.telefericoteide.com/altavista). We

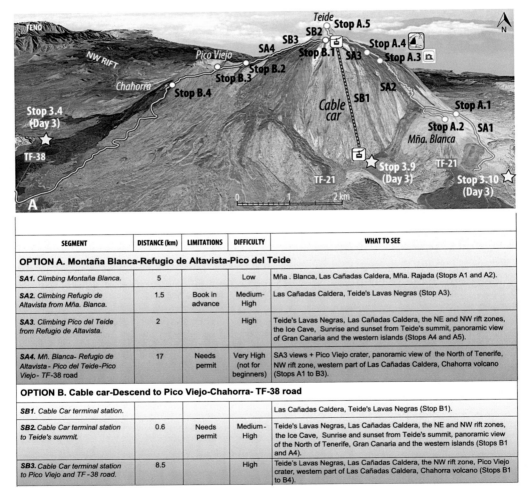

SEGMENT	DISTANCE (km)	LIMITATIONS	DIFFICULTY	WHAT TO SEE
OPTION A. Montaña Blanca-Refugio de Altavista-Pico del Teide				
SA1. Climbing Montaña Blanca.	5		Low	Mña . Blanca, Las Cañadas Caldera, Mña. Rajada (Stops A1 and A2).
SA2. Climbing Refugio de Altavista from Mña. Blanca.	1.5	Book in advance	Medium-High	Las Cañadas Caldera, Teide's Lavas Negras (Stop A3).
SA3. Climbing Pico del Teide from Refugio de Altavista.	2		High	Teide's Lavas Negras, Las Cañadas Caldera, the NE and NW rift zones, the Ice Cave, Sunrise and sunset from Teide's summit, panoramic view of Gran Canaria and the western islands (Stops A4 and A5).
SA4. Mñ. Blanca- Refugio de Altavista - Pico del Teide-Pico Viejo- TF-38 road	17	Needs permit	Very High (not for beginners)	SA3 views + Pico Viejo crater, panoramic view of the North of Tenerife, NW rift zone, western part of Las Cañadas Caldera, Chahorra volcano (Stops A1 to B3).
OPTION B. Cable car-Descend to Pico Viejo-Chahorra- TF-38 road				
SB1. Cable Car terminal station.				Las Cañadas Caldera, Teide's Lavas Negras (Stop B1).
SB2. Cable Car terminal station to Teide's summit.	0.6	Needs permit	Medium-High	Teide's Lavas Negras, Las Cañadas Caldera, the NE and NW rift zones, the Ice Cave, Sunrise and sunset from Teide's summit, panoramic view of the North of Tenerife, Gran Canaria and the western islands (Stops B1 and A4).
SB3. Cable Car terminal station to Pico Viejo and TF -38 road.	8.5		High	Teide's Lavas Negras, Las Cañadas Caldera, the NW rift zone, Pico Viejo crater, western part of Las Cañadas Caldera, Chahorra volcano (Stops B1 to B4).

FIGURE 5.88

Different hiking excursions are possible to inspect the Teide and Pico Viejo stratovolcanoes (image from *Google Earth*). Note, all of them need to be well planned and should only be attempted with appropriate equipment, maps, and supplies.

recommend booking the permit, the "Refugio," and, if desired, also the cable car (€25 for adults and €12.50 for children for a round trip) well ahead of your visit because these activities are very popular with tourists, geologists, and locals alike.

SA1. Climbing Montaña Blanca

The track to climb to Mña. Blanca (SA1 in Fig. 5.88) and to Teide's summit (SA2 and SA3) commences 2 km west of *Stop 3.10*. Look for a small parking place and signs indicating "Montaña Blanca" and "Refugio de Altavista." Park the car there and use the wide dirt track that heads north and up-hill. The track crosses a thick bed of phonolitic pumice of the Mña. Blanca (~ 0 AD) eruption. Large pumitic scoria (Fig. 5.89D) and bread crust bombs (Fig. 5.89C) are concentrated here because of the proximity of the vent.

Stop A.1. Halfway to the top of Mña. Blanca, the track passes among large accretionary lava balls that detached from the front of a black lava flow visible on Teide's slopes

FIGURE 5.89

(A, B) Large accretionary lava balls ("*Huevos del Teide*," meaning Teide's eggs) that detached from the front lobe of the medieval Lavas Negras eruption that is visible in the background. Images (C and D) show bread crust bombs from Mña. Blanca and felsic (phonolitic) scoria.

(Fig. 5.89B). The formation of these lava balls, particularly spectacular here, has been discussed above (see, eg, Fig. 5.71G).

Stop A.2. As the track winds its way up to the top of Mña. Blanca, the views gradually extend over the eastern part of the Las Cañadas Caldera (Fig. 5.90A) and a magnificent set of volcanic features becomes visible. This includes the characteristic lava dome eruptions on the south flank of Mña. Blanca that produced viscous phonolitic coulées (in the foreground) and the Mña. Rajada dome (mid-ground), which erupted extremely viscous lava flows that piled up over the vent, thus building an endogenous dome with a pronounced "rosette" structure (Fig. 5.90B).

Once you reach the top of Mña. Blanca, have a look around. Following inspection of Mña. Blanca's summit, return to the main track at the place where the trail to *Refugio de Altavista* continues (SA2 in Fig. 5.88).

SA2. Climbing to Refugio de Altavista from Mña. Blanca

This segment of the hike (SA2 in Fig. 5.88) is relatively short (1.5 km), but the steepness and high altitude (2725 masl) make it strenuous. Walk slowly and steadily when breathing gets difficult and gradually make your way to the *Refugio*.

Stop A.3. The *Refugio* (Fig. 5.91A) was built in 1892 on the initiative of the English explorer Graham Toler, who lived on Tenerife, to meet the demand of the increasing number of visiting scientists in the 18th and 19th centuries. It was refurbished in 1997 and again in 2007. At the *Refugio* you can enjoy spectacular views and impressions.

Stop A.4. We recommend that you visit the "ice cave" after having had a little rest at the *Refugio*. The sunset should be stunning at this altitude (125 m higher in altitude than the Refugio), but it is nothing compared with the view from the summit of Teide if you

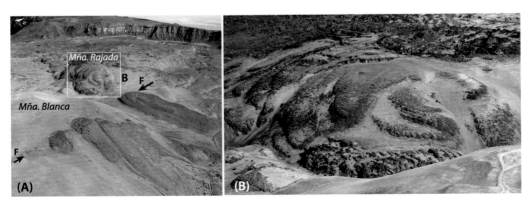

FIGURE 5.90

(A) *Google Earth* view of the eastern part of the Las Cañadas Caldera, with Mña. Blanca and Mña. Rajada in the middle ground. (B) Detailed view of Mña. Rajada from the summit of Mña. Blanca, showing a distinct "rosette" structure.

FIGURE 5.91

(A) The eastern flank of Teide shows the medieval Lavas Negras that were deposited over Mña. Blanca pumice. The *Refugio de Altavista* is indicated by a white circle (image from *Google Earth*). (B) Louis Feuillée's drawing of Teide (1708) with the location of the ice cave "Cueva del Hielo" (image courtesy of *Cabildo Insular de Tenerife*). (C) Facilities and main features to inspect during the ascend to the Teide summit (image from *Google Earth*). (D) View into the "Cueva del Hielo."

have the energy (Fig. 5.92A). The original trail to Teide's summit was laid out by the "*neveros*," local people who extracted ice from the "*Cueva del Hielo*" (the ice cave) and carried it on mules to *La Orotava*. There it was sold to rich people to prepare sorbets and ice creams. As Louis Feuillée quoted in 1724 "*the men that come to the Pico to get ice, that only exist in the foot of the sugar loaf* (the summit cone), *have made this trail*" (Fig. 5.91B). The legendary "ice cave" (Fig. 5.91C) is part of a lava tube with a skylight that allows access to the interior of the tube. Year-round ice forms inside the tube because the interior is largely protected from the sun while being at high altitude, and therefore it is exposed to relatively high rainfall and low temperatures (Fig. 5.91D).

FIGURE 5.92

Spectacular scenery can be seen from Teide summit. The shadow of Teide extends beyond Gran Canaria at sunset (A), and beyond La Gomera at sunrise (B). Sunrise from the summit of Teide rivals Maui's most spectacular sunrises on Haleakala (the House of the Sun).

SA3. Climbing Pico del Teide from Refugio de Altavista

Please note that you can only stay in the refugio for one night. You "must not miss" climbing the summit during the early morning of the second day (*Stop A.5*). From the summit you can see a spectacular sunrise (Fig. 5.92B), with the shadow of Teide silhouetted against the bright crimson sky extending over the island of La Gomera and beyond.

Around the summit crater (note that walking inside the actual crater is not allowed), which is still residually active, you might see fumaroles of steam and sulfur of up to ~90°C, which is the boiling temperature of water at this altitude (Fig. 5.93).

FIGURE 5.93

(A) Fumarols in the summit crater of Teide. Inside the crater, which is still residually active, you will see (and smell) fumaroles of steam and sulfur at approximately 86°C (the boiling temperature of water at this altitude). (B) Sulphur crystals encrusting a fumarole vent. This is the effect of sublimation, as the sulfur transitions directly from gas to solid. (C) The original summit crater was considerably deeper and with large amounts of sulfur. During World War I, the sulfur was extracted, however, reducing the crater to its present appearance. (D) Sulfur at the rim of the "Old Teide," the site of Teide's former summit prior to the medieval Lavas Negras summit eruption.

After the early morning summit climb, you can either hike back down to Mña. Blanca, take the cable car to the foot of Teide, or continue toward Pico Viejo and the TF-38 road.

Note also that you can technically proceed to the summit from the *Refugio* without a permit if you make it back to the cable car terminal before 9:00 am because the National Park Rangers guard the direct path from the cable car terminal to the summit during the day. However, you should, of course, have a valid permit with you when attempting to summit Teide, irrespective of the time of day.

OPTION B. CABLE CAR TERMINAL—ASCENT TO TEIDE SUMMIT—DESCENT TO PICO VIEJO-CHAHORRA AND TF-38

A different approach to summit the Teide is to take advantage of the cable car and be brought to the foot of the medieval cone that makes Teide's summit. You may climb the path to the summit directly from the cable car terminal (0.5 km and 175 m of altitude) if you have a valid permit.

Please keep in mind that it can get very windy, and gloves, warm clothing, and wind protection are essential. Also, note you are rapidly elevated from approximately 2350 masl to 3550 masl at the upper cable car terminal station and to 3718 masl if you climb the summit cone. This can be tricky for some, and mild symptoms of altitude sickness occur frequently (dizziness, headache, trouble breathing, etc.).

Stop B.1. The cable car travels up 1200 m to the summit cone in approximately 8 minutes. During the ascent, the views of Las Cañadas Caldera get progressively more spectacular, while right below the cable car you can see the phonolitic flows that were deposited on the steep flanks of Teide volcano.

Once you arrive at the summit terminal station, you can see beyond Tenerife to the neighboring islands of Gran Canaria, La Gomera, La Palma, and El Hierro, at least on clear days. For Teide peak hike up the trail and see the summit geology above. Return to the cable car station. Either return to the foot of Teide with the cable car now or walk from the cable car terminal westward on a well-paved path (*Telesforo Bravo*). On this path you will soon pass a levee of the Lavas Negras and you will see a patch of white rock with a spot of bright yellow. The smell and the color indicate the presence of sulfur (Fig. 5.93B,D). You are at the rim of one of the craters of the "Old Teide" volcano (pre-Lavas Negras eruption), which is intensely altered by fumarolic activity within Teide's older crater complex of pre-medieval times.

The shape of Teide volcano was quite different prior to the medieval Lavas Negras event, and it was approximately 200 m lower (Fig. 5.94A). The summit was made up of two concentric craters (Fig. 5.94B). The medieval summit cone that formed during the Lavas Negras eruption is nested in these Old Teide craters and gives Teide its characteristic present-day profile (Fig. 5.94C).

The trail then guides you through Lavas Negras for approximately another 450 m until you reach the end of the paved trail. From there, a simple path on top of the Lavas Negras lava flows starts, trodden by the millions of feet of Teide's previous visitors. Some unstable blocks are a hazard to watch, so please walk carefully.

In front of you, as you look toward the west, there is a magnificent view of Pico Viejo volcano and its spectacular 850-m-wide summit crater (Fig. 5.95A).

After another 1.3 km, the path leaves the Lavas Negras and runs on smooth pumice from nearby lava domes, such as Los Gemelos (The Twins) and La Mancha Ruana.

Stop B.2. The trail then crosses the saddle between the Teide and Pico Viejo volcanoes. The twin lava domes of Los Gemelos show prominent pressure ridges, and the high viscosity of the lava prevented it from flowing very far from the vent, solidifying in short (~300 m) coulées (Fig. 5.95B).

Continue on the main path westward. The trail climbs gently onto the rim of Pico Viejo crater and then offers a view of the entire Pico Viejo summit crater complex (Fig. 5.96).

Stop B.3. Walking on the caldera rim westward, a wide plateau of sub-horizontal bedded basaltic lavas appears soon. These form a large block in clear discordance with the steep outward-dipping lava flows that form the flanks of Pico Viejo (see Fig. 5.96). This evidence points to the occurrence of two vertical collapse events that have likely

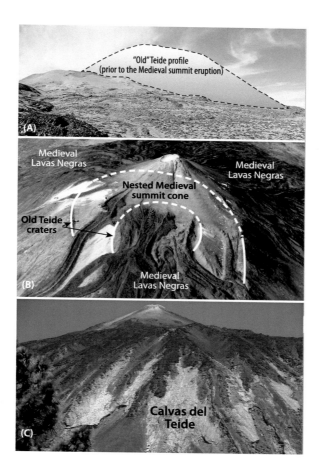

FIGURE 5.94

(A) The outline of Teide was quite different prior to the development of the present summit cone during the latest summit eruption (Lavas Negras). This new cone grew nested in the Old Teide summit crater that was likely a 1-km-wide double crater complex as shown in (B) (image from *Google Earth*). "Calvas del Teide" (Teide's bald patches) at the northern flank of Teide (C). These indurated and thick white volcaniclastic slabs correspond to phreatomagmatic surges, involving the opening of a lateral vent during a period when water was available at this altitude and had access to magma, likely when the stratocone was covered by a thick cap of snow or ice.

affected the Pico Viejo summit. The first collapse basin was filled with lava to form a lava lake, which subsequently collapsed again, leaving behind a large block of sub-horizontal bedded lavas that you can inspect today.

Moreover, a gray pyroclastic surge deposit covers the plateau and mantles the flanks of Pico Viejo. It is apparently related to the explosion crater that opened at the southern part of the summit caldera (see Fig. 5.96). This explosive activity is related to the interaction of magma with water from the local aquifer or, more likely, from

FIGURE 5.95

(A) View of Pico Viejo volcano, with the southwest coast of Tenerife in the backdrop (B) Los Gemelos, which are two lava domes with short but spectacular coulées located in the saddle between Teide and Pico Viejo.

FIGURE 5.96

Inside view of the 1-km-wide Pico Viejo summit crater. The crater was filled with successive sub-horizontal flows. Only a part of these remain, forming a block on the western side of the caldera that is discordant to the outward-dipping lavas of the crater wall. The more recent explosion pit in the foreground produced the explosion breccia deposits that cover the block of the caldera-filling lava flows and the flanks of the summit crater.

snow-melting water. Note that similar explosive (phreatomagmatic) episodes from meltwater have occurred on Teide also (see, eg, Carracedo and Troll, 2013). The northern flank of the volcano is mantled by thick indurated white volcaniclastic slabs devoid of any vegetation (see Fig. 5.94C), locally known as "Las Calvas del Teide" (Teide's bald patches). These deposits have formed as phreatomagmatic surges, likely involving the opening of a lateral vent during a period when water was available at this altitude with the stratocone covered by a thick cap of snow or ice.

FIGURE 5.97

Panorama view of the northwest rift zone from the summit of Pico Viejo volcano. The Miocene Teno massif is visible in the background, and La Palma and La Gomera can be seen in the far distance.

FIGURE 5.98

(A) Trail connecting the Pico Viejo and Chahorra volcanoes. The road TF-38 in the distance crosses the Chahorra lava flows. (B) The vents of the 1798 Chahorra eruption are aligned along a northeast–southwest eruptive fissure that is radial to the Pico Viejo summit and originates from the flank of Pico Viejo. Image from Google Earth.

Continue for approximately 300 m down the western flank of Pico Viejo to have a superb view of the northwest rift zone (Fig. 5.97). On clear days the view extends to the Miocene Teno massif and the islands of La Gomera and La Palma.

Continue on the main trail and proceed toward TF-38. After a very steep and approximately 1.4-km-long stretch, the trail arrives at Chahorra volcano (Fig. 5.98A,B).

Stop B.4. This spot offers you a close-up view of the 1798 Chahorra eruptive vents, aligned along a radial fissure on the western flank of Pico Viejo volcano (Fig. 5.98B). The trail encircles the main cinder cone, where large accretionary balls (similar to the Teide's eggs) have accumulated. The trail continues for another 4.5 km on Pico Viejo lava flows toward the TF-38 road and thus back to civilization.

END OF DAY 4.

DAY 5: ANATOMY OF THE DEEPLY ERODED NORTHEAST RIFT AND THE MIOCENE−PLIOCENE ANAGA MASSIF

Day 3 was, in part, dedicated to the active northwest rift zone, with its internal structure concealed by recent volcanism (see Figs. 5.26 and 5.27). In contrast, lower rates of eruptions on the northeast rift zone (NERZ) coupled with intense erosion and mass-wasting by giant landslides have uncovered the lower parts of this particular rift, revealing aspects of the internal structure of an otherwise inaccessible geological formation (see Figs. 5.24−5.25). The current day aims to inspect this geological situation and then continues with an option to tour the Miocene−Pliocene Anaga shield volcano after the NERZ itinerary (Fig. 5.99).

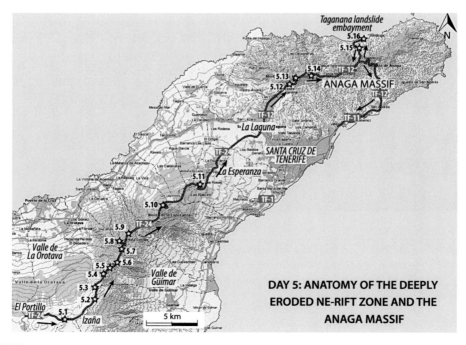

FIGURE 5.99

Map of the northeast rift zone (*Dorsal de La Esperanza*) and the Anaga massif, with the stops of *Day 5* marked. Map from IDE Canarias visor 3.0, GRAFCAN.

FIGURE 5.100

Oblique aerial view of the northeast rift zone (NERZ) from the southeast, with Teide volcano in the background and the main stops of *Day 5* indicated. The NERZ is approximately 35-km-long and forms one of the structural backbones of the island. Its activity goes back to the Miocene, and thus the NERZ is one of the long-lived geological features of the island.

Starting from *Los Cristianos*, pass through *Vilaflor* and into the Las Cañadas Caldera at *Boca de Tauce* (repeating the route of the first part of *Day 3*). Continue on TF-21 through to the cross-roads after the Teide National Park Visitor Center (last stop of Day 3). There, join TF-24 (Figs. 5.99 and 5.100) until you reach *Mirador Corral del Niño* (km 38 on TF-24). The road from *El Portillo* crosses a series of cinder cones and lavas of the northeast rift zone that have been dated to between 31 and 37 ka (Carracedo et al., 2007b) and that were fed by a swarm of dyke intrusions that forms the backbone of this northeast long-axis of Tenerife.

Stop 5.1. Volcán de Fasnia (1705 AD) (N 28.3028 W 16.5194)

From the mirador you can see the astronomical observatory of *Izaña* in the distance. If you look to the south, a cluster of relatively small, intensely black cinder cones from the 1705 eruption should be visible (Fig. 5.101). These cones are the central vents of a northeast–southwest-trending, 10-km-long fissure eruption (ie, from an approximately 10-km-long dyke). The other vents are located to the northeast (Volcán de Güímar) and to the southwest (Volcán Siete Fuentes), probably marking the distal edges of the 1705 eruptive fissure.

The 1705 eruption is the first to occur on Tenerife since the Spanish conquest (after more than 213 years of repose). The eruption was preceded by intense seismicity, but no casualties were reported. The eruption lasted from December 31, 1704 to March 27, 1705. Specifically, the Siete Fuentes volcano erupted from December 31, 1704 to January 5, 1705; Fasnia volcano followed from January 5 to January 16, 1705, and Güímar volcano opened up from February 2 to March 27, 1705. This event was followed only a little more than a year later by the Garachico

FIGURE 5.101

Volcán de Fasnia, the central vent of the 1705 fissure eruption. This volcanic episode lasted 12 days, from January 5 to 16, 1705 AD. The eruption formed a chain of small cinder cones and emitted short basaltic flows. The longest flow from Volcán de Fasnia was arrested after flowing down-slope for approximately 4 km.

eruption in May 1706 (eg, Romero, 1991; Solana and Aparicio, 1999). A visit to the Fasnia vents is optional and can be done via a dirt track that commences a little further from the turn for *Izaña*.

Continue on TF-24 and pass *Izaña*. Right after you pass the hill on which *Izaña* is constructed, a prominent ankaramite dyke protrudes on the right side of the road, just before a panoramic view of the Güímar Valley opens up in front of you approximately 1 km further. Park the car along the road here and walk along the road for a few steps until you get a full view of the valley (provided that clouds are not an issue).

Stop 5.2. Valle de Güímar (N 28.3193 W 16.4938)

This 10-km-wide depression is the result of a giant landslide that occurred approximately 830 ka ago (see Figs. 5.25A and 5.102). A geological section of the valley shows a partially filled collapse basin carved into a Quaternary basaltic sequence that crops out in the walls of the embayment. The southwest wall, the "Pared de Güímar," has been deeply cut by the Bco. de Badajoz in the contact between the precollapse and post-collapse sequences (red box in Fig. 5.102). The youngest age of the pre-landslide sequence, at the bottom of the barranco, and the oldest age from the filling sequence help constrain the age of the giant landslide to between 831 ± 18 ka and 860 ± 18 ka (Carracedo et al., 2011b; Kissel et al., 2014). The core of the northeast rift is formed by older Pliocene lavas, which are accessible through *galerías* (see, eg, Fig. 5.25B and Delcamp et al., 2012).

Once you have taken in the views of the valley, continue on TF-24. Soon the road starts to descend. Look for one of the paved parking spaces to the left, which provide spectacular views of the Orotava valley. Continue for a short distance until you reach a tight U-bend and park the car in the parking lot on the right side just before the bend or a few meters further down if this parking area is full.

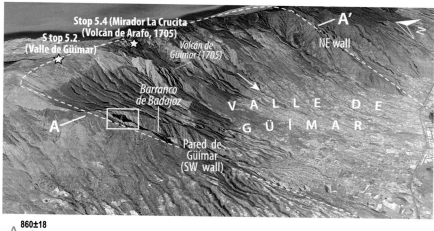

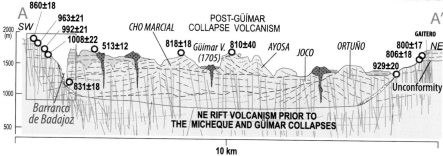

FIGURE 5.102

(A) *Google Earth* image of the Güímar valley. (B) Associated geological cross-sections (after Carracedo et al., 2011a).

Stop 5.3. La Tarta (The cake) (N 28.3344 W 16.4902)

Walk along the road inside the wooden barrier (please note, this is a popular stop and a lot of traffic passes here at times) to observe interbedded phonolitic pumice and basaltic lapilli (Fig. 5.103). This sequence implies that, at that time, felsic and mafic eruptions were taking place semi-simultaneously on Tenerife. Because the individual units are separate (ie, no mixed facies are seen within the package), the two magmas are unlikely to represent a single eruption. It appears that phonolitic tephra erupted after basaltic tephra, which was followed by yet another basaltic eruption. The basaltic/phonolitic sequence originated from the independent eruption of a deep basaltic magma that fed the NERZ and a phonolitic one from the LCV, which would be derived from inside the LCC Indeed, this bimodal tephra sequence is characteristic of the transition zone between the mafic rift zones and the central felsic volcanic system (eg, Wiesmaier et al., 2012).

Continue your descent on TF-24. The road soon crosses numerous northeast−southwest-trending sheet intrusions, which are the feeder dykes to the northeast rift (see Fig. 5.22). After approximately 8 km, you will arrive at *Mirador de La Crucita*. Park the car to have a closer view of Güímar valley from here.

FIGURE 5.103

(A) "La Tarta" (the cake) bend. Felsic and mafic layers of pyroclastic ash and scoria in the transition zone between the felsic central complex and the mafic northeast rift reflects two alternating geological regimes. (B) Enlarged red square from (A) shows the sequence of basaltic lapilli (Strombolian events) and phonolitic pumice (Plinian eruption) in a close-up view. Note the sharp boundary between the felsic and mafic deposits.

FIGURE 5.104

(A) Basaltic dyke parallel to the road TF-24 that runs along the NERZ. This type of exposure allows insight into the structure of a rift. The host rock is pahoehoe plagioclase basalt. (B) View from *Mirador La Crucita* (Stop 5.4), looking south. In the background is the Volcán de Güímar (also known as Volcán de Arafo), another eruptive vent of the 1705 eruption besides Fasnia.

Stop 5.4. Mirador La Crucita *(Volcán de Güímar, 1705 AD) (N 28.3452 W 16.4818)*

A little further from the parking area, the road exposes a basaltic dyke that runs parallel to the road, allowing insight into the structure of such a sheet intrusion (Fig. 5.104A). Walk down to the mirador on the other side of the road to enjoy a view of the 1705 AD Volcán de Güímar that is nested in the Güímar valley (Fig. 5.104B). This eruption started on February 2, 1705, and lasted until March 27, 1705. The eruption, a characteristic basaltic fissure event, was preceded by a year of intense seismicity (1704 is known on Tenerife as "the year of the earthquakes"). Several of the 1705 lava flows ran down-hill, and the longest came to rest at approximately 300 m from the coast (see Fig. 5.24D). Luckily, no casualties were reported from this eruption.

Continue on TF-24 for a short distance and park on the side of the road just past the km 29 road sign and close to a thick mafic dyke.

Stop 5.5. Dykes of the NE rift zone (N 28.3528 W 16.4742)

After inspecting the thick dyke, walk back past the km 29 sign toward the km 30 sign. In the road cuts here, an excellent section of thin pahoehoe plagioaclase basalt flows is exposed that has been intruded by numerous dykes of somewhat similar composition. Sections of the dykes show a wealth of internal features such as crystal concentrations (Fig. 5.105A) and jointing (Fig. 5.105B). Crystal

FIGURE 5.105

Characteristic features of basaltic dykes. (A) Concentration of olivine and especially pyroxene crystals in the center of a dyke. (B) Cooling joints in an aphyric basalt dyke.

concentrations are frequently observed in these dykes, with fine-grained margins and a zone of large crystals in the central part. These differences are caused because cooling of the external parts of the dyke is relatively quick due to the rapid loss of heat to the surrounding rock, and the resulting texture is correspondingly fine-grained at the "chilled margin". The chilled margin then acts as an insulator, allowing the central core of the dyke to cool slower, resulting in a coarser texture (larger crystals) in the center of the dyke. An alternative idea is "flow differentiation," in which already formed crystals "bounce" off the solid margin and thus cluster in the center of the flowing magma (see also Delcamp et al., 2012, 2014; Carracedo and Troll, 2013).

Joints in dykes (and lava flows) are formed by shrinkage during cooling and are perpendicular to the cooling fronts (eg, contacts). In lava flows, which are generally sub-horizontal, jointing generates semi-vertical columns, frequently hexagonal in structure (eg, at Los Órganos, chapter: The Geology of La Gomera). In dykes, cooling occurs dominantly at the sides, and thus jointing is often seen as a horizontal feature here.

Note the abundant small holes in dykes and flows drilled to extract oriented cores for palaeomagnetic studies, principally to determine geomagnetic polarities and internal flow patterns (Delcamp et al., 2010, 2012; Carracedo et al., 2011b; Kissel et al., 2014).

From here, you may want to look north. The view in front of you offers a glimpse of the dyke swarm that forms the axis of the northeast rift (Fig. 5.106).

FIGURE 5.106

Dyke swarm on the axis of the northeast rift zone at the eastern wall of the Orotava valley. These likely fed cinder cones at the surface. Eruptive vents are densely packed along the crest of the NERZ, whereas lava flows dominate the flanks.

Continue on TF-24 for approximately 1.5 km. The road crosses a dense cluster of basaltic dykes that belong to the NERZ dyke swarm. Park the car at *Mirador de Ayosa* that will soon appear on your left.

Stop 5.6. Mirador de Ayosa. *Orotava Valley (N 28.3560 W 16.4681)*

This mirador is at the center of the northeast rift zone and offers a spectacular view of the La Orotava landslide basin and the exhumed axis of the northeast rift zone. Walk back on the road to have a closer look at the dykes and their structures.

This stop also offers a closer view of the head of the Orotava Valley, showing the dense swarm of dykes parallel to the axis of the northeast rift (Fig. 5.106).

The deep structure of rift zones is rarely accessible, except through galerías and in cases like this, in which a giant landslide mass-wasted the northern flank of the ridge to expose its internal structure (see also Figs. 5.22, 5.23, and 5.25).

Continue on TF-24 for approximately 3.3 km until a sign for *Fuente de Joco* appears. Park the car on the right side of the road opposing "Joco's spring."

Stop 5.7. Fuente del Joco *(N 28.3689 W 16.4632)*

The high permeability of volcanic rocks is one of the main reasons for the lack of permanent rivers or lakes on Tenerife. The only other freshwater resource is groundwater. Since the 1960s, hundreds of horizontal tunnels (*galerías*) were excavated to access the main insular aquifers (Fig. 5.107A) to supply potable water to an increasing population and a booming tourism industry (at present Tenerife hosts 0.9 million inhabitants and approximately 5 million annual visitors).

Before galerías were available, natural springs were vitally important. Thus, topographic maps have abundant toponyms referring to permanent springs that existed before extraction through the galerías. Artificial extraction through galerias has now drained the main aquifers considerably and, by now, depleted most of the natural springs of the old days.

Natural springs such as at *Fuente de Joco* (Fig. 5.107B) provide an excellent example of how these springs were formed. A thick but fractured lava flow with a basal breccia layer of very porous scoria drains the water from up-hill. The vertical water seepage is then stopped at a contact with an impermeable baked soil (*almagre*) below the lava, forcing the water to migrate horizontally and surface in a spring that is derived from such a "perched aquifer" (Fig. 5.107C). Today, the majority of natural springs on the island are only names on maps. They have dried out or are by now contaminated with heavy soil, minerals, or volatile contents due to low fluxes. The water supply on Tenerife is thus not sustainable in the long run, and innovative long-term solutions seem highly needed.

Continue down-hill on TF-24 for 0.5 km and enter a road to the left signposted *Chipeque* and *Chimague*. From these miradors, a general view of the Güímar and the La Orotava valleys is allowed on a good day.

Stop 5.8. Mirador de Chipeque *(N 28.3740 W 16.4637)*

Only approximately 200 m after the turn on your right, *Chimague* mirador appears, residing above the central part of the Güímar landslide scar. From this mirador you have a superb view of the northern wall of this lateral collapse embayment. Only a short distance further along the road, you

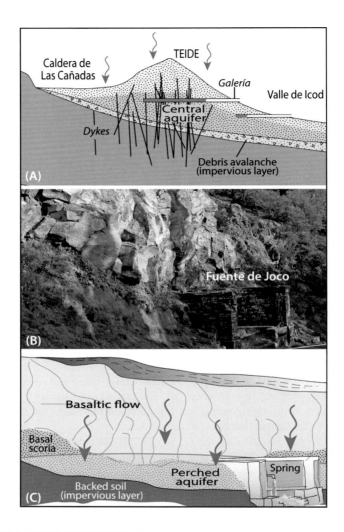

FIGURE 5.107

(A) Sketch of the main aquifer system of Tenerife that also provides most of the water used on the island. (B) *Fuente de Joco*. (C) Sketch of the perched aquifer supplying the Fuente de Joco spring.

will reach the *Mirador de Chipeque*, one of the best points for a panoramic view of the Orotava Valley and Teide volcano (Fig. 5.108).

From here you can observe the structure of the La Orotava Valley, which originated in a giant landslide that mass-wasted an estimated volume of approximately 57 km³ at approximately 600 Ma (between 690 and 566 Ma ago). The landslide left a characteristic depression with rectilinear walls, whereas post-landslide volcanism has now partly filled the La Orotava Valley, thus producing a flat but mildly inclined floor in places.

FIGURE 5.108

Spectacular view of the Orotava landslide valley from *Mirador de Chipeque* (Stop 5.8), with Teide volcano in the backdrop. Note the recent lava fill in the upper parts of the Orotava basin.

Note that the paving stones here at the mirador are from the 1736 tholeiite eruption on Lanzarote, which has become a very popular building stone in the Canaries and is characterized by bubble channels that mark intense degassing of the lava (see also Fig. 7.32).

Return to TF-24 and continue down-hill. Pass the km 25 road sign and park on the left side of the road. On your right, the road crosses a thick basaltic flow with spectacular spheroidal alteration.

Stop 5.9. Spheroidal weathering (N 28.3814 W 16.4586)

Alteration of basaltic lava, particularly those with columnar jointing, begins when weathering agents (eg, slightly acidic rain- or groundwater containing dissolved CO_2) first react chemically with the minerals of the rock, which ultimately leads to the conversion of rock to soil (Fig. 5.109A–C). Water attacks the rock along cracks and hydrolysis transforms anhydrous minerals into clays and other sheet silicates, broadly following the susceptibility series: glass $\geq$ olivine > plagioclase > pyroxene > opaque minerals.

During weathering of basalt, Ca, Mg, K, Na, Rb, and Sr show generally rapid early loss, which is a function of the alteration of volcanic glass to smectite, the alteration of olivines to iddingsite, and the early weathering of plagioclase.

Rainwater seeps progressively through these fractures toward the interior of the hexagonal prisms, leaving rounded lumps of less weathered rock behind (Fig. 5.109D). Ultimately, the entire rock is weathered down and transformed to a mixture of iron oxide-hydroxides and clay minerals that are stable at surface conditions.

Continue down-hill on TF-24, which follows the crest of the northeast rift. Just after the km 20 road sign, you will encounter *Mirador de Ortuño*, where you can park to have another magnificent view of Teide volcano.

Stop 5.10. Mirador de Ortuño (N 28.4058 W 16.4238)

Large-scale faulting on Tenerife is usually gravitationally induced (normal faulting). Giant land-slides are the extreme example, whereas smaller normal faults are the more common type (eg, at *El Risco*, chapter: The Geology of Gran Canaria).

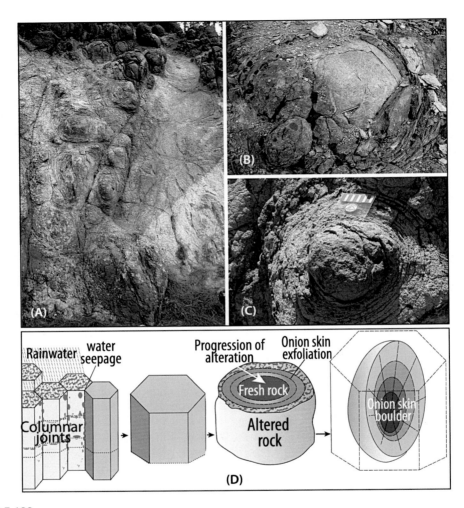

FIGURE 5.109

(A) Lava flow with columnar jointing in an advanced state of weathering. Note the onion skin boulders are almost completely transformed from the original rock and are pervasively altered and disintegrated. (B) An exfoliated boulder showing the onion skin structure in detail. (C) Only the very interior of this boulder remains. (D) Cartoon showing the stages of alteration of basaltic lava flows to form spheroidal boulders and, ultimately, soils. Because of "morphological similarities," these exfoliated onion skin boulders have initially been confused with pillow lavas.

Yet smaller normal faults are observed when differential settlement has occurred, such as when a thick lava flow settles over altered lapilli or similar. The road cut at the mirador here shows a normal fault of this type. Here, the lava flow sank down-hill due to gravity because it was deposited on an altered lapilli bed that was not sufficiently resistant (Fig. 5.110).

FIGURE 5.110

Normal fault caused by the load of the top lava flow over altered lapilli and scoria beds at *Mirador de Ortuño*. Such small gravity faults are frequent in sequences of alternating lava flows and layers of altered lapilli.

Contrasting the Hawaiian Islands, no earthquakes on active faults have been recorded on the Canaries in the past 20 years, although persistent seismicity has been recorded in the past two decades below the ocean floor between Tenerife and Gran Canaria. This place was interpreted as a regional and highly active fault by some workers, although, more probably, the seismicity in that area is associated with an active submarine volcano (eg, Krastel and Schmincke, 2002b; Carracedo et al., 2015a,b).

Continue down-hill on TF-24. Pass the sign for *Mirador Montaña Grande* and park the car on the left side. The *Mirador de Mña. Grande* provides a panoramic view of the eruptive centers on the edge of the northeast rift zone and, in the distance, the Pliocene shield volcano of Anaga while being opposite a reddish scoria cone that has been partially excavated in the past.

Stop 5.11. Mirador de Mña. Grande (N 28.4284 W 16.3798)

This mirador overlooks the eruptive vents at the end of the northeast rift (Fig. 5.111), where the control of the rift wanes and the clustering of vents along the axis of the rift relaxes (ie, the vents fan out). Many of these vents have been mined to use the lapilli for building and road construction or to complement agricultural efforts.

The Anaga massif in the distance (Fig. 5.111) is, to a large degree, covered by the *Laurisilva* (Laurel forest), a floral association of the rain forest type that extended in the late Miocene and early Pliocene over much of Southern Europe and Northern Africa but survived only in isolated spots such as the Canaries (Tenerife, La Palma, and La Gomera) and on Madeira.

You can close the day here and return to your base once you reached *La Laguna* at the end of TF-24, or continue to inspect Anaga.

FIGURE 5.111

Cinder cones at the end of the northeast rift zone viewed from *Mirador de Montaña Grande*. The Pliocene Anaga massif is visible in the distance.

If this is your closing stop of Day 5, then continue on TF-24 to *La Laguna* and successively take TF-5 in the direction of *Santa Cruz de Tenerife*, then TF-2 for *Aeropuerto del Sur*, and, finally, TF-1 to your base in the south.

For a tour of the Anaga massif (to inspect the main geological features and enjoy the spectacular flora), continue on TF-24 to *La Laguna* and cross the town toward the North. Maybe consider a stop in La Laguna if time permits; otherwise, continue on the road to *Las Mercedes* (TF-12) (see Fig. 5.99).

After an approximately 7-km drive, first across the *Vega de La Laguna* plane and then through the *Las Mercedes laurisilva* rain forest, you will see a mirador in a sharp bend on your right. Park the car there for a magnificent panoramic view of *La Laguna*, the northeast rift, and Teide volcano in the far distance.

Stop 5.12. Mirador de Jardina (N 28.5240 W 16.2881)

This mirador provides a view of the Anaga formations as scattered paleoreliefs in the foreground of the northeast rift zone (middle ground of your view).

The Anaga shield volcano developed during the Pliocene (see Fig. 5.112) on top of an extension of the Miocene central shield (see Fig. 5.3B,C) and has been inactive over the past approximately 4 Myr. During this period of eruptive quiescence, erosion carved a dense and deep network of barrancos. Dyke ages define the timing of a major northward sector collapse at approximately 4.2 Ma, which formed the Taganana embayment (Walter et al., 2005).

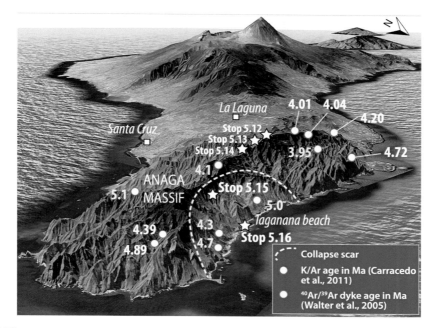

FIGURE 5.112

Google Earth view of Tenerife from the northeast, with the Anaga massif in the foreground. Most of the Anaga shield developed during the Pliocene and has been inactive since, explaining the intense erosion and deeply incised barrancos (eg, Walter et al., 2005).

The volcanic stratigraphy of the Anaga volcano is relatively simple, with two major geological series. The lower basaltic series is made of strongly eroded basaltic sequences of lava flows and pyroclasts and is densely traversed by numerous dykes that form well-defined dyke swarms (Fig. 5.113A). The upper series is made up of capping sub-horizontal basaltic and felsic lava flows, many of which are more resistant than the older units and therefore cause the steep profiles of the hillsides here (Fig. 5.113B).

The city of *La Laguna* (The pond) takes its name from a large lake that existed there and around which the former capital of Tenerife was established (Fig. 5.114A,B). The city was constructed on an extensive and fertile plane that formed when the natural drainage of the Barranco de Las Mercedes was obstructed by the growth and development of the northeast rift zone. The area then filled with fluvial and alluvial sediments (Fig. 5.114C).

Continue up-hill on TF-12 to *Mirador Cruz del Carmen* (2.4 km).

Stop 5.13. Mirador de la Cruz del Carmen (N 28.5308 W 16.2800)

The view is similar to that at the *Jardina mirador*, but here you have an informative public exhibition illustrating the endemic flora of the area and a bathroom facility.

FIGURE 5.113

(A) *Roque Taborno*, an erosion remnant of the lower basaltic series of the Anaga massif. Note the dykes intruding the basaltic sequence. (B) Posterosional lavas, some of them phonolitic, moved along the barrancos. The subsequent long period of eruptive inactivity and erosion left these lava flows forming the crests of the interfluves in a spectacular inversion of relief.

Continue on TF-12. After approximately 1 km, take a right turn at the upcoming fork. Continue for approximately 1 km until you reach the parking area of *Mirador Pico del Inglés*.

Stop 5.14. Mirador Pico del Inglés (N 28.5331 W 16.2641)

This 990 masl viewpoint offers a 360° panoramic of the Anaga massif. The Anaga shield volcano is predominantly made of basaltic pyroclastic material and lavas, whereas the upper part of the shield is capped by mafic lava flows and felsic rocks, with the latter predominantly as intrusions in the form of phonolitic plugs and necks.

FIGURE 5.114

(A) The city of *La Laguna*, with the northeast rift and Teide in the background. (B) *La Laguna* (The pond) takes its name from a large pond that existed and around which the former capital of Tenerife was established (courtesy of *Cabildo Insular de Tenerife*). (C) The drainage of the southwest flank of the Anaga massif was interrupted by the construction of the northeast rift zone during the Pliocene to Quaternary, causing the creation of a lake, subsequently partially filled with sediment. The lake existed until a few decades ago, particularly during intense rain periods, but has been effectively drained over the past decades via the modern sewage system of *La Laguna*.

Evidence of terminal felsic eruptions is provided by phonolitic lava flows that generally rest unconformably on the basaltic sequences. These phonolite flows filled barrancos and now appear as crests by relief inversion (see Fig. 5.113B).

Return to TF-12 and drive eastward for approximately 12 km, following several road signs for *Taganana*. Join TF-134, still following the signs for *Taganana*, and pass through a large tunnel. On the other side of the tunnel you will find yourself inside the Taganana collapse embayment. Continue for 1 km after the tunnel until you reach a very sharp bend and a small mirador with a parking area (near road signs for km 2 on TF-134).

Stop 5.15. The Taganana collapse embayment (N 28.5582 W 16.2057)

Once through the tunnel, a wide horseshoe-shaped embayment opens up toward the north. This sector is structurally and characterized by a unit of brecciated polylithological materials inclined toward the north that separates the intra-embayment formations from the younger series. Intense erosion has, moreover, exhumed the phonolitic plugs that are frequently seen as up-standing masses inside the embayment (Figs. 5.115 and 5.116).

FIGURE 5.115

Taganana embayment looking eastward (photograph taken from Stop 5.15). A substantial portion of the rock mass here is dykes, dominantly inclined to the south (d in the figure). Felsic plugs in this part of Tenerife (Roques de Anaga) experienced intense erosion and coastal regression and remind us of La Gomera.

FIGURE 5.116

Phonolitic plugs in the Taganana embayment. The plugs show concentric contraction joints (onion skin layers) similar to the famous "*Roques*" of La Gomera.

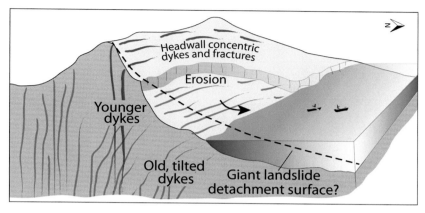

FIGURE 5.117

Block diagram showing main types of faulting and dyke deformation recorded in the Taganana collapse embayment (after Walter et al., 2005). A period of creep is inferred before the N-flank of the massif failed at approximately 4.2 Ma.

The rocks here appear intensely disturbed and shattered, and dykes that locally form up to 90% of the rock mass are brecciated throughout and tilted toward the north (Fig. 5.117). These features resemble those of debris avalanche deposits (Walter et al., 2005).

The Taganana embayment is likely the result of destabilization, displacement, and collapse of the northern flank of the Anaga massif. The intensely sheared character of the Taganana sector suggests that, prior to the collapse, this volcano sector was creeping toward the sea, which thus fractured the rocks in this area.

The road continues through the village of *Taganana*, and from there to the coast at the *Taganana* or *Roque de las Bodegas* beach (Fig. 5.118A). There you can park the car for a view of the coast and the cliffs of the Anaga massif.

Note the abundant dykes and phonolite plugs in inversion relief. This is clear evidence of the erosive retreat of the coast that has been exposed for approximately 4 Ma to the aggressive marine erosion on the windward side of the Anaga peninsula.

Stop 5.16. Taganana beach (N 28.5698 W 16.2046)

Here, the older Anaga rocks are overlain by coarse palaeo-screes, which are presently being undercut by the sea. This exposure offers an interesting insight into the likely structure and state of alteration of the interior of comparable younger volcanoes at a depth-level of 1 to 2 km below their summits (eg, the northeast rift zone on Tenerife and also the Cumbre Vieja on La Palma). Much of

FIGURE 5.118

(A) View of the Taganana coast looking South. (B) *Google Earth* image of the eastern part of the Taganana landslide embayment. Erosion caused the regression of the coast by at least 1 km, as indicated by the dykes that have resisted erosion. Consider doing the coastal walk around one of the dyke intrusions at the far end of the beach.

the rock is altered to clay minerals that have been oxidized and hydrated upon exposure, hence the widespread rusty colors. Intrusions, like dyke and plugs, tend to be somewhat more resistant to erosion and form upstanding rocks here in the area (like the coastal walk around a dyke intrusion at the end of the beach; Fig. 5.118).

FIGURE 5.119

(A) View of the Taganana beach. (B) Under the sand, rounded phonolite beach pebbles occur, which often contain pyroxenite and amphibole-rich xenoliths, indicating that the cap of phonolitic rocks on top of Anaga was once considerably more extensive and was supplied by a deep-reaching plutonic plumbing system.

Make a note of the abundant rounded phonolite beach pebbles, which often contain pyroxenite and amphibolite xenoliths and may indicate that the cap of phonolitic rocks on top of *Anaga* may once have been considerably more extensive (Fig. 5.119).

The *Roque de las Bodegas* area comprises a number of fish restaurants and is a good place for a simple but nice meal. Return to TF-12 and drive toward *Santa Cruz* to join TF-1 and return to your base.

END OF DAY 5.

THE GEOLOGY OF GRAN CANARIA

The Geology of the Canary Islands. DOI: http://dx.doi.org/10.1016/B978-0-12-809663-5.00006-2

357

THE ISLAND OF GRAN CANARIA

Gran Canaria is the central and the third-largest island (1560 km²) of the Canarian archipelago, after Tenerife and Fuerteventura, and was the second-largest in population in 2014 (851,157), after Tenerife (889,936) (*ISTAC, Estadísticas de la Comunidad Autónoma de Canarias-Demografía*).

The island is circular in shape, approximately 45 km in diameter, and rises to an altitude of 1949 meters above sea level (masl). This configuration led to the formation of a set of deep radial barrancos and canyons and a mountainous interior (Fig. 6.1).

Roads on Gran Canaria will take you virtually anywhere; however, the only fast highways go around the coast, semi-encircling the island from *Agaete* in the northwest to the capital *Las Palmas*

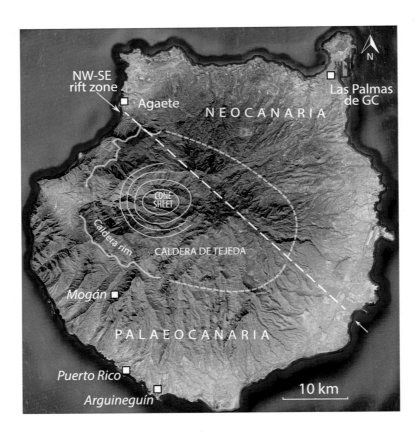

FIGURE 6.1

Google Earth image of Gran Canaria showing the main geomorphological units of the island. The northwest−southeast geographical divide separates the old (Miocene) southwest half of the island (Palaeocanaria), from the younger (Plio-Quaternary) part, or Neocanaria. The Miocene Tejeda Caldera (white solid line indicates the visible caldera margin, broken line is inferred) and the central cone-sheet swarm (in yellow) are shown.

and from there to the southern tourist resorts of *Puerto Rico* and *Mogán*. In the western part of the island, from *Agaete* to *Mogán*, roads are narrow and bendy and run parallel to hundred-meter-high cliffs (eg, *Andén Verde*), likely representing the scar of a westbound Miocene giant landslide. Roads in the island's interior are frequently steep and bendy and require an experienced driver.

The main geomorphological features of Gran Canaria are related to the volcanic evolution of the island, and are divided in two equal parts by a northwest—southeast line that coincides with a Pliocene rift zone (see Fig. 6.1). The southwest older part (Palaeocanaria) is formed by the Miocene volcanics, whereas the younger northeast portion (Neocanaria) concentrates the rejuvenation and recent volcanism (Plio-Quaternary) (Bourcart and Jérémine, 1937).

The central part of Gran Canaria is occupied by what can be considered the most distinct geological feature of the island, the Miocene Tejeda collapse caldera with its central cone-sheet swarm (eg, Schmincke, 1967; Crisp and Spera, 1987; Clarke and Spera, 1990; Schirnick et al., 1999; Troll et al., 2002).

Gran Canaria is likely one of the most comprehensively studied oceanic islands in the world. Following Tenerife, Gran Canaria attracted early scientific interest (eg, Bourcart and Jérémine, 1937), but the most relevant studies for our modern thinking commenced in the second half of the 20th century (eg, Schmincke, 1967, 1969, 1973, 1974, 1976, 1979a,b; Fúster et al., 1968c).

As shown in the geological map (Fig. 6.2), Gran Canaria not only presents an interesting combination of basaltic shield volcanism and caldera-forming felsic eruptions, with abundant intracaldera and extracaldera ignimbrites and a spectacular cone-sheet swarm, but also exhibits these geological features in superb outcrops, as erosion carved deep radial canyons that now allow detailed inspection of the successive volcanic episodes of the island.

THE SUBMARINE (SEAMOUNT) STAGE OF GRAN CANARIA

Submarine (seamount) deposits do not crop-out on Gran Canaria, yet available information on this stage has been provided by the Ocean Drilling Program (ODP), Leg157 (eg, Schmincke et al., 1995; Schmincke and Segschneider, 1998). Seismic profiles confirm that this stage amounts to approximately 90% of the total volume of the island (Schmincke and Sumita, 1998; Krastel and Schmincke, 2002a). Cores from five offshore sites (Fig. 6.3) recovered hyaloclastite tuffs and debris-avalanche deposits, interpreted as the products of shallow (<500 m deep) submarine eruptions (Schmincke and Segschneider, 1998). These deposits appear interbedded with those of the subaerial eruptions, indicating semi-continuous volcanic activity from submarine to subaerial volcanism, implying that this transition is mainly a change in eruptive mechanism, but not necessarily of the underlying magmatic cause.

Erosion outpaced volcanic construction of Gran Canaria likely from the Middle Miocene, leading to the development of a wide marine abrasion platform encircling the island (Fig. 6.4). This platform is particularly extensive in the southern "Paleocanaria" that lacks any volcanic rejuvenation after the eruption of the Fataga ignimbrites approximately 8—9 million years (Ma) ago (although a Pliocene Roque Nublo cover once existed at ~4 Ma). At that time, both the climate and the marine fauna of the Canaries were tropical, similar to the present climate of the Caribbean or the Gulf of Guinea (Meco et al., 2007). Thus, changes in the sea currents and the colder climate

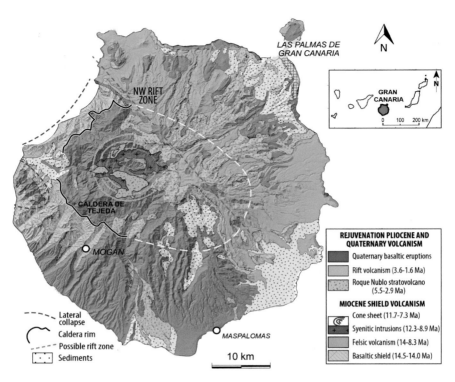

FIGURE 6.2

Simplified geological map of Gran Canaria after Fúster et al. (1968); Schmincke (1976); Ballcels et al. (1992); and Carracedo et al. (2002). Note the differences in age and composition of the southwest and northeast parts of the island caused by an extensive resurfacing of the northeast by younger volcanism (green colors).

during the Pliocene (since ~4 Ma ago) caused the extinction of this marine tropical fauna and provided the source of the organic aeolian beaches and dunes that abound on the island, especially on the southern and southeastern leeward coasts, such as the famous *Dunas de Maspalomas*.

THE SUBAERIAL GROWTH OF GRAN CANARIA

The two main subaerial phases of growth of an oceanic island, the shield and posterosive (rejuvenation) stages, are superbly represented on Gran Canaria, with a long period of volcanic quiescence between these two major constructive episodes (Fig. 6.5).

On Gran Canaria, there is little doubt about the short duration of the shield stage (~1 Ma), followed by an extended period of postshield volcanism. On the other Canary islands, particularly the eastern ones, this definition is lacking, likely because of the poor resolution between the shield stage and postshield stage volcanism that "artificially" expands the duration of the former. This

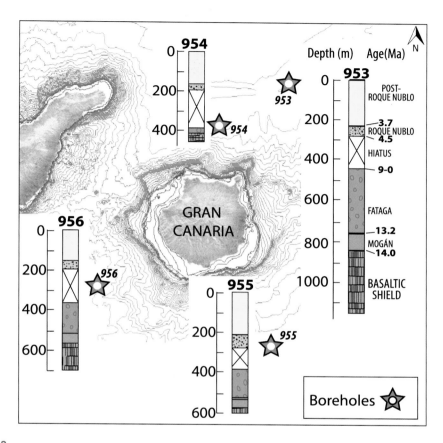

FIGURE 6.3

Boreholes drilled during ODP Leg 157 into the volcanic apron of Gran Canaria. The drilled offshore deposits can be correlated with the time-equivalent onshore stratigraphy (after Schmincke and Sumita, 1998).

problem is increased by the need to sample "fresh" lavas for radiometric dating, leading to a "propensity" to select samples of the younger postshield stage lavas (see also chapter: The Geology of Fuerteventura, Fig. 8.14). It is easier to avoid this issue on Gran Canaria because the shield stage volcanism is capped by predominantly felsic volcanic deposits, contrasting the eastern islands where this stage is mainly basaltic also, and hence very similar to the shield stage volcanics.

AGE OF GRAN CANARIA VOLCANISM

Gran Canaria and Tenerife are the best-dated islands in the Canaries and probably the best dated ocean islands in the world. At least 180 K/Ar, ^{40}Ar/^{39}Ar, and ^{14}C ages from lavas of Gran Canaria (Figs. 6.5 and 6.6) have been published (Abdel Monem et al., 1971, McDougall and

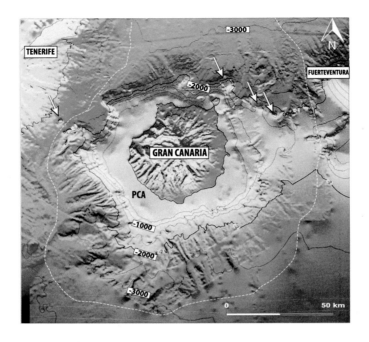

FIGURE 6.4

The island of Gran Canaria towers above the 3-km-deep ocean bed to almost 2 km above sea level. Note the large shallow marine platform that encircles the island, caused by marine abrasion and changes in sea level by successive glaciations. The arrows indicate submarine volcanoes (from Masson et al., 2002).

Schmincke, 1976; van den Bogaard and Schmincke, 1998; Guillou et al., 2004b; Rodríguez-Gonzalez et al., 2009).

The dating of ignimbrites and lava flows from the Mogán and Fataga groups, as well as from many tephra fallout layers from the Roque Nublo group, is particularly detailed, and these age determinations provide a high-resolution chronostratigraphic framework for Gran Canaria, particularly for the Miocene felsic period (Fig. 6.7).

MIOCENE SHIELD BASALTS

More than 1000 km^3 of basaltic lavas were erupted on Gran Canaria during the shield stage in a relatively short time interval and thus at very high eruptive rates. Although the shield stage of Gran Canaria was dominated by the eruption of basalts, the extended postshield period comprises two

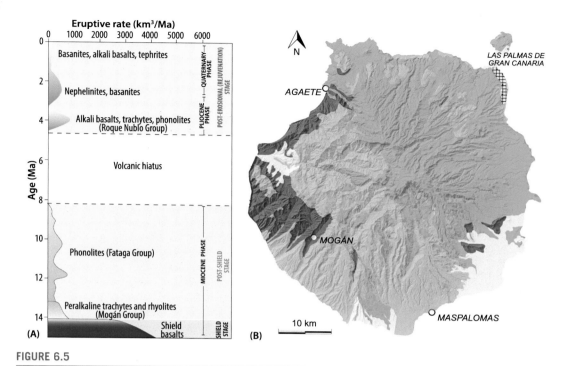

FIGURE 6.5

(A) Stratigraphy, eruptive rates, and ages of the main volcano-stratigraphic units of Gran Canaria. (B) Miocene outcrops (dark blue, shield stage basalts; pink, felsic postshield volcanism) and the rejuvenation Plio-Quaternary volcanism (green). Modified from Schmincke and Sumita, 1998, 2010.

main pulses of very distinct felsic volcanism. The composition of the eruptives changed, and so did the eruptive mechanisms, leading to a widespread cover of explosive ignimbrite deposits.

Radiometric dating of the shield basalts indicates that a complex 2000-m-high shield volcano, larger than the present island of Gran Canaria, was likely constructed between 14.5 and 14.0 Ma (McDougall and Schmincke, 1976; van den Bogaard and Schmincke, 1998). This continuous phase of eruption of low-viscosity basalts led to the accumulation of thick sequences of lava flows of Hawaiian-type fissure eruptions (Fig. 6.8A), with scant interbedded pyroclastics (Fig. 6.8B) but characterized by the intrusion of multiple generations of dykes and sills (Fig. 6.8B,C).

A simplified model section through Gran Canaria at the end of the shield stage ($\sim$14 Ma ago) shows a core of plutonic rocks replacing the older oceanic crust with a large basaltic magma chamber system that resided at the base of the crust or within the island and represents the source of the shield basalts (Fig. 6.9, Schmincke, 1993). A second smaller magmatic chamber became established at a shallow depth (4–5 km deep), where differentiation prepared the felsic (rhyolitic and phonolitic) magmas that would generate the highly explosive events of large volumes of

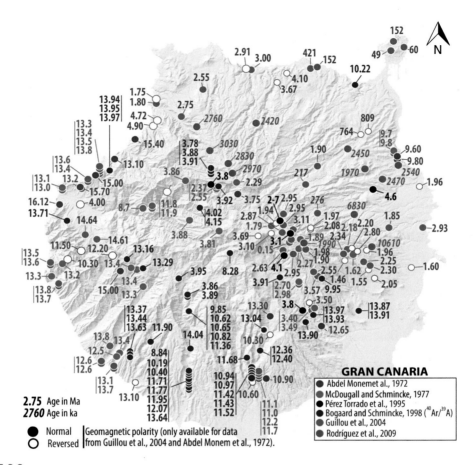

FIGURE 6.6

Compiled radiometric (K/Ar and ^{40}Ar/^{39}Ar) and geomagnetic polarity ages of lavas and volcanic deposits from Gran Canaria, likely one of the best dated oceanic islands on the globe (see text for details).

ignimbrites that led to the formation of the Mogán and Fataga group deposits and the Tejeda collapse caldera (eg, Freundt and Schmincke, 1992; Troll and Schmincke, 2002; Troll et al., 2002; Hansteen and Troll, 2003).

Present-day outcrops of the shield basalts are located in the western and southwestern parts of the island (see Figs. 6.2 and 6.5B), forming steep cliffs that expose more than 1000-m-thick basaltic volcanic sequences (Fig. 6.10) topped by an extensive felsic cover.

The extent and morphology of the Miocene shield volcano cannot be defined accurately, but they are inferred. Schmincke (1993) proposed up to four distinct shield volcanoes, the Agaete, Arucas, Agüimes, and Güigüí-Horgazales eruptive centers (see lower left inset in Fig. 6.10), but more likely there was a main shield volcano located in the central western part of the island, as

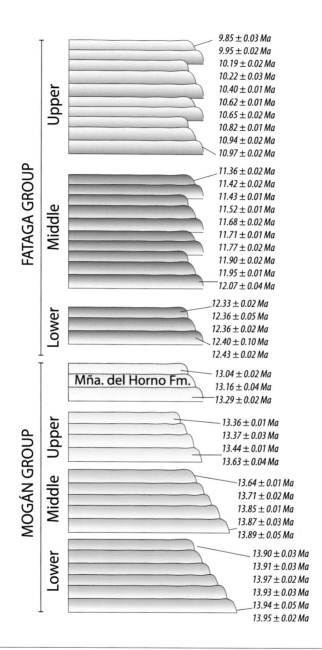

FIGURE 6.7

Chronostratigraphy of the Miocene felsic volcanism of Gran Canaria. Ages are from single crystal $^{40}Ar/^{39}Ar$ data (after van den Bogaard and Schmincke, 1998).

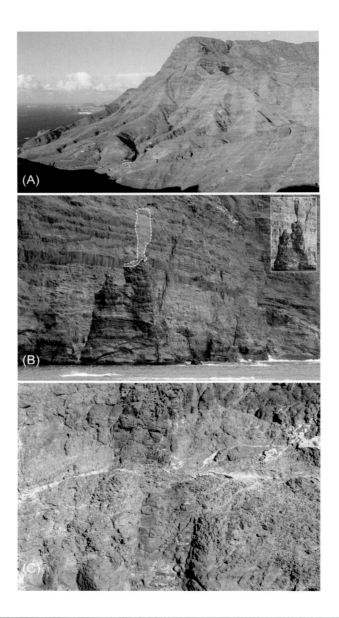

FIGURE 6.8

Characteristic aspects of the Miocene shield volcanism of Gran Canaria. (A) Basaltic sequence topped by Miocene ignimbrites forming the El Risco cliffs on the northern coast of the island. (B) Detail of the basaltic shield sequence at *Puerto Agaete*. The monolith in the inset, the *Dedo de Dios* (God's finger), partially collapsed during a storm in 2005. (C) Faulted basaltic dyke in altered basaltic lava flows of the northern shield volcano (near *Andén Verde*).

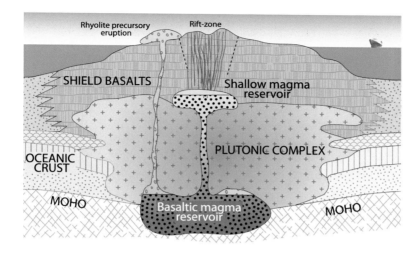

FIGURE 6.9

Model cross-section of Gran Canaria approximately 14 Ma ago, depicting the principal geological features. A plutonic core is gradually replacing the older oceanic crust where two magma chambers existed: a deep basaltic reservoir located at the base of the crust and a shallow one at 4–5 km (ie, relatively high in the crust). The latter chamber is where the highly evolved magmas of the Mogán formation differentiated (after Schmincke, 1993).

suggested by Balcells et al. (1992) and Troll et al. (2002) following the analysis of the distribution and orientation of the Miocene dykes.

A major unconformity is visible in Bco. de Güigüí, which separates the older Güigüí formations from the younger Horgazales basalt formations. The presence of a thick layer of breccias between both formations and their small age differences suggest that a giant landslide may have affected the Güigüí volcano but that the collapse depression was rapidly filled again by the Horgazales volcanism (Fig. 6.11).

MIOCENE FELSIC ROCKS

A shallow level, approximately 4–5 km deep, reservoir was established at the late stages of the shield volcano (see Fig. 6.9), evolved to rhyolitic magma compositions, and was intermittently replenished by the lower basaltic magma chamber (Schmincke, 1969, 1973; Crisp and Spera 1987; Clarke and Spera, 1990; Cousens et al., 1990; Freundt and Schmincke, 1992, 1995; Troll and Schmincke, 2002). This rhyolitic magma system produced the first highly explosive eruptions on the island, with the ejection of large volumes of ignimbrites of the Mogán group (Fig. 6.12). This violent episode began with the eruption of one of the most widespread ignimbrite units on the entire island, ignimbrite "P1", which extends over 400 km^2 and represents a bulk volume of 60 km^3 or more (Freundt and Schmincke, 1992; Schmincke, 1993; Schmincke and Sumita, 1998).

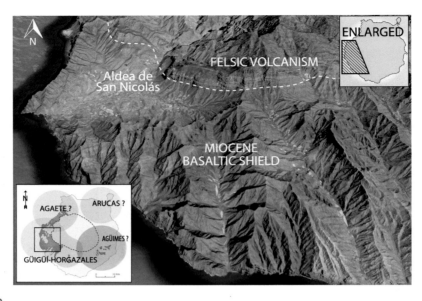

FIGURE 6.10

Deeply eroded Miocene basaltic shield of the southwest part of Gran Canaria (image from *Google Earth*). Barraconos cutting the lava sequence facilitate inspection of the oldest lavas. Schmincke (1993) postulated the construction of the Gran Canaria shield from three different eruptive centers: one near *Agaete*, in the northwest of the island; one at *La Aldea*, in the west; and one close to *Agüimes*, in the southeast. Schmincke and Sumita (1998) discussed a fourth emission center near *Arucas*, in northeast Gran Canaria (inset lower left). Based on the analysis of dyke distribution, Balcells et al. (1992) and Troll et al. (2002) proposed the existence of only one center, located in the central part of the island. This latter configuration seems to also fit better with the subsequent formation of the central collapse caldera at Tejeda.

The eruption of such a large volume of rhyolitic magma emptied the shallow magmatic chamber and caused the collapse of the volcano forming a huge caldera (the Tejeda Caldera), which is believed to have formed synchronously with the emission of ignimbrite P1 (eg, Freundt and Schmincke, 1992).

Felsic Miocene extracaldera volcanism has been divided into two groups (Schmincke, 1969, 1993), the lower trachytic to rhyolitic Mogán group (~300 m thick) and the younger and thicker (>500 m) trachytic-to-phonolitic Fataga group (Fig. 6.13).

THE MOGÁN AND FATAGA GROUPS

Intermittent interaction between the lower basaltic magma chamber and the shallow level reservoir (see Fig. 6.9) led to repeated cycles of differentiation processes that produced batches of evolved magmas with similar, but different, diagnostic compositions that now facilitate the correlation of individual cooling units.

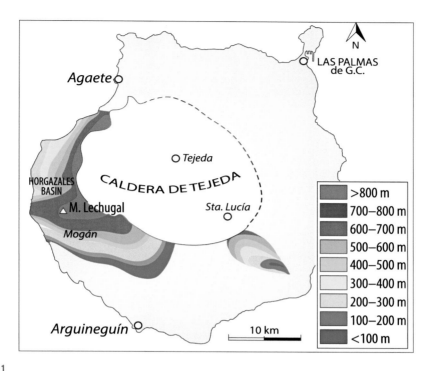

FIGURE 6.11

Elevation of the Miocene shield volcano before being capped by P1 ignimbrite (after Freundt and Schmincke, 1992 and Schmincke and Sumita, 2010).

Of all the ignimbrites of the Mogán group, the most peculiar is ignimbrite P1. It is only slightly younger than the Miocene shield basalts (see Figs. 6.12, 6.13, and 6.14). This cooling unit, located at the base of the sequence, had a total eruptive volume of 60 km^3 or more, indicating the onset of a voluminous and highly evolved eruption when construction of the basaltic shield was ending. The 14.1-Ma-old P1 unit is compositionally zoned from a felsic lower part to a basaltic upper part. The partitioning in three units of different bulk composition is easily recognized in the field due to their different colors, comprising a lower pink porphyritic rhyolite, a central mixed rock, and an upper basaltic unit (Fig. 6.14; Schmincke, 1969). The origin of this outstanding composite flow has been proposed as due to mixing of rhyolite, trachyte, and basalt magmas erupted from a zoned magmatic chamber (eg, Schmincke, 1974, 1979b; Freundt and Schmincke, 1992, 1995). The P1 unit extends over much of the southern and western part of the island (Fig. 6.14), from *Agaete* and *Andén Verde* in the north and west to Bco. de Arguineguín and *Agüimes* in the south and southeast, and it has been identified in offshore deposits (eg, Freundt and Schmincke, 1998).

During the eruption of P1, ring fractures opened at the caldera rim, erupting felsic products toward the interior and toward the outside of the caldera (Schmincke, 1967, 1969) (Figs. 6.15 and

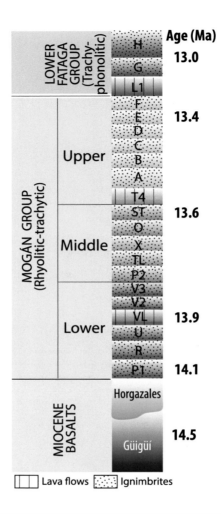

FIGURE 6.12

Simplified stratigraphic section of Miocene eruptive episodes on Gran Canaria (modified from Schmincke and Sumita, 1998 and Hansteen and Troll, 2003).

6.16). Many more such events followed and erupted welded ignimbrites, originating from the collapses of the hot eruptive columns of pyroclastic materials emitted (Mogán group).

The first phase (14.0−13.3 Ma) of extracaldera volcanics comprises, besides the P1 unit, 15 rhyolitic−trachytic cooling units (the Mogán group, Figs. 6.12, 6.13, and 6.16), with some of them more than 100 m thick in places (Schmincke, 1993) and a combined volume of approximately 800−1000 km^3 (Schmincke and Sumita, 1998). Radiometric ages and the lack of interbedded

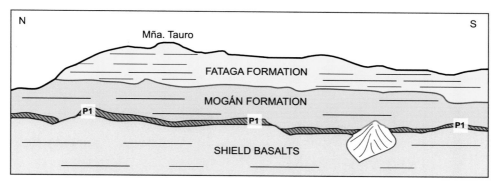

FIGURE 6.13

View of the Horgazales Fm. (Miocene basalts) overlain by the trachyte-rhyolite Mogán group (with basal rhyolite-basalt ignimbrite P1), Bco. Mogán. The sequence of Mogán units is capped by Fataga group trachyphonolite ignimbrites and lava flows (simplified from Schmincke, 1993).

epiclastic deposits (eg, sediments, palaeosols) point to an emission of the consecutive units in relatively quick succession, and the average repose period is currently estimated at approximately 40−50 kyr (van den Bogaard and Schmincke, 1998).

The second felsic eruption phase (13.3−8.3 Ma) produced 600−800 km^3 of trachyphonolitic ignimbrites (the Fataga group, Figs. 6.12, 6.13, and 6.15). Lava flows and epiclastic deposits are frequently interbedded with these younger ignimbrites, particularly at the top of the sequence, likely due to progressively longer periods of eruptive quiescence between the explosive events.

Although eruptive vents probably continued to be ring fractures at the caldera rim, the presence of thick lava sequences (eg, at Mña. de Amurga), abundant intrusions (dykes and lava domes), and areas of intense hydrothermal alteration suggest the existence of a felsic stratovolcano in the central part of the island in the late Miocene (Schmincke, 1993; Donoghue et al., 2010).

THE TEJEDA COLLAPSE CALDERA

The development of the Caldera de Tejeda and the enormous volume of associated ignimbrites ($\geq$1500 km^3) likely represent the most dramatic volcanic features in the geological evolution of Gran Canaria. Two previous events set the necessary conditions for the generation of the caldera: (1) the emplacement of a shallow level reservoir that facilitated magma differentiation and accumulation of highly evolved magmas and (2) the eruption of the massive P1 ignimbrite, which emptied this reservoir and initiated a complex interplay of repeated inflation and deflation cycles (Troll et al., 2002). This gravitational and vertical collapse of the reservoir roof opened ring fractures at the caldera perimeter that acted subsequently as conduits for further the massive eruption of ignimbrites (Fig. 6.17) (eg, Schmincke, 1967, 1969; Troll and Schmincke, 2002).

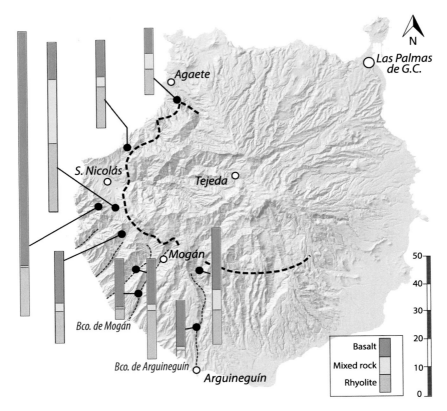

FIGURE 6.14

Exposed sections of the P1 ignimbrite (after Freundt and Schmincke, 1992, 1995).

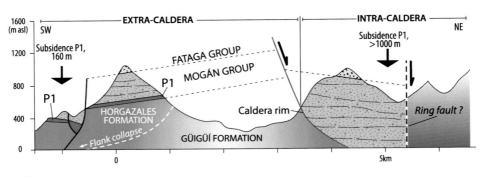

FIGURE 6.15

Schematic cross-section showing the Tejeda caldera to the right. The intracaldera and extracaldera Mogán and Fataga group deposits fill the caldera and cap the Miocene shield basalts (from Schmincke 1993; Schmincke and Sumita, 2010).

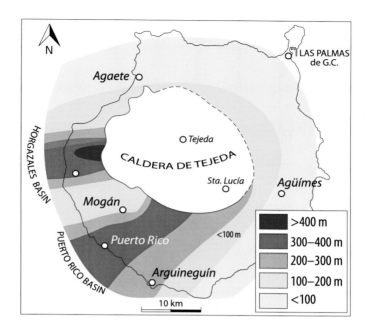

FIGURE 6.16

Thickness of the Mogán group ignimbrite sequence. Maximum thickness data reveal two main palaeo-basins. The Horgazales and Puerto Rico former valleys were filled with successive ignimbrite eruptions of the Mogán group (after Schmincke and Sumita, 2010).

The 20×17-km-wide elliptical Caldera de Tejeda, with a vertical displacement of its interior of 1 km or more, can be inspected in detail in the western part of Gran Canaria (see Figs. 6.1, 6.2, 6.15, and 6.18). Along a 30-km stretch, the caldera rim is marked by an eye-catching bright green to pink deposit formed by hydrothermal alteration of intracaldera pyroclastic rocks along the caldera's marginal ring fracture system (Fig. 6.19). These hydrothermalized rocks (eg, Donoghue et al., 2008), locally known as the "Azulejos" (from Spanish *azul*, meaning blue), are similar to the Tenerife Azulejos in the Las Cañadas Caldera (see chapter: The Geology of Tenerife, Fig. 5.78C,D).

Once caldera collapses began, the shallow level magma chamber was repeatedly replenished with basaltic magma from the lower reservoir (eg, Crisp and Spera, 1987; Leat and Schmincke, 1993; Freundt and Schmincke, 1995; Sumner and Wolff, 2003), triggering explosive eruptions. A mechanism that likely accelerated and maintained the massive explosive volcanism was the gravitational load of the sinking block (cauldron block), repeatedly forcing the expulsion of ignimbrites through the ring fractures along the caldera perimeter (see Fig. 6.17A).

The intra-caldera deposits that almost entirely filled the depression were erupted from the same ring fractures, although inspection is not easy because they are largely concealed by later volcanics and were affected by subsequent intrusive episodes.

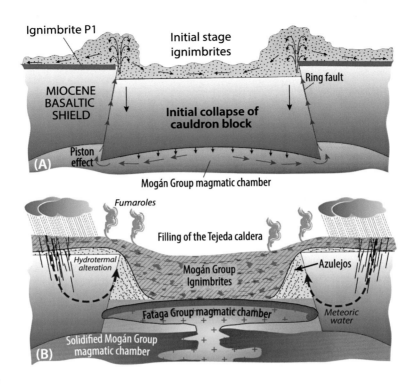

FIGURE 6.17

Sketch of the processes involved in the formation of the Miocene Tejeda Caldera. Modified from Troll et al., 2002 and Donoghue et al., 2008.

INTRACALDERA INTRUSIVES: THE TEJEDA FORMATION

The Tejeda formation is a large intrusive complex that formed inside the caldera basin as the depression was being filled by rhyolitic and trachytic pyroclastic flows and lavas of the Mogán group, and later by the trachyphonolitic ignimbrites and associated sill intrusions of the Fataga group. The intrusive complex was divided by Schmincke (1967) into three concentric zones approximately centered within the Tejeda Caldera: (1) a 3 × 5-km central core of intracaldera breccias intruded by syenite stocks and less than 20% dykes (Fig. 6.20); (2) a cone-sheet zone approximately 10 km in diameter in which the older sheet intrusions are trachytic and the younger ones are phonolitic, and where cone-sheet density increases up to more than 90% in places; and (3) a wide (2−3 km) marginal belt where dyke density is less than 20% and virtually absent within 2 km of the caldera margin (see also Schirnick et al., 1999; Donoghue et al., 2010).

FIGURE 6.18

Pliocene and younger lavas overlie the Tejeda caldera in the far distance. In the foregroud is the Roque Nublo monolith (image courtesy of Foto aérea de Canarias, Gobierno de Canarias).

FIGURE 6.19

(A) Topographic caldera margin at *Fuente de Los Azulejos*. Shield basalts to the lower left are unconformably overlain by steeply dipping intracaldera Mogán-aged tuffs. Note the progressive shallowing of the intracaldera deposits to a flat geometry on top of the pile. (B) The bright-colored tuffs are characteristic for low-temperature hydrothermal alteration due to extensive fluid flow in the faulted caldera marginal areas (see Donoghue et al., 2008) (images courtesy of S. Burchardt).

The total volume of intrusions of the Tejeda Formation has been estimated at 250 km³ (Hernán and Vélez, 1980), and these authors estimated a depth of 2 km below sea level for the top of the reservoir that fed the cone-sheet dykes (ie, shallower than the previous Mogán reservoirs at ~5−6 km) and inferred an original vertical extent of at least 3 km for the cone-sheet swarm. The average dip of the cone sheets also decreases from 50° to 30° toward the periphery, and syenite screens between the cone sheets are replaced by screens of intracaldera fill toward the caldera margin.

Cone-sheet intrusion usually occurs during periods of caldera resurgence (eg, Burchardt et al., 2013). It appears that after magma chamber evacuation and associated collapse (Fig. 6.21A), the

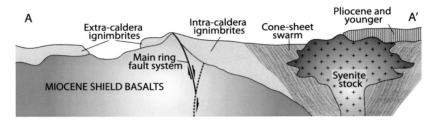

FIGURE 6.20

Generalized cross-section through Gran Canaria from *San Nicolás* (A) to approximately the center of the island (A′), approximately west−east, showing the Miocene shield basalts, the felsic ignimbrite suite, and the subsequent intrusion of the cone sheets and syenite stocks that mark the terminal stages of the Miocene postshield volcanism (after Troll et al., 2002).

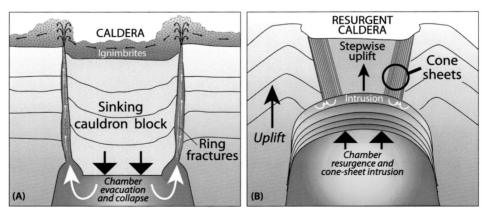

FIGURE 6.21

(A) Caldera-forming ignimbrite eruptions that were initiated by P1 ignimbrite (Mogán group) change with time to (B) resurgence of the caldera, leading to a stepwise uplift of the cauldron roof through injection of dykes as magma pushes upward again. This episode constructed a dense conical dyke swarm (the cone sheets) during the late Fataga period.

FIGURE 6.22

View of the deeply eroded cone sheets from *Mesa de Acusa* (looking south). The cone sheet swarm is unconformably overlain by the Roque Nublo lava flows (A) and massive Roque Nublo breccia deposits (B). Image courtesy of L.M. Schwarzkopf.

magma chamber became sealed again and also overpressured, and therefore exerted enough upward force to uplift and fracture the overlying crust (eg, Walter and Troll, 2001).

Each time the upward pressure lifted up the cauldron block, the intruding magma filled the resulting conical cracks and solidified to form a dyke intrusion (Fig. 6.21B). After a long period of repeating this process many times (from ∼7.3–11.7 Ma; Schirnick et al., 1999), many thousands of conical dykes were intruded, creating the spectacular cone-sheet swarm exposed today (Fig. 6.22).

The intrusion of the Tejeda cone-sheet swarm caused structural uplift of the core zone, probably by a piston-like uplift mechanism. The emplacement of the dyke swarm caused uplift of the host and roof rocks by at least 1400 m (Schmincke, 1993; Schirnick et al., 1999), and likely a large stratovolcano grew inside the caldera at this stage (eg, Donoghue et al., 2010). Cone sheets have recently gained renewed interest for their relevant role in constructing the interior of long-lived volcanoes (eg, the Paleogene Ardnamurchan intrusive complex in northwest Scotland, Burchardt et al., 2013, and the Zarza intrusive complex in Baja California, Johnson et al., 1999) and are likely the key conduit type to feed volcanic flank eruptions.

VOLCANIC QUIESCENCE AND SEDIMENTATION

Eruptive activity in the late stages of the first cycle of volcanic construction of Gran Canaria (the basaltic shield, the caldera, and the postcaldera phases) progressively declined and, finally, the island entered a long period of eruptive quiescence (5.3–8.8 Ma). During this episode the island was intensely eroded. This repose period affected the entire island, but it is most evident in the southwest part of Gran Canaria, where the Miocene volcanics have not been as intensely resurfaced (see Fig. 6.5A). The northeast part, in turn, is almost entirely covered by Pliocene to Quaternary volcanism, reaching 500 m of total deposit thickness in some places (eg, Guillou et al., 2004b; Pérez-Torrado et al., 1995).

A radial network of barrancos is incised in the Miocene felsic deposits, frequently reaching the substratum of basaltic shield lavas in the south of the island. This erosive configuration will later control the distribution of rejuvenated volcanism that is often expressed as intracanyon lava flows and deposits.

Products of erosion accumulated during the Upper Miocene and Pliocene, mainly at the northeast, north, and south coastal regions (the Lower Member). When posterosional eruptive activity recommenced (rejuvenation), large volumes of these sediments were remobilized by the renewed volcanism and deposited as alluvial fans, forming the thick "Las Palmas Terrace" (Lietz and Schmincke, 1975), which was more recently relabeled as the Las Palmas Detrital Formation (LPDF) (Gabaldón et al., 1989; Fig. 6.23). This massive littoral sedimentation that formed the LPDF coincided with the renewed volcanism of the rejuvenation stage approximately 5.5 Ma ago, which eventually led to the construction of the Roque Nublo stratovolcano in the center of the island.

The return of eruptive activity and intense sedimentation concurred with a phase of marine transgression in the island that formed fossiliferous marine deposits corresponding to the Middle Member of the LPDF (Fig. 6.23), interbedded with epiclastic formations (the Upper and Lower Members of the LPDF). Early Pliocene littoral and marine deposits contain a characteristic fossil assemblage of tropical climate: *Strombus coronatus*, *Nerita emiliana*, *Gryphaea virleti*, *Patella intermedia*, and *Rothpletzia rudista* (Lietz and Schmincke, 1975).

Evidence of the Pliocene marine transgression is provided by well-preserved basaltic pahoehoe lava-fed deltas exposed in exceptional outcrops at the north−northeast coastal area of Gran Canaria

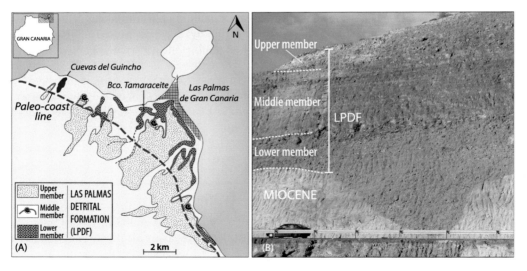

FIGURE 6.23

(A) Simplified map of the Las Palmas detrital formation (LPDF) (after Gabaldón et al., 1989). (B) Close-up view of a representative section in which the three members of the LPDF unconformably overlie a Miocene ignimbrite substratum.

FIGURE 6.24

The northern cliffs of Gran Canaria expose Miocene and Pliocene strata in spectacular semi-vertical sections at *El Rincón*. A thin white band (the "marine band") is visible high up in the cliffs. The white arrows follow the base of the Roque Nublo pillow lavas, indicating the Pliocene sea level before the post-Pliocene sea-level drop.

(Fig. 6.24). These lava-fed deltas correspond to the Pliocene initial volcanic activity of the Roque Nublo stratovolcano and crop out at different heights above the present sea level. The transition from subaerial lavas to submarine pillows (Fig. 6.25A; up to 120 m high in some places) marks the precise location of the sea level at that time (~4 Ma ago), before a post-Pliocene sea-level drop. This is likely the result of a combination of eustathic and isostatic processes, with the latter probably caused by the load of the neighbor island of Tenerife (Menéndez et al., 2008).

This situation is similar to that described in Fuerteventura. There, Pliocene lava flows entering the sea (eg, at *Ajuy*) define the sea level at that time (see chapter: The Geology of Fuerteventura, Figs. 8.32 and 8.33).

Polygonal feeder tubes filled with hydroclasts (Figs. 6.25B and 6.26) formed as thick lava flows entered the sea under the influence of a tidal regime. In the LPDF, they appear interbedded between submarine and subaerial lithofacies, thus marking the passage from subaerial to submarine conditions. These tubes show sharp contacts between external polygonal walls composed of coherent lava and an internal filling of densely packed angular hyaloclastite material, thus representing a fascinating petrological and evolutionary record for Gran Canaria (Pérez-Torrado et al., 2015).

PLIOCENE TO QUARTERNARY REJUVENATION VOLCANISM

This stage comprises the volcanic activity of the past 5.5 Ma and involves two main phases. First, the construction of a central felsic stratovolcano (the Roque Nublo volcano); second, the post-Roque Nublo volcanism. Notably, the composition of lavas changed from phonolitic (in the latest phases of the Roque Nublo period) to basaltic and basanitic (in the Holocene). Some eruptions

FIGURE 6.25

(A) Closely packed pillow lavas above white marine sediments of the Middle Member of the LPDF. (B) Polygonal feeder tubes filled with hydroclasts in pillowed lava flows of the Pliocene Roque Nublo group (from Pérez-Torrado et al., 2015).

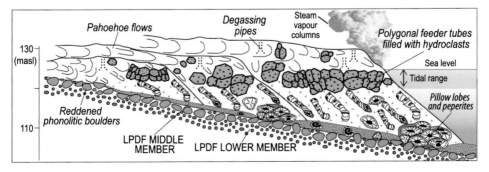

FIGURE 6.26

Interpretative sketch of the formation of the lithofacies association of the northern Roque Nublo Pliocene lava delta (RNPLD) (from Pérez-Torrado et al., 2015).

FIGURE 6.27

View of the Roque Nublo monolith at sunset from *Pozo de las Nieves*, with Teide volcano in the far distance. The full Roque Nublo volcano probably would have been of similar shape and size as Teide before it collapsed and eroded away (dotted line).

contain pre-Hispanic artefacts and other physical remains, such as, the 1970-year Before Present (BP) *Caldera de Bandama* eruption (Rodríguez-Gonzalez et al., 2012), implying that these most recent volcanic events were in fact witnessed by the indigenous population of the island.

The construction of the Roque Nublo volcano shares notable similarities with the development of the Teide stratocone (Fig. 6.27). Although more speculative, a comparison of the Caldera de Tejeda ignimbrites and cone sheets that culminated in the construction of a central felsic volcano also bears similarities with the Las Cañadas Volcano on Tenerife, which also shows a sequence of successive caldera-forming explosive events and likely a similar intrusive complex. However, a main difference is the late-shield stage of island development for the Tejeda intrusions, where the Las Cañadas Volcano developed during the rejuvenation stage of Tenerife. This circumstance may also be reflected in the total duration of the eruptive periods of approximately 8.8 Ma for the Tejeda complex but only approximately 3.5 Ma for the Las Cañadas Volcano.

THE ROQUE NUBLO GROUP

Construction of the Roque Nublo (RN) stratovolcano commenced approximately 5.5 Ma ago. Initial strombolian eruptions were located at the center and in southern Gran Canaria and consisted of a northwest–southeast lineament of basaltic cinder cones (Pérez-Torrado et al., 1995; Pérez-Torrado, 2000).

Approximately 4.6 Ma ago, eruptive activity concentrated at the center of the island for a period of 1.5 Ma, first as basaltic effusive eruptions that reached the sea along the radial fluvial network

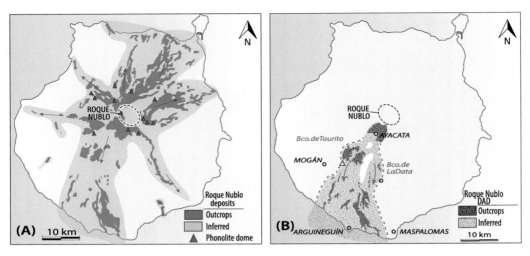

FIGURE 6.28

(A) Outcrops and former extent of the Roque Nublo group deposits. (B) Collapse of the flanks of the Pliocene Roque Nublo stratocone led to large debris-avalanche deposits (DAD) that traveled far beyond the coastline. The deposits comprised at least 14 km³ and covered an area of approximately 180 km² in south Gran Canaria prior to its partial erosion (after Pérez-Torrado et al., 1995; Mehl and Schmincke, 1999).

(eg, the thin flows in the Mesa del Junquillo, topped by thick agglomerate, ignimbrite, and breccias deposits, see Fig. 6.22), where some of these developed pillowed coastal lava deltas (Figs. 6.26 and 6.28A).

The persistent residence of magmas in shallow magma reservoirs at that time then led toward more explosive trachytic and phonolitic compositions, culminating approximately 3.9 Ma ago with the eruption of thick deposits of ignimbrites and breccias associated with relatively large phreato-magmatic and vulcanian events (Pérez-Torrado et al., 1995; Pérez-Torrado, 2000).

The distribution and thickness of the Roque Nublo deposits indicate a volume of the volcano of approximately 200 km³ and a former altitude of at least 2500 masl. Persistent northeasterly trade winds likely had an effect on the asymmetrical configuration of the stratovolcano, showing gentle northern slopes but rather steep southern flanks (Fig. 6.27; Pérez-Torrado et al., 1995; Pérez-Torrado, 2000). The growth of the volcano induced a progressive instability that finally triggered several northbound and southbound lateral collapses (Figs. 6.28B and 6.29). The most voluminous landslide removed the steeper and unstable southern flank, spreading thick debris-avalanche deposits that are now exposed along approximately 25-km distance from the volcano and that are present in many parts of the island (Fig. 6.30; Brey and Schmincke, 1980; García Cacho et al., 1994; Mehl and Schmincke, 1999). These deposits have now been detected in boreholes drilled during the ODP Leg 157 in the volcanic apron of Gran Canaria (see Fig. 6.3) and according to Mehl and Schmincke (1999), the corresponding debris avalanche(s) entered the sea at velocities of more than 100 m/s (Fig. 6.28B).

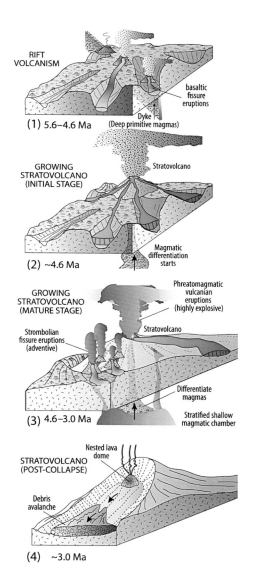

FIGURE 6.29

Cartoon depicting the evolution of the Roque Nublo stratovolcano from the initial stage of construction (1) to the final lateral collapse (4). At its peak of growth, approximately 3 Ma ago (3), the Roque Nublo volcano towered over the island with an altitude ≥2500 masl, likely similar to the Teide stratocone on Tenerife today. At that point, the Roque Nublo volcano was probably the highest peak in the Canary Islands (modified from Pérez-Torrado et al., 1995).

FIGURE 6.30

Panorama view of central and western Gran Canaria from *Pico de las Nieves*, the highest point on the island. Roque de Bentaiga, in the center, and Teide volcano in the far distance are visible. The lateral collapse of Roque Nublo volcano formed the rocks that make the Roque Nublo monolith and the thick slabs below it (left of image).

Reconstruction of the original distribution of the Roque Nublo Debris Avalanche Deposits (RNDAD) shows that they covered approximately 180 km^2, with an estimated volume of approximately 14 km^3.

The most spectacular feature of the Roque Nublo group is the large megablock with the 80-m-high Roque Nublo monolith, which gave the name to the entire Roque Nublo formation (Fig. 6.31A). Mehl and Schmincke (1999) describe the megablock moving above dyked in situ Roque Nublo pyroclastics and lava flows, as shown by striations that indicate transport of the megablock from north to south (Fig. 6.31B).

ROQUE NUBLO INTRUSIVE FACIES

Evolved magmas appeared in the later stages of the Roque Nublo cycle and were very viscous. They forced their way to the surface by intruding and deforming the earlier Miocene and Roque Nublo volcanics and formed lava domes that cluster around the periphery of the Roque Nublo stratovolcano (Figs. 6.28A and 6.32). The arrangement is remarkably similar to the peripheral lava domes surrounding the Teide stratocone (see Figs. 5.34 and 5.39), or the felsic domes of La Gomera that cluster around a central complex (see Figs. 4.6 and 4.14). There are approximately a dozen Roque Nublo–aged felsic lava domes on the island, with the most spectacular being the Roques de Tenteniguada and the approximately 3.9-Ma-old hauyne phonolite plug of Risco Blanco that is located in the northern wall of Bco. de Tirajana (Fig. 6.32B).

POST-ROQUE NUBLO VOLCANISM, THE LATEST CONSTRUCTIONAL PHASE OF GRAN CANARIA

A deep level of understanding of the geological evolution of Gran Canaria during the Miocene shield and postshield stages was obtained through decades of dedicated and detailed on-shore and off-shore studies, for the most part by Schmincke and co-workers.

These pioneering studies proposed a period of volcanic repose (~ 0.5 Ma) between the end of the Roque Nublo group and the onset of the post-Roque Nublo volcanism. The synthesis of the latest results obtained from combined mapping, magneto-stratigraphy, and the most recent K–Ar age

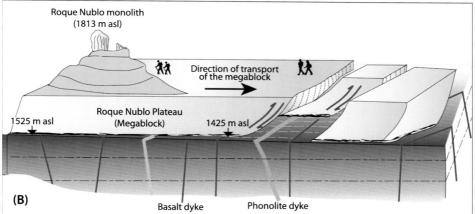

FIGURE 6.31

(A) Roque Nublo monolith is the remaining part of a thick slab of debris-avalanche deposits (image courtesy of *Foto aérea de Canarias, Gobierno de Canarias*). (B) The 80-m monolith is part of a large megablock that shows a slight inclination (~5°) from left to right in the sketch. Mehl and Schmincke (1999) interpreted this inclination and the detachment surface at the base of the Roque Nublo monolith to indicate a listric *'decollement'*.

dating (Guillou et al., 2004b; Pérez-Torrado et al., 2006) for this period, however, indicate rather more continued volcanism in the transition from the Roque Nublo to the Quarternary volcanic events (see Figs. 6.33 and 6.34).

Specifically, the combination of radiometric dating and stratigraphic correlation with geomagnetic reversals led to a detailed reconstruction of the eruptive history of the island during this period (Fig. 6.34). The new ages obtained by Guillou et al. (2004b) for the Miocene stage confirm the very detailed study of van den Bogaard and Schmincke (1998) and are consistent with the ages

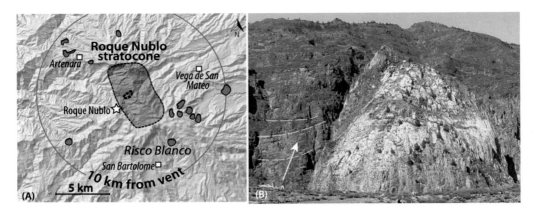

FIGURE 6.32

(A) Felsic lava domes around the Roque Nublo stratovolcano form a peripheral cluster. This arrangement is similar to the peripheral lava domes surrounding the Teide stratocone or the group of felsic domes in the center of La Gomera. (B) Risco Blanco, a hauyne-bearing phonolite intrusion. The ~3.9-Ma-old Risco Blanco dome intruded Roque Nublo lava flows. Note the upturned lava flows around the intrusion (eg, to the left of the dome, indicated with an arrow).

published by McDougall and Schmincke (1976), whereas the work of Guillou et al. (2004b) also improved the geochronology of the Pliocene to Quaternary rejuvenation phase.

Eruptive activity in this period began approximately 3.5 Ma ago, coinciding with the latest, intrusive period of growth of the Roque Nublo stratovolcano. Basaltic fissure eruptions developed a lineation of eruptive vents in a northwest−southeast direction in the northeast flank of the island (Fig. 6.33A). Progressive concentration of these vents led to the establishment of a vaguely defined rift zone located below the crest of the island (with a slight deviation toward the northeast). This circumstance forced the flow of recent lavas toward the northern and northeastern parts of Gran Canaria (Fig. 6.33B). At later stages, eruptive vents dispersed, but mainly on the northeast flank of the island. The recent volcanic resurfacing of the northeast flank of Gran Canaria is thus consistent with the early proposal by Bourcart and Jérémine (1937) of a *Neocanaria* (the northeast Pliocene and Quaternary part) and a *Paleocanaria* (the southeast Miocene part).

RECENT (QUATERNARY) VOLCANISM

In this period, volcanic activity is spatially and temporally disperse and, with the exception of the volcanic field of *Tafira* (reverse polarity), the recent eruptive activity belongs to the Brunhes period (<0.78 Ma), and most of it is likely younger than 0.5 Ma.

Eruptions during this period were mainly basanitic or nephelinitic, and the erupted lavas were dominantly channeled through barrancos, frequently creating considerable disruption of "normal" fluvial processes (Rodríguez-Gonzalez et al., 2009, 2012).

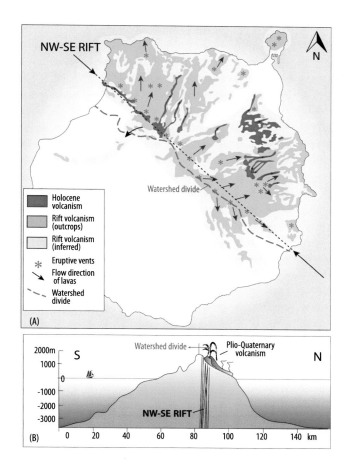

FIGURE 6.33

(A) Distribution of post-Roque Nublo volcanism, probably initially related to a northwest−southeast fractue or rift zone, while, during the Holocene, eruptions migrated eastward. The broken blue line indicates the watershed, which is consistently higher than the Holocene vents. This feature confined the recent eruptions mainly to the northeast half of the island (see cross-section in B) (from Pérez-Torrado et al., 1995; Carracedo et al., 2002).

The most recent period of eruptive activity (Holocene) was poorly dated and understood prior to the significant improvement provided by the recent work of Rodríguez-Gonzalez et al. (2009), in which 13 individual eruptions were dated with ^{14}C (Fig. 6.35).

During the past 11 ka, 24 monogenetic basaltic eruptions occurred in the north and northeast sector of Gran Canaria. Archeological studies have shown that the more recent eruptions affected prehistoric human settlements on the island, which is relevant in respect to the assessment of present-day eruptive hazards, particularly in densely populated areas. For instance, several of these

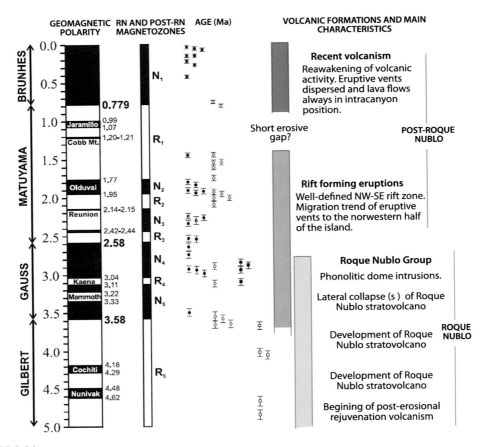

FIGURE 6.34

Graph of available K−Ar data versus the Astronomical Polarity Time Scale (APTS; from Berggren et al., 1995). Note the overlap of the late Roque Nublo eruptions (yellow) with the initial rift volcanism (in orange) (from Perez Torrado et al., 1995; Guillou et al., 2004b).

recent eruptions took place close to the island's capital, Las Palmas, which currently hosts more than 400,000 inhabitants (Fig. 6.36).

Field studies by Rodríguez-Gonzalez et al. (2009) showed that these eruptions typically constructed strombolian cones (up to 250 m in height) and erupted relatively long lava flows (up to 10 km). The total volume of these eruptions is approximately 0.388 km^3 (46.1% as tephra fall, 41.8% as cinder cone deposits, and 12.1% as lava flows). The relatively low eruption rate (0.04 km^3/ka) during the past 11 ka is consistent with Gran Canaria being in a late stage of island evolution, as is implied by the regional volcano−tectonic framework of the Canary Islands.

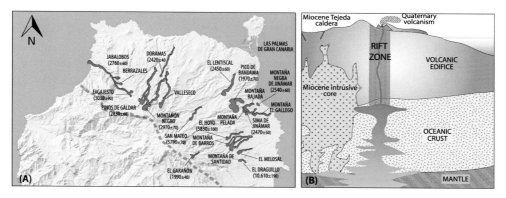

FIGURE 6.35

(A) Sketch map showing the distribution of Holocene eruptions on Gran Canaria and their radiocarbon ages. The stippled blue line is the watershed divide. Note that the Holocene eruptions are consistently located east of and below this divide (after Rodríguez-Gonzalez et al., 2009). (B) Displacement of Quaternary volcanism to the northeast of the dense Miocene intrusive core (after Ye et al., 1999; Hansteen and Troll, 2003).

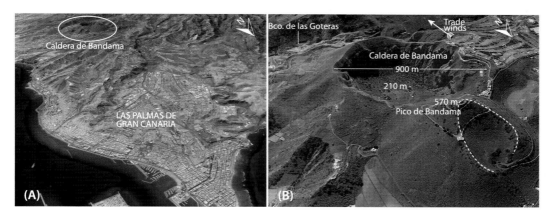

FIGURE 6.36

(A) Location of the 1970-year BP Caldera de Bandama eruption, which is close to *Las Palmas de Gran Canaria* (~7 km), the capital of the island. (B) View of the cinder cone of Pico de Bandama and the Caldera de Bandama, which started as a strombolian cone but evolved to explosive pulses that ended with a partial collapse that formed the present-day Caldera de Bandama (images from Google Earth).

THE 1970 BP BANDAMA EXPLOSIVE ERUPTION

Holocene volcanism on Gran Canaria is rather monotonous (ie, predominant basaltic strombolian eruptions). However, one of these eruptions, the 1970-year-old Bandama event, is particularly interesting because of its proximity to Las Palmas and its evolution toward highly explosive events (Hansen et al., 2008).

The Bandama volcanic complex is formed by a strombolian cone (*Pico de Bandama*) and a caldera-like depression of phreatomagmatic collapse origin (named *Caldera de Bandama*; Figs. 6.36B and 6.37). The cone has an asymmetric crater that is open to the northeast due to the influence of the trade winds (Figs. 6.36B and 6.37), whereas the caldera presents an elliptical outline with a diameter of 930×860 m and a maximum decrease of 269 m (Hansen, 1993; Hansen et al., 2008).

The phreatomagmatic eruptions leading to the caldera genesis were preferentially oriented to the southwest and their deposits covered an area of approximately 11 km^2 (Figs. 6.37 and 6.38). The distance/fragmentation values determined for these deposits show that the most explosive pulses of the eruption probably reached a subplinian magnitude.

A ^{14}C age of 1970 ± 70 BP indicates that the Bandama volcanic complex ranks among the most recent eruptions on Gran Canaria, which is consistent with the earlier discovery of a millstone of the pre-Hispanic age under the Bandama pyroclastic deposits.

Remarkably, the area covered by the Bandama deposits is at present one of the most densely populated on Gran Canaria. The capital of the island, *Las Palmas de Gran Canaria*, with more than 375,000 permanent residents, is only 7 km from the Bandama volcanic vents (see Figs. 6.36A and 6.38A).

Is the Bandama event the last or merely the latest eruption on Gran Canaria? With up to 24 Holocene eruptions, there is little reason to assume that after the Bandama eruption volcanism became extinct, and we must prepare for future eruptions in the region (eg, Carracedo et al., 2015a). The eruptive hazard in this densely populated area will be, in part, conditioned by the eruptive mechanisms (eg, purely strombolian or strombolian with phreatomagmatic episodes, like during the 1949 Hoyo Negro eruption on La Palma or the Los Erales eruption on Tenerife, see Figs. 3.59C, 3.60A, and 5.66) and by the proximity of the vent to the city of *Las Palmas*. Finally, the direction of the wind during the eruption will also play a significant role and in the event of unfortunate circumstances phenomenal mitigation measures would be required (see, eg, Carracedo et al., 2015a).

A VOLCANIC HISTORY OF GRAN CANARIA

The analysis of volcaniclastic rocks drilled during ODP Leg 157 at sites 953 and 954 to the north and sites 955 and 956 to the south of Gran Canaria (see Fig. 6.3), with a cumulative penetration of almost 3000 m and an overall recovery rate of 75%, correlates surprisingly well with the geological evolution established for the onshore island (Fig. 6.39). These geological data, combined with a wealth of published geochronological and volcanological information, allowed the detailed reconstruction of the development of one of the main islands of the Canarian archipelago (eg, Schmincke and Sumita, 1998). According to this scheme, the evolutionary stage of Tenerife is a step behind that of Gran Canaria (~ 3 Ma), whereas Lanzarote and Fuerteventura are a step ahead.

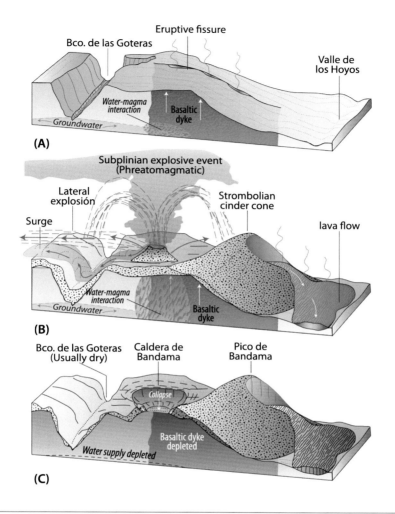

FIGURE 6.37

(A) The Caldera de Bandama eruption started as a basaltic fissure eruption that constructed two aligned strombolian vents (B). One of the vents interacted with groundwater, thus changing to phreatomagmatic explosive behavior that generated pyroclastic falls and flows that ended in the formation of (C) the present caldera complex (modified from Hansen, 1993).

Comparing Tenerife and Gran Canaria, the peak of the construction of both islands is represented by a large stratovolcano developed at the top of the island (Roque Nublo on Gran Canaria and Teide on Tenerife). The growth of Roque Nublo volcano ended approximately 3 Ma ago with catastrophic destruction of the stratocone, whereas the Teide volcanic complex is still in a late stage of development. Therefore, the question is whether both stratovolcanoes will share a similar *grande finalé* or if Teide will merely cease activity and disappear by slow and gradual erosion.

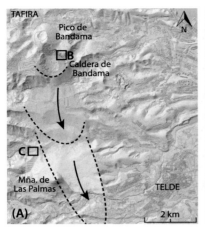

FIGURE 6.38

(A) Isopach map indicating the dispersion of pyroclastic fall from the Bandana precaldera strombolian vent. (B) Pyroclastic deposits of the Bandama eruption at *Cueva de Los Canarios*. (C) A sequence of pyroclastic eruptions occurred during the explosive phase of the Bandama eruption (photograph courtesy of C. Moreno).

GIANT LANDSLIDES AND MAGMATIC VARIABILITY

Giant lateral collapses on Gran Canaria have been inferred from onshore observations but have been principally studied off-shore through geophysical means and through the data provided by the boreholes drilled during the ODP Leg 157 in the volcanic apron of Gran Canaria (Schmincke and Sumita, 1998). This information was complemented with high-resolution seismic reflection data afforded by seismic profiles around Gran Canaria obtained during the pre-site survey by the RV

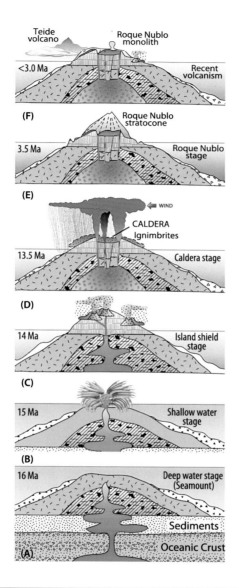

FIGURE 6.39

Cartoon showing the volcanic history of Gran Canaria (modified after Schmincke and Sumita, 1998). The evolution of this island is approximately 3 Ma ahead of Tenerife. Although Gran Canaria has by now lost the Roque Nublo stratovolcano, Teide volcano on Tenerife is still growing. Currently available geological information is insufficient to foresee whether or not Teide stratocone will experience a catastrophic *finalé* similar to the one seen at the Roque Nublo volcano, or if it will simply end its eruptive activity without major collapse events and gradually succumb to erosion.

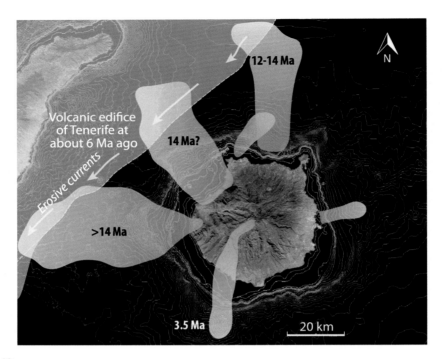

FIGURE 6.40

Bathymetric map showing several large landslides detected by morphological and seismic observations. Note that the volcanic apron of Tenerife overlies the giant landslides from the northwest flank of Gran Canaria (from Funck and Schmincke, 1998; island image from *Google Earth*).

Meteor (Schmincke and Rihm, 1994). Using this approach, a number of large mass-wasting events are identified (Fig. 6.40), interbedded with the pelagic background sedimentation, and are interpreted to represent the deposits associated with slope failures at the flanks of Gran Canaria (Funck and Schmincke, 1998; Krastel et al., 2001).

In relation with the formation of the Tejeda Caldera, it is noteworthy that a general destabilization of the island by caldera volcanism may have favored the formation of this felsic shallow level magma chamber, which, in turn, favored lateral landslides (eg, Troll et al., 2002).

As described in chapter "The Geology of Tenerife," this apparent link between lateral collapses and emplacement of shallow felsic caldera systems has been the case on Tenerife for the lateral collapse that formed of the Las Cañadas Volcano that produced the current Las Cañadas Caldera. It may therefore be a common case in Atlantic ocean island evolution that felsic caldera episodes destabilize island edifices to induce higher rates of lateral collapse events.

On-shore evidence for massive landslides on Gran Canaria has been found for the Pliocene (3.5 Ma) lateral collapse of the Roque Nublo volcano, but it is less direct for the Miocene landslides that have been confirmed from off-shore records (eg, Fig. 6.40; Funck and Schmincke, 1998). Interestingly, proof of a giant landslide has recently been obtained in the Agaete valley, but

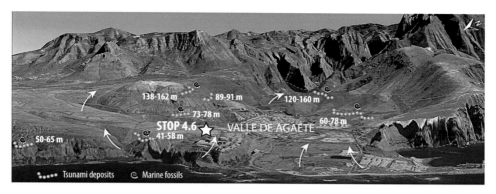

FIGURE 6.41

3-D view of the "flooded" study area (Valle de Agaete) reconstructed from marine conglomerate deposits (gray dots) that occur at altitudes up to 160 masl (from Pérez-Torrado et al., 2006).

it is not related to Gran Canaria. The deposit probably relates to the much younger giant landslide of Güímar on Tenerife which formed at approximately 830 ka (see Figs. 5.28 and 5.102) and directly faces the Agaete Valley (see also Fig. 1.32).

Marine deposits of conglomerates with fossil rhodolites and marine shells crop out at up to 188 masl along the Agaete valley walls (Fig. 6.41). These deposits have been interpreted as the result of the tsunami caused by the Güímar lateral collapse, that invaded the Agaete valley. This situation provided favorable conditions, maybe unique in the Canaries, for the inland deposition of high-energy tsunami deposits (Pérez-Torrado et al., 2006).

THE TIRAJANA MULTI-SLIDE CALDERA

The Barranco de Tirajana is a large (35 km^2) concavity formed by erosion of the barranco headwall (Fig. 6.42). What makes this feature particularly interesting is the multiple slope failures.

The formation of this "caldera" has initially been related to volcanic processes, such as explosion, lateral collapse, and vertical collapse. However, Fúster et al. (1968c) and Araña and Carracedo (1980) interpreted this depression as exclusively erosive, without any volcanic or tectono-volcanic implication.

Up to 28 recorded slumps have occurred in the depression (Fig. 6.43), separated in successive generations. The extent of the displaced materials ranges up to 6.5 km^2, and the volume of the slumps varies from 0.18 to 1.35 km^3 (Fig. 6.43A). The slipped materials consistently moved toward the present barranco bed (Fig. 6.43B), indicating that its course was similar to when the slides occurred (Lomoschitz and Corominas, 1997).

The deposits of these gravitational instabilities are different from volcanic debris avalanche accumulations, and the failure surfaces have generally affected solid Miocene and Roque Nublo pyroclastics, lavas, and ignimbrites.

FIGURE 6.42

Google Earth image of the headwall of Bco. de Tirajana. Sedimentary deposits (red) fill part of the depression and are mostly related to slumps from the steep barranco walls.

This erosive depression incised a previous shallow valley since approximately 0.6 Ma to recent times. During this period, several pulses of instability occurred, with the latest being the 'Rosiana slump'.

THE ROSIANA SLUMP

A 850×400 m landslide scarp gave rise to a volume of approximately 3×10^6 m^3 of collapsed material in February 1956 (Fig. 6.44). Major previous slides involved volumes exceeding 10^6 m^3 and occurred in 1879, 1921, and 1923, usually after intense rainfall. The last catastrophic movements in 1956 destroyed houses, infrastructure, and communication lines (an arterial road and a bridge), and prompted the evacuation of 250 people (Fig. 6.44).

According to Solana and Kilburn (2003), the fast rates of movement, reaching maximum values of approximately 7.3 cm/hour near the top of the slide and 0.2 cm/hour at the toe, were accompanied by meter-long ground cracks 0.5−1 m wide and several meters deep that caused additional damage to several buildings.

SEDIMENTARY FORMATIONS

Most of the erosion of Gran Canaria occurred during the long period (8.8−5.3 Ma) of eruptive quiescence. Products of this erosion accumulated during the Upper Miocene mainly in the

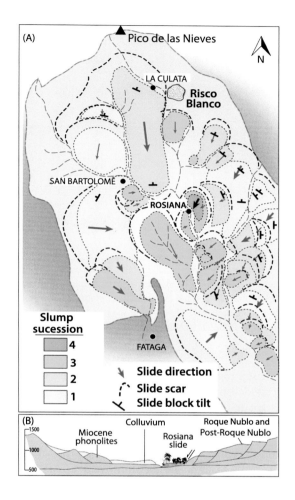

FIGURE 6.43

(A) Multiple slide blocks inside the headwall of Bco. de Tirajana. (B) The slides and the transported blocks tend to converge toward the barranco bottom (after Lomoschitz et al., 2002).

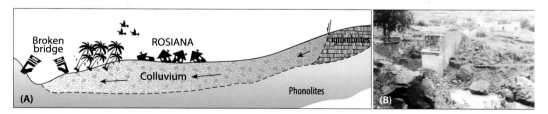

FIGURE 6.44

(A) The youngest major slump on Gran Canaria occurred in 1956 and destroyed a bridge and several houses of the small village of *Rosiana* (B). Image (B) courtesy of Cabildo Insular de Gran Canaria.

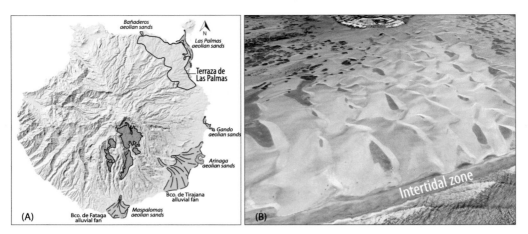

FIGURE 6.45

(A) Sediment that accumulated during the long period of eruptive quiescence (8.8–5.3 Ma) was remobilized by the renewed Pliocene eruptive activity and was redeposited in large volumes as alluvial fans onto the northeast sector of the island, forming the Las Palmas Terrace as well as the mouths of the main barrancos of the southeast sector of Gran Canaria (Tirajana and Fataga). (B) The coastal delta of Bco. de Fataga has also been occupied by an extensive field of aeolian sand dunes (Dunas de Maspalomas), one of the main tourist attractions of the island. Image from Google Earth.

northeast, north, and south coastal regions. The Pliocene renewal of eruptive activity (rejuvenation) deposited large volumes of eroded products as alluvial fans at the northeast sector of the island, forming the Las Palmas Terrace (eg, Menéndez et al., 2008, see Fig. 6.23). However, the northeast sector of Gran Canaria was resurfaced to a great extent by the Roque Nublo and post-Roque Nublo volcanism, capping a large part of these deposits.

The most relevant outcrops of recent sedimentary formations are alluvial deposits of the largest barrancos in the southeast (Tirajana and Fataga) that form wide coastal fans (Fig. 6.45A).

Aeolian sand dunes top these alluvial fans, forming the extensive dune field of Maspalomas (Fig. 6.45B). The recently formed dunes rest on top of consolidated marine deposits with *Strombus bubonius*, a characteristic fossil of the "Senegalese fauna" (Meco et al., 2007), indicating the presence of dunes in this area since at least the Pliocene.

ROCK COMPOSITION AND VARIATIONS

Gran Canaria presents the greatest variability of composition of igneous rocks of the entire Canarian archipelago, being the richest of all oceanic islands in magmatic diversity. Besides the characteristic evolution of Oceanic Island Basalt (OIB) magmas of the Canaries, producing lavas of the basanite–basalt to trachyte–phonolite series (Fig. 6.46), Gran Canaria exceptionally presents other types of magma, such as tholeiite basalts and also rhyolites. The latter rock type is only present in the Canaries on the island of Gran Canaria (see, eg, Fig. 5.43).

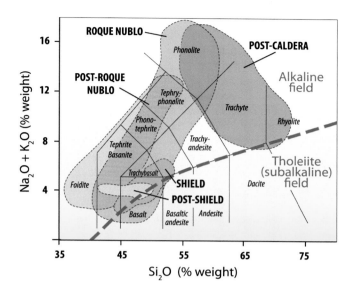

FIGURE 6.46

Total alkali versus silica (TAS) diagram of more than 500 lava compositions from Gran Canaria (data from Balcells et al., 1992; Schmincke, 1993; Pérez-Torrado, 2000; and Guillou et al., 2004b). Note the rich compositional diversity when compared with a similar diagram of rock composition from Tenerife (Figure 5.43A).

The geological evolution of Gran Canaria is thus separated into three main evolutionary stages: the Miocene shield stage with its late felsic phase, the Pliocene posterosional stage of Roque Nublo volcanism, and the Pleistocene, Holocene, and recent alkaline volcanic activity in north and northeast Gran Canaria. Each stage has its own distinct magma compositions and types related to discrete magmatic processes at depth, and likely related to three successive mantle "blobs" of variable chemical composition (eg, Hoernle and Schmincke, 1993; Schmincke and Sumita, 1998). However, recent dating now supports the Pliocene to Holocene events on Gran Canaria to represent a single activity cycle and, hence, discussion on the detailed causes of magmatism on Gran Canaria will probably continue for many years to come.

In any case, the exceptional range of magma compositions and ample variety of eruption products and styles render Gran Canaria likely the geologically most complete Canary island. Thus, it rightly deserves its place among the most spectacular "natural laboratories for ocean islands volcanism" worldwide (see, eg, Figs. 1.33 and 1.34).

GEOLOGICAL ROUTES ON GRAN CANARIA
DAILY ITINERIES

This chapter aims to provide a general background on the volcanic geology of Gran Canaria and suggests a number of scenic drives to key geological units and features that will help to glean an insight into volcanological processes and temporal geological evolution of the island. All excursions are designed to commence in *Maspalomas/Playa del Inglés* because most visitors are likely to give preference to the southern sun instead of the northern clouds. There are five itinerary days (Fig. 6.47) that can be followed in detailed maps (eg, Fig. 6.48).

The present series of drives aims at building a swift and useful overview of the volcanological history of the island by inspection of accessible outcrops near major roads. Where appropriate, other localities are suggested where more or better exposures can be inspected.

The basaltic shield makes up the majority of the island's edifice, with all younger rocks (those that make up a major part of the exposed island) representing perhaps 2—5% of the island's total volume (Krastel and Schmincke, 2002a,b). Of these, however, the pyroclastic deposits of the closing shield stage are particularly prominent, forming thick blankets of ash that must have once covered the entire island (Days 1 and 2). These deposits were erupted from a major caldera volcano (the Tejeda Caldera) that is superbly exposed, and a scenic traverse will take us from the extracaldera deposits through the hydrothermally altered caldera margin into the fill of the caldera basin (Day 2). A late resurgence of the caldera can be investigated through numerous sheet intrusions

FIGURE 6.47

Geological itineraries in Gran Canaria (map from *Google Earth*). See Figures 6.48, 6.64, and 6.83 for detailed maps of the geological excursions (eg, STOPs and names of the roads, etc.).

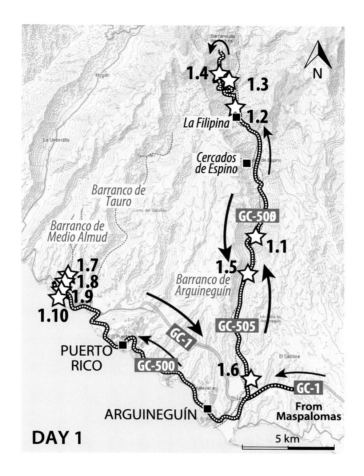

FIGURE 6.48

This day starts by inspecting the Barranco de Argueneguín with shield basalt lavas, P1 ignimbrite on shield basalts, and zoned ignimbrites and concludes the day in Bco de Medio Almund. Map from IDE Canarias visor 3.0, GRAFCAN.

that take the shape of an inverted cone (called cone sheets) and associated stocks of coarse syenite (an alkali-version of granite). At the end of Day 2, a major unconformity that is marking the decline of the Miocene Tejeda volcano is inspected. Day 3 starts with a drive to the extracaldera Fataga ignimbrites and into the Tejeda Caldera once more to inspect the Miocene intracaldera rocks and also the Pliocene Roque Nublo—type breccias that rest unconformably on the Miocene formations. After a volcanic quiescence of several Ma, the Roque Nublo era saw the evolution of a large stratovolcano, probably similar in size and character to present-day Teide on Tenerife. Much of this volcano is now eroded and we are only left with some remnants exposed in the central highlands of Gran Canaria. The Roque Nublo rocks themselves were later covered—at least in part—by yet younger Pleistocene to Holocene basaltic deposits mainly in the north and northeast of the island. The second part of Day 3 focuses on Pleistocene and Holocene cinder cones, maars, and scoria deposits that are of very mafic alkali-rich character. These fresh volcanic landforms are vivid reminders of the last eruption on Gran Canaria dating back only approximately 1500—2000 years. An inspection of the Las Palmas Volcanic field follows, which is a cluster of eruptive vents of this recent geological period, shaping the northeast of the island by producing a characteristic "nobbly" terrain. Day 4 is a circum-island trip that first inspects uplifted pillow lavas and marine sediments in northern Gran Canaria, followed by inspection of coastal instability in the steep north and north-west of the island. A summary day is planned for Day 5, traversing the island in a south to north fashion, with spectacular exposure of Miocene, Pliocene, and recent volcanic rocks. The day closes in the lush north of the island and offers an opportunity to visit scenic *Teror* or *Las Palmas*.

LOGISTICS ON GRAN CANARIA

A note on supplies and logistics. You will need to rent a car, and plenty of companies are available offering medium-sized vehicles for between €150 and €200 for a full week. Remember, horse power is an advantage on bendy mountain roads. You should, moreover, have a detailed roadmap with you in the car. The ones supplied by car hire companies are frequently insufficient (precise maps can be obtained from visor.grafcan.es). Besides this, clothes for warm as well as cold weather (eg, high in the mountains) should be carried at all times (including a wool hat, a raincoat, and gloves). Further, a first aid kit, plenty of water, sun block, and a sun hat are good companions on Gran Canaria. Finally, it is always recommended to fill the cars with fuel before heading into the islands' interior. More and more petrol stations crop up these days in even the smaller settlements on the island, but these may not be open during weekends and public holidays, or they can have seemingly random business hours.

DAY 1: MIOCENE ROCKS OF GRAN CANARIA: SHIELD BASALTS AND IGNIMBRITES

Starting point is *Maspalomas* (Fig. 6.48). From there, drive south toward *Mogan* on GC-1. Turn off after approximately 7 km for *Arguineguín* (GC-500; Exit 56).

Just outside *Arguineguín*, turn off the main road to the right and drive over a little bridge that goes over the road on which you just came. After approximately another kilometer, turn left again at a roundabout and follow signs toward *Cercados de Espino* (GC-505). Soon you will pass a brown to cream-colored, pumice-rich pyroclastic deposit of Fataga age, overlain by a more typical green Fataga welded ignimbrite deposit.

After approximately 1.4 km into the barranco, you will pass a major quarry on the right. This quarry delivers pumice-rich rock aggregate as raw material to the cement works in *Arguineguín* village (from nonwelded, pumice-rich ignimbrite). We will stop here briefly on our return trip. Continue up *Barranco Arguineguín* and note greenish Fataga phonolite ignimbrites on either side of the valley now.

Note, you are driving along the bottom of a major barranco now and the Fataga deposits on either side dip gently toward the sea. This implies that you are driving through progressively older units as you are driving along the subhorizontal barranco floor. Soon we will stop for inspection of the shield basalts, for example, right after the Bar *El Pinillo*. Park on the left side of the road on the open ground next to the bar.

Stop 1.1. Miocene Shield Basalts (N 27.8285 W 15.6623)

Walk approximately 100−150 m to the west from the parking spot to get good exposure on ground level. Here, a multitude of thin (1−5 m thick) basalt lava flows are visible, often with top and bottom breccias that mark the boundaries between the individual flows (Fig. 6.49A,B). These rocks are among the oldest on the island and are part of the basaltic shield on which all younger rocks rest. The basaltic shield is the most voluminous unit, but it only makes up a small part of the onshore exposures. Note that some of the basalt lavas here are characterized by a high percentage of large plagioclase crystals (Fig. 6.49B).

Continue inland on GC-505. A little further into barranco Arguineguín, you will meet a fork. Go left toward *Soria* and pass through the village of *La Filipina*.

Approximately 1.5 km after *La Filipina*, park the car at the side of the road in a sharp bend that is running up-hill so that your car is visible by road users from both sides.

Stop 1.2. Barranco Arguineguín: P1 Ignimbrite (N 27.8756 W 15.6714)

The rock exposed in the road section is "P1" ignimbrite and vitrophyre (Fig. 6.50). The ignimbrite shows a notable horizontal variation because of changes in density of the composite flow with increasing distance from the source (Fig. 6.51).

At the base of this outcrop, just slightly above the tarmac, you will find a black vitrophyre, where there is an initially chilled horizon from when the hot flow met the cold substrate (Schmincke and Swanson, 1967). The main body of the ignimbrite, in turn, is made of a pinkish ash-rich matrix and a very high percentage of feldspar crystal.

The high crystal content is astonishing, and it seems plausible that strong winds have removed fine ash from the eruptive cloud(s), producing a crystal-enriched area near the vent deposit (fines-depleted, i.e loss of particles <2mm due to high gas flux or winds during the eruption). Note the dark lumps of basalt scattered throughout the outcrop. These are enclaves of a more basaltic magma that was mixed into the erupted material at depth.

When done, walk down toward *La Filipina* village for approximately 300 m. Note the basalts underlying P1 are shield basalts again (Fig. 6.52A,B). Enter a little road toward the valley bottom to see more ignimbrite (sharp left turn).

After approximately another 300 m, at a little bridge, you will see P1 in front of you. It is dipping toward the valley floor, suggesting it to be a ground-hugging valley infill deposit into an older Miocene valley during the P1 eruption.

Follow the little road to the left for a short while. Take the left road when reaching a small crossroad. Here, the top of the shield basalt sequence is overlain by pink P1 a little higher up on the hill (on your left). Continue up-hill for a few steps until you come to a small but wide cave in the rock face (Fig. 6.53).

Stop 1.3. Bottom of Upper Barranco Arguineguín: P1 Ignimbrite (N 27.8791 W 15.6709)

At this location you will find P1 vitrophyre again. Walk up-hill from the cave and observe changes in P1 composition. Note an increasing amount of basaltic enclaves that is present on the way up until P1 grades into a fully basaltic whole-rock composition. No direct boundaries

FIGURE 6.49

(A) Massive Miocene shield basalts in *Bco. Arguineguín*. (B) Some of the lava flows here contain abundant large plagioclase crystals. Images courtesy of L.K. Samrock.

between the various compositions are observed, implying that the different lithologies come from the same compositionally zoned eruption and, thus, magma chamber. When inverting the eruptive sequence back into the source chamber, the low-density felsic magma would reside high up in such a chamber and erupt first. Denser, mafic magma would, in turn, reside lower in this chamber and erupt later.

Return to the car and continue up-hill on GC-505 to drive along the extensive Mogán ignimbrite succession exposed in the barranco wall.

FIGURE 6.50

Barranco Argueneguín. (A) Massive deposit of P1 rhyolite ignimbrite. Note the considerable thickness of the deposits. (B) P1 ignimbrite with mafic blobs. Also note the crystal-rich nature of the main ignimbrite body.

Approximately 2 km after the last stop, you will find a spectacular exposure of a now altered and clay-dominated former vitrophyre on your right side in the passing road section. This vitrophyre separates ignimbrite "O" from the one below. Note the undular nature of the now yellow contact, suggesting an uneven substrate onto which the ignimbrites were originally laid down (Fig. 6.54).

Continue on this road up-hill until you find a larger open area on a plateau on your right side (~4 km after Stop 1.3). Park your car here.

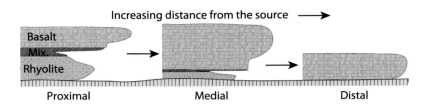

FIGURE 6.51

Horizontal variation in the progression of P1 composite flow with increasing distance from the source in the center of the island. Modified from Freundt, Schmincke, 1992, and Schmincke, 1993.

Stop 1.4. Barranco Arguineguín: Overview Mogán-Fataga Succession (N 27.8856 W 15.6777)

From the plateau, an extracaldera fault in Mogán units is visible in the Eastern barranco wall. Fataga units (top-most) are undisturbed. This suggests that caldera movements largely stopped before the Fataga period (Fig. 6.55), implying that the Tejeda Caldera was essentially formed during Mogan times. Also, note the remnants of Roque Nublo valley fill to your left (arrow in Fig. 6.55C).

Turn around and return back the same way you came to the barranco bottom! Approximately 8 km after *La Filipina*, look for a flat area on the left side where a water pipe almost meets the road.

Park a little further on in a small lay-by and walk back for approximately 100−150 m onto the flat ground next to the road. There, composite ignimbrite P1 is exposed again, displaying spectacular mixing of the individual layers (Figs. 6.56 and 6.57).

Stop 1.5. Lower Barranco Arguineguín: Zoned Ignimbrite P1 with Mixing Features (N 27.8083 W 15.6675)

Note the veins of the lower units intrude into the upper basaltic units. This relationship implies that the rhyolite was still plastic during the time the later basalt was deposited, suggesting them to be coeruptive, and thus likely from the same eruptive source and event. Magma mixing during the eruption of ignimbrite P1 caused evacuation of the magma chamber through ring fractures while deep basaltic magma was injected at the base of the reservoir (Fig. 6.56; Freundt and Schmincke, 1992, 1995).

Continue toward the coast and stop at the large quarry to your left, a little further down the barranco (just before you pass under the motorway bridge again; Fig. 6.58).

Stop 1.6. Fataga Ignimbrite Quarry, Cantera Ignimbrite (N 27.7714 W 15.6659)

The beige-brown Cantera ignimbrite that is quarried here is rich in pumice clasts and thus is a major ingredient for the cement blending at the cement works in *Arguineguín*. The quarry has been

FIGURE 6.52

Upper Barranco Arguineguín. (A) Down in the valley, walk toward the inclined P1 deposits on the left of the valley slope. (B) To your right, P1 is seen again in the valley floor, underlain by shield basalts (image courtesy of L.M. Schwarzkopf).

in operation for approximately 50 years, and quarrying began practically at the road, giving you a feel for how much rock has been removed in that time.

Continue on GC-500 and into *Arguineguín* village. A quick break can be taken at the fuel station right at the main road. It offers a coffee shop and simple bathroom facilities.

From *Arguineguín*, continue on GC-500 toward *Puerto Rico*. This coastal route offers some breathtaking exposures of Mogán and Fataga ignimbrites (Fig. 6.59), but also a large number of vacation resorts.

Just after *Puerto Rico* and below a major cliff, you will pass superb exposures of the gray-blue ignimbrite "E" (Upper Mogán group) with very flat and long fiamme. You will also see a

FIGURE 6.53

Just at the little cave, you will pick up the base of ignimbrite P1 again.

FIGURE 6.54

Basal vitrophere of ignimbrite "O," Barranco Argueneguín. The undulated nature of the contact suggests an uneven substrate onto which the ignimbrite was deposited.

FIGURE 6.55

(A) View of the east wall of Barranco Arguineguín with the full succession of shield basalts and Mogán and Fataga ignimbrites. (B) Close-up of Mogán and Fataga succession. Note, the offset of units in the lower part of the photo (Mogán group) that is absent in the higher units (Fataga group). Image courtesy of L.M. Schwarzkopf. (C) View of Upper Barranco Arguineguín. Note, the hill on the far left is a Pliocene Roque Nublo type of valley infill. (D) Behind your parking spot towers Mt. Tauro, which is made of Fataga deposits all the way from here to its very top.

beautifully sandy beach, *Playa de Amadores*. Virtually all of the sand at this beach has been shipped in from the Sahara though.

Around the next cliff nose you will enter the mouth of Barranco Tauro. Note the pebbly and boulder-rich beach here. This appearance is the more natural look of beaches in the central and western Canaries.

Soon after entering Barranco Tauro you will meet a roundabout. Go left toward *Mogán* when coming from the coastal road.

Continue on GC-500 for approximately 2.5 km and park at the mouth of Barranco Medio Almud in a lay-by on your right, just when the road reaches the lowest point in the valley. Please wear your safety vest for the next few stops.

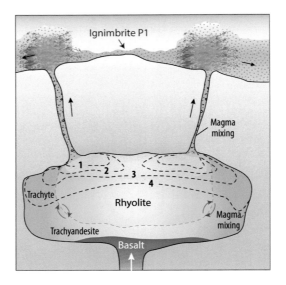

FIGURE 6.56

Model explaining magma mixing during eruption of ignimbrite P1 and evacuation of the magma chamber through ring fractures. Simultaneously, deep basaltic magma is injected at the base of the reservoir. Numbers refer to the evacuation isochrons, controlling magma extraction and mixing in the conduit during ascent (from Freundt and Schmincke, 1992).

FIGURE 6.57

(A) Outcrop of zoned P1 deposit in lower Barranco Arguineguín. (B) Plan view of the lower rhyolite intruding into the more mafic (reddish) middle part of the deposit. (C) Offshoot from lower rhyolite layer of P1 into the more mafic layers above, indicating the felsic and mafic compositions were co-eruptive (ie, both were liquid at the same time and thus cooled together; a cooling unit). Images in A and C courtesy of L.M. Schwarzkopf.

FIGURE 6.58

Cantera quarry near *Arguineguín* (image courtesy of *Excavaciones PEDRO ESTÉVEZ, S.L.*). This quarry provides *pozzolana* as the basic ingredient for the manufacturing of cement. The needs for infrastructure in the fast developing tourist area in the south and southwest of Gran Canaria started in the second half of the 20th century. This quarry, for example, provided cement for the construction of the Soria dam, at the head of the Bco. de Arguineguín, which was constructed over more than 10 years (1962−1972), and is the largest in the Canaries (120 m high with a maximum capacity of 32 hm^3) (see http://www.pestevez.com/cantera.htm).

Stop 1.7. Barranco Medio Almud: Zoned Ignimbrite deposit "A" (N 27.8081 W 15.7380)

The reddish-brown unit to the right of the road is ignimbrite "A" of the Upper Mogán group (Figs. 6.60A−D). This ignimbrite is characterized by dark and light fiamme in a much finer ash matrix. Feldspar content is highly variable, with more complex and abundant types in the dark fiamme. Likely crystal settling in the zoned magma chamber is the cause of this crystal variation (see Troll and Schmincke, 2002) when assuming that the now "reddish" rhyolite magma was originally located above the darker trachyte magma in the former storage chamber.

Also, note the major "flow unit boundary" within the exposure (Fig. 6.60). Although very similar in composition above and below this boundary, a pause in eruption and a new high-energy pulse is required to explain this phenomenon.

Walk up along the roadside toward the tunnel (backwards and in the direction of *Arguineguín*) and note the greenish densely welded ignimbrite above ignimbrite "A." This is ignimbrite "B," separated from "A" by a pronounced former vitrophyre. This vitrophyre is now largely changed to clay (Fig. 6.61A). Note the rotated clasts and the flow fabrics in ignimbrite "B" and the sparcity of crystals it contains (Fig. 6.61B).

Walk further up-hill along the road to the next vitrophyre (base of ignimbrite "C"). At that point, cross to the other side of the road. Look at the exposure from there and you should see

FIGURE 6.59

(A)−(C). Impressions of coastal exposures of Miocene Mogán ignimbrites along GC-500.

vertical reddish stains, probably fossil fumaroles that have degassed volatiles from the hot deposit, similar to those in the valley of 10,000 smokes in 1912, perhaps (Fig. 6.61C).

Walk further up the road to just before the tunnel and continue walking up-hill on the dirt track that commences to the right of the tunnel entrance.

Stop 1.8. Barranco Medio Almud; *Dirt track beside tunnel: Ignimbrite "D" (N 27.8050 W 15.7383)*

The ignimbrite here is ignimbrite "D." Note the rotational clasts, dense welding, and the fair number of plutonic fragments such as syenites and gabbros that were ripped out from the chamber and conduit walls of the original magma reservoir at depth (Fig. 6.62). The rotated clasts are especially noteworthy because they also document post-depositional (rheomorphic) flow of the hot ash pile once it was laid down from the original pyroclastic current (eg, Kobberger and Schmincke, 1999).

Note the horizontal flow unit boundary exposed in ignimbrite "D." Locally this boundary is seen to truncate fiamme, suggesting several eruptive pulses that made up the final ignimbrite deposit here. These would then cool together, and we thus refer to such an ignimbrite as a cooling

FIGURE 6.60

(A) Ignimbrite "A" exposure at Stop 1.7, Barranco Medio Almud. (B) Note the internal boundary within ignimbrite A. (C) Close-up of the internal flow unit boundary within ignimbrite A. (D) Dark- and light-colored fiamme in ignimbrite A (see text for details; images in *A−C courtesy of L.M. Schwarzkopf*).

unit, implying a link in time but allowing for a pulsing eruptive sequence with successive individual "flow units". Continue up the track until you reach the next vitrophyre (ignimbrite "E").

Stop 1.9. Barranco Medio Almud: Ignimbrite "E" (N 27.8044 W 15.7385)

This ignimbrite unit displays dense welding and abundant and intensely flattened white pumices with abundant tension gashes, which are also a rheomorphic flow feature indicating postdepositional down-slope movement of the still hot and plastic mass (Fig. 6.63; see Leat and Schmincke, 1993).

Continue walking on the track to the cliff nose. Note the trachytic unit (the dark peaks above you) in the upper part of ignimbrite "E" and the lack of a "vitrophyre" between the gray rhyolitic part and the dark trachyte. This ignimbrite is another example of vertical compositional zoning in

FIGURE 6.61

(A) Ignimbrite "B" rests on ignimbrite "A". Note the former vitrophyre (now clay-rich) at the base of ignimbrite B. (B) Ignimbrite B with solid clast now weathered out. Note B is very crystal-poor but shows a high degree of internal welding. (C) Vertical, reddish staining suggests fossil fumarole activity in ignimbrite "C". Images courtesy of L.M. Schwarzkopf.

FIGURE 6.62

Syenite clast in densly welded ignimbrite D. Note compaction and rotation features around the syenite clast, implying postdepositonal (rheomorphic) flow of the still hot ignimbrite deposit (image courtesy of L.M. Schwrazkopf).

magma chambers (Fig. 6.64), which must have been a major factor of magmatic evolution for these explosive eruptions back in the Miocene.

Stop 1.10. Barranco Medio Almud: end of path (N 27.8037 W 15.7387)

If your timing is good, you will now find yourself with a beautiful silhouette of Teide volcano on Tenerife on the far western horizon, but of course only if the weather plays along.
END OF DAY 1.

DAY 2: THE MIOCENE TEJEDA CALDERA

Today, you can either take the motorway from *Maspalomas* to *Mogán* (GC-1, blue) or the more scenic route along the coast from *Maspalomas* to *Arguineguín*, and from there to *Puerto Rico* and *Mogán* (GC-500 green; Fig. 6.65).

If you choose to drive along the coastal road (recommended), then you will drive through a golf resort soon after *Maspalomas*, before extensive Miocene conglomerates, breccias, and ignimbrites give a glimpse into the dynamic erosional and eruptive processes that shaped Gran Canaria in the late Miocene (Fig. 6.66A,B).

We come back to the relevance of this major upper Miocene erosion episode later in the day. Note that the beaches here are very popular with tourists and locals alike. These are natural beaches; the boulders and pebbles are not imported ones.

FIGURE 6.63

(A) Very long fiamme with tension gashes in ignimbrite "E" (see Leat and Schmincke, 1993). (B) Ignimbrite E has flown toward the sea! Note the opening of the tension gashes. Top flow to the right is needed. (C) Tension gashes in fiamme of ignimbrite E. Note that this fiamme is now largely broken up. Rheomorphic deformation (ie, postdepositional plastic flow) in ignimbrite E has locally broken up fiamme entirely and rotated them due to stretching and down-slope movement.

FIGURE 6.64

(A) Looking up-hill, one can see the dark trachyte top of ignimbrite E that stands in strong contrast to its light-colored rhyolitic base. This is another vertically zoned ignimbrite. (B) Note the thin erosional remnant of dark trachyte that rests on top of the rhyolite (ignimbrite E). Images courtesy L.M. Schwarzkopf.

The barren landscape here in the very south of Gran Canaria shows remnants of intense terracing. This is the result of one of the island's biggest recent environmental disasters. Gran Canaria was a dominant deliverer of tomatoes to the United Kingdom in the 1950s and 1960s. However, brackish and saltwater was used for irrigation, producing tasty and large tomatoes but causing the soil to become highly saline and thus unusable for many decades due to extreme salinity (Fig. 6.66C,D).

After a few kilometers along the barren coast here, the *Arguineguín* cement works will become visible in the distance (front left). Around the level of the cement works, a complex intercalated sequence of conglomerates, Fataga-age pyroclastic deposits, and local debris avalanches is exposed in the road cuts to your right (Fig. 6.67).

Once you approach *Arguineguín*, join the motorway to *Mogán* (GC-1). Once on the motorway (GC-1, blue), continue toward *Puerto Rico* and *Mogán*.

Soon you will pass a spectacular series of Fataga ignimbrites to the left and right along the motorway in a set of relatively new road sections. Also, note extensive debris avalanche and debris flow deposits of similar ages that are intercalated with the pyroclastic units.

After driving through several short tunnels, you will reach the termination of the motorway. Keep right here and follow signs for *Mogán* on GC-200 (not *Puerto de Mogán*).

At the km 63 road sign on GC-200, pass through the village of *Los Palmitos*. After another 3−4 km (km 60 on GC-200 road sign), pass through *Las Casillas* and *Molino de Viento* villages. Note the giant furniture and kitchen utensils on display here that were designed by a German ex-patriot who has lived on the island for many years. The display is set up next to a traditional windmill on your left (the *Molino*).

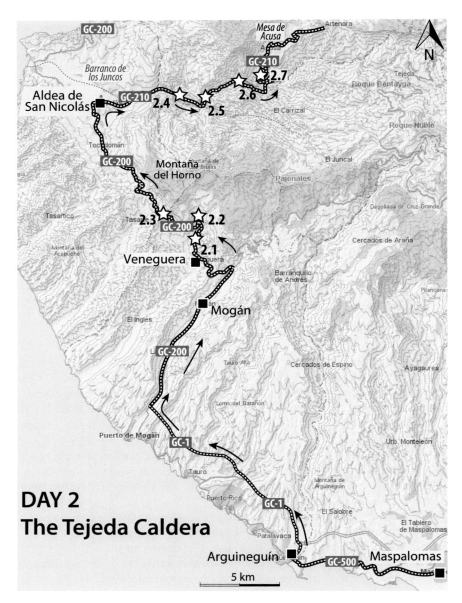

FIGURE 6.65

On Day 2, a visit to Los Azulejos, the caldera margin, and hydrothermal alteration sites. The intracaldera cone-sheet swarm, caldera-fill deposits, and the central syenite intrusions are our main focus. Map from IDE Canarias visor 3.0, GRAFCAN.

FIGURE 6.66

(A) Extensive Miocene conglomerates along GC-500. (B) Close-up of Miocene conglomerates with mainly pebbles of ignimbrite and some basaltic rocks. (C) and (D) The barren landscape here is the result of using brackwater for irrigation in the 1960s, with disastrous long-term consequences.

FIGURE 6.67

Once the Arguineguín cement works come in sight, you are within Miocene Fataga group pyroclastic deposits that are intercalated with Miocene conglomerates.

At approximately km 59 on the GC-200 road signs, you will reach *Mogán* village itself. This is the locality that gave the name to the Mogán group ignimbrites of the afternoon of Day 1 (eg, the cliffs to your right and left are almost exclusively made of Mogán ignimbrites here).

Follow GC-200 until signs for *Veneguera* and *La Aldea* appear at approximately the km 56 road sign on GC-200. You are now back in the old shield basalt rocks that underlie the Mogán ignimbrites, because you were once more driving inland on reasonably flat ground while the Mogán units dip gently toward the sea.

Continue on GC-200 toward *Veneguera* and *La Aldea*. The road soon gets very bendy and will bring you up-hill again. You are advised to drive with caution in this part of the island.

Continue for a while on GC-200 until you see some spectacular views of the colorful hydrothermally altered rocks at *Fuente de Los Azulejos* in the distance in front of you.

A little after the km 51 road sign on GC-200, a major dyke intrusion is cut by the road and a lay-by to your left opposes the outcrop. Park here for today's first major stop.

Stop 2.1. Southwest Gran Canaria: Mogán Rhyolite Dyke (N 27.9140 W 15.7298)

The dyke is rhyolitic in composition and of Mogán age ($\sim$13.5−14 Ma), and strikes at an angle to the center of the island (ie, to the central caldera). Note the abundant feldspar crystals and the steep flow banding that is preserved in the rhyolite dyke. The dyke (and the associated plug further up-hill) are likely feeders to one of the Mogán ignimbrites of Day 1 (Fig. 6.68).

Continue on GC-200 and after a few more kilometers, another major stop of the day approaches. Park your car at *Fuente de los Azulejos*, which offers spectacular geology and the opportunity for some simple refreshments (Fig. 6.69).

Stop 2.2. Fuente de Los Azulejos: Hydrothermal alteration at caldera margin (N 27.9240 W 15.7273)

Looking toward the northwest from the parking place, you will see semi-horizontal shield basalts in unconformable contact with Mogán age ignimbrites to the right. Note the steep dip of the Mogán tuffs near the contact and their increase in thickness further to the right. This is the topographic margin of the central Miocene Tejeda Caldera. Note it is not the caldera fault itself (see Fig. 6.15), but an erosional margin of the caldera perimeter that eventually was covered with eruptive materials. For example, higher on the right side, the dips of the units tend to shallow out, implying the basin was progressively filled with tuffs at that time. The sequence is covered by massive green to brown Fataga ignimbrites that are flat-lying and seemingly unaltered, giving an age constraint to the alteration event (12.5 Ma in this case; see, eg, Troll et al., 2002; Donoghue et al., 2008).

Two large brownish units appear within the altered rocks that pinch out toward the west. These are likely sill intrusions of Fataga age or major ignimbrites that have escaped alteration (perhaps due to high compaction). Their convex upward shape indicates intrusive inflation, and the former idea may be more probable.

The entire site is a classic case of relief inversion, where the younger rocks are found at higher altitudes even though they had previously been in a lower topographic position (the basin fill). The vivid colors of the tuffs are characteristic for low-temperature hydrothermal alteration due to extensive fluid flow in the faulted caldera marginal areas (Fig. 6.70). A variety of hydrothermal minerals are present, such as clays, zeolites, micas, and chlorite, as well remnants of the original mineral assemblage (eg, Donoghue et al., 2008).

FIGURE 6.68

(A) Mogán rhyolite dyke in the road section. Please be careful when crossing the road. (B) Dyke rock in detail. Note the steep banding, indicating magma flow, and the abundant feldspar crystals. (C) Looking up-hill, where the intrusion thickens. Images courtesy of L.M. Schwarzkopf.

FIGURE 6.69

Topographic caldera margin at *Fuente de Los Azulejos*. Shield basalts to the lower left are unconformably overlain by steeply dipping intracaldera Mogán-aged tuffs. Note the progressive shallowing of the intracaldera deposits to a flat geometry on top of the pile (Fataga Formation). See also Figure 6.19.

Ignimbrite relics that escaped this alteration event can be found a few steps further along the road from the refreshment kiosk. Here, in a drainage pit, an ill-defined lens of still glassy ignimbrite is found that escaped extensive overprint (maybe due to locally high compaction and welding) and the transition from altered to fresh rock can be studied here in detail (Fig. 6.71).

Walk to the north—northwest along the road for approximately 300 m. Here, you will see the contact between the shield basalts and the caldera-infill tuffs. Note the large number of "pick up" clasts here and a spectacular pinkish fiamme near this spot (set in a green ash matrix). This fiamme displays internal flow folding. Both way up (out of caldera) and way down (into caldera) are indicated, implying that many of these tuffs had a hard time escaping from the caldera, with only the bigger eruptions producing extracaldera outflow deposits as well (Fig. 6.72). If you are willing, several small footpaths lead up these cliffs in the area here for some great views, and we recommend a little stroll.

Continue on GC-200 toward *San Nicolás/La Aldea*. Soon the road will climb up-hill again. Only a short distance further, at approximately the km 45.5 road sign on GC-200, right on the ridge where you pass into Barranco de Tasarte, you may view the dyke from the earlier road cut in the distance. Park in a small lay-by on the right side just after the crossover into the next barranco.

FIGURE 6.70

(A) Hydrothermally altered caldera margin tuffs show a variety of vivid colors and internal textural features. (B) Take a short walk into the small gorges to inspect the full range of features exposed (image courtesy of *S. Burchardt*). (C) Close-up of hydrothermally altered intracaldera tuffs at *Fuente Los Azulejos*. Note how the colors grade into each other, indicating a complex alteration mineralogy (see Donoghue et al., 2008).

FIGURE 6.71

Specimen of glassy obsidian lens in hydrothermally altered tuffs from *Fuente de Azulejos*. From Donoghue et al., 2008.

FIGURE 6.72

(A) Steeply inclined greenish and cream-colored ignimbrites at the caldera margin. (B) Close-up of rock outcrop near the topographic caldera margin. Note the different alteration pattern of the host tuff (green) and the more solid fiamme at some sites. This particular fiamme shows flow structures that indicate back-flow into the caldera. Images courtesy of L.M. Schwarzkopf.

Stop 2.3. West Gran Canaria: Mogán dyke intrusion viewed from a distance (N 27.9257 W 15.7476)

Walk back to the little hill to view the intrusions. The dyke system in the distance ($\sim 1.5-2$ km to the South) shows giant magma mixing features (mafic enclaves) that seem to have been caught in the process of disintegration during conduit transport (Fig. 6.73). Unfortunately, most of the mafic material has now weathered away, leaving big holes in the dyke. Also note the late plug in the far distance from the exposed dyke system (Fig. 6.74), which seems to have channeled the final episode of the eruption. Plug flow is considered to be less energy-demanding than dyke flow, and thus waning eruptive pressure likely facilitated the change from a fissure-fed to a plug-fed system.

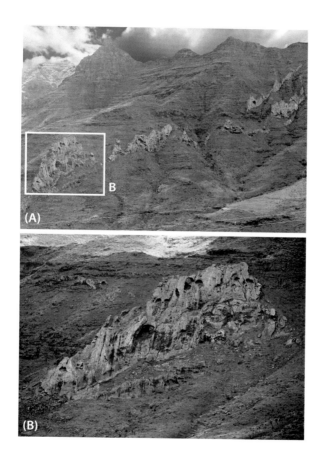

FIGURE 6.73

(A) Rhyolite dyke with large holes from weathered out, large mafic enclaves. (B) Enlarged part of the white box area in (A). Images courtesy of L.M. Schwarzkopf.

FIGURE 6.74

(A) Looking back to the southeast, the dyke can be traced all the way to the caldera rim. (B) The dyke developed a late, semi-cylindrical plug close to the caldera, implying that magma flow eventually focused along individual spots of this fissure. Images courtesy of L.M. Schwarzkopf.

Continue on GC-200 for some time now. After the next major ridge you will descend again, now into Barranco La Aldea. If the weather permits, you might catch a glimpse of northern Tenerife (Anaga) in the far distance from here. The rocks you are passing now are shield basalts, but in places they are covered by recent rock-fall debris.

Continue until GC-200 eventually meets a T-junction in the outskirts of *San Nicolás* (km 35 on GC-200 road sign), turn right onto GC-210 (green), and follow the signs for *Artenara*.

If you consider having lunch in the old village core of *San Nicolás*, then drive straight across the next roundabout and then toward the village church and park near there. Otherwise, continue toward *Artenara* on GC-210.

After only a few blocks you will meet another T-junction. Go right, and only approximately 10 m after that go left again (still following GC-210 signs!). Note that some road signs are old and give obsolete road numbers, which can be a little confusing at times. Once you have successfully maneuvered through *San Nicolás*, you will find yourself on the relatively narrow road into Barranco La Aldea (still following signs for *Artenara*). After approximately 1 km into the Barranco (ie, after *San Nicolás*), a bridge crosses a small river that is generally dry in summer (just after the km 33 road sign on GC-210). After only 0.5 km you will cross the same river again, and then once more after another 200 m.

Continue now on the left bank of the barranco and river up-hill. Soon after the third little bridge over the usually dry river, you will encounter a large overhanging block on the valley side (Fig. 6.75). This block looks as if it is ready to fall at any point. It is best to pass quickly because in this area here much of valley walls to the right and left are covered with scree and coarser block deposits that formed from instabilities of the barranco walls. After approximately another 2 km into the barranco, the very vivid colors of the hydrothermally altered rocks occur again. However, not all of this material here is *in situ* (in place). Recent debris avalanche deposits, possibly as young as 4 Ma, have redistributed much of the original rock here. Note that various large blocks of altered rock are partially disintegrated within these younger debris deposits. Higher up the slopes, however, in situ caldera infill with original hydrothermal alteration colors can be observed and is hence the most probable source of the debris flows and avalanche deposits at your level.

FIGURE 6.75

Large overhanging block over the road into the caldera, Barranco La Aldea.

Continue into the caldera along GC-210. The road gets very narrow soon and driving is not for the faint-hearted. Note the massive ignimbrite packages and intrusive sills while continuing your drive into the caldera.

At approximately the km 30 road sign on GC-210, you will note coarse caldera infill breccias to your left, implying that you are moving into the deeper parts of the intracaldera succession. Continue down-hill, passing a mixture of caldera infill sediment and pyroclastic rocks cut by an increasing amount of cone sheet−type intrusions.

The landscape of this inner part of Gran Canaria was once referred by the Spanish playwright and philosopher Miguel de Unamuno as a "*Tempestad petrificada*" (thunderstorm frozen in rock). Accordingly, the road is very adventurous, much like the description suggests.

Just after a steep and narrow drive up-hill, you will reach a plateau on the right side of the road at approximately the km 28 road sign on GC-210. Park your car here on the open ground.

Stop 2.4. Inside the caldera: The cone sheet swarm (N 27.9858 W 15.7372)

Just opposite the parking space, on the other side of the road, follow a little dirt track for approximately 500 m. Usually a chain is mounted so that one cannot drive up this track. Approximately 50 m before a little finca, walk straight up-hill (no path) and walk over rough ground with numerous boulders of dark rock until you reach the top of a little ridge (~200 m).

Walk until approximately N 27.9895 W 15.7367. Here, you will get a superb view of the caldera infill with abundant cone sheets cutting through the caldera infill at angles of approximately 40−50°. With a bit of good light, one can even work out a time sequence of the intrusions from the cross-cutting relationships of the individual sheets (Fig. 6.76). Note that this spot is part of the transitional zone between the almost cone sheet−free outer belt of the caldera infill and the inner portion that is extensively intruded by these sheets (Fig. 6.77).

Walk back to the car. From the parking place on the little plateau, one can see a cone sheet lifting the strata it cuts by nearly its own thickness right above the lake in the cliff face on the opposite side of the barranco. It is assumed that the combined intrusion of hundreds of cone sheets here in the center of the island caused considerable vertical growth of the Miocene island edifice, pushing the island's roof upward like a gigantic hydraulic jack for perhaps 1 km or more (Schmincke, 1993; Schirnick et al., 1999; Donoghue et al., 2010).

Continue to drive along the lake side and further into the interior of the island. A major dam in the lake will appear on your right. Note also that the number of cone sheets that cut into the caldera fill has been steadily increasing. A short while after the km 27 road sign on GC-210, park your car in a small lay-by on your left side, which is located under an overhang of rock (Fig. 6.78A).

Stop 2.5. Cone sheet intrusions: close up inspection (N 27.9820 W 15.7243)

In this area, close to 100% of the rock mass consists of cone sheet intrusions. This is an excellent location to inspect intrusive contacts, chilled margins, and cross-cutting relationships of the cone sheets with each other because the cavernous outcrop at the lay-by here gives good 3-D insight into structure and geometry of these sheets (Fig. 6.78B).

Continue along the GC-210 into the caldera. You will now pass numerous cone sheets along the bendy road. They are of variable freshness and coloration, with virtually no country rock relicts

FIGURE 6.76

(A) Take a position on the grassy slope opposite a magnificent hillside of cone-sheet intrusions. The cone-sheets cut intracaldera deposits in the rock face opposite your position. (B) Close-up view of the cone-sheet dyke swarm at Stop 2.4. Images courtesy of L.M. Schwarzkopf.

between. So, they have effectively made their own host rocks and "cooked" each other, giving rise to the many degrees of freshness and coloration seen here (eg, Donoghue et al., 2010).

Continue on GC-210 for another short stretch until you arrive at the base of a major water dam.

Stop 2.6. Central Gran Canaria: Syenite intrusions (N 27.9924 W 15.7070)

Just to your left is a jagged rock surface, dipping toward you at approximately 45° (Fig. 6.79). It is cut by several dykes. This is a small syenite intrusion with a sanidine crystal foliation (Fig. 6.79A). Location: N 27.9926 W 15.7071.

FIGURE 6.77

Stop 2.4: Remnants of older caldera fill are still visible to your left but are displaced relative to each other by the numerous cone-sheet intrusions. Image courtesy of L.M. Schwarzkopf.

At the rock slab, crystals are seen to plunge to the west (30°) and east (20°) on either end of the outcrop, but they are fairly horizontal in the top part of the rock mass. This suggests a curved foliation that mimics upward flow of magma and a push from underneath (ballooning). In places one can also see syenite crystals being caught in the process of incorporation into one of the cross-cutting dykes (Fig. 6.79B), giving an example of how xenocrysts can enter ascending magmas (see, eg, chapter: The Geology of La Gomera).

On the other side of the road (Location: N 27.9920 W 15.7070), one will find a more coarsely crystalline syenite (Figs. 6.79C,D). Big sanidine crystals of varying morphology can be seen there in contact with domains of lower crystallinity, probably reflecting migration of variously solidified crystal mushes at this structural level. This is the very top of the plutonic heart of Gran Canaria and represents a window into the usually hidden magmatic system of Miocene Gran Canaria.

Continue up the bendy serpentine-style road on the right side of the dam. Once you are up on the flank of the dam, follow the lake outline for the next little while. Note, the Roque Nublo

FIGURE 6.78

(A) Cone sheets intruded by cone sheets at Stop 2.5. (B) Close-up of cone-sheet contacts. Chilled margins are frequent, implying that some time must have passed between the individual intrusions. Images courtesy of L.M. Schwarzkopf.

FIGURE 6.79

(A) Syenite intrusions at side of the dam. This warrants a closer inspection. (B) Dyke intruding syenite. Have a close look and see if you can find xenocrysts. Image courtesy of *L.M. Schwarzkopf.* (C) and (D) Coarse syenite of the plutonic core of the island is exposed just on the other side of the road. The syenite contains large potassium feldspar crystals (sanidine) and records interaction of former coarse and finer crystal mushes.

monolith is now visible in the far distance along the valley axis and Roque Bentayga, a similar Pliocene erosion remnant, is visible somewhat closer. Also, a little windmill to the top left of your position is now becoming visible. The little windmill is our next stop.

For there, continue to follow GC-210 into the barranco until you meet a fork at the km 22 road sign. At the fork, continue to follow GC-210 (green) toward *Artenara*, which is initially down-hill.

On the way, you will encounter numerous more cone sheets intruding into each other (a cone sheet swarm), many of which are hydrothermally altered with green and beige alteration colors.

These rocks, however, are not as strongly altered as the rocks near the caldera margin (eg, those at *Fuente Los Azulejos)*. This difference is likely the result of less severe alteration near the top of

FIGURE 6.80

View of the cone-sheet swarm around the *Presa del Parralillo* (at the windmill; Stop 2.7). In the distance and on the right, you can see the flat Roque Nublo deposits.

a major hydrothermal cell or represents the final and weakest gasp of repeated pulses of hydrothermal activity.

Black staining is frequently visible and is the result of low-temperature precipitation of manganese oxides from running (and evaporating) ground and surface waters.

Continue up-hill toward the windmill, which will provide us with a major viewpoint and park your car right in front of the windmill (Fig. 6.80).

Stop 2.7. Central Gran Canaria Miocene–Pliocene Unconformity (N 27.9926 W 15.6942)

Looking south, you will see flat units overlying the extensive cone sheets in the rock faces opposing the parking site on the other side of the barranco. These are flat-lying Roque Nublo lavas and pyroclastic rocks, separated from the cone sheets by very irregular angular unconformity (Fig. 6.81).

This major unconformity implies a major erosional hiatus between the late Miocene Tejeda volcano and the Pliocene Roque Nublo volcanic episode. During this time "gap," erosion must have stripped away much of the entire Miocene volcanic edifice, literally to its roots. If you recall the extensive late Miocene conglomerates from early in the morning, this is where it all went.

Note also the intensely colored cone sheet rocks near Stop 2.7 (Fig. 6.82A) and the variable degrees of alteration, indicated by the colors. In addition, note that the strike of cone sheets below you swings around when you revolve around your own axis (Figs. 6.82B–D), giving you a good 3-D sense of the overall cone sheet swarm (ie, the inverted cone geometry).

From Stop 2.7, continue up-hill (on GC-210) toward *Acusa* and *Artenara*. Drive through more cone sheets for several kilometers until the massive flat-lying Roque Nublo sequence above the village of *Acusa* appears (Fig. 6.83).

At approximately the km 17 road sign on GC-210, one can view the cone sheet–Roque Nublo contact especially well. This place is a beautiful stop for taking photos, but watch the traffic, please.

When entering *Acusa* village, you are driving through Roque Nublo group sedimentary rocks as well as flat lavas and pyroclastics. Continue on this road until you see a sign for *Vega de Acusa*.

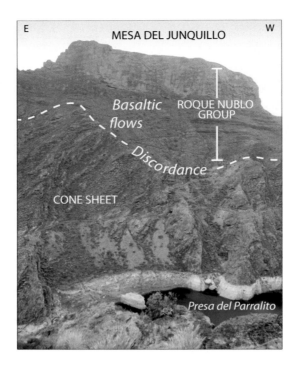

FIGURE 6.81

View from the parking site at the traditional Canary windmill above the lake (Stop 2.7). Looking south, the major Miocene-Pliocene unconformity is visible. Note the irregular topography of the Miocene rocks.

Enjoy some spectacular views over central Gran Canaria from here. To the south, you can see Roque Nublo, Roque de Bentayga, and various scattered mountain settlements. A little road to *Acusa Seca* goes off to the right, offering some of the best views, but it is a rather narrow drive and you might need to watch your time.

After enjoying the impressions, drive until you reach a major crossroad. There, follow GC-210 (green) to *Artenara*. When you reach the roundabout in *Artenara*, you have several options to return to your base.

1. Turn here and drive back the same route that you came over the day (spectacular driving!), approximately 2.5 hours.
2. Continue via *Tejeda* GC-210 (green) to *Maspalomas*, approximately 2−2.5 hours, partly tracing the route of Day 3 (mountainous driving!).
3. Go to the north (*Las Palmas*) via GC-211, and from there to *Maspalomas* via the coastal motorway, but note that this is a very long drive, albeit beautiful, as it takes you through the lusher parts of N-Gran Canaria (3−3.5 hours).

END OF DAY 2.

FIGURE 6.82

(A) Intensely colored cone-sheet rocks near Stop 2.7. Note the variable degrees of alteration are indicated by colors. (B) View to northwest, with inclined cone sheets in foreground. (C) Note inclined cone sheets down at lake level when looking west and southwest. (D) View of central Gran Canaria from your site. The Roque Nublo monolith is usually visible from here when looking to the south and southeast. Images in (B) and (C) courtesy of *L.M. Schwarzkopf.*

DAY 3: ROQUE NUBLO VOLCANO AND RECENT VOLCANISM

From *Maspalomas*, drive up to *Playa del Inglés* and find Avenida de Tirajana (one of the main roads). Drive up-hill and follow signs for *Fataga* (Fig. 6.84). Go straight across several roundabouts until you approach the end of the village.

At the last roundabout, follow signs for *Fataga* (GC-60 red). Pass through the village of *San Fernando*, which almost directly follows after *Playa del Inglés*. There is a pleasant restaurant near the village end (*Casa Vieja*) that is possibly worth an evening visit.

Continue to follow GC-60. Soon the road will start to become a little bendy (near the km 47 road sign on GC-60).

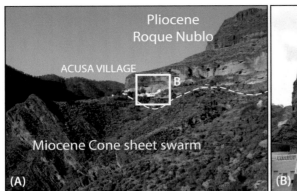

FIGURE 6.83

(A) *Acusa* village at the boundary between the Miocene cone sheets and the Pliocene Roque Nublo deposits. Use of the caves as dwellings dates back to pre-Hispanic times (image coutesy of *L.M. Schwarzkopf*). (B) After the conquest, the aboriginal caves were in continued use by the new population until the present day.

After a short while you will note a quarry in the valley below on your left. The quarry breaks Fataga rock for road gravel and building stones. Cathedral de Santa Ana in *Las Palmas* is constructed from Fataga rock, for example.

Continue on GC-60, and soon you will drive through Pliocene Roque Nublo—type breccia of a former valley fill deposit (inspection of similar rocks follows later in the day).

Approximately 5 km after *San Fernando* (a little after the km 43 road sign on GC-60), you will pass a Guanchee interpretation center (*Mundo Aborigen*). If you are interested in the way of life on the island prior to the Spanish conquest, then stop here. Otherwise, follow GC-60 up-hill for another 2 km.

Make a brief stop on the little plateau and mirador *Degollada de las Yeguas* (480 m altitude), which will appear on your left side soon.

Stop 3.1. Degollada de las Yeguas: *View of the late Miocene Fataga ingnimbrites and lava succession (N 27.8193 W 15.5792)*

If weather permits, then you will have some scenic views of the *Fataga* lava and ignimbrite succession to the northwest of your position (Fig. 6.85).

Continue on GC-60, now down-hill. Drive to the valley bottom along the bendy road with Fataga-age lavas, tuffs, and debris-avalanche deposits in the sections next to the road. Once on the valley floor, note the large but recent rock fall to your left (near the km 38 road sign on GC-60).

Continue on GC-60 toward *Fataga* village and drive along spectacular Fataga exposures on your left and right. Optional stops between the km 34 and 35 road signs on GC-60 are very much recommended.

Inspect the succession and try to distinguish between lavas and ignimbrites. Lavas often show irregular top surfaces, whereas ignimbrites tend to form flat top surfaces, at least in this succession. This valley and the village of *Fataga* represents the type locality for the Fataga group.

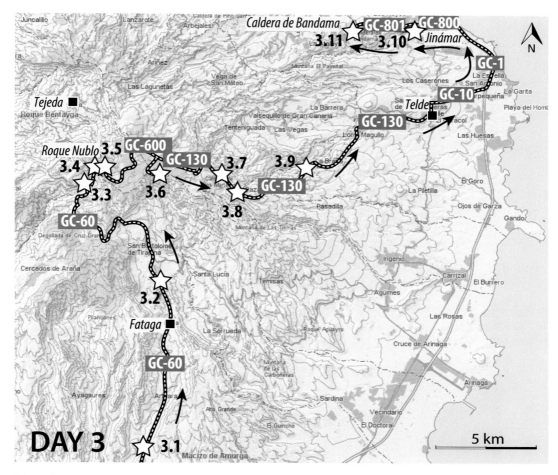

FIGURE 6.84

Itinerary for Day 3. During this day excursion, we inspect the Fataga succession at its type locality and drive to the center of the island with a spectacular view of Pliocene lavas and the Pliocene Risco Blanco intrusion. For the Roque Nublo debris-avalanche breccias, we visit the mountain village of *Ayacata* and hike to the Roque Nublo monoliths. From there, drive to the highest point of the island with an overview of Gran Canaria (*Los Pechos*). Finally, after inspecting the Caldera de Los Marteles and phreatomagmatic deposits at recent maar volcanoes and Holocene bomb-beds in northeast Gran Canaria, we end the day exploring recent cinder cones (1500–3000 years) in Jinámar and Bandama and with an overview of Las Palmas volcanic field. Map from IDE Canarias visor 3.0, GRAFCAN.

FIGURE 6.85

Barranco Fataga. Fataga ignimbrites and lava succession. This is the type locality of the Miocene Fataga group.

FIGURE 6.86

Hydrothermal colors at the caldera margin just after *Fataga* village, Upper Barranco Fataga.

Continue up-hill and pass through the village of *Fataga*. Follow signs for *San Bartolomé*. Just after *Fataga* village, you will see the typical colors of the hydrothermal intracaldera rocks to your left (Fig. 6.86). Much of the material is reworked here and is part of younger slump and slide deposits, and it marks that the caldera margin is not far from here.

Continue up-hill for several kilometers along another bendy stretch. At the km 28 road sign on the GC-60, you will enter a pine forest. Just after the next bend, drive into the parking area of a larger restaurant. This will be our next stop and viewpoint.

Stop 3.2. Scenic view onto Risco Blanco phonolite intrusion (N 27.9071 W 15.5691)

Park in front of the restaurant and walk a few steps toward the north until you reach a small cliff ($\sim$100 m). On the far hillside, the Risco Blanco phonolite intrusion (white and pointy) will be visible, unless you picked a bad weather day. The Pliocene basalts are bent upward at the sides of the intrusion, but it appears that the Risco Blanco magma became stuck at this level because no phonolite penetrates beyond the pointy top (Fig. 6.87).

Also note the inclined and rocky slope to your left (northwest), where no houses are found. It is made up of a massive landslide deposit that has collapsed from the headwall of the cliff into the barranco.

Return to your car and continue on GC-60, but now down-hill. Approximately 2 km after Stop 3.2, follow signs for *St. Bartolomé* and *Tejeda*, which point up-hill (still on GC-60).

Drive through the village of *San. Bartolomé* and continue on GC-60 toward *Tejeda*. The road quickly climbs up-hill again, allowing some closer views of the Risco Blanco intrusion and also down the full length of Barranco Tirajana.

A little after the km 23 road sign on GC-60, you will also get some close-ups of the headwall of the barranco that is made up of Pliocene lavas and sediments. At this point you should also see the first brown road signs for *Roque Nublo*.

Drive over the pass into the next barranco just after the km 20 road sign on GC-60 (*Cruz Grande*). Now you will drive along Miocene caldera infill on which the Pliocene deposits rest (note the altered colors again), but some outcrops here are in part reworked materials.

Drive on until you hit a major fork. There, follow GC-60 for *Ayacata* and *Roque Nublo*. To your left (in the distance) you can see Miocene caldera infill and cone sheets of the cone-sheet swarm of Day 2, which are overlain by massive Roque Nublo group debris avalanche deposits. Drive further and into the mountain village of *Ayacata*.

Park the car in front of the small restaurant *Casa Melo*. We recommend a "coffee stop" here for sustenance before the Roque Nublo hike!

After your break, walk west along the road (GC-60) for maybe 350 m, until the km 14 road sign on GC-60 appears. Walk toward the rock exposure of steeply dipping Roque Nublo breccia to your right (Fig. 6.88).

Stop 3.3. The Roque Nublo breccia deposits (N 27.9580 W 15.6107)

Note the deposit looks massive on first glance, but it contains many volcanic fragments that are variably altered. Clast alignment might be detected and seems very steep. A peculiar jointing (perhaps related to depositional bedding planes) is observed (Fig. 6.88). One would think that a giant volcano must have given rise to such a large volume of breccia deposits. Note that mafic volcanic

FIGURE 6.87

(A) The Pliocene Risco Blanco phonolite intrusion in Roque Nublo—age basalts seen from the viewpoint at Stop 3.2 (looking North). (B) Close-up of the Risco Blanco intrusion.

rocks seem to dominate the clasts here, with virtually no other lithologies present, but there is notable variation among these mafic volcanic clasts also.

Return to your car and follow GC-600 (green) up to the Roque Nublo monolith (direction *San Mateo*). Note that you have to go back on GC-60 for a few meters from your parking place. Do not follow GC-60 any longer!

After a short while in Miocene intracaldera sequences, the road will bring you into Roque Nublo breccia units. After approximately 3.5 km on GC-600, a lay-by on your right marks the start

FIGURE 6.88

Massive Roque Nublo breccia deposit. Here, the clasts are mainly mafic volcanic rocks. Note the steep jointing present in the rocks.

of the Roque Nublo track. Park the car here. Parking may be limited in the high season and you may need to park a few hundred meters further up the road.

Stop 3.4. Roque Nublo hike (N 27.9657 W 15.6015)

Bring clothes for cold weather, water, some snacks, and, ideally, a first aid kit. You will be out now for approximately 1.5–2 hours. Follow the hiking path on the left side of the road up to the monolith. Just a few 100 m up the path you will see more Roque Nublo breccia to your right. Walk on the path through the pine forest for perhaps 20–30 minutes (Fig. 6.89) and stop in the second major bend, just before reaching a little plateau. Location: N 27.9680 W 15.6101.

FIGURE 6.89

(A) The lower parts of the path are wide and well laid out (image courtesy of *L.M. Schwarzkopf*). (B) Higher on the Roque Nublo path, it can get a little rougher, so good equipment is advised in case the weather changes unexpectedly (image courtesy of *C. Karsten*). (C) Look at the volcanic deposits here along the trail. They dip differently than those earlier in the day.

Just before the plateau, Roque Nublo breccia and other volcanoclastic layers with medium dips approximately 40° to the northeast are exposed. Note that these units dip opposite to the ones you inspected this morning. This material here is widely held to represent an intra-crater facies of the long-lost Roque Nublo volcano.

Walk up to the first plateau. Note the blocks of Roque Nublo ignimbrite in the stairs next to the plateau. Although these rocks are not directly exposed on the path, they occur nearby, highlighting the in part rather explosive nature of the Pliocene Roque Nublo volcano.

Make your way up to the main (second) plateau (Fig. 6.90A). Keep your eyes on the ground when walking toward the Roque Nublo monolith. Once your eyes have adjusted, abundant clasts of gabbro, diorite, large amphibole crystals and aggregates, and even amphibole megacrysts can be found on the plateau (Figs. 6.90B,C, see Frisch and Schmincke, 1969). *Please do not hammer and only take loose material with you!*

Contrasting the first Roque Nublo breccia from this morning, which showed an occurrence of mafic volcanic clasts only, there is a notable abundance of plutonic clasts in these rocks here. This realization has been used to suggest progressive collapse of the Roque Nublo volcano down to its plutonic roots, which happened through three or more collapse events that gave rise to the three major flow units of the Roque Nublo breccia on the island (Pérez-Torrado et al., 1995).

On an altitude of approximately 1750 masl, cross the plateau and walk toward the monolith (Fig. 6.91). Walk on the western side of the monolith onto a small area where the monolith shows an internal boundary. The boundary shows breccia on either side but displays features of typical plastic deformation in the form of a shear zone right at the foot of the monolith (Fig. 6.92). Location: N 27.9706 W 15.6130 (1753 masl). This observation implies that the rock mass was deposited on a shallow slope but during high-energy movement to allow for plastic deformation of the base, which was likely caused by frictional heating or similar type of processes.

In the distance you can see *San Nicolás* to the west and the windmill on cone sheets from Day 2. If it is cloudy, be very careful here because it is a little steep in places.

Walk around to the east side of the monolith also, where you will find the shear zone again just behind a little plateau, which also lends itself for a short rest. The view to the east down into the barranco yields a glimpse into the roots of the Roque Nublo volcano. A major gabbro intrusion and a number of dykes can be seen across the valley, suggesting that the center of the Roque Nublo volcano was probably located around this part of Gran Canaria.

The Roque Nublo monolith is also popular with climbers, and you might find a bunch of them attempting to make their way up the monolith while you are there. Usually they use elaborate climbing equipment, but we advise watching for falling rocks, ropes, or climbers!

Return to the car. Once you are there, continue up the road for approximately 400 m from the car park, until you come to a sharp right bend. This is already our next stop.

Stop 3.5. Roque Nublo dyke with gabbro xenoliths (N 27.9661 W 15.5978)

The dyke is in the second sharp road bend after the Roque Nublo car park, and the dyke crops out on both sides of the road, defining an overall strike of northeast—southwest to north—south (Fig. 6.93).

The dyke is phonolitic in composition and belongs to the Roque Nublo radial dyke swarm. Note the large gabbro fragments (xenoliths) within the dyke and the striations on the dyke's margins (in the chilled part). The striations plunge approximately 30° toward the south and southwest,

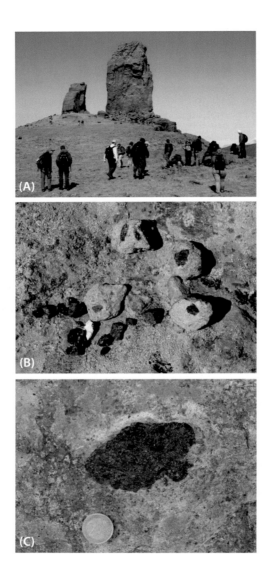

FIGURE 6.90

(A) Main (second) plateau before Roque Nublo monolith. (B) Large amphibole crystals in Roque Nublo deposits on main plateau. (C) Amphibole crystals and aggregates in Roque Nublo deposits. Some can reach quite respectable sizes (eg, Frisch and Schmincke, 1969). Images courtesy of L.M. Scharzkopf.

FIGURE 6.91

The Roque Nublo monolith in its full beauty. Walk onto the left (west) side first to inspect the exposed boundary at the base of the monolith. Encircled climber gives the scale.

indicating not only a vertical flow component but also a lateral one for this dyke! The dyke can be traced for a distance and you may want to follow it for a few steps.

Return to your car and continue the drive up-hill on GC-600, with Roque Nublo breccia exposed in the road sections on your right.

After approximately 3 km you will pass some cavernous weathering Roque Nublo deposits on either side of the road. Similar weathering patterns have frequently been used as shelter and seasonal settlements by aboriginal Guanches in lower altitudes, and they were used at these higher altitudes as temporary shelters during the hunting season.

FIGURE 6.92

(A). At the base of Roque Nublo monolith. (B) At the western foot of the Roque Nublo monolith, a shear zone is exposed. (C) Close-up showing shear sense indicators. These suggest the former movement of the top toward the south.

Continue through the dense pine forest and pass a larger barbecue place in the forest that is usually packed on summer weekends. Soon after, you will pass a number of fields on both sides of the road. Note the dark scoria that forms the top layer on the fields. This fine scoria and lapilli materials are derived from Holocene volcanoes; their local name is *picón* (Spanish for medium-grained

FIGURE 6.93

Dyke of Roque Nublo—age with large gabbro xenoliths, a short bit up the road from the Roque Nublo parking site. Image courtesy of L.M. Schwarzkopf.

charcoal, eg, for burning in braziers that would be similar in form and size to the basaltic lapilli). Picón has the ability to hold water for much longer than the top soil due to its internal porosity (vesicles), making it an extremely useful helper in local agriculture where actual rainfall can be very limited (see also chapter: The Geology of Lanzarote, Figs. 7.15 and 7.21).

Soon after passing the lapilli-covered fields, you will come to a road junction. Turn right toward *Pozo de las Nieves* (*Los Pechos*) onto GC-130 (green).

Continue on the bendy road for approximately 2 km and then turn right for Mirador *Pozo de las Nieves* using GC-134 (yellow). Drive up to the mirador and the radar station, and use the small road just to the right of the entrance of the radar station. Drive up this small road for 2 minutes until you reach a parking place.

Stop 3.6. Pozo de las Nieves *(Los Pechos) 1949-m altitude: The highest point on the island (N 27.9621 W 15.5720)*

This is the highest point of the island, reaching almost 1950 m in altitude. On a clear day you will get some amazing views of Miocene rocks in the south, the Roque Nublo monolith, and even the active Teide volcano on Tenerife in the far distance (Fig. 6.94).

Walk upstairs behind the parking place for the best views, including Roque Bentayga in the west and the top of the Risco Blanco intrusion below you to the east.

FIGURE 6.94

(A) View from *Pozo de las Nieves* (1949 masl) toward Tenerife, the highest point on the island. Note the Pliocene deposits with Roque Nublo monolith in the foreground and Miocene and Pliocene rocks in the distance to the right. (B) Pico de Teide on Tenerife is sometimes visible in the far distance. Try to imagine the full Roque Nublo volcano (red dashed line) using Teide on Tenerife as an analog.

Return to the junction of GC-134 (yellow) with GC-130 (green) and follow the signs for *Telde* (GC-130 green east). This stretch now marks the end of Pliocene and the start of the Holocene deposits in the area.

After approximately 1 km, you will encounter scoria deposits to your right, originating from a maar volcano that is located just behind the trees.

Continue down-hill. Approximately 5 km after the last junction you will find Caldera de Los Marteles (another maar) to your right. A lay-by on your right offers safe and comfortable parking.

Stop 3.7. Caldera de Los Marteles: Maars and maar deposits (N 27.9609 W 15.5354) at 1544 masl

Walk back on the road (up-hill) for approximately 200 m to the next bend to inspect early maar deposits that are mainly made up of country rock fragments (Roque Nublo lithologies), grading upward into more juvenile magmatic deposits (higher proportion of scoria; Fig. 6.95A).

Return to the car and inspect the Los Marteles crater itself, which now hosts several fertile fields. The crater is dated at approximately 90,000 years before present. Local concentric settling fractures are still obvious (Fig. 6.95B), and a larger post-maar or syn-maar dyke is visible near the parking site (Fig. 6.95C).

Continue on GC-130 and drive up-hill. A little after the next bend, the Marteles deposits are exposed again in the road section to your left (pass slowly), or consider another stop here if you are willing.

Some 1.5 km after Los Marteles, park your car in a sharp right bend with a lay-by on the left side.

Stop 3.8. Northwest Gran Canaria: Holocene lapilli and bomb deposits (N 27.9529 W 15.5282)

Walk back approximately 50 m to inspect a Holocene bomb deposit (Fig. 6.96A−C). The dark volcanic bombs are beautifully spindle-shaped, indicating that they experienced ballistic transport while still plastic. This is not a protected site and you are free to take a souvenir. If you are lucky, you might also find some white, foamy ocean crust sediment xenoliths (xeno-pumice), too (Fig. 6.96D−F), but they are rare here.

Continue down-hill toward *Telde*, but note frequent Holocene lapilli scoria and bomb deposits along the way.

Drive through the village of *Cazadores* in the direction of *Telde*. A short break might be appropriate here.

Continue on GC-130, and approximately 2 km after the village of *Cazadores* you will get some nice views over the Las Palmas volcanic field. The volcanic field comprises clustered and isolated cinder cones and maars that are aligned in places, although this is hard to make out from here and the distribution of vents appears a bit random from our position.

At approximately the km 16 road sign on GC-130, you will pass a little cinder cone on your left that is still actively quarried for lapilli (picón). Stop on the right side of the road and walk over to the quarry.

Stop 3.9. Northwest Gran Canaria: Trees versus fumarole pipes in Pleistocene cinder cone deposits (N 27.9636 W 15.4889)

Some beige-to-brown vertical colors that are frequently associated with decimeter-sized holes can be seen in the road sections of the scoria deposits here. The suggestion that these represent remnants of former trees, however, is unlikely. Some are mere animal burrows, others are plant root discoloration, and yet others may be old fumaroles or degassing pipes. So, there are more plausible and simpler explanations for these features than burned tree trunks (Fig. 6.97A).

Follow GC-130 down-hill, and after 3−4 km yet another cinder cone with bombs and scoria is passed (El Hoyo, approximately 6000 years old), and several more cones become visible in the distance, also.

FIGURE 6.95

(A) Caldera de Los Marteles; tuff ring deposits. (B) The Pleistocene Caldera de Los Marteles is approximately 90 ky old (image from *Google Earth*). (C) Radial dyke at Los Marteles, near parking site. Images in A and C courtesy of L.M. Schwarzkopf.

FIGURE 6.96

(A) Scoria and bomb deposits near Los Marteles (Stop 3.8). (B) Closer view of scoria layer and loosely consolidated bombs. (C) Individual bombs of basanite composition show twisted spindle shapes. (D) Frothy xenolith sample (xeno-pumice) of sedimentary deviation (see Hansteen and Troll, 2003). (E) Further example of frothy xeno-pumice from El Hoyo cinder cone, which you will pass a little further along the road from Stop 3.9. (F) Example of xeno-pumice in scoria deposits from La Isletta, located in the far distance beyond Las Palmas city. Sample courtesy of F.J. Pérez-Torrado.

FIGURE 6.97

(A) Cinder cone that is actively quarried for picon (stop 3.9). (B) Jinámar cinder cone lurking behind a derelict pumping houses. Park on the opposite site of the fuel station for Stop 3.10. (C) Walk toward the exposure behind the fuel station to the outcrop of the Jinámar scoria cone deposits or via a pass near the old pump house. The deposits here are poorly consolidated, making the slopes of the cone rather unstable (images in A−C courtesy of *L.M. Schwarzkopf*). (D) Vintage impression of Jinámar cinder cone in approximately 1906 viewed from the northwest. Note that the pumping house was not yet erected and the western flank of the cone had not yet degraded. Building activity subsequently destabilized the toe of the cone, leading to high levels of erosion and flank instability over the past 100 years. Image courtesy of Cabildo de Insular de Gran Canaria.

Continue on GC-130 toward *Telde*, and from there make your way to the motorway (GC-1) near the coast. Enter the motorway in the direction toward the north (*Las Palmas*). Note at *Telde* that you will need to join GC-10 (red) first, which will bring you to GC-1 (blue).

Once you are on the motorway, drive toward *Las Palmas* for approximately 3 km. In the distance (front right), La Isleta appears, which is another product of the most recent volcanic episode. The area is in part closed off, however, because it hosts a naval facility.

Exit the motorway after approximately 3 km at *Valle de Jinámar* and turn right (inland) at the next round about, which you will reach shortly after the blue-green motorway bridge.

Continue toward *Valle de Jinámar* and *Marzagán*. Drive on for a short while until you see a BP fuel station to your right that has a cinder cone behind it. Park on the free space parallel to the road (Fig. 6.97).

Stop 3.10. Jinámar volcano: Holocene cinder cone deposits (N 28.0366 W 15.4176)

Walk toward the exposure for a few hundred meters to inspect the cinder cone behind the petrol station. A little path leads to the outcrop (Fig. 6.97). Note the freshness of the deposits and the often blue shine on it from surface oxidation. The scoria carries pyroxene crystals as well as olivine, and it is generally of very alkaline character, like all the other Holocene volcanics of the island. This eruption here is dated to only approximately 1500 years before present (Rodríguez-Gonzalez et al., 2009). Note the steep slope of the deposits, making it quite difficult to walk up-hill (Figs. 6.97C,D).

Return to the car and continue toward *Marzagán* (GC-800, green). After *Marzagán* village, follow signs for *Bandama* (GC-801, yellow). After approximately 4 km on GC-801, a sharp turn to the left will bring you to *Bandama* (GC-821, yellow).

Soon the road climbs steeply and joins GC-802, and then GC-822, before it circles around Bandama hill, leading up to Bandama peak. Park at the vantage point on the peak of Pico Bandama (elevation 569 m).

Stop 3.11. Pico de Bandama: Holocene activity in northwest Gran Canaria (N 28.0377 W 15.4578)

To the north, you can inspect low-lying hummocky deposits and a collapse scarp in the Bandama cinder cone. These two features are related.

To the south, you will look into the Caldera de Bandama (Fig. 6.98), a maar with a Roque Nublo basement found as country rock in the maar deposits in the crater's flanks.

Bandama is only approximately 2000 years old and is another example of the younger volcanic activity on the island (Hansen et al., 2008; Rodríguez-Gonzalez et al., 2009). The cone is a little older than the Bandama maar (Fig. 6.37), which implies that an early strombolian cinder cone phase was followed by subsequent phreatomagmatic activity (eg, Hansen, 1993).

On a good day you will also get some spectacular views onto Las Palmas volcanic field to the northeast, with La Isleta in the far distance. Almost all topographic rises in this area are little Holocene volcanoes, and one can count up to a dozen of them in the wider field of view.

Because of this splendid location, Pico de Bandama was used as a German radio station to help coordinate their Atlantic U-boat fleet during the Second World War. For that reason, a bunker system was constructed inside the top of Bandama hill by the German Navy that would allow hosting radio personnel and mounting an antiaircraft gun onto the peak on short notice. Luckily, the Pico Bandama position was never required to help fend off an allied assault; otherwise, its geological beauty might not have been preserved in quite the same way.

Looking at the deposits in some more detail, the alkaline basaltic scoria that forms the Bandama cone also carries some rare nodules of mantle origin (peridotite fragments) with big (up to 5 mm) olivine crystals and spinel present. The mineral spinel is stable under upper mantle conditions between 20 and 80 km depth, thus providing us with an estimate of the depth of mantle

FIGURE 6.98

(A) View from the west onto the Pico and Caldera de Bandama (*Google Earth image*). (B) Pyroclastic deposits of the Caldera de Bandama rim, *Llano de los Canarios*. (C) View into the Caldera de Bandama from the road (image courtesy of *A. Hansen*). (D) and (E) Sequence of alternating fall and surge deposits in the tuff ring of the Caldera de Bandama. Images in D and E courtesy of L.K. Samrock.

melting and magma generation beneath the recent volcanic field (ie, ≤ 80 km). In addition, if you have a good eye, some small beige-to-white frothy fragments of oceanic sediments (xeno-pumice) can be found here as well (see, eg, Fig. 3.69).

Once you have taken in the fine views of Bandama, return to GC-1 and join GC-1 in the direction toward *Telde*, *Aeropuerto*, and *Maspalomas* to return to your base in the south of the island.

END OF DAY 3.

DAY 4: COASTAL GEOLOGY AND DESTRUCTION OF OCEANIC ISLANDS

From *Maspalomas*, drive north toward *Las Palmas* on the main motorway GC-1 (blue) for approximately 40 minutes (Fig. 6.99). Approximately 10 km before reaching *Las Palmas*, the motorway splits into GC-1 and GC-3. We join GC-3 keeping left and follow signs for *Tamaraceite*. Drive on GC-3 for approximately 8 km and through several tunnels until you reach a right exit into GC-340 to Tamaraceite. Follow this road for another 3 km, and after two sharp bends take a left onto a small road named Carretera del Cuartel Manuel Lois (an old disused navy base) (N 28.1207 W 15.4479). *Note that the ongoing building activity in 2015 may change this junction significantly.*

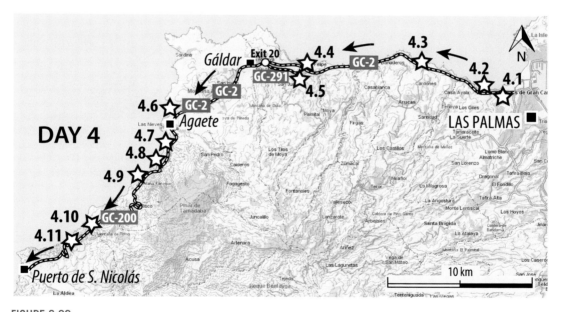

FIGURE 6.99

Itinerary Day 4. This day excursion is dedicated to inspecting the north and northwest coast, including uplifted marine rocks and pillow basalts in the northern cliffs, Pico de Gáldar cone, *Agaete*, and the "bite in the apple" tsunami deposits in Barranco Agaete, and a small-scale coastal landslide just before the village of *El Risco*. The day will end in *St. Nicolás*, with views of La Aldea valley followed by a visit to the romantic fishing port of *Porto de St. Nicolás*. Map from IDE Canarias visor 3.0, GRAFCAN.

FIGURE 6.100

(A) Pillow lavas on marine marls at Stop 4.1. (B) and (C) Close-up of disturbed marine marl layers. (D) Close-up of the pillows. Pillow lavas form from lava erupting into water, implying these rocks, now at approximately 97 masl, were once at or just below sea level. Images in A and B courtesy of L.M. Schwarzkopf.

Follow this small road until you encounter an outcrop with pillow lavas. Park and walk the road. There, you will also encounter very light-colored marine sediments overlain by pillow lavas of Roque Nublo age on the left side of the road section (Fig. 6.100).

Stop 4.1. Las Palmas outskirts and Barranco Tamaraceite: uplifted pillow lavas (N 28.1139 W 15.4599) at 97 masl

Here, beach boulders and cobbles are overlain by whitish carbonates and marls of shallow marine origin, which in turn are overlain by extrusive pillow lavas. Higher up in the outcrop, the pillows are followed by hyaloclastites (ie, fragmental rocks from submarine eruptions).

Pillow lavas are characteristic for submarine eruptions where hot extruding lava chills against cool seawater and rapidly forms a solid crust. Within this crust, the hot lava is thermally insulated and can continue to flow, forming tubular structures that appear as pillows in most cross-sections. Eventually, the magma in the tube hardens as well, forming radial cooling cracks (as is the case here), or alternatively it drains out, leaving an empty shell.

The occurrence of pillow lavas at this elevation (97 masl) implies uplift on this order of magnitude since the time of original formation of the pillows (the pillows are Pliocene and belong to the Roque Nublo group).

The reason for this uplift, which is only seen in northeast Gran Canaria, could be that the recent magmatic activity (eg, Bandama-type volcanism) is fed from a larger magma storage area underneath this part of the island. Alternatively, it is conceivable that the younger and larger island of Tenerife, with the active volcano of Pico de Teide, exerts flexural strain onto the ocean crust in the whole region, causing the entire island of Gran Canaria to become slightly tilted toward Tenerife (eg, Menéndez et al., 2008). It is hard to distinguish between these two hypotheses for now, and the two are also not mutually exclusive; therefore, both may act together.

Return to GC-340 and continue down-hill; from there, you can soon join the motorway GC-2 in the direction toward *Arucas*, *Agaete*, and *Gáldar* (ie, westward). Approximately 2 km after entering GC-2, drive along an impressive cliff exposure to your left that is made up of Miocene and Pliocene ignimbrites, lavas, breccias, and conglomerates (see Fig. 6.24). After only a few minutes on GC-2, stop in a large lay-by on your right side close to a large statue (*Monumento al Atlante*).

Stop 4.2. Northern cliffs of Gran Canaria: Miocene and Pliocene rocks (N 28.1277 W 15.4623)

Approximately two-thirds up the cliff, a thin but persistent white layer is visible for kilometers, but it is locally down-faulted into blocks, forming a horst and graben arrangement (eg, looking toward the southwest from your position). This is the marine layer beneath the pillow lavas from our last stop (STOP 4.1). It can be seen to form a widespread and easily traceable marker horizon for N-Gran Canaria.

Continue on GC-2 westward. After approximately another 2 km, you will see a recent cinder cone on your left that deposited scoria over the steep cliff ahead of you (the Arucas volcano). It is presently quarried for picón (scoria and lapilli for road construction and agriculture).

Continue on GC-2 toward *Agaete* and *Gáldar* for another 2 km and leave the motorway at the *Arucas* exit. Park directly in the bend when reaching the bridge over the motorway (eg, just next to the usually locked barrier on the right side).

Stop 4.3. Arucas: Uplifted marine platform (N 28.1439 W 15.5098)

From your parking site just before the motorway bridge, walk down-hill for a few minutes until you see a coastal platform with a light-colored cover rock. Walk down-hill until the road splits and take the left turn. Once you reach the coastal platform on the right side of a little bay, progress to the light-colored thin cover on top of the darker rocks that make up the cliffs (Fig. 6.101A).

The outcrop here (the Arucas platform) is a coastal abrasion platform where Pliocene shells in marine sediment are found on top of Miocene basaltic lavas. The platform is approximately 350,000 years old, and the Arucas volcano (the cinder cone in the background) is approximately 125,000 years old (Fig. 6.102A). *Patella* is the main fossil in the raised marine sediments (Figs. 6.101B,C), and frequent wasp burrows can be observed, too, suggesting that this light

FIGURE 6.101

(A) Walk down to the coastal platform. Where the path splits, walk to the left until you reach the light-colored marine sediment layer with invertebrate remains resting on top of older basalt (image courtesy of *L.M. Scharzkopf*). (B) Pliocene shell beds and marine boulders on uplifted Arucas platform (~45 masl). (C) Close-up of Pliocene shells in raised marine sediments.

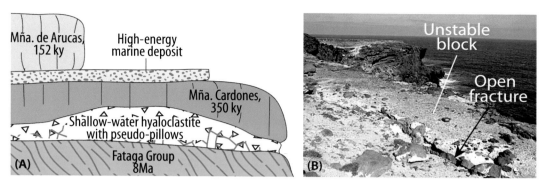

FIGURE 6.102

(A) Sketch indicating the structure of the northern wall of the little bay near *Arucas* (image courtesy of *L.M. Scharzkopf*). (B) A steep fracture that detaches a major sliver of the headwall of the little bay is visible (in 2014). Do NOT step on it because you are approximately 50 m above sea level!

sedimentary layer formed in a shallow marine setting or a tidal area before being covered by the younger Arucas lavas.

Here, at an elevation of approximately 45–50 masl, the Pleistocene marine sediment comprises not only shell debris from *Patella* but also larger remnants of bivalves and gastropods that are seen frequently (mussels and snails). The rock is, in fact, not "fossilized" and the animal remains are effectively still carbonate in their composition (ie, they are not lithified!).

The insect burrows in the upper part of the marine deposit suggest a fair bit of time between the younger lavas on the top formation of the platform, although exact time spans cannot be accurately derived.

The wider implication of these sea creatures being found at a level above the shoreline is that they underline the concept of uplift discussed previously at Stop 4.1. The evidence seen at the present location would support an uplift of approximately 45 m since the Pleistocene. If this information is taken at face value, then it would imply an acceleration of uplift of north Gran Canaria in the more recent geological record (ie, relative to the total uplift of 95 m since the Pliocene).

If this is correct, then the recent volcanic activity in northeast Gran Canaria may be a more dominant factor than flexural tilt of oceanic crust because Tenerife is already several Ma old. However, sea level was higher in the Pleistocene, making this estimate a little uncertain.

If you have a packed lunch, then this may be a good spot for a picnic! Note: At the time of writing (2014), the headwall of the bay displayed a major arcuate fracture that destabilizes a sliver of rock approximately 5 m wide (Fig. 6.102B). Do not step on the headwall and do not picnic there!

This type of fracture is symptomatic of intense coastal erosion that undercuts steep rock cliffs in the north and northwest of the island. Once this rock falls, it will likely cause a splash that will create a mini-tsunami within the confines of the little bay itself.

Return to your car and continue on GC-2 toward *Gáldar* (West). To turn the car, you must cross the bridge first. Note that GC-2 (blue) turns into GC-2 (red) at some point along this stretch.

The steep cliffs and the strong waves here in the north are very popular with surfers. For example, Björn Dunkerbeck, a multiple world champion of surfing, has a popular surfing school here on the north coast.

The north of Gran Canaria also grows a lot of bananas and sugar cane because this is the weather side of the island, and bananas and sugar cane need a lot of water. Perhaps not the most sensible crops on an archipelago that continuously suffers from water shortage, but bananas and rum made from sugar cane provide regular income for the local farming communities.

After a while, "Pico de Gáldar" appears in the distance. Canarios lovingly call this cone "little Teide." It is a Pleistocene (approximately 1.2 Ma old) pyroclastic cone that experienced several small-scale to medium-scale landslides prior to erection of the village that now creeps up its slopes.

After a few more kilometers on GC-2, a thick succession of (Pliocene to Peistocene) Roque Nublo and post-Roque Nublo lavas is exposed in the cliffs above the road. The road now climbs up-hill and you will drive through a series of small tunnels.

As soon as the road descends, turn right into GC-291 (exit 20) and follow brown signs for *Cenobio de Valerón* for a few kilometers.

Stop 4.4. Cenobio de Valerón *archaeological site (N 28.1435 W 15.6124)*

After a short drive on GC-291, you will arrive at the famous archaeological site *Cenobio de Valerón* (entrance fee €2.50, closed on Mondays). Park in front of the entrance stairs and walk up to the little office to be admitted. The site likely represents a pre-Hispanic grain deposit dug into a fossilized Pliocene cinder cone (Fig. 6.103). Although legends about housing young maidens in

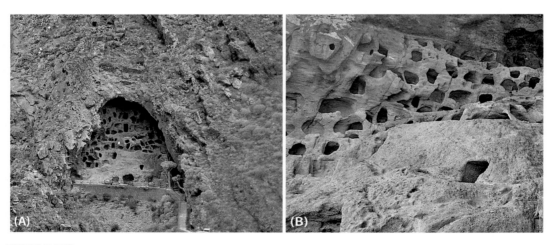

FIGURE 6.103

(A) *Cenobio de Valerón*, a pre-Hispanic grain deposit dug into a Pliocene cinder cone (Mña. del Gallego) (image courtesy of *Foto aérea de Canarias*). (B) Close-up of the individual niches carved into the lapilli and fine scoria. Likely, these were used as grain stores by the aboriginal population. Image from IDE Canarias visor 3.0, GRAFCAN.

these cells are widespread, the grain store concept strikes modern archeologists as the more realistic option. Note that xeno-pumice is present in some of the scoria here, also.

Return to GC-2 via GC-291, but make sure to briefly stop on the way to admire Pico de Gáldar from the distance before heading back on the motorway (ie, in the parking place of the larger shopping center and restaurants before joining the GC-2 motorway again).

Stop 4.5. Distant view of Pico de Gáldar (N 28.1435 W 15.6124)

Note the angle of the slope is approximately 33—35°, which is approximately the angle of repose for stable accumulation of loose granular material such as sand. This indicates that this cone's architecture is prone to instability. Also, note the spoon-shaped collapse scars on both sides of the cone.

Although Pico de Gáldar is unlikely to erupt again, it still represents a considerable landslide hazard for the village at its base and hints at the fate of all large volcanic structures; they ultimately decay.

Return to and continue on GC-2 toward *Agaete*. Note that GC-2 is blue again now (ie, an official motorway). When driving past Pico de Gáldar, note the onion-like appearance of the layers excavated by erosion and the houses built within these more wind-sheltered depressions (Fig. 6.104).

After another 10 km on GC-2, and when descending into the next valley (Barranco de Agaete), you will get some nice views of Tenerife to your right (if the weather permits), dominated by Pico Teide in its center. *Teide* is more than 3700 m high and the tallest peak in Spain. Its name is probably derived from "Echeide," the aboriginal word for "hell," because it had several eruptions in pre-Hispanic time that were likely witnessed by the Guanches.

Just before the village of *Agaete*, you will be able to see the concave northwestern coastline of Gran Canaria (the "bite in apple"), which is believed to have sourced a major Miocene landslide (Fig. 6.105; Funck and Schmincke, 1998).

FIGURE 6.104

(A) Pico Gáldar (1.3 Ma) pyroclastic cone in northwest Gran Canaria. The town of *Gáldar* has developed in the southern flank of the volcano, protected from the predominant winds (the trades). (B) *Google Earth* map view of Pico de Gáldar, showing the radial pattern of gullies carved by torrential rains.

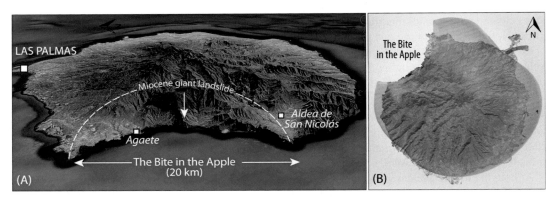

FIGURE 6.105

(A) Aerial view of the northwest coast of Gran Canaria. The large embayment formed through a giant landslide in the Miocene. (B) The landslide scarp resembles a "bite in the apple."

Enter *Agaete* village and drive across the first roundabout in the direction toward the cemetery (*cementerio*). Drive up-hill for approximately 300 m, pass the cemetery on your right, and park at the sign for *Playa de La Caleta*.

Stop 4.6. Playa de La Caleta, tsunami deposits, and parking site (N 28.1051 W 15.7048)

From the parking site, walk toward the northwest along a dirt track (approximately 5 minutes) until you meet the coast. Then, turn right onto a terrace, where you will find a marine conglomerate deposit that rests on 1.8-Ma-old lavas (Fig. 6.106, but see also Fig. 6.41). The main outcrop is at location N 28.1077 W 15.7049.

Here, rounded and angular boulders occur together with marine shells, which thus represent a high-energy deposit that is believed to originate from a giant wave, a tsunami that originated from a giant collapse on Tenerife (the Güímar collapse) at 840 ky (Paris et al., 2005a; Pérez-Torrado et al., 2006). The landslide scar of Güímar is approximately 80 km away from Gran Canaria and formed approximately 840,000 years ago (see chapter: The Geology of Tenerife).

These unusual deposits can be found up to an elevation of 160 m above sea level here in northwest Gran Canaria, and there it looks very similar, suggesting a wave approximately 20–30 m high that rushed to 160 m above sea level (tsunami run-up) rather than a stable water column (Pérez-Torrado et al., 2006). Locally, a soil is underlying the tsunami deposit, supporting that it is not a "regular" marine layer. Two distinct facies are visible, believed to represent the initial in-flow and later back-flow of the giant wave (Fig. 6.41).

Return to the car and continue toward *San Nicolás* (GC-200, green). To get there, drive toward *Puerto Agaete*, but turn left before entering the port area (ie, enter GC-200 green), unless you want a short break in *Puerto Agaete* for a coffee or snack.

After approximately 0.7 km on GC-200, right in the first major bend after the turn off, you will again encounter the conglomerate deposits from previously, which are believed to be tsunami

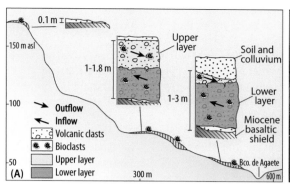

FIGURE 6.106

(A) Marine conglomerates outcrop along the southern wall of Bco. Agaete and along the road to La Aldea valley. The cobble fabrics in the sedimentary deposit are believed to represent the successive in-flow and out-flow of a giant tsunami wave (after Perez-Torrado et al., 2006). (B) The tsunami deposit along GC-200.

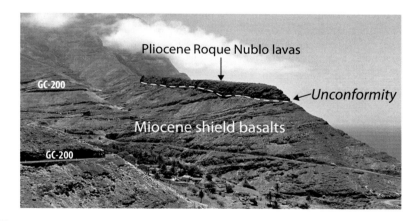

FIGURE 6.107

Miocene shield basalts overlain by Pliocene (Roque Nublo) intracanyon lavas. This major unconformity is a spectacular example of inverted relief. Likely, the Pliocene Roque Nublo lavas filled a barranco carved into the Miocene shield lavas. Erosion then inverted the relief because the younger Roque Nublo lavas are more erosion-resistant than the old and deeply altered Miocene sequences, and now there is a ridge where previously there was a valley. This setting gives an idea of the intense erosive activity on this windward flank of Gran Canaria. The picture was taken from N 28.0819 W 15.7058.

deposits (see Fig. 6.106). These deposits here are from the same event and can be found throughout Barranco Agaete (Paris et al., 2005a; Pérez-Torrado et al., 2006).

Continue further up-hill through shield basalts that somewhat surprisingly dip toward the center island (Fig. 6.107). Geophysical evidence suggests that a positive gravity anomaly exists off-shore

Agaete and that a separate volcanic edifice may have existed there, adding substantial weight to the crust and thus promoting landslides. The inward-dipping lavas thus may have originated from a long-lost volcanic edifice formerly located out at sea.

Approximately 2.5 km after *Agaete* (still on GC-200), you will encounter massive Miocene Mogán and Fataga pyroclastic flows on top of the shield basalts. As you pass, note the strong alteration in the basalts. Zeolites, calcite, chalcedony, amygdales, and gypsum are frequently found in these rocks and originate from hydrous alterations within the buried rocks of the shield basalt sequence back in the Miocene.

After approximately 4 km on GC-200, a major unconformity of Roque Nublo group lavas that rest on Miocene shield basalts will appear in front of you (Fig. 6.107). The Roque Nublo rocks here show a seaward dip (most likely lava flows filling a paleobarranco that are now in topographic inversion), whereas the shield basalts still dip toward the island center, implying a large Miocene basalt volcano once stood out there in the sea.

Drive up to the hill with the unconformity in sight, pass Barranco de Guayedra and several sharp bends, and park in a little lay-by next to the unconformity.

Stop 4.7. Miocene Roque Nublo unconformity (N 28.0747 W 15.7102)

Walk up to the unconformity for a closer inspection and for some nice views of the Atlantic Ocean. Note that it can be windy out here, so be prepared. In addition, the remains of a small prehistoric shelter or settlement are found here, and you may want to take a minute to inspect these also. Rumors indicate that this might have been an ancient type of lighthouse where a large fire would have guided sea-going vessels back to Agaete bay.

Continue on GC-200. Just after the little lay-by, note a massive phonolitic dyke in the road section. After another approximately 600 m, you will see an erosional remnant of a dyke (at N 28.0752 W 15.7153) sticking out of a cliff like a wall (Fig. 6.108). In the Canarian tongue, these features are referred to as *taparuchas* and are particularly well developed on the island of Gomera (see Figs. 4.13, 4.30, and 4.31).

Continue on GC-200, along the wild and rugged coastline of northwest Gran Canaria. Basalt flows and younger dykes of mostly late Miocene (Fataga) age are now passing by. Approximately 10 km after *Agaete*, you will get to the El Risco landslide (Fig. 6.109A). You will need to park after a larger left bend, where you will find an open space to your left. From here, you need to walk back onto the road to inspect the head of the slide, but you can also hike down the dirt track opposite to the parking space to investigate the toe of the slide (Fig. 6.109B).

Stop 4.8. El Risco landslide (N 28.0600 W 15.7296)

Let's do the head of the landslide first. Walk back on the road around the last major bend and approach the gully that separates the stable from the unstable rock mass (see Longpré et al., 2008b). On the way you will have passed two major phonolite dykes that cut through the shield basalts. The gully at the landslide margin follows a radial fault (radial to the Miocene caldera), and a lithified fault breccia is visible up on the slope above the road. The head wall of the slide, in turn, is a recent concentric open scarp and is the result of a more recent movement. Continue to walk along the road around the landslide mass, but please watch the traffic (Fig. 6.109).

FIGURE 6.108

(A) Prominent dyke remnant along GC-200 (*Google Earth image*). (B) The weathering-resistant phonolitic dyke stands proud, like a wall (dyke comes from Scottish for a stone wall), a feature locally referred to as a *taparucha*.

To view the toe of the landslide, walk back to the car and then down the dirt track opposite the parking site. Pass a large upstanding Fataga phonolite dyke on the way and walk out, along the dyke, onto the nose that sticks into the sea (*Laja del Risco*). Turn back and view the landslide from below (Fig. 6.109).

Take a rest here and enjoy the sea breeze. Consider inspecting the dyke in more detail on your walk back.

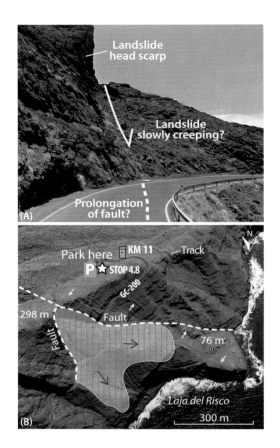

FIGURE 6.109

(A) El Risco landslide approached from the north. Note the large detached mass of material and the steep overhanging headwall (courtesy of *M.-A. Longpré*). (B) Faults (broken lines) and landslide area. White arrows indicate prominent basaltic and phonolitic dykes (after Longpré et al., 2008b).

Once back at your vehicle, continue your drive until the village of *El Risco* approximately 10 minutes further. Right at the main road is a small coffee shop with outside seating. This is a pleasant lunch spot with an excellent inspection opportunity of another Fataga phonolite dyke north of the little bar.

Stop 4.9. Village of El Risco: concentric phonolite dyke (N 28.0467 W 15.7282)

Walk back down the road toward *Agaete* for a few minutes to admire the phonolitic dyke, cutting shield basalts, while enjoying a break in El Risco. Unlike most of the other phonolitic dykes, this one is concentric and not radial to the caldera, which is a little unusual, but all the more noteworthy.

After your inspection, leave *El Risco* village and follow GC-200 up-hill toward *San Nicolás*. The road becomes reasonably narrow and bendy soon, with a steep drop (up to 600 m vertical cliffs) to your right for several kilometers and we recommend that *an experienced driver should do this leg.*

Approximately 2—2.5 km after *El Risco* village, the road intersects the transition between intracaldera breccias and pyroclastics and the shield basalts. The felsic pyroclastic units and the sedimentary breccias often display vivid alteration colors in a very similar style to those at *Los Azulejos* inspected on Day 2, although this can be difficult to see if vegetation is thick.

Drive a little further through a small tunnel and look back to the port of *Agaete*. You will note the overall curved coastline of the Miocene giant landslide and, in addition, the shallow inward-dipping caldera margin becomes very clear from here. Note plenty of radial phonolitic dykes and other intrusions on the way. Continue on GC-200, but please drive carefully here.

Approximately 8 km after *El Risco* village, the first ignimbrites of the Mogán succession can again be seen to overlie shield basalts. At *Andén Verde*, a stop can be taken (Fig. 6.110A). Stop at the lay-by on the right just on top of the ridge. At the time of writing (2014), a mirador was in preparation but was not completed yet.

Stop 4.10. The Andén Verde, northwest Coast (N 28.0279 W 15.7695)

Climb up the small rock ridge behind the lay-by for some spectacular views of the coast and Tenerife on a good day, but again, please be careful! If you look around in a calm manner you might see one of the large Canary lizards that are sometimes sunbathing around here. However, stray dogs have also been reported in this location, and you may need to offer a bite of food to buy your peace from the latter species.

Behind you (inland), a stack of Mogán ignimbrites is resting on top of shield basalts again (at Andén Verde hill), marking the transition to explosive activity inspected on Day 1. When looking toward the south (ie, toward *San Nicolás*), numerous white covers of tomato plantation greenhouses are seen to be creeping up the valley.

This valley is exceptional on Gran Canaria because it is the only major barranco that is concentric to the caldera, rather than radial. Interestingly, it almost exactly follows the margin of the Miocene Caldera, and a link is quite possible.

Continue your journey down to *San Nicolás*. Note P1 ignimbrite to your right, only a few hundred meters after the *Andén Verde* stop. A little further, a quick stop at *Mirador El Balcón* is possible if you want to admire the coastline of northwest Gran Canaria another time (Fig. 6.110B).

FIGURE 6.110

(A) The GC-200 road manages the sheer cliffs of the "bite in the apple" in the northwest landslide scar. The cliff here is almost 600 m high. (B) View onto the western part of the bite in apple cliffs of Gran Canaria from *Mirador El Balcón*. (C) From the mirador, if weather permits and you have good luck, then you will see Tenerife and Teide in the distance and may enjoy one of the glorious sunsets that are relatively frequent in this part of Gran Canaria.

Stop 4.11. Mirador El Balcón, *northwest Coast (N 28.0192 W 15.7853)*

From *Mirador El Balcón*, you may enjoy one of the best view of the western part of the "bite in apple" cliff (Fig. 6.110B). If weather permits and you have good luck, you will see Tenerife and Teide in the distance, and you may enjoy one of the glorious sunsets relatively frequent in this part of Gran Canaria (Fig. 6.110C).

Continue down-hill toward *San Nicolás*. Some of the open hills to your left have produced hearsay reports of high proportions of sea shells and beach pebbles, probably where the current greenhouses extend up to the valley. One may risk speculation on potential tsunami deposits again, with a wave originating from some of the neighboring islands (eg, Güímar on Tenerife) that could have had the energy to transport beach sediment inland for several kilometers and into the open valley of La Aldea (see Giachetti et al., 2011).

A short time after Stop 4.11, and approximately 30 km after *Agaete*, you will meet a fork. Follow GC-173 (yellow) to visit *Puerto de la Aldea* (*San Nicolás*'s harbor) and its beach, which offers a number of very pleasant seafood restaurants and bars.

Otherwise, continue to the left on GC-200 (green) to return to *Maspalomas*. From La Aldea valley, one can see the Roque Nublo monolith in the center of the island when looking up the Barranco, provided the weather permits. Drive into the outskirts of *San Nicolás* and turn right toward *Mogán* just before reaching the village core, but stay on GC-200 (green) and complete the round trip of the island. The leg to come was previously used (Day 2), and you will find that you have probably become slightly familiar with the route already.

END OF DAY 4.

DAY 5: TRAVERSING THE ISLAND SOUTH TO NORTH

Start the day by making your way to *Puerto Rico* and *Mogán* either via the new stretch of motorway in the South (GC-1) or via the coastal road GC-500 (Fig. 6.111). Once you arrive in Barranco de Mogán, continue on GC-200 (green) inland through a series of villages, including the village of *Mogán* itself. At the end of the settled valley, you will reach a sharp bend in the road that also offers the opportunity to join GC-605 (yellow) in the direction of *Tejeda* and *Presa de las Niñas*. Join GC-605, but note that this route should not be taken in poor weather or during heavy rain!

After approximately 1.5 km on GC-605, you will note dark Miocene shield basalts that make up the rocks in the valley flanks. Take a stop at either of the locations below to inspect the shield basalt flows and their associated breccias. Some related cinder and scoria deposits within the shield basalt sequence here point toward fossil vents in the region that were subsequently buried ("fossilized") by Miocene basalt flows.

Stop 5.1. Miocene Shield Basalts, Upper Barranco Mogán (N 27.9082 W 15.7039 or N 27.9074 W 15.7040)

The shield basalts display some massive lava flows here, and scoria-rich areas define former near-vent facies (Fig. 6.112). Some crystal-rich basalt varieties do locally occur.

Continue on GC-605 up-hill, and after several kilometers on a winding and narrow mountain road (Fig. 6.113A) you will drive through the hydrothermally altered ignimbrites of the Mogán succession again. We are very close to the caldera margin here, and the hydrothermal overprint in the rocks is relatively intense (see, eg, Donoghue et al., 2008).

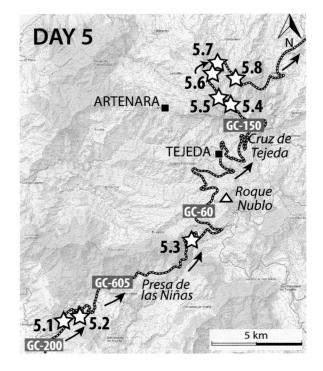

FIGURE 6.111

Day 5. This excursion traverses the island from south to north. The drive will reach the highland lakes (dams) within Miocene intracaldera rocks and the central Roque Nublo deposits before passing *Ayacata* and *Tejeda* mountain villages and the Roque de Bentayga monolith. After visiting *Cruz de Tejeda*, you will further inspect the Holocene volcanism, such as eruptive vents, cinder cones, lapilli beds, lava flows, and hornitos. Finally, we descend into the lush valleys of north Gran Canaria and return via *Teror* and *Las Palmas* to the south. Map from IDE Canarias visor 3.0, GRAFCAN.

FIGURE 6.112

The shield basalts display some massive lava flows here in the southwest, but scoria-rich areas occur that define former vent areas. Image courtesy of C. Karsten.

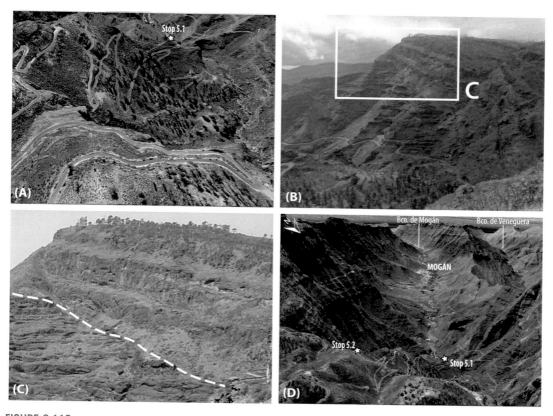

FIGURE 6.113

(A) The bendy mountain road climbs up from Bco. de Mogán and passes the Miocene caldera margin. (B) View of the caldera margin unconformity between the shield basalts and the inclined intracaldera ignimbrites that rest unconformably above. (C) Close-up of the unconformity (image courtesy of *L.M. Schwrzkopf*). (D) View down Barranco Mogán (*Google Earth image*).

A little further, a mirador on GC-605 on the right side appears. Park your car there. The rocks exposed here in the wall behind you are the Upper Mogán ignimbrites again, but these are in bad shape due to intense hydrothermal alteration.

Stop 5.2. Head of Barranco Mogán and Miocene ignimbrites (N 27.9084 W 15.6961)

Looking to the northwest, one can see the caldera margin unconformity between the shield basalts at the bottom and the inclined intracaldera ignimbrites that rest unconformably above (Figs. 6.113B,C). On a good day one should also have a splendid view down the long axis of Barranco Mogán, whose head region we are at now (Fig. 6.113D).

Continue up-hill through hydrothermally altered Miocene intracaldera ignimbrites, now of Mogán-Fataga transitional age. After a few minutes, a small erosional remnant of Roque Nublo deposit stands proud in the distance to your right (ie, when looking southwest), a mini–Roque Nublo monolith if you will (Fig. 6.114A).

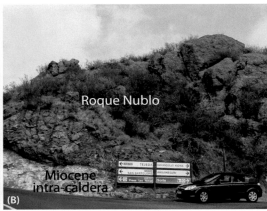

FIGURE 6.114

(A) A small erosion remnant of Roque Nublo breccia is visible from the road (to your right), resembling a mini-Roque Nublo monolith. (B) Unconformity between Miocene and Pliocene (Roque Nublo) rocks at the crossroads. Turn left here toward *Ayacata*.

A few kilometers further you will reach a junction with GC-505. Stay on GC-605 and continue toward *Tejeda* (ie, turn left). Here at the junction, Roque Nublo deposits overlie the Miocene intra-caldera rocks and you can put your finger on the unconformity (Fig. 6.114B).

Pass more Roque Nublo breccias and note that after a while, the rocks become more coherent and several lava flows with top and bottom breccias appear in the road sections.

Drive by *Cruz de San Antonio* and a splendid highland landscape will unfold before you (if it is not cloudy). Take a quick stop at *Cruz de San Antonio* if you like, or continue to drive inland (approximately northward).

The rocks along the road are now fresh-looking Roque Nublo basalt flows, and you may make an optional stop to enjoy the views and inspect the lava exposures.

Continue on GC-605 and pass several small to medium highland lakes on your right. Note that just before the km 11 road sign on GC-605, you will encounter Roque Nublo breccias again for a short while before you drive within basalt flows once more. The freshness of these flows is rapidly diminishing, however, as you drive further inland.

Near the km 8 road sign on GC-605, you will note a drastic change in the rock type exposed as you have just made your way into the late Miocene cone-sheet swarm that rests beneath the Roque Nublo cover (compare to Day 2).

Continue inland. The boundary between the Miocene cone sheets and the Roque Nublo deposits is now undulating along the side of the road, and you will see a bit of both rock types coming and going for a little while. Soon, however, you will drive fully in hydrothermally altered cone-sheet units for several kilometers. The majority of the cone sheets here dip toward the north (approximately toward the center of the island) as we are now approaching the central highlands from the South.

FIGURE 6.115

(A) Massive Miocene cone sheet swarm with a northerly dip at Stop 5.3. (B) The majority of the intrusions of the cone-sheet swarm dip toward the center of the island, approximately toward the north here, as we are now approaching the central highlands from the south. Image courtesy of C. Karsten.

FIGURE 6.116

Drive along the near 300-m-high Roque Nublo breccia cliffs here that represent another erosional remnant similar to the Roque Nublo monolith itself. The road (GC-60) follows the unconformity between Roque Nublo (top) and Miocene intracaldera rocks (bottom) for a while after leaving *Ayacata*.

Stop 5.3. Hydrothermally altered cone sheets (N 27.9493 W 15.6312)

Soon after the km 4 sign on GC-605, stop in a lay-by on the right (in a sharp bend) to inspect the hydrothermally overprinted cone-sheet swarm (Fig. 6.115).

Continue on GC-605, and at the end of GC-605, just after the km 1 road sign, you will see a small lay-by on the right. Park to inspect the unconformity between the Miocene cone-sheet swarm and the overlying dark and massive Roque Nublo breccias once more, which are now exposed pronounced cliffs in front of you (eg, Location: N 27.9541 W 15.6185) (Fig. 6.116).

Continue into the village of *Ayacata* for a short refreshment stop (ie, turn right onto GC-60 red). In *Ayacata*, park outside the little coffee shop that we used on Day 3.

You may now consider doing the Roque Nublo hike (~1 hour) if you did not do it on Day 3 (eg, because of weather or time constraints). If you choose this option, continue up-hill on GC-600 (green) toward *San Mateo* for approximately 10 minutes and fall back on instructions for Day 3.

To continue the north–south traverse of the island, however, join GC-60 (red) toward *Tejeda*, which will take you further along the steep cliffs of Roque Nublo breccia on your right for a while (Fig. 6.116). Above you and to your right are still Roque Nublo-type rocks, but below you are the Miocene intracaldera cone sheets, which have much lighter colors.

The road then descends for a stretch. Near the km 9 road signs on GC-60, you will find yourself next to Miocene cone sheets again. Note the dip of the cone sheets has changed, reflecting that our position relative to the focal point of the Miocene cone-sheet swarm has changed also. After the km 8 road sign, you will once more traverse the unconformity and Roque Nublo lavas will appear to your right for approximately 1 km, followed by Roque Nublo pyroclastics and breccias.

Yet a little further, you will encounter a fork. To the left is Roque Bentayga (via GC-607 and GC-671, yellow), another erosional remnant of the Roque Nublo period, similar to the Roque Nublo monolith itself. Take a short drive there (10 minutes) and park at the end of GC-671 at a well laid out mirador to stretch your legs. Stunning views are the reward and a short hike up the Roque there is recommended (Fig. 6.117). Note, however, that the entrance road to this mirador is closed on Sundays!

Return to GC-60 and continue toward *Tejeda*. The Roque Nublo lavas and pyroclastics on your right pass once more into the Miocene intracaldera rocks as the road descends toward *Tejeda* (eg, after km 6 on the GC-60 road signs). This underlines the irregular topography of the Miocene–Pliocene unconformity, as you now see Roque Nublo deposits below the altitude level of some of the Miocene cone-sheet rocks we passed only a few kilometers earlier.

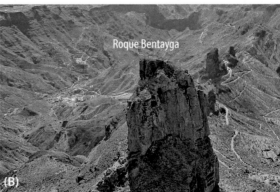

FIGURE 6.117

(A) Roque Bentayga with Roque Nublo in the background. (B) Roque Bentayga deposits are aligned with the other Roque Nublo monoliths, suggesting it to be part of a former valley-fill deposit that was later eroded to now form an inverted relief. Images courtesy of Fotos aéreas de Canarias.

A little closer to *Tejeda*, several rock faces can be seen to have protective shot-crete implemented and also wire fencing, reflecting the crumbly and rather unstable nature of the exposed hydrothermally overprinted Miocene intracaldera rocks here in the central parts of the island.

In *Tejeda* village you have several options for refreshments, and there is also a fuel station in the center of the village in case you are running low. Right at the fuel station, you will meet another fork. Make a right turn toward *Las Palmas* and *Artenara* (ie, stay on GC-60).

After *Tejeda* village, continue to follow signs for GC-60 (up-hill). A left turn follows after a short while, but continue on GC-60 to *Las Palmas* and *Artenara*. Signs for *Cruz de Tejeda* should appear now, and after less than 0.5 km, you will meet a roundabout. Drive up-hill, and follow signs for *Cruz de Tejeda* on GC-15 (red) now. Follow GC-15 up-hill for a while.

At the next fork, go left so that you stay on GC-15 (red). Now you are in Roque Nublo group rocks once more. When arriving in *Cruz de Tejeda*, a major tourist attraction, please drive carefully because many people may be walking on the streets outside the cafés and souvenir shops. An optional stop is recommended.

A small road (GC-150) goes off to the left side at the end of *Cruz de Tejeda*. Follow GC-150 (green) toward Pinos de Gáldar and continue up-hill for approximately 4 km through Roque Nublo strata.

Once the road descends again, you will encounter Holocene scoria and lava deposits on your left (near km 9 road sign on GC-150). These are very reddish (oxidized) from contact with air in a hot condition and are locally seen to overlie weathered Roque Nublo deposits. The Holocene scoria deposits here were erupted from a series of vents and cones around Montañón Negro (big black mountain). As the name implies, black scoria and lavas of fresh appearance are also found here in the area, and an inspection follows in a few kilometers.

Continue on GC-150 a little further. A Holocene feeder dyke can be seen to have cut into the surrounding scoria (Fig. 6.118A). Make a brief stop for an inspection.

Stop 5.4. Holocene dyke intrusion into scoria (N 28.0247 W 15.6102)

Although strongly weathered, pyroxene and feldspar plus altered olivine can still be recognized in the dyke intrusion. Notably, the country rock next to the intrusion is reddened and was likely "fritted" from the influence of the hot magma close by (Fig. 6.118B).

Continue on GC-150. Just before you see a long stonewall going up-hill to a large cinder cone, you will need to park where the walk starts, which is in a relatively steep left bend. Try to park a little off the road if possible, or drive a little further and walk back.

Stop 5.5. Montañón Negro: Lapilli and scoria deposits (N 28.0257 W 15.6122)

Follow the stone wall for a short while and then walk to the left to make your way to the fresh exposures on the northwest face of Montañon Negro (Fig. 6.119A,B). Note it can be slippery when walking on lapilli.

The age of the cone is 2970 years BP (Rodriguez-Gonzalez et al., 2009), so it is geologically very young. Note the white frothy xenoliths (xeno-pumice) that can be recovered with ease and in relatively high numbers from this site (Fig. 6.119C−H). These comprise igneous rocks that were recycled and also oceanic sediments (eg, Hoernle, 1998; Hansteen and Troll, 2003) that were entrained in ascending magma similar to occurrences on La Palma, Lanzarote, and, of course, the ones that erupted near El Hierro in 2011 (eg, Carracedo et al., 2012a,b, 2015a).

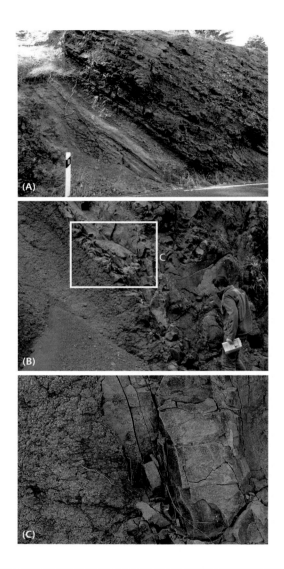

FIGURE 6.118

(A) Reddish Holocene scoria along GC-150. (B) This Holocene sheet intrusion in scoria shows baked contacts in the scoria next to the intrusion margin. (C) Close-up of the baked contact. Images courtesy of L.M. Schwarzkopf.

Return to the car and continue on GC-150. After the km 11 road sign on GC-150, you will leave the forest. Layers of dark (fresh) scoria and lapilli will be visible in good weather and the Montañón Negro cinder cone will appear in full beauty on your right again. You are now fully engulfed in the Holocene eruptives that cover large areas of the north and northeast of Gran Canaria.

After another 1−2 km, you might be rewarded with a stunning view of Teide volcano on Tenerife, at least on a good day and soon after you will meet a fork with GC-21. Right across the

FIGURE 6.119

(A) The young Monatñon Negro cinder cone next to GC-150. (B) The former quarry displays dark and well-bedded scoria layers and quite abundant xeno-pumice (white specs on the ground). Xeno-pumice is relatively frequent here, or at least it is easily spotted among the black juvenile scoria deposits (images in A and B courtesy of *L.M. Schwarzkopf*). (C)−(F) Frothy sedimentary xenolith (xeno-pumice) close-up, recording vesiculation, thermal cracking, and partial melting, although original bedding structures are sometimes still recognizsed (eg, in F). (G) and (H) A larger specimen that is in the early stages of partial melting, as seen in the bubbly outer zone and thin melt film that line the fracture walls in the interior of the speciment.

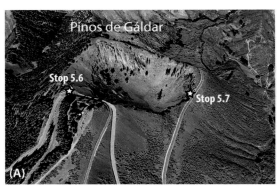

FIGURE 6.120

(A) View of the Pinos de Gáldar eruptive vent and the location of Stops 5.6 and 5.7 (image from *Google Earth*). (B) View of the interior of the crater at stop 5.6; image courtesy of L.M. Schwarzkopf.

fork of GC-150 (green) with GC 21 (red), you will find a lay-by and mirador with a small parking space. Park here for the next stop.

Stop 5.6. Pinos de Gáldar and Holocene Eruption Crater (N 28.0386 W 15.6186)

From here you can get a birds-eye view of the Holocene (~3000 year old) "Pinos de Gáldar" eruptive crater (Fig. 6.120). The crater is almost the same age as the Montañon Negro cinder cone, our last stop, but the color of the basaltic lapilli here is red, likely due to intense oxidation from prolonged exposure to heat in the crater area.

Continue on GC-21 toward *Valleseco* and *Las Palmas*. After only 1 km, you will meet a T-junction with GC-70-(red). Turn left onto GC-70 and follow this road toward *Moya* and *Gáldar*.

After only approximately another 1 km, the larger Holocene Pinos de Gáldar crater opens on your left again. This site is worth a photo stop, but please be careful here; traffic can be heavy and the lay-by is slim.

Stop 5.7. Pinos de Gáldar: Holocene crater (2830 BP) (N 28.0401 W 15.6157)

Beautiful exposures of crater rim facies are seen here (Fig. 6.120B). The deposits overall are oxidized, compatible with a near-vent facies deposit where the combination of heat and air contact allow rapid uptake of oxygen, producing a rust-like color.

Turn your car, drive back to the previous junction, and continue on GC-21 toward *Las Palmas*. After the km 3 road sign on GC-21, you will drive by a Holocene lava field to your left and right. This makes a very useful stop and is usually shady in the afternoon.

Stop 5.8. Holocene "aa" lava flows with hornito (N 28.0350 W 15.6065)

Here, bottom and top breccias of several lava flows are superbly exposed, displaying more massive flow centers, the typical architecture of viscous "aa" or "blocky" lavas. Unfortunately, the rocks are not the freshest ones on the island and have seen significant low-temperature hydrous alteration. Remember, this is the wet side of the island and these rocks have presumably seen a lot of moisture (Fig. 6.121A).

FIGURE 6.121

(A)−(C) Holocene lavas with large hornito at Stop 5.8 (images courtesy of *L. Samrock*). Note the strong overgrowth here on the wet side of the island.

These lavas are part of the eruptives of Montañón Negro that you visited earlier and are approximately 3000 years old. Walk down the road for a few hundred meters until a large "hornito" becomes visible. From here, the valley-filling nature of the lava also becomes apparent. The top surface of the flow looks very fragmented and rough; it looks almost unmodified apart from the plants that slowly conquer the lava surface (Fig. 6.121B,C).

After the Holocene lavas, the road rapidly brings you to lower altitudes and back into the underlying post-Roque Nublo and Roque Nublo sequences. Continue on GC-21 (red) toward *Vallesceco*, *Teror*, and *Las Palmas*. Soon, you will be able to see *La Isleta*, the volcanic arm of the island to the northeast of *Las Palmas* in the far distance. Continue on GC-21 down-hill toward *Las Palmas*. A cultural stop in *Teror*, which you will pass, is highly recommended.

You will note that Gran Canaria is more lush here on the Atlantic weather side due to considerably higher rainfall. This gives rise to a series of springs. The spring water is bottled and available

FIGURE 6.122

Santa Ana cathedral in the old town of Las Palmas de Gran Canaria (the *Vegueta*) is constructed of local stone, dominantly of Fataga ignimbrites, and makes for a beautiful stop in the late afternoon (image courtesy of C. Karsten).

to you in all supermarkets on the island (eg, *Agua de Teror*, *Firgas*, *El Toscal*). Chances are that the bottled water you have in the car with you is originally from here somewhere!

Further down-hill, join GC-3 (blue) and follow signs for *Las Palmas*, *Telde*, and *aeropuerto*. Consider a visit to *Las Palmas*, especially the old town Vegueta, while you pass. Santa Ana cathedral, for example, is worth a visit, especially for its Fataga-aged gray-green building stone (Fig. 6.122).

From *Las Palmas*, make your way to the north—south motorway (GC-1, blue). Follow the motorway GC-1 toward the south (*Sur*) to *Telde*, *aeropuerto*, *Playa del Inglés*, and *Maspalomas*. From here, it will be approximately 40 minutes back to your base.

END OF DAY 5.

THE GEOLOGY OF LANZAROTE

The Geology of the Canary Islands. DOI: http://dx.doi.org/10.1016/B978-0-12-809663-5.00007-4

THE ISLAND OF LANZAROTE

Lanzarote, the most easterly island of the Canary archipelago, is only 140 km off the African coast. Lanzarote is 60 km long, 20 km wide (862 km^2), and is elongated in a northeast—southwest fashion; thus, it is parallel to the African continental margin (see Fig. 7.1). The topography of the island is characteristic of mature islands, with the landscape dominated by deeply eroded volcanic massifs (eg, the Los Ajaches and Famara massifs), U-shape barrancos, and high vertical cliffs, separated by a wide central plain that is covered with organic aeolian sands. Recent volcanoes are grouped in a northeast—southwest trending central rift zone, with the 1730—36 and the 1824 vents and deposits being expressions of the island's "active" volcanic status.

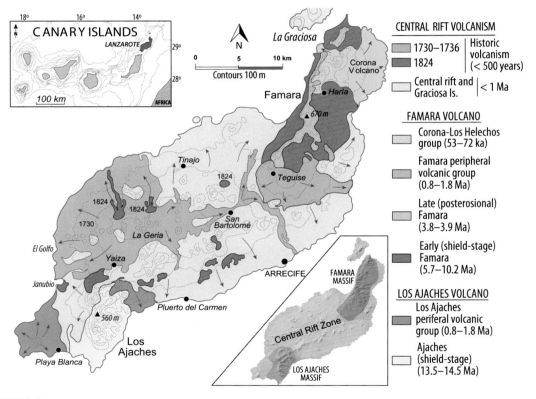

FIGURE 7.1

Geological map of Lanzarote. The inset shows the main geomorphological features of the island (modified from Fúster et al., 1968 and Carracedo and Rodriguez-Badiola, 1993). Lower inset image from Google Earth.

From a geological point of view, Lanzarote is the northern prolongation of Fuerteventura, because the two islands are only separated by a shallow stretch of sea, '*La Bocaina*' (40 m water depth), likely a land bridge that connected both islands during previous glacial periods of marine regression.

In comparison with other oceanic islands (eg, Kauai, the oldest emerged island of the Hawaiian archipelago, which is only 6 million years (Ma) old), the 15-Ma-old island of Lanzarote remains emerged above sea level because of the lack of significant subsidence in the Canaries. Instead, most of the original subaerial volume of the island was removed through catastrophic mass wasting and gradual erosion.

The record of volcanic activity on Lanzarote extends from the mid-Miocene to the present day. The island was formed by two independent shield volcanoes: the southern Ajaches volcano, now reduced to 560 meters above sea level (masl), and the northern Famara volcano (now 670 masl). According to Coello et al. (1992), both massifs seem to have grown together in the late Miocene because flows of the younger Famara volcano encircle the older Los Ajaches massif (Figs. 7.1 and 7.2). Later marine erosion must have removed most of the Famara sequences that connected the two Miocene massifs,

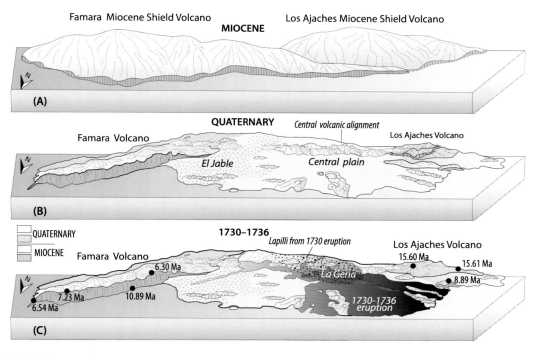

FIGURE 7.2

Cartoon showing different stages in the development of Lanzarote. Although the Los Ajaches and Famara Miocene massifs may be considered independent shield volcanoes, they grew into a single island during the Late Miocene and early Pliocene. Based on Coello et al. (1992) and new radiometric ages from H. Guillou, personal communication.

leaving sparse outcrops of this volcanism surrounded by more recent lava fields (known as "kipukas" in Hawaiian terminology).

After several million years of eruptive quiescence, volcanism on Lanzarote went through a rejuvenation phase, erupting mafic lava, scoria, cinder, and lapilli from a northeast—southwest central rift zone (Fig. 7.2B). This late phase culminated with the eruption of the Corona volcano in the Famara area, approximately 21 ka ago, and the historical (<500 years) eruptions of 1730—36 and 1824 (Fig. 7.2C). These historical eruptions likely represent the only Holocene volcanism on the island. The widely accepted idea among the population of the Canaries that Lanzarote (the "island of the 1000 volcanoes") is the most volcanically active in the archipelago is therefore not strictly justified.

THE 1730—36 LANZAROTE ERUPTION

The eruption products of the 1730—36 events represent some of the most vivid geological features of Lanzarote. The 1730—36 eruption is the third-largest basaltic fissure eruption in historical times (after the 934 AD Eldgja and 1783 AD Laki eruptions in Iceland). Although most fissure eruptions have not lasted long in eruptive duration, with the exception of the continuing Pu'u O'o eruption on Kilauea in Hawaii (30 years at the time of writing), the 1730—36 events differ greatly, with an eruptive duration of 6 years. In contrast, typical cinder cone events last on average only a few months in the Canaries (see Tables 3.1 and 5.1).

Over the 6 years of the 1730—36 eruption, between 3 and 5 km^3 of basaltic pyroclasts and lavas were emitted, covering approximately 225 km^2 (one-third of the island). The consequences of the prolonged volcanism on the island's economy and population were devastating at the time. A great part of the most fertile farmland was destroyed, along with 26 villages. The subsequent famine eventually forced most of the inhabitants to leave the island during the time of the ongoing eruptive events (Fig. 7.3).

A useful side effect is that eye-witness accounts are available for the reconstruction of this complex eruption and the successive eruptive vents and lava fields that were produced. In particular, detailed entries in the diary kept by the Parish Priest of *Yaiza* (included in the work of von Buch, 1825) and the official reports to the Royal Court of Justice of the Canary Islands by the authorities appointed to manage the crisis (Real Audiencia de Canarias 1731). The latter document, found as original in the Spanish "Archivo General de Simancas" (Carracedo et al., 1990; Carracedo et al., 1992), includes an oil painting that depicts the progress of the eruption in November 1730, when the governor of Fuerteventura, alarmed by the reported destruction, sent an artist to Lanzarote to "precisely map the places lost by lavas and those lost by sands (lapilli)" (Fig. 7.4A). Today, we distinguish at least 10 main eruptive centers and five eruptive phases along the 14-km-long volcanic fissure of the 1730—36 eruption (Fig. 7.4B,C; after Carracedo et al., 1992), and the old reports have been a vital tool in establishing the sequence of events.

THE 1824 LANZAROTE ERUPTION

A period of 88 years of eruptive repose on Lanzarote followed after the 1730—36 eruption. Eye-witness accounts mention the occurrence of earthquakes as early as 1813 that may have been a

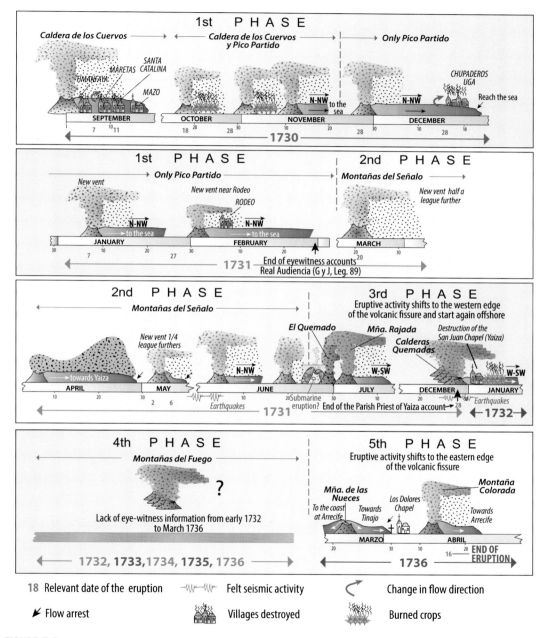

FIGURE 7.3

Reconstruction of the 1730−36 eruption of Lanzarote with information provided by eye-witness accounts such as the Real Audiencia de Canarias (1731) and the diary of the Parish Priest of *Yaiza*, as well as through geological observations (after Carracedo et al., 1992). Contemporaneous reports ceased in early 1732, unfortunately, because the authorities and most of the population had evacuated the island at that point.

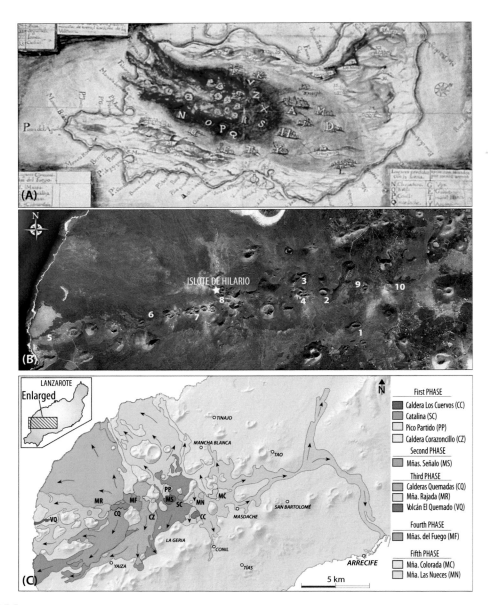

FIGURE 7.4

(A) Map of Lanzarote from November 1730 (courtesy of the *Archivo General de Simancas, Valladolid*, Real Audiencia de Canarias, 1731) showing the distribution of eruptive vents, the flow of lavas, and the extent of "sand cover" (lapilli beds). (B) Aerial view (*Google Earth image*) of the vents of the 1730–36 eruption on Lanzarote. Except the initial vents (1–3), emplaced in a northwest–southeast trend, most of the 1730–36 eruptive vents opened along a west-to-east alignment. (1) Caldera de los Cuervos (initial vent). (2) Caldera de

(*Continued*)

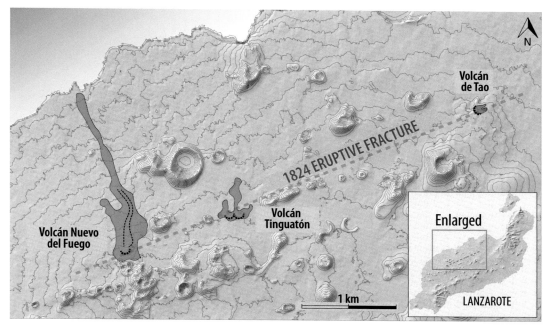

FIGURE 7.5

Location of the three eruptive vents of the 1824 eruption of Lanzarote. Note that the vents are aligned on a northeast—southwest fracture, probably induced by a dyke that fed the eruption (background image from *GRAFCAN*).

precursor to the subsequent 1824 eruption. The 1824 eruption was shorter in duration than the 1730—36 eruption, lasting only 86 days, from July 31 to October 25, 1824.

The 1824 eruption constructed three strombolian cinder cones, Volcán Nuevo del Fuego, Volcán Tinguatón, and Volcán de Tao, along a 14-km northeast—southwest eruptive fissure, likely related to the intrusion of a dyke (Fig. 7.5). The evolution of this eruption has been described by Hernández-Pacheco (1910) based on a detailed contemporaneous account by the Parish Priest of

◀ Santa Catalina. (3) Pico Partido. (4) Montañas del Señalo. (5) Volcán El Quemado. (6) Montaña Rajada. (7) Calderas Quemadas. (8) Montañas del Fuego. (9) Montaña de las Nueces. (10) Montaña Colorada (last vent of the eruption). (C) Geological map of the 1730—36 eruption. The eruption is separated into five phases characterized by changes in the eruptive dynamics and style, in magma composition, and in its relationship to structural features (Carracedo et al., 1992). PP = Pico Partido; SC = Caldera de Santa Catalina; MS = Montañas del Señalo; VQ = Volcán El Quemado; MR = Montaña Rajada; CQ = Calderas Quemadas; MF = Montañas del Fuego; MN = Montaña de las Nueces; MC = Montaña Colorada. After Carracedo et al. (1992).

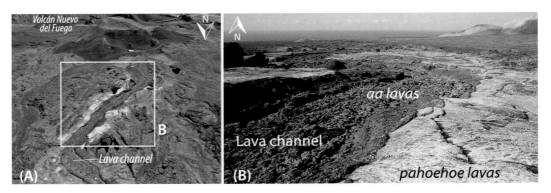

FIGURE 7.6

(A) Lava flow of Volcán Nuevo del Fuego, the westernmost eruptive center of the 1824 eruption. The lava flows formed lava channels and a combination of "aa" and pahoehoe structures. (B) Close-up view of the lava channel.

San Bartolomé (Baltasar Perdomo), that was, however, attributed by Rumeu de Armas and Araña (1982) to Ginés de Castro, the governor of the island. This account describes the main events of the 1824 eruption, which are surprisingly similar to those from other Canarian fissure eruptions (eg, the 1705 eruption on Tenerife, see Fig. 5.41B). Lava flows of the 1824 eruption are short (eg, none of them reached the sea) and low-volume. The lavas flowed on top of the 1730−36 volcanics, and thus the economic damage was small. However, very fluid lavas from the westernmost vent, Volcán Nuevo del Fuego, formed a pahoehoe lava field with spectacular features of combinations of "aa" and pahoehoe forms (Fig. 7.6).

One of the most interesting events of the 1824 eruption occurred during the activity of the Tinguatón volcano (Fig. 7.7A), the last vent to open and the one with the shortest activity in duration (October 17−25). Its active phase started with an eruptive column of incandescent pyroclasts *"…much higher than the surrounding mountains and the night sky was lit up over most of the island…"*. The vents then poured out three lava flows, with the longest advancing for approximately 5 km. The following morning the crater was full of water, and later that day a new dense and intensely black eruptive column rose, indicating the interaction of magma and water. In the afternoon, the load of the water and increasing gas pressure breached the crater at its southern rim and formed a river of hot salty water with the "color of bleach" that carved a channel still recognizable to this day (Fig. 7.7B).

Water was flowing continuously from October 19, and on October 21 and 22 pressurized water formed geysers more than 30 m high. On October 25, the flow of water ceased together with the 1824 eruption.

PHREATOMAGMATISM ON LANZAROTE

Phreatomagmatism is a frequent eruptive mechanism on the Canary Islands, as seen during the 1949 eruption on La Palma or at the Caldera del Rey vent system on Tenerife. However,

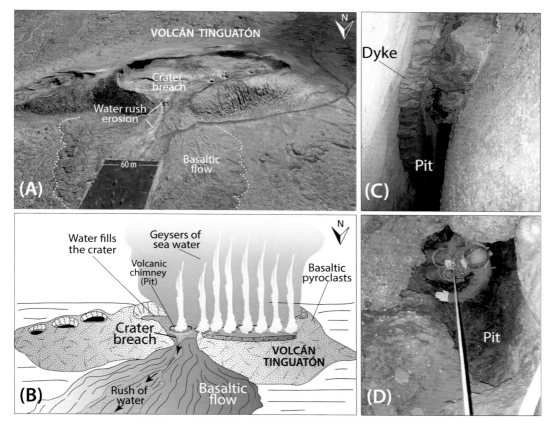

FIGURE 7.7

(A) *Google Earth image* of the Tinguatón volcano, which was located in the center of the 1824 eruptive fissure and was the last vent to erupt. Note the wide crater that shows several pits, locally referred to as the "chasm of the Devil." (B) Ejection of pressurized water is reported to have formed geysers. The load of the water filling the crater eventually breached its northern rim and formed a rush of hot salt water with "the color of bleach" that carved a channel still recognizable today. (C) Dyke intruding one of the pits. (D) The eruptive conduits were "cleaned" during the emission of water, leaving some 'clean' pits with more than a 100-m depth.

phreatomagmatic eruptions are particularly abundant on Lanzarote, with spectacular volcanological features and proportions. The low elevation of the island of Lanzarote, with the greater part of the island below 300 masl, has seemingly favored hydrovolcanism, generating frequent tuff cones and tuff rings such as El Golfo and Caldera Blanca. Many of the eruptive vents can be observed to have started as hydrovolcanic and ended as strombolian, leading to the large, wide-cratered volcanic cones (see also Fig. 5.20). This explains the frequent use of the toponym "caldera" in the recent fissure eruptions of Lanzarote, although the features thus named are actually volcanic cones (see Clarke et al., 2009).

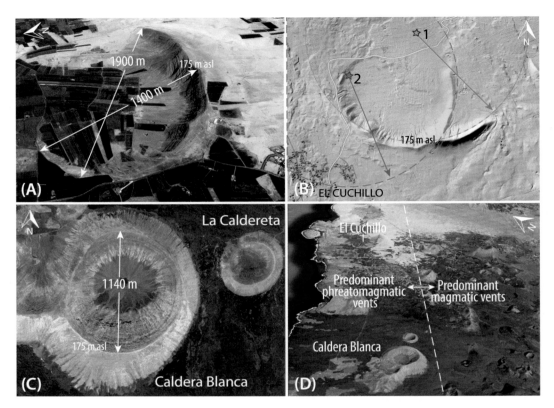

FIGURE 7.8

(A) View of the El Cuchillo maar (1400 m wide, 175 m high). (B) Centers of the successive vents (1 and 2) of the phreatomagmatic eruption (from Aparicio et al., 1994). (C) Another phreatomagmatic volcano, the Caldera Blanca, an 1140-m-wide and 450-m-high tuff cone, has been extensively colonized by lichens, which give it its conspicuous white color, hence its name. (D) Close to the shoreline, the water table is shallow and the eruptions are predominantly phreatomagmatic, whereas inland vents are, for the most part, magmatic.

One of the most spectacular hydrovolcanic vents is the maar of El Cuchillo, near *Tinajo*. (Fig. 7.8A). Another equally outstanding phreatomagmatic volcano is the Caldera Blanca tuff ring. This 1140-m-wide, 450-m-high tuff cone ring has been extensively colonized by lichens that give it its conspicuous white color, hence its name (Fig. 7.8B).

THE CLIMATE OF LANZAROTE

The central and western Canaries, presently in a comparatively early stage of development, show high relief and therefore trap the humidity of the tradewinds, which brings abundant rainfall on their northern seaward flanks. Conversely, the tradewinds pass unchecked over the lower eastern

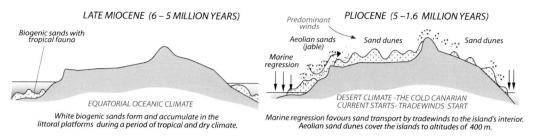

LATE MIOCENE (6 – 5 MILLION YEARS)

Biogenic sands with tropical fauna

EQUATORIAL OCEANIC CLIMATE

White biogenic sands form and accumulate in the littoral platforms during a period of tropical and dry climate.

PLIOCENE (5 –1.6 MILLION YEARS)

Predominant winds

Aeolian sands (jable)   *Sand dunes*   *Sand dunes*

Marine regression

DESERT CLIMATE -THE COLD CANARIAN CURRENT STARTS- TRADEWINDS START

Marine regression favours sand transport by tradewinds to the island's interior. Aeolian sand dunes cover the islands to altitudes of 400 m.

FIGURE 7.9

Origin of the organic aeolian sands associated with a Pliocene climatic change in the Canaries. The drastic climatic change was ultimately caused by the closure of the isthmus of Panama in the Pliocene, thus initiating the cold Canary ocean current. The extinct tropical fauna gave rise to organic sands, later transported by winds to the island's interior. Modified from Meco et al. (2007).

islands, explaining the sparser rains and desert-like conditions. This effect is exemplified on Lanzarote by the layer of light-colored sand that covers the central plains of the island (locally known as *jables*, from the French *sables*). Aeolian sands, formed by accumulation of organic detritus in the littoral platforms of the island, are derived from fragmented shells of a marine tropical equatorial fauna that was thriving in the eastern Canaries during the Miocene (Fig. 7.9A). The remains of this tropical fauna, which became extinct when the climate changed, gave rise to organic sands that were transported by wind into the island's interior (Fig. 7.9B).

The drastic climate change in the Canary Islands was caused by the closure of the isthmus of Panama in the Pliocene that initiated the North Atlantic current. This clockwise North Atlantic ocean current system branches south and flows southwestward along the northwest coast of Africa as far south as Senegal. As the current flows around the Canary Islands, it brings cold water from the north and the humid tradewinds, which help to lessen the heating effect of the Sahara to the east.

An interesting geological aspect is that although Canarian igneous rocks do not contain any free (primary) quartz crystals, this mineral is relatively abundant in aeolian sediments and is a main component of some sedimentary basement xenoliths, but it is not directly related to the Canary magmatic suite. A potential source of the quartz crystals found in surface sediments here could be sand plumes that originate from large sand storms in the Sahara desert (Figs. 7.10 and 7.11) and that can transport considerable quantities of aeolian dust that is deposited in the Canarian Archipelago (Criado and Dorta, 2003). Notably, quartz-bearing xenoliths (xeno-pumice) are frequently present in the 1730−36 and the 1824 eruptives and represent recycled materials from preisland sedimentary strata that were originally also sourced from Africa (Rothe and Schmincke, 1968; Aparicio et al., 2006, 2010; Carracedo et al., 2015a).

GEOLOGICAL ROUTES

The island of Lanzarote is relatively small (length 58 km), relatively flat, and has abundant high-quality tarmac roads. We offer a 3-day field program as a first approach to the geology of the island. Each day focuses on one of the main geological features: (1) the Miocene Ajaches shield;

FIGURE 7.10

Massive sandstorm blowing off fine dust from the northwest African Sahara desert toward the Canary Islands. The dust is loaded with crystals of quartz, abundant in the continent but virtually absent on the Canary Islands in primary form. Image courtesy of NASA.

(2) the Famara Miocene shield; and (3) the central plateau hosting the spectacular 1730−36 AD and 1824 AD eruption vents and lava fields (Fig. 7.12).

We assume that most people will find accommodations in the area around *Puerto del Carmen* or *Arrecife*; therefore, the trips are arranged to commence from there. It is strongly advised not to leave valuables in the car, even if you only leave the car for a few minutes. Please also make sure you always use safety triangles and high-visibility vests when operating near roads or on road sections.

DAY 1: AJACHES SHIELD, SALINAS DEL JANUBIO, EL GOLFO

Drive from *Puerto del Carmen* or *Arrecife* northward to join LZ-2 west toward *Yaiza* (Fig. 7.13). Once you are on LZ-2 and beyond *Puerto del Carmen*, you will drive past some large cinder cones on your right (eg, Caldera de Gairía), which reach between 500 and 600 masl (Fig. 7.14A,B). After only a few kilometers, a fresh road-cut in basalt becomes available on the right side. Park the car in the open ground in front of the outcrop to inspect the Miocene shield basalt series.

Stop 1.1. Miocene shield basalts (N 28.9415 W 13.7086)

The exposed rocks here are part of the Miocene shield series and belong to the oldest rocks exposed on the island. A massive but jointed olivine basalt lava is seen here (Fig. 7.14C). The olivine has largely decayed and is represented by brown alteration colors that occur as specks throughout the

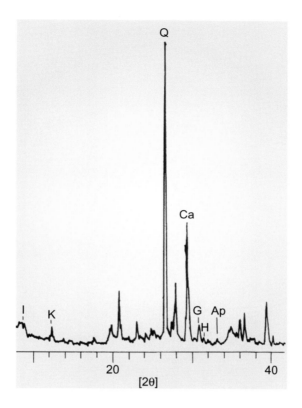

FIGURE 7.11

X-ray diffractogram of a sample of dust taken near the northern airport of Tenerife (*Guamasa*). The chemical composition of the dust is consistent with a source of granite and carbonate compositions, which are widespread rock types on the African continent, but not on the Canaries. From Criado and Dorta (2003).

rock. The joints are filled with calcite from the thick caliche cover above (eg, top of outcrop). The olivine basalt here belongs to the southern extension of the Famara shield volcano that gradually engulfed the older Ajaches shield to create the island we see today (see Figs. 7.1 and 7.2).

Continue on LZ-2. A short while further, a large open cone, Caldera Riscada (approximately 1 Ma old), appears in the distance before you. Make a quick stop at the side of the road near the prominent camel and donkey park on top of a small height on the road.

The "caldera," as it is known here, is rather wide, a characteristic feature of many cinder cones on the island (Fig. 7.14B). It is likely that shallow groundwater was involved in the eruption and added external water. This will increase a volcano's explosive potential, which is likely to carve a wider vent complex (Carracedo et al., 1992). The phenomenon of magma−water interaction is known as "phreatomagmatic" activity (eg, Clarke et al., 2009). The low elevation of the island, often below 300 m, has favored interaction of ground or marine water with magma to produce

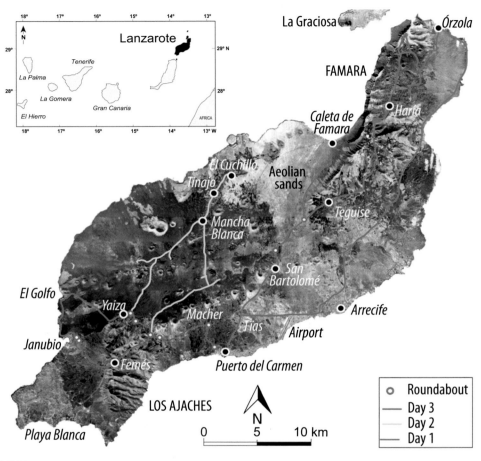

FIGURE 7.12

Main roads and stops of the geological itineraries on Lanzarote. Image modified from Google Earth.

abundant explosive (phreatomagmatic) eruptions with volcanic cones generally larger, flatter, and with wider craters than in the remainder of the Canary Islands. Local tradition probably explains the frequent use of the toponym "caldera" in the recent fissure eruptions of Lanzarote, although these vents are actually volcanic cones.

Continue on LZ-2 until you reach the turn for *Femés* (LZ-702, yellow), and follow LZ-702 toward the south-southwest.

After only a few hundred meters, you will drive up-hill and through the deposits of the Caldera Riscada that we have inspected from a distance a few miles earlier. A small collapse on the left flank in the "caldera's" rim shows a rather attractive onion pattern of individual layers that make up the interior of the caldera's tuff ring. Note that each layer formed from a discrete explosive

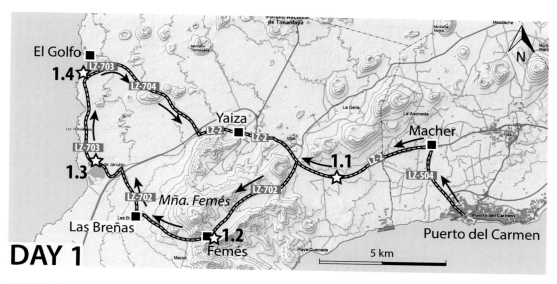

FIGURE 7.13

Map indicating the Stops of Day 1. The day is dedicated to the Miocene Los Ajaches shield volcano. Map from IDE Canarias visor 3.0, GRAFCAN.

FIGURE 7.14

(A) *Google Earth* map view of volcanic cones of the central rift near *Uga*. The use of the term "caldera" for cinder cones is frequent in Lanzarote, likely due to their particular morphometry, showing lower than average height-to-width ratio (ie, wider and lower profiles than cinder cones). This is mainly due to a pronounced central crater area. (B) Caldera Riscada. (C) Miocene shield basalt series at Stop 1.1.

FIGURE 7.15

(A) Valle de Femés, carved by deep erosion into the Miocene Los Ajaches massif, near the village of *Femés*. The U-shaped profile originates from filling of a V-shaped valley with erosional debris and lapilli from Pleistocene eruptions in the vicinity. (B) Section of valley fill showing altered lapilli layers and soils in succession. This material is frequently quarried and used on the island as a soil supplement.

outbursts. If you are willing, a quick stop at the side of the road is optional, but make sure to use safety triangles and high-visibility vests.

Continue toward the south through *Casitas de Femés*. After the settlement, a wide valley opens up (Valle de Femés). The flanks of the valley are made of caliche-covered Miocene basalts, and the valley itself is originally of Miocene age also. Formerly a V-shaped valley (a barranco), it is now filled in the bottom with thick lapilli and aeolian sand deposits (Fig. 7.15A); thus, the valley appears rather flat in the center and, hence, U-shaped (inset in Fig. 7.15A). In the valley center, aeolian sands and altered lapilli are quarried to provide fertile soil supplements for agricultural efforts elsewhere on the island. An optional stop is quite possible if you would like to inspect the workings (Fig. 7.15B). Otherwise, drive into *Femés* and follow signs for *Mirador de Femés*. Park at the mirador for an overview stop.

Stop 1.2. Mirador de Femés (N 28.9134 W 13.7798)

At the mirador (~ 360 masl), Miocene shield basalts can be seen to your left and right (when looking southwest) (Fig. 7.16A,B). On the right, and somewhat behind you, is a more recent cinder cone that has a set of radio and TV poles mounted on its top (Montaña Atalaya de Femés). Looking west, reddish deposits that belong to Montaña Femés (0.2 Ma) are visible in the foreground (Fig. 7.16C). In the far distance, you can see another cone (Montaña Roja, 1.17 Ma) that fed extensive lavas onto the plain in front of you and these reached the sea in places. In this way, these types of lava eruptions create new land on ocean islands, especially if many eruptions follow in quick succession.

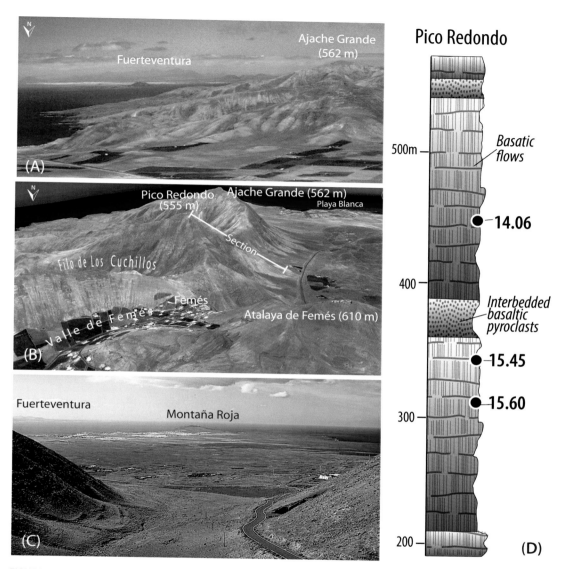

FIGURE 7.16

Perspective views of the Miocene Los Ajaches massif. (A) View from the northeast, with Fuerteventura in the background. (B) *Google Earth* view of Los Ajaches with Mña. Atalaya de Femés in the foreground. The morphology of Los Ajaches massif (> 14 Ma) is characteristic of very old volcanic sequences. U-shaped valleys and sharp (knife-type) interfluves have formed from intense erosion, and the white color is due to a surface layer of soil encrusted with calcium carbonate (caliche). (C) View toward the west from *Mirador de Femés*. In the background, the 1.17-Ma-old Montaña Roja cinder cone and associated lava fields. (D) Stratigraphic section of Los Ajaches massif (indicated in B), with K-Ar ages in Ma (ages from *Guillou, personal communication*).

The church at the mirador was, in fact, built during the 1730−36 eruption. It carries a carved plate at its frontal face commemorating a visit by Bishop Dávila y Cárdenas who was sent from Madrid to report on the eruptive events. Incidentally, he was also the person to suggest the use of lapilli (*picón*) for agriculture, now seen almost everywhere in the Canaries and also right outside *Femés*. The Bishop noted that where thick lapilli cover (*picón*) accumulated on fields, plants would die quickly, whereas fields with a thin lapilli cover would actually experience a boost in plant growth. The function of the *picón* is to hold moisture for longer in the small vesicles than can be accomplished by regular soil, thus providing better irrigation to the growing plants. This concept is so widely used these days in pot plants and gardening efforts all over the world that it is fair to say that the Bishop's little invention here had global consequences.

The bar next to the mirador is usually a good place for having some refreshments. Otherwise, continue south toward *Salinas del Janubio* on LZ-702. Drive down-hill through the Miocene shield basalts and the overlying recent cinder deposits from Montaña Femés. At the base of the hill you will encounter a roundabout, where you go straight across in a westerly direction (direction *Las Breñas*).

Soon after the roundabout, another recent cinder cone appears to your right (Caldereta de Maciot) with a quarried southern face that exposes the interior sequence of a scoria cone extremely well. A stop here is optional. Continue into *Las Breñas* and turn right after only a few hundred meters into the village. Drive through the outskirts of *Las Breñas* village, and leave the village on LZ-703 in a northerly direction.

A considerable lava field will open up before you that is part of the 1730−36 Timanfaya eruption deposits. We inspect the 1730−36 deposits in more detail on *Day 2*. Continue on LZ-703 until you reach the *Mirador de las Salinas* just before reaching the 1730−36 lavas. Park here for an overview stop.

Stop 1.3. Mirador Salinas (N 28.9420 W 13.8190)

The salt plant in the bay below the mirador is, in fact, built into a little lagoon (Fig. 7.17A). A shallow sand bank further out makes this site ideal for salt production because the water here is already somewhat elevated in salinity to start with. The method of salt extraction was introduced in 1895 and has changed little since last century. Large wooden staves known as *palancas de madera* are used, with sea water passing through narrow channels into ponds where the water simply evaporates and the salt crystallizes and thus is concentrated (Fig. 7.17B). The residue is then washed in clean salt pans, where the process is completed when leaving behind bright sparkling crystals of salt only (Fig. 7.17C). Salt production on Lanzarote reached up to 13 million kilograms yearly at maximum production (mainly used for dry-salting fish), but it is currently reduced to approximately 2.5 million kilograms per year.

The cliffs beyond the salt plant are formed by a Middle Miocene basaltic sequence in the lower part, overlain by a Miocene beach with tropical fauna and Upper Miocene basalt lavas resting unconformably on top of the Miocene beach. These probably originate from the Famara Shield volcano in the northeast of the island (Fig. 7.18A). The entire sequence is overlain by Pleistocene lavas from Montaña Atalaya de Femés (0.2 Ma, Zazo et al., 2002). Three marine terraces incise the upper sequence, corresponding to +20 m, +1.5 m, and +1 m, with the +1.5 m and +1 m terraces probably representing the "Last Interglacial" and the "Holocene," respectively (Fig. 7.18B, Zazo et al., 2002).

Continue on LZ-702 toward *El Golfo*, as you are now driving through the "malpaís" (badlands) of the 1730−36 lavas. Drive until you reach the El Golfo vent. Park at the little mirador *Charco de los Clicos*.

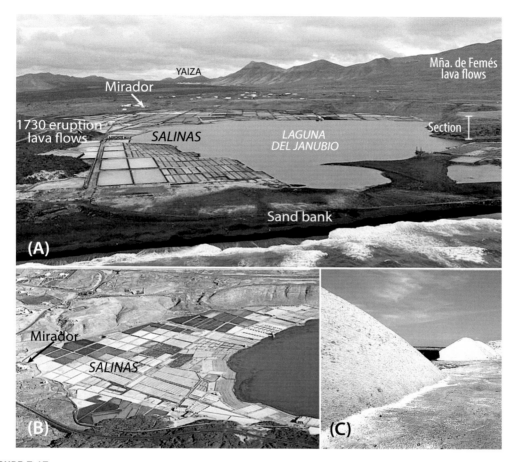

FIGURE 7.17

(A) *Salinas del Janubio.* Salt extraction was introduced here in 1895 and has changed little since then. Sea water is channeled into ponds where the water simply evaporates, leaving crystals of salt (photo courtesy of *Foto aerea de Canarias*). (B) The *Salinas* in full production. Note the various colors of the basins reflecting different stages of the production process (image from *Google Earth*). (C) Approximately 2.5 million kilograms of salt are commercially produced on Lanzarote every year, contrasting to the up to 13 million kilograms during the times of maximum production.

Stop 1.4. El Golfo (N 28.9736 W 13.8313)

El Golfo volcano corresponds to a phreatomagmatic cone half-removed by syneruptive explosions and subsequent marine erosion (Pedrazzi et al., 2013) showing the internal structure of this type of volcano (Fig. 7.19A).

At the parking site a spectacular face is exposed (Fig. 7.19B,C), with lower brownish deposits that grade into darker deposits up-section. The El Golfo vent had contact with seawater and erupted

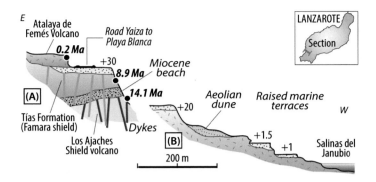

FIGURE 7.18

Stratigraphic sections of the cliffs around *Salinas de Janubio*. (A) Coastal cliff showing a Middle Miocene sequence of lavas from the Ajaches shield at the base of the cliff, unconformably overlain by Upper Miocene lavas of the Tías formation (Carracedo and Rodríguez-Badiola, 1993), probably originating from the Famara shield. Ages from *H. Guillou (personal communication)*. (B) Sequence of raised marine terraces (altitude in meters) in the margin of the *Salinas de Janubio* lagoon. According to Zazo et al. (2002), the +1.5 m corresponds to the last interglacial maximum, whereas the +1 m level corresponds to the Holocene time period. Modified from Carracedo and Rodríguez-Badiola (1993) and Zazo et al. (2002).

in a phreatomagmatic fashion, hence the fine ash and small lapilli (ie, higher eruptive energy results in smaller particles) (Fig. 7.19D). The brown colors, in turn, are from fluid circulation within the El Golfo edifice, leaving behind certain areas that are more oxidized. Note the wealth of peridotite xenoliths (mantle nodules) that occur in some horizons as well as crustal xenoliths from the preisland ocean crust (limestone and siliciclastic xenolith) that are, in part, light-colored and frequently frothy (ie, xeno-pumice; Troll et al., 2012) (Fig. 7.19E). Bomb sags are abundant (Clarke et al., 2009). These form when fragments of rock ejected from a vent eventually land on a wet and ductile substrate (Fig. 7.19F,G).

Take a walk down to the beach along a well-paved coastal path (approximately 1.5 h) to inspect the El Golfo deposits and the heart of the vent complex in detail.

When reaching the broad bay at the end of the paved walkway, you can either turn back or continue your walk across the bay into the very reddish and more oxidized beds of the El Golfo volcano.

Return to the parking site once you have had enough sea air and impressions of phreatomagmatic depositional features.

When leaving the parking area, continue toward *El Golfo* and *Yaiza* (ie, turn left after the parking site and follow LZ-703). At the next junction, join LZ-704 toward *Yaiza*.

Drive up-hill for a short while and over a small saddle between two recent cinder cones before going down-hill. The plains in front of you are almost completely covered with 1730–36 lavas forming extensive badlands (*malpaís*). Consider a quick stop when convenient to get a taste of the program of *Day 2*.

After crossing the "malpaís," join the road to *Yaiza* and take a peek at the hills in the far north. This will be the goal of *Day 3*.

After passing through *Yaiza*, return to *Puerto del Carmen/Arrecife* via LZ-2.

END OF DAY 1.

FIGURE 7.19

Geological features of the El Golfo phreatomagmatic vent complex. (A) Aerial view of El Golfo volcano.
(B) Exposed face of the internal part of the El Golfo volcano. (C) Panorama view of the interior of the crater.
(D) Higher eruptive energy in phreatomagmatic eruptions results in smaller and more intensely fragmented
particles, many from the local basement rock. (E) Frothy crustal xenoliths (xeno-pumice) from the pre-island
ocean sediments (image courtesy of *S. Wiesmaier*). (F) Trails of white and dark xenoliths as well as an internal
fault with a downward displacement on the left side of the photo. (G) Bomb sag implying soft and wet substrate
of the layers at the time of deposition (see Clarke et al., 2009).

DAY 2: THE 1730—36 ERUPTION

Make your way to the roundabout close to Montaña Riscada, where LZ-2 crosses LZ-30 (see start of Day 1). From here, continue toward *La Geria* on LZ-30 (that is to the right and up-hill) (Fig. 7.20). Soon you will pass agricultural fields with many individual, small, half-circle walls (Fig. 7.21). These are there to protect plant seedlings from wind and airborne lapilli (*picón*) and are a trademark of Lanzarote. The consequences of the prolonged volcanism (from 1730 to 1736) on the economy of the island and the local population were devastating; great parts of the most fertile land were destroyed along with 26 villages, and the subsequent famine eventually forced the

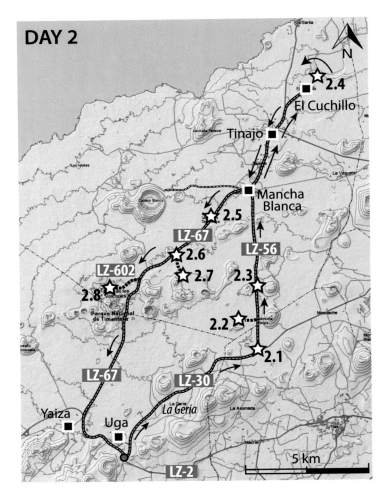

FIGURE 7.20

Map showing the Stops of Day 2. This day is mainly dedicated to the 1730—36 eruption. Map modified from IDE Canarias visor 3.0, GRAFCAN.

FIGURE 7.21

(A) Oblique view (from the west) of the *La Geria* area that is mantled with dark layers of lapilli from the 1730–36 eruption. (B) Vintage impression of *La Geria* (courtesy of *Cabildo Insular de Lanzarote*). (C) The 1730–36 lapilli cover, initially considered to be catastrophic for the island's farmland soon became a traditional mulching technique on Lanzarote because it protects and nurtures vegetation when present in limited quantities. In thick lapilli beds, holes are dug to reach the underlying soil to grow the malvasia (malmsey) grapevines. In these holes, the lapilli (*picón*) trap the atmospheric humidity, thus avoiding the need for irrigation. The holes also protect the plants from airborne particles and drought during times of strong winds, particularly when hot, dry easterly winds (*calima*) blow from the Sahara desert.

majority of the inhabitants to leave the island. However, soon after the eruption, the island's agricultural resources improved again, especially with the introduction of the lapilli cover as a mulching technique. This approach dramatically reduces evaporation losses of nocturnal condensation of water from the atmosphere. The technique is particularly useful in the valley of La Geria, where a large portion of the 1730–36 lapilli accumulated (Fig. 7.21). Notably, the very old and well-known Malvasia grape is still grown here and it is grown in this particular fashion!

Continue on LZ-30. You will see several *"bodegas"* on both sides of the road, where you can taste the Malvasia wine. It is said to originate from an ancient family of grapes believed to be of Greek provenance (from the Greek town of *Monemvasia*). Continue a few kilometers on LZ-30 until you reach another turn. Here, you should turn left onto LZ-56 (direction *Tinajo*), but first park the car at an open ground just at the right side of the junction.

Stop 2.1. The 1730–36 eruption (N 28.9830 W 13.6834)

The exposures on either side of the road are the 1736 tholeiite flows (Fig. 7.22A). Note that the rather crystal-poor medium gray rock contains darker vesiculated domains that also sometimes carry siliciclastic and limestone fragments and small, partly digested relict xenoliths, suggesting the gas-rich darker magma to be charged with H_2O and especially CO_2, potentially with a significant contribution from crustal recycling of sedimentary rocks (Aparicio et al., 2006).

The tholeiite came rather late in the eruptive sequence of the 1730–36 events, allowing for some interaction time with the crust. Although some scientists consider a mantle phenomenon for this late tholeiite pulse (Sigmarsson et al., 1998; Thomas et al., 1999; Lundstrom et al., 2003), recent work by Aparicio et al. (2006, 2010) has shown that crustal recycling was extensive and that the tholeiite magma represents the culmination of magma residence and associated differentiation at crustal levels. There, the magma interacted with the sedimentary rocks below the island and the island's interior to produce the more silica-rich "tholeiites." This is supported by the relatively frequent sedimentary xenoliths in the earlier 1730–36 deposits that occur alongside the easily spotted mantle xenoliths and frequently show a range of disintegration textures, underlying active magma–crust interaction processes during the 1730–36 events (see also Stop 2.7).

The tholeiite lavas flowed dominantly east, encircling the Montaña Negra volcano just north of the road (Fig. 7.22A). If you walk along LZ-30 for a few steps, you will see what looks like mud cracks in the upper crust of the lava flows (pressure ridges; see also chapter: The Geology of El Hierro, Figure 2.48). These are made by internal inflation of the lava flow by progressive liquid supply to the unhardened interior of the flow and the resulting stretching of the already hardened upper crust (Fig. 7.22B).

Return to your car and join LZ-56 (left turn). Drive for approximately 1 km on LZ-56 through more of the extensive 1730–36 lavas. To your right, an old but entirely lapilli-covered cone appears (Montaña Negra), and to your left is the very initial vent of the 1730–36 eruption (Caldera de los Cuervos). To inspect Caldera de los Cuervos, park the car on the open ground just after the municipality mark.

Stop 2.2. Caldera de los Cuervos (N 28.9928 W 13.6849)

To take the hike to Caldera de los Cuervos and to inspect the intravent and vent facies in detail, consider approximately 1–1.5 h for the hike and inspection. Bring water and sun protection.

When approaching the vent (a path exists), you can see a broken-out edge in the crater wall and a large jumble of rock in the lava field a short distance further (Fig. 7.23A). This feature is crucial for the identification of this cone as the initial vent of the 1730 eruption. The contemporary diary of the eruption by Andrés Lorenzo Curvelo, the Parish Priest of *Yaiza*, describes the event: "*On the seventh day of September (1730) a great rock burst outwards with a thunderous sound, and by its*

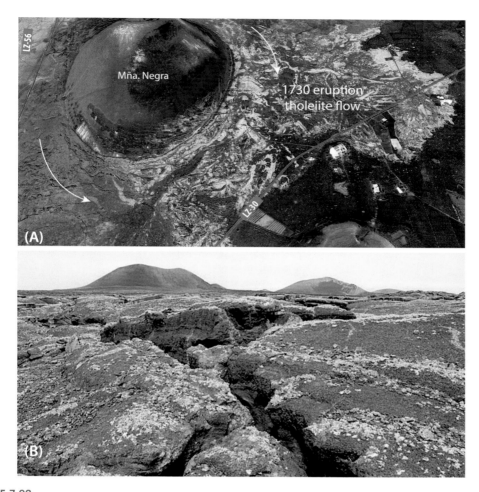

FIGURE 7.22

(A) *Google Earth map* of Mña. Negra encircled by the tholeiite flows of Mña. de las Nueces. Note the distinct wake effect in the flow pattern of the lava. (B) Close-up view of a pressure ridge in the lava flow originating from lateral compression and increased internal pressure of the formerly liquid lava. Pressure ridges usually develop on the surface of flows that are ponded in flat areas or in depressions. The brittle crust frequently buckles to accommodate the migrating and, in part, inflating core of the flow, thus creating a central crack along the length of a tumulus.

pressure forced the lava going northwards to change direction, flowing then to the northwest and west-north west...". This rock now lies 150 m north of a breach in the flank of the cone (Fig. 7.23A), coinciding with the missing part of the volcano that was detached by the pressure of the lava filling the crater. The block, shifted by the lava, did indeed divert the former northbound

FIGURE 7.23

Caldera los Cuervos, the initial vent of the 1730−36 eruption. (A) This vent was identified by the detached and rafted block, described in detail in contemporary eye-witness accounts (see Carracedo et al., 1992). (B) The eruptive conduit of the cone is lined with lava that includes abundant peridotite xenoliths, as shown in (C) These "mantle nodules" represent fragments from the melting region beneath the island, where the bulk of the magma in the Canaries is produced (images in (B) and (C) courtesy of S. Wiesmaier).

flow to the northwest (Carracedo and Rodríguez-Badiola, 1991; Carracedo et al., 1992). A similar event occurred during the 1824 eruption, but that time it occurred because of the pressure of water that was filling an eruptive crater (Fig. 7.7).

Make your way to the gap in the crater wall through a little moat. When reaching the gap in the wall, a path leads down into the vent cavity. The wall is largely made of mafic agglutinates that dip into the crater (ie, intracrater facies).

Once inside the crater, a fresh rock face is exposed to your left (Fig. 7.23B). The rock there is packed with large mantle nodules of variable angularity and size (Fig. 7.23C). This is a spectacular outcrop, so please do not damage this face by taking samples! Instead, pick a sample from the ground, where you will find plenty of mantle xenoliths available as fallen blocks and scree! These peridotite xenoliths are messengers from the upper mantle, maybe from approximately 50−80 km below the island, and they tell us where and how the mantle below Lanzarote began to melt to produce the magmas that we see as lava flows, scoria cones, and feeder dykes today. Note that you are right inside one such outlet, in the vent of a young volcano!

Return to the car when convenient.

Stop 2.3. View on Montaña Colorada and Montaña de las Nueces (N 29.0046 W 13.6843)

Back at the car, look north from your parking site and Montaña Colorada (a reddish hill) can be seen to the right of LZ-56 (Fig. 7.24A,C) and Montaña de las Nueces (the mountain of nuts, for the noise it makes when walking on the lapilli and bubbly lava fragments) is to the left of LZ-56. The latter cone is the vent that produced the tholeiitic flows toward the end of the 1730−36 eruption (Fig. 7.24A,B).

Continue your drive northward and after passing between the two cones, you will drive through extensive lava fields that are derived from Montaña de las Nueces. The main flow-field of tholeiite pahoehoe lavas displays a discontinuous system of lava tubes. Montaña Colorada erupted after the Las Nueces event and closed the Timanfaya activity in April 1736 (Carracedo et al., 1992; Solana et al., 2004).

The spectacular eruption of 1730−36 extended for approximately 14 km along a west-to-east propagating fracture that seems to have acted like a zip, suggesting that the source of magma feeding the fissure eruption was located off the west coast of Lanzarote (Fig. 7.25).

A particularly spectacular feature of the 1730−36 eruption is the *Cueva de los Naturalistas* (Fig. 7.24D), first described by Hernández-Pacheco (1910) and then again by Martín and Díaz (1984). The tube is close by and has elliptical sections with a height up to 8 m and a width of approximately 20 m, and it shows delicate lava stalactites and stalagmites (Fig. 7.24E). Note that access to the public is not advised and we must discourage a visit due to structural instabilities.

You may, however, want to consider visiting the former house of César Manrique at some point during your stay on Lanzarote. César Manrique was a famous local artist and his former home is built inside a lava tube with skylights (Fig. 7.24F). The Fundación César Manrique (www.fcmanrique.org), Calle Jorge Luis Borges 10 in *Tahiche* (just north of *Arrecife*), is open daily from 10:00 to 18:00. An entrance fee of €8 applies for adults, but it is free for children to visit (2015 rate).

Continue on LZ-56 to the village of *Mancha Blanca*. There, take the road north to *Tinajo* ("Avenida de los Volcanes"). Continue through *Tinajo* and leave *Tinajo* village to the north on "Avenida la Cañada." A great view onto El Cuchillo volcano will now open up for you. Continue north on "Calle del Cuchillo" to the village of *El Cuchillo* ("the knife," named so because of the sharp edge of the nearby tuff ring). Cross *El Cuchillo* village in the northeast direction (on "Calle los Arenales") and then follow a dirt track to your right for approximately 400 m until you reach the foot of the El Cuchillo volcano. Park there and walk to the crater rim to inspect this spectacular double maar.

Stop 2.4. El Cuchillo maar (N 29.0861 W 13.6568)

The El Cuchillo volcano is a 175-m-high (see Figs. 7.8 and 7.26A), and the floor of this 1400-m-wide maar crater is only approximately 40 m above present sea level. The maar formed from at

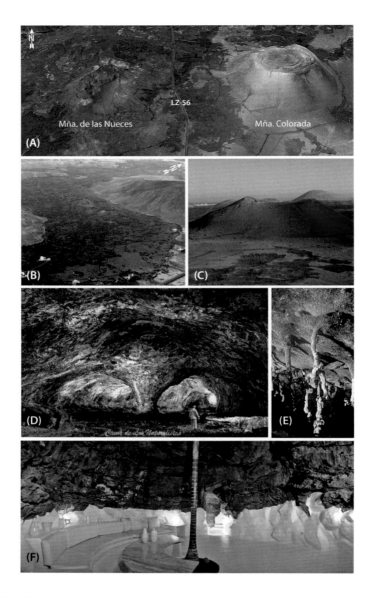

FIGURE 7.24

(A) *Google map* view of Mña. de las Nueces and Mña. Colorada, the final vents of the 1730−36 eruption. The former of the two vents has been intensely quarried for the ornamental value of its shelly pahoehoe lavas, and the latter for its lapilli. (B) The Mña. de las Nueces tholeiite lava was very fluid and was emitted at high rates. It flowed on slope angles of ∼1 degree for approximately 20 km, reaching the coast near the capital *Arrecife* (see Solana et al., 2004). The main flow-field has a discontinuous system of lava tubes (eg, the *Cueva de los Naturalistas*) described in detail by Hernández-Pacheco (1910). (C) Montaña Colorada, which erupted after the Las Nueces event and closed the Timanfaya eruptive activity in April 1736. (D) *Cueva de los Naturalistas* on a vintage postcard. (E) Lavacicles (lava drip structures) likely result from remelting through heating of the roof of the lava tube. This type of formation results in a very irregularly shaped stalactites (images (D) and (E) courtesy of *Cabildo Insular de Lanzarote*). (F) The house of the famous local artist César Manrique was built in 1968 in the Montaña de las Nueces lava tube. Image courtesy of Fundación César Manrique, Lanzarote.

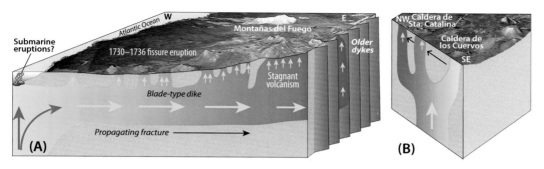

FIGURE 7.25

(A) Sketch illustrating the association of the 1730−36 fissure eruption with a propagating fracture. (B) The plumbing system of the initial stage of the eruption may have been independent and related to a different (southeast−northwest trending) fissure. This two-step volcano-tectonic evolution probably lengthened the 1730−36 eruption well beyond the usual duration of the historical eruptions in the Canaries.

least two different eruptive phases; the most recent one created the deeper western explosion crater (Figs. 7.8A,B and 7.26B).

The walls of the cone are made of thinly bedded palagonitic tuff breccia (see Fig. 7.26B−D), with abundant peridotitic xenoliths and angular blocks of pre-existing rocks. Features typical of high-energy hydrovolcanic explosions are abundant, such as cross-bedding and ballistic impact sags. The surface is very slippery; therefore, it is safer to observe details of these features from the floor of the crater. Once you have had sufficient insight, return to *Tinajo* and *Mancha Blanca*. Very soon after the village entrance to *Mancha Blanca*, turn left onto LZ-67, following the signs for *Timanfaya National Park*.

Continue on LZ-67, and once you leave *Mancha Blanca* village, a great view onto Caldera Blanca will open up for you (Fig. 7.26E, F). An optional visit to Caldera Blanca is recommended. For that, a 5-km track starts from LZ-67 just west of *Mancha Blanca* village (at N 29.0435 W 13.6942). Follow this track for approximately 400 m and park at the available parking site. Then, continue on foot. The track traverses fresh 1730−36 lavas on the way to La Caldereta and Caldera Blanca (Fig. 7.26F). Climbing to the rim of Caldera Blanca, the trail allows the inspection of this spectacular ca. 1-km-wide hydrovolcanic caldera and offers magnificent panoramic views of the 1730−36 and the 1824 vents and lava flows.

Return to the car and to LZ-67, and continue west through agricultural plantation. Just when reaching the first 1730−36 lavas again, the "*Timanfaya Visitor Center*" appears on you right. Park here.

Stop 2.5. Timanfaya Visitor Center (N 29.0331 W 13.7040)

Take a stop here for refreshments and a wealth of background information on the 1730−36 events, including rock samples, maps, and a brief audio visual simulation of an eruption (very attractive for kids and those young at heart). Note that an entrance fee applies (€9 for adults).

After the visitor center, continue on LZ-67 for only a short drive (1 km). Now, you are passing the 1824 vent of Tinguatón volcano on your left. An optional stop is quite possible. For this, you

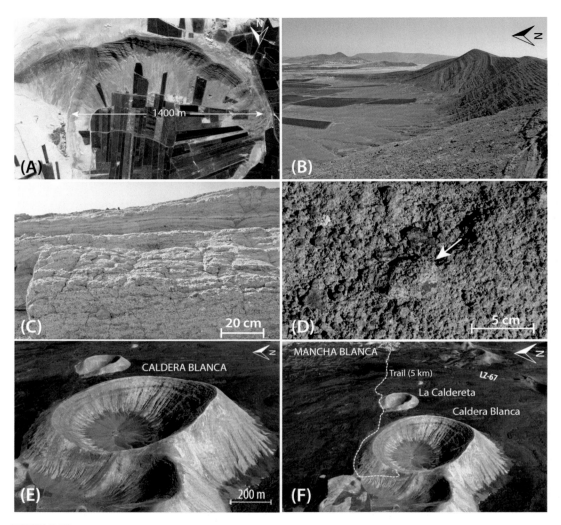

FIGURE 7.26

(A) *Google Earth* map view of the El Cuchillo maar, a spectacular hydrovolcanic eruption that formed a double crater now partially covered with aeolian sands. (B) View from the western end of the crater rim. In the far distance, the Famara shield volcano can be seen. (C) Close-up view of the explosion breccia forming the hydrovolcanic cone. (D) Detail of the breccia showing a mixture of fragmented crystals, predominantly olivine and pyroxene, and an altered peridotite nodule (white arrow). (E) Caldera Blanca, a spectacular 1140-m-wide and 450-m-high phreatomagmatic tuff-cone, has been extensively colonized by lichens. This gives it its conspicuous white color. A tour encircling the rim of the volcano offers splendid views of the 1730 and 1824 vents and lava fields in the surrounding area. (F) Caldera Blanca is easily accessible on foot via a dirt track starting just after the km 8 sign post on LZ-67. After 400 m, the track ends in a parking space at the beginning of a 5-km trail.

have to park in the entrance of a dirt track on your left, just 700 m from the visitor center (there are limited places available to park at N 29.0331 W 13.7040). Walk from there for approximately 400 m to reach Tinguatón volcano. There you can inspect the interesting features derived from magma—water interaction that occurred during the 1824 eruption (see Fig. 7.7 and the summary of the 1824 Lanzarote eruption in the opening section of this chapter).

Return to the car and to LZ-67, and drive through 1730—1824 lava flows. Very soon, you will see a larger sign with *Timanfaya* and the local "devil symbol" (*El Diablo* by César Manrique) right in front of a larger cinder cone (Montaña Tíngafa). Park here for a hike to Pico Partido on the other side of the road (1—1.5 h in total).

Stop 2.6. Hike to Pico Partido (N 29.0189 W 13.7202)

A small and, at first, barely noticeable path (made by people's footsteps, locally known as *señalo*) takes you through the lava "malpaís" in the foreground. After that, walk up-hill over lapilli beds before reaching a spectacular lava channel that emerged from the 1730—36 eruption crater above (Figs. 7.27A and 7.28) and that drained the lava lake that was inside the crater during its active phase. Take a few minutes to inspect the inside of the now drained lava channel where spectacular "shark teeth" are visible. Shark's tooth lava stalactites are drip structures that form during the rise and fall of flowing lava against the walls of the channel.

Overbank flows are clearly recognizable (Fig. 7.27). These structures form inside lava channels while they are still active. The volume of lava moving down a channel will fluctuate so that the channel may be full or overflowing at times, but nearly empty at other times.

Continue up-hill and inside the crater. Once there, you will see the remains of a former lava lake that shows a solid crust on top (now partly collapsed). This lake fed the lava channel we have just seen (Fig. 7.28).

Stop 2.7. Pico Partido lava lake (N 29.0111 W 13.7184)

Walk around a little, but please be extremely careful; the ground is hollow under the chilled upper crust because the lava has drained out in many places (which puts your legs in real danger in case of an involuntary breakthrough).

On the other side of the crater, the lava channel that filled the lake is visible and lava seemingly cascaded down from another vent above.

Looking north, Isla Graciosa and Isla de Montaña Clara should be visible and, if the weather permits, Isla de Alegranza can be seen in the far distance. These three islands form the most norther-ly land of the Canary archipelago. In the foreground, slightly to the right, is the 1824 (Tinguatón) vent again, which pierces through the 1730—36 lavas.

Return to the car the same way you came, and note the larger blocks along the way. Some of them are large gabbro and peridotite xenoliths (Fig. 7.29A,B) that were brought up by the 1730—36 and by the 1824 eruption as well. The peridotite nodules represent fragments from the melting region beneath the island where the bulk of the magma in the Canaries is produced. Also, note an abundance of small frothy pumiceous fragments (xeno-pumice) in the 1730—36 lavas and lapilli beds (Fig. 7.29D—H). These xeno-pumice fragments seem especially abundant in the lapilli beds; check Mña. Tíngafa, where your car is parked, and you will find many small xeno-pumice frag-ments among the loose Tíngafa deposit. It thus seems reasonable to hypothesize that xeno-pumice

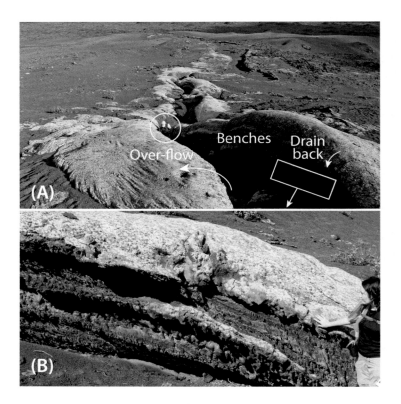

FIGURE 7.27

(A) Close-up view of the lava channel at Pico Partido and its main features. Circled person shown for scale. (B) Ornamental structures inside the lava channel. The "shark's tooth" lava stalactites shown in the photo are drip structures that form during the rise and fall of moving lava against the walls of the channel. Image B courtesy of S. Wiesmaier.

may be a factor in making eruptions more explosive (Aparicio et al., 2006; Troll et al., 2012; Carracedo et al., 2015a).

Continue on LZ-67 and then join LZ-602 toward the *Timanfaya National Park*.

Stop 2.8. Islote de Hilario, Timanfaya National Park (N 29.0061 W 13.7533)

Enter the *Timanfaya National Park* (LZ-602) if you would like to inspect the core area of the 1730–36 events. The entrance fee is €8 for adults and €4 for children (2014 rate). Drive to the visitor center and restaurant, which is also the starting point for the bus tour that takes you around the National Park (Fig. 7.30A). Note that you cannot walk or drive around in the National Park beyond the main road by yourself. The bus tour is, however, included in the fee for the National Park, and you will get to see the heartland of the 1730–36 eruption. Near the restaurant, an artificial geyser

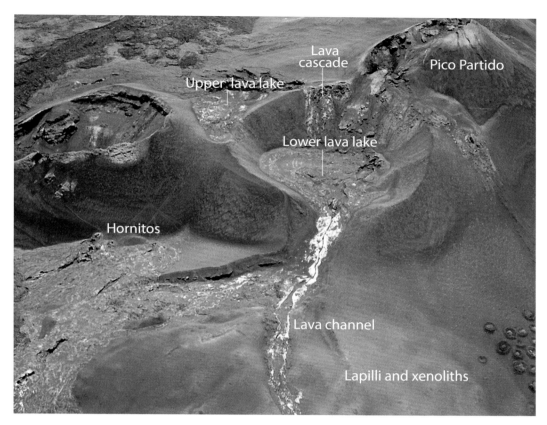

FIGURE 7.28

Oblique *Google Earth* view onto Pico Partido volcano, a spectacular cluster of basaltic cinder cones, lava lakes, and lava channels that formed during the 1730−36 eruption.

is produced by pouring water over a geothermal heat source (Fig. 7.30B). Sometimes dry grass or straw is put on top of this site and is shown to spontaneously combust from the heat of the hot gases that are still emitted. Up to 400°C is reached a few meters below the surface here (Fig. 7.30C).

The gases are, however, dominantly made up of air that is sucked into the system elsewhere, gets heated, and escapes at this site (Fig. 7.30D). Note that the restaurant barbecues their chicken with "hot air" from channelled gas emissions too, in case you would like to take in a geothermal energy kick. Once you have enjoyed the splendor of the National Park facilities, return to LZ-67, and from there to your base by following LZ-67 toward the south and then first the road signs for *Yaiza* and then *Puerto del Carmen* and *Arrecife*.

END OF DAY 2.

FIGURE 7.29

Various xenolith types in the area of *Pico Partido* and *Mña. Tingafa*. (A) Gabbro. (B) Peridotite. (C) Olivine crystals in sand from the disintegration of peridotite xenoliths. (D) Xeno-pumice that has been partially melted and vesiculated while entrained in magma. Examples of crustal xenolith types in 1730—36 lavas. (E) Quartzite. (F) Claystone. (G) and (H) Siliciclastic (formerly sedimentary) xeno-pumice showing intense degassing features and, in places, relict internal bedding is still preserved (images in (G) and (H) from Carracedo et al., 2015a). The various degrees of partial melting and vesiculation observed in xeno-pumice are likely a function of residence time in the basanite magmas, with longer residence causing melting and vesiculation to become more advanced.

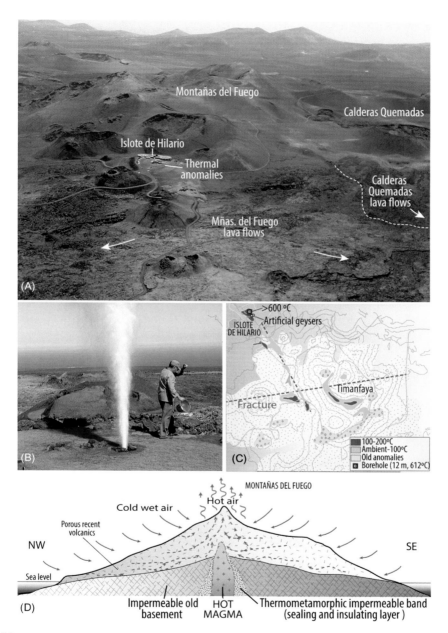

FIGURE 7.30

(A) Aerial view of the core area of the *Timanfaya National Park*. (B) An artificial geyser is produced by pouring water over a geothermal heat source. (C) Map of the geothermal anomalies in Timanfaya. (D) Model explaining the origin of the thermal anomalies by a modest remnant of hot magma at a depth that still remains from the 1730−36 eruption. Modified from Calamai and Ceron (1970).

DAY 3: THE FAMARA MIOCENE SHIELD VOLCANO

Take the LZ-35 toward *San Bartolomé* and continue on LZ-20 north until you reach a round-about with a modern sculpture (Monumento al Campesino, by César Manrique). Here, turn to the right onto LZ-30 (Fig. 7.31). The lavas of the 1730−36 (mainly tholeiite) eruption appear again on your right. Tholeiite lava (also referred to as tholeiite basalt) is similar in composition to alkali basalt but richer in silica and poorer in total alkali element content (Na + K) (Fig. 7.32A). Subalkaline tholeiitic magmas are very rare in the Canary Islands (ie, only seen

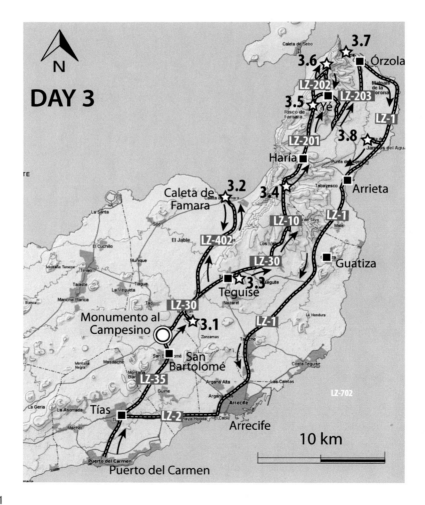

FIGURE 7.31

Map of Day 3 with the route through the central plateau of recent volcanism and the Famara shield volcano. Map modified from IDE Canarias visor 3.0, GRAFCAN.

FIGURE 7.32

(A) Alkali versus silica diagram for lavas of the different phases of the 1730−36 eruption. The latest phases fall into the field of subalkaline tholeiites, which is unusual for the Canary Islands and may be linked to assimilation of silica-rich sedimentary materials (Aparicio et al., 2006, 2010). (B) The tholeiite lava of the 1730−36 eruption is a popular building stone in the Canaries and is obtained in this quarry near *Teguise* (images in (B) and (C) courtesy of *C. Karsten*). The massive and compact lava is used as cladding (C) and paving stone (D) in all the Canaries. The tholeiite lava is easily recognized by the large and bubbly degassing features that appear darker and hint at complex petrogenetic processes for this lava in particular (eg, in (C)).

among recent products in the 1730–36 final eruptive episodes and in the older rocks of El Hierro), and mildly alkaline to highly alkaline basalts are usually predominant. However, tholeiites are abundant in other geological contexts, especially at mid-ocean ridges and in the shields of the Hawaiian Islands.

The unusual tholeiite lavas of Lanzarote are a popular building stone in the Canaries and are quarried here in this particular flow near the road in a large quarry that is not obvious from the road, but it is accessible through a dirt track off to the right just before a large farm (*Complejo Agro-Industrial de Teguise*). This is an active quarry, and we advise that you should ask for permission before attempting to enter.

Stop 3.1. Quarry near Teguise, 1736 tholeiite lava (N 29.0270 W 13.5931)

This massive and compact lava is used as paving and cladding stone in all the Canaries (Fig. 7.32B–D). Note, for example, that the airport of Lanzarote is decorated with this rock too! The base of the lava here shows intense striations and rapid flow of the tholeiite lava is indicated, likely a function of very low viscosity (Solana et al., 2004). The intense degassing features frequently found in the tholeiite lavas suggest that it had a very high gas content, which may be a key reason for this exceptional behavior (see also above).

Return on LZ-30 to the roundabout of the "Monumento al Campesino" and turn left toward *Caleta de Famara* on LZ-402 for a view of the aeolian sands and the spectacular cliffs of *Famara*. At this point, the 1730–36 eruption flow diverts in two branches, one directed toward Arrecife's harbor (*Los Mármoles*) and the other one southward to the sea at *Caleta de Famara* (see Figs. 7.1 and 7.24B). We will follow the course of the second one, crossing aeolian sands that mantle the final course of the flow.

The aeolian sands formed by accumulation in the littoral platforms of the island (see Fig. 7.9B) of a pre-Pliocene, 50–100-m water depth marine tropical fauna, with abundant foraminifera. The climate in the Canaries changed to cold and arid conditions during the Pliocene, giving rise to deposits of finely fragmented shells, exposed during a marine regression and transported inland by wind action (see also chapter: The Canary Islands: An Introduction).

For approximately another 5 km, the road continues through the sand-covered 1730–36 flows, leaving the foothills of the Famara cliff on your right. On arriving to the entrance of *Caleta de Famara*, park the car and follow the footpath to enjoy the spectacular view of the Famara cliff.

Stop 3.2. The Famara Cliff (N 29.1148 W 13.5589)

The Famara cliff is a 600-m-high escarpment formed by a sequence of subhorizontal basaltic lavas erupted in the Upper Miocene between approximately 10.9 and 6 million years ago (Fig. 7.33). The present cliff is the result of intense erosion, likely accentuated by a lateral collapse of the west flank of the massif during this period, followed by subsequent marine erosion, as evidenced by the wide Pleistocene-Holocene abrasion platform (inset in Fig. 7.33).

Return to LZ-30 and follow signs for *Teguise*. Soon, a castle on the hills in the far distance appears. *Castillo de Santa Bárbara* is guarding over the town of *Teguise*, which was the capital of Lanzarote until 1847. *Teguise* has a violent history because it was frequently ransacked

FIGURE 7.33

Panorama view of the Famara cliff. The cliff in the photo is approximately 2 km side-to-side and is 340 m high here. The entire Famara cliff is approximately 15 km long. The inset shows the formation of the cliff by marine erosion, likely after initiation by a giant landslide off the western flank of the old Famara volcano.

and devastated by invasions from pirates and privateers (eg, by Sir Francis Drake, who was known and feared here as "El Draque"), despite the presence of the *Castillo de Santa Bárbara*. Note that the Castillo is built on the top of Mount Guanapay, a volcanic cone of Pleistocene age.

When reaching *Teguise*, you will get to a major fork where you should turn left onto LZ-10 in the direction of *Haría*. Note the castle is now towering over the valley on your right. *Teguise* also hosts the *Casa Museo Palacio Spinola*, a family home for nobles from the mid 1700s that provides a sense of that long-gone period, and you may wish to pay a visit.

Just before exiting the village, a sharp turn to the right brings you up to the castle, which also hosts the *Museo de la Piratería* (www.museodelapirateria.com/en/). The castle (*Castillo de Santa Barbara*) was used in the past as a military stronghold and artillery base that would oversee large stretches of central Lanzarote and as a refuge in case of pirate attacks. The crater of the volcano provided a natural depression to hold rainwater (locally known as *maretas*) that could supply the castle. These *maretas* were used around the island in pre-Hispanic times, and the early Spanish settlers likely "learned the trick" from the aboriginal population.

Stop 3.3. Castillio de Santa Barbara and Museo de la Piratería *(N 29.0578 W 13.5504)*

The castle itself sits on top of an old volcanic crater (~1 Ma) but is covered by weathered brownish lapilli, contrasting the "imported" lapilli at the parking site. Looking southwest, Fuerteventura is visible in the far distance. In the foreground, two types of cinder cones are seen. One type has wide craters of light colors. These are 100−200 ka old. Directly west (between the light-colored craters) are the darker cinder cones of the 1730−36 eruption. To the north, you can see the late Miocene rocks of the northern shield (the Famara shield), now with a radar station on top, which are 10−5 Ma old and are part of the island's oldest rock series. In the far north, the small islands of Graciosa, Montaña Clara, and Alegranza are likely visible again.

Note the light-colored craters in the southwest and west all seem to open to the north and northeast. This phenomenon is probably a function of the regionally predominant tradewinds that come from this direction and cause less ash to be deposited on the northerly flanks of erupting cinder cones. It is this lower side that is also preferentially eroded.

If you are interested, the Pirate Museum is worth a visit. The entrance fee is €3 per person (2014 rate).

When finished here, return to LZ-10 and continue northward toward *Haría*. Soon you will be driving through Miocene basaltic shield stage rocks locally covered with recent lapilli blankets. A short distance after you have passed the military radio station on your left, turn right into *Mirador de Haría*, which has a large parking site and a restaurant.

Stop 3.4. Mirador de Haría (N 29.1276 W 13.5135)

Consider a quick stop for refreshments and inspect the layered stratigraphy of the Miocene rocks exposed below the mirador in the canyon (Fig. 7.34A). The canyon is U-shaped, again due to younger infill accumulating in the bottom of the originally V-shaped valley (Fig. 7.34B). Continue on LZ-10 down-hill for a closer inspection of the Upper Miocene succession ($\sim$7−6 Ma). These rocks all belong to the Famara shield that forms the northern part of Lanzarote.

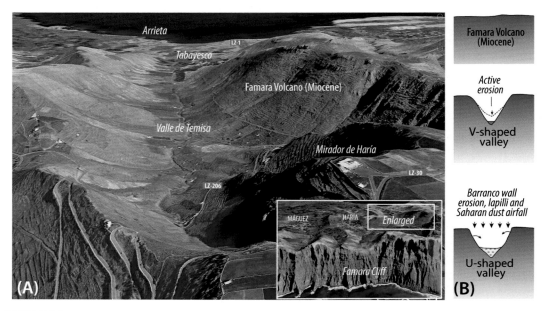

FIGURE 7.34

View from *Mirador de Haría*. (A) The U-shaped Valle de Temisa is a characteristic feature of the Miocene Famara massif. The inset shows the valley head truncated by the Famara cliff (images from *Google Earth*). (B) Sketch explaining the evolution of V-valleys to U-valleys through erosion, volcanic deposition of lapilli, and from air-fall dust from the Sahara dessert.

Drive through *Haría* in a northerly direction, following signs for *Mirador del Rio* ($\sim$7–8 km from here) and for the village of *Máguez* (LZ-201). Pass through *Máguez*, and then pass by several smaller cinder cones (Los Helechos $\sim$80 ka, and Corona volcano $\sim$20 ka, Carracedo et al., 2003) just after leaving *Máguez* village.

Continue to follow signs for *Mirador del Río*. Note that the signs will take you onto LZ-201 and then LZ-202. Stay on LZ-202 for a short stretch and pass the *Finca de Corona* before you reach a small road to the parking place for *Mirador las Rositas* on you left. This is our next stop.

Stop 3.5. Mirador las Rositas (N 29.1965 W 13.4924)

Walk down the footpath from the parking site for approximately 5 minutes until you reach a small wall and paved area along the path before the downward stairs commence. This is our viewpoint. You don't need to go further down.

Looking southwest, columnar jointed lavas of the Famara shield (Miocene) are visible. In the foreground, the Corona lavas are seen to have cascaded down over the Miocene cliff (Fig. 7.35). Looking northwest, the island of Graciosa is visible, with a phreatomagmatic cone at its left tip and with the small islands of Montaña Clara and Alegranza in the background. Looking northeast, the Miocene basalts of the Famara shield are visible again. Note the reddish cinder cone deposits in the lower part of the cliff to the northeast, marking a former vent site in the Miocene basalt succession.

Once you have taken in the view, walk back to the car, but note the steep wall that runs semi-parallel to the path. This is a former bit of lava crust, or a "levee," that was upturned here at the Corona flow periphery. If you look on the valley side of the wall-like feature, you will see it lacks the sharp edges that are typical for a dyke.

Return to LZ-202 and continue toward *Mirador del Río*, which is a little further to the north. Soon you will be on a high coastal road that follows a spectacular cliff and will bring you to the next mirador (*Mirador del Río*). Note the thick caliche cover on top of the cliffs to your right that are formed from hardened aeolian sands with a high organic (shell and coral) component.

Stop 3.6. Mirador del Río (N 29.2140 W 13.4811)

At the mirador you can either walk around a little to catch the views for free or enter the facilities to enjoy the views in comfort and with refreshments (entrance fee €4.50 per person). From there, you will have a spectacular view of the "*isletas*" of La Graciosa, Montaña Clara, and Alegranza (Fig. 7.36).

From the parking place, looking south, you see the open northern rim of Montaña Corona, whose lavas we had seen at the last stop. Take in some of these breathtaking views and enjoy the likely breeze because, from here, it is down-hill toward the coastal regions again.

Continue your journey toward *Yé* and, once there, join LZ-201 again, going toward *Arrieta*.

Soon you will drive down-hill through the lava fields of the Corona volcano (the Corona vent is now to your right). Note the larger rafts within the lavas to your left (at the toe of the flow). The lavas have transported these rafts from the former crater walls all the way down to the low ground near the shore.

Continue through agricultural plantations for a short while. Turn left into LZ-203 toward *Órzola* and pass the large blocks and rafts that we saw from a distance (Fig. 7.37). Note that there are quite a few of them! Drive into the town along the main street and make your way to

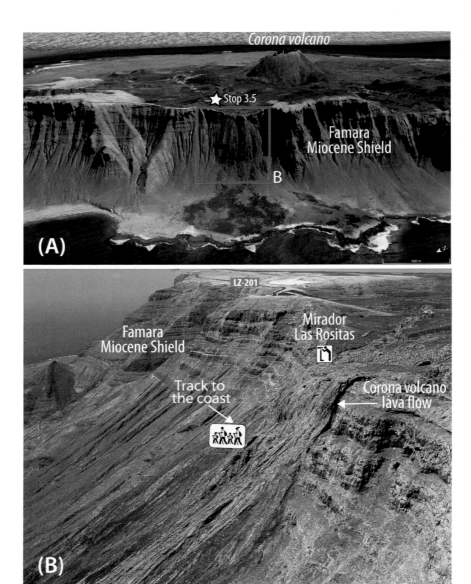

FIGURE 7.35

(A) *Google Earth* view of the eastern Famara cliffs and the Corona volcano lavas that cascaded over the cliff. (B) Closer view of the lava flows of the Corona volcano that cascaded down the Famara cliff. The main lavas of the Corona volcano flowed toward the eastern coast, but a small branch detached westward from the main flow to cascade down the Famara cliff here at the *Mirador de las Rositas*. From here, a trail follows the lavas to the western coast, approximately 340 m below.

FIGURE 7.36

View from *Mirador del Río* looking north. The small islands (isletas) of La Graciosa, Montaña Clara, and Alegranza are visible. The isletas are island volcanoes constructed on a shallow platform carved by marine erosion. The oldest eruptive sequences, resting unconformably over the Miocene Famara massif, have been dated to approximately 45 ka (de la Nuez et al., 1997) and represent the post-Miocene eruptive rejuvenation observed around Lanzarote.

the harbor at the end of the main street. From here you can view, or walk to, the Miocene Famara volcano's eastern cliffs.

Stop 3.7. Fossil beach at Órzola (N 29.2255 W 13.4646)

Park the car and walk along the path heading north to a white sand beach. After a short walk you will arrive at the foot of the cliff and a conspicuous, up to 7-m-thick, white sand layer is seen. This paleo-aeolian dune is interbedded in Upper Miocene basaltic flows (Fig. 7.38A). Large eggshells found in these dunes (Fig. 7.38B), initially attributed to a nonflying fossil ostrich, raised an interesting biogeographical debate because it would suggest a land connection (a land bridge) between the Canary Islands and the African continent between the late Cretaceous and the Miocene. This hypothesis is at odds with the regional geology because it is now well known that all the Canaries rest on oceanic crust, and the stretch of ocean between

FIGURE 7.37

Gigantic lava rafts "floating" in the former Corona lavas. These large blocks of solid lava grew by accretion while they were dragged down-slope by the pressure of the flow. In some eruptions, such as the 1730—36 event, rafts were even witnessed to have rotated while floating in lava.

Lanzarote and the continent has been a deep, continuously active sedimentary basin since the Cretaceous. New evidence seems to solve the contradiction, pointing to these eggs as being related to paleo-flying birds that were some type of large oceanic bird probably resembling modern albatrosses, but larger. These birds likely nested in the aeolian sands, thus allowing some eggs to be preserved (García Talavera, 1990).

Return to the harbor in *Órzola* for a rest and/or refreshment stop and to view the spectacular cliffs of north Lanzarote once more. Also, note the Corona lavas that reach all the way to the sea here in the port area.

Once you feel you have collected sufficient views, pictures, and information, return the way you came into *Órzola* village, but take a left turn at the end of the town and follow the signs for *Cueva de los Verdes* (Cave of the Greens) (LZ-1) in the direction of *Arrecife*. Drive along the coast for several kilometers through the lavas of the Corona volcano (the volcano is still in the background). Then, turn right for 1 km to visit the *Cueva de los Verdes* (entrance fee €8 per person in 2014). This cave is the longest and most spectacular lava tunnel in the Canaries.

FIGURE 7.38

(A) Fossil beach interbedded within Miocene lava flows of the Famara cliff near *Órzola*. (B) Fossil egg of a large oceanic bird, probably resembling modern albatrosses but larger, that nested in the aeolian sands (photo courtesy of *Cabildo Insular de Lanzarote*). (C) Eggshell of *Puffinus* (a mid-Pleistocene shearwater seabird) in a sand dune, shown for comparison. From Meco et al. (2007).

Stop 3.8. Cueva de los Verdes lava tube (N 29.1604 W 13.4385)

The *Cave of the Greens* (the former owner's family name was "Los Verdes"), is part of a 7.5-km-long and up to 35-m-high lava tube that originated from Corona volcano and is one of the world's longest and largest lava tunnels. A 40- to 45-min guided tour shows spectacular views of the complex lava tube that is wide enough at places to host a conference room today. In contrast, several hundred years ago, it served as a refuge for local people from pirate attacks and slave hunters (Figs. 7.39A–C, 7.40, and 7.41).

After the visit, return to LZ-1 and follow the sign to the nearby *Jameos del Agua* (optional). Both cave openings are natural "skylights" (locally called *jameos*) produced by collapses in the roof of the lava tube (see Figs. 7.39A,C and 7.40A,B). A main difference is, however, that the *Jameos del Agua* has a 50-m-long, 10-m-wide tidal seawater lagoon (Fig. 7.39D) that is semiilluminated by the partial collapse of the ceiling and that has been developed as a tourist attraction. Diving investigations found a 1.6-km-long submerged marine prolongation of the lava tunnel (*Túnel de la Atlántida*), and the seaward section of the tunnel terminates at 64 m below sea level in

FIGURE 7.39

The Corona lava tube. (A) Map of skylights in the lava tube. (B) View inside the tube (photo courtesy of *S. Socorro*). (C) Swimming pool inside a large skylight in the Corona lava tube. (D) Tidal seawater lagoon in the lava tube near the coast (*Jameos del Agua*). The inset shows the eyeless and depigmented endemic *Munidopsis polymorpha* crab that is adapted to this niche environment.

a *cul-de-sac*. Although this was initially interpreted as a submarine extension of the onshore tunnel of the Corona eruption, the length and depth of the submarine part of the tube are, however, physically unfeasible. Recent ^{40}Ar/^{39}Ar dating of the Corona lavas determined that the Corona eruption occurred at 21,000 ± 6500 years ago, potentially corresponding with the last glacial maximum at 18,000−21,000 years ago. Accordingly, the lava tube may have formed under entirely subaerial conditions, but it was flooded during postglacial sea level rise. Therefore, it may not be a submarine lava tube *per se*, but it likely represents a flooded onshore feature (Fig. 7.40C; Carracedo et al., 2003).

The eyeless and depigmented endemic *Munidopsis polymorpha* crab lives in this submarine lava tunnel and can be observed in the lagoon inside the lava tunnel (inset in Fig. 7.39D). This, together with a series of notable volcanological features, makes the *Jameos del Agua* worth a visit.

Once your interest is exhausted, we close our visit to Lanzarote here. Return to LZ-1 in the direction of *Arrecife* and return to your base.

END OF DAY 3.

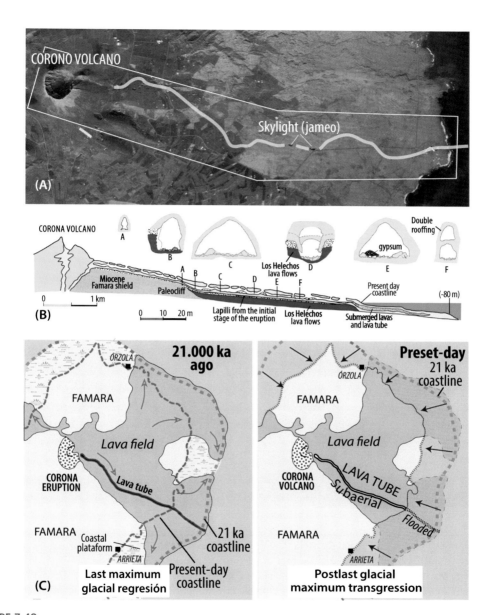

FIGURE 7.40

(A) Map of the Corona lava tube (image from *Google Earth*). (B) Sketch of the Corona lava flows and lava tube. (C) Formation of the lava tube happened under subaerial conditions during the last glacial maximum (ie, low sea level) between 18,000 and 21,000 years ago. (D) The tube was partially flooded and submerged during the subsequent deglaciation and associated marine transgression (from Carracedo et al., 2003).

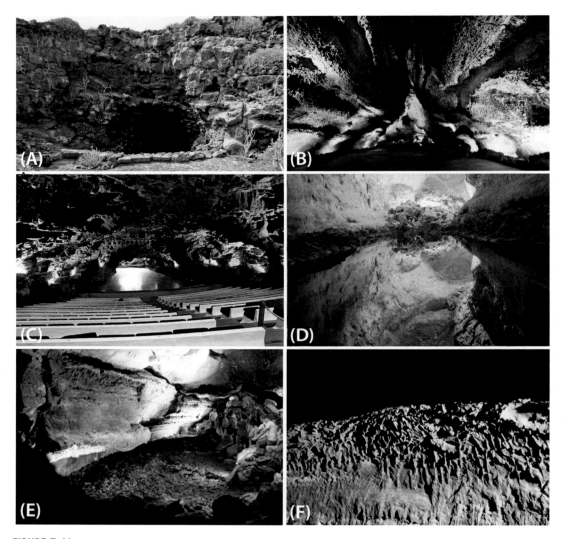

FIGURE 7.41

Different views and features inside the *Cueva de Los Verdes* lava tunnel. (A) The entrance of the cave uses a skylight (*jameo*) that formed by collapse of the roof of the tube. Note the pile of pahoehoe flows. (B) The tunnel is very wide in places, with some sections with an angular ("gothic") roof, typical of levee overgrowth. It is large enough to host a conference room (C). (D) Partitions divide the lava tube horizontally into "double levels"; however, water ponds sometimes reflect the roof of a single-level tube to simulate a double-level cave. (E) The walls of the tunnel are often lined with a layer of hardened lava left by intermittent flow, sometimes deformed by gravitational slumping and sliding down the walls while still plastic. The floor of the tube is usually formed by the congealed surface of the last lava flow and can be a smooth pahoehoe or spinose aa-type, depending on whether the lava was stagnant or moving when it hardened. (F) The roof of the tunnel frequently shows primary ornamental features, such as lavacicles and lava stalactites. These features document secondary melting and dripping from the roof during the active period of the lava tunnel when molten lava was flowing just a little under the roof (ie, where you are now).

THE GEOLOGY OF FUERTEVENTURA

8

The Geology of the Canary Islands. DOI: http://dx.doi.org/10.1016/B978-0-12-809663-5.00008-6

THE ISLAND OF FUERTEVENTURA

As the bathymetry of Fuerteventura and Lanzarote shows (Fig. 8.1), the two islands form a continuous ridge, from the El Banquete seamount in the south to the Isletas north of Lanzarote. The platform between the islands likely formed by marine abrasion during previous glacial periods, when

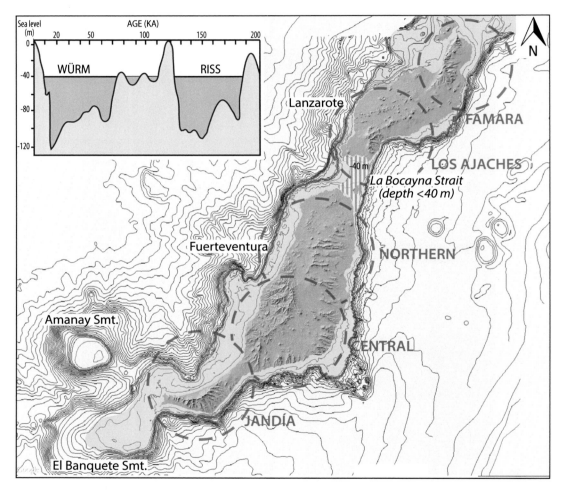

FIGURE 8.1

The islands of Fuerteventura and Lanzarote are separated by a narrow (11 km) and shallow (≤40 m deep) strait, La Bocayna (Bathymetry map from *IDE Canarias visor 3.0, GRAFCAN*). The two islands formed a single island during periods of previous low sea-level standing (eg, during glaciations). The inset illustrates the periods during which the Bocayna strait emerged and both islands were connected, forming a <200-km-long and ridge-like land mass. The red circles indicate the main shield volcanoes.

the sea level dropped by 100 m or more. Thus, the "connected" islands of Fuerteventura and Lanzarote formed by the construction of five overlapping shield volcanoes (red circles in Fig. 8.1) that were aligned in a southwest—northeast direction. Although Fuerteventura and Lanzarote are at present separate islands from a geographical point of view, over considerable periods of geological time the islands were connected and formed a single, elongated, ridge-like land mass more than 220 km long (see inset in Fig. 8.1).

MAIN GEOMORPHOLOGICAL FEATURES

The main geomorphological features of Fuerteventura were originally defined by Fúster et al. (1968a) and comprise: (1) the Jandía peninsula, separated from the rest of the island by an isthmus largely covered with organic aeolian sands (Fig. 8.2A,B); (2) the Betancuria massif, formed by mainly plutonic rocks that have resisted erosion to form a compact group of mountains; (3) the central plain, a 25-km-long depression, interpreted as a fault system (Hausen, 1967) and then later as the scarp of a giant landslide (Stillman, 1999); and (4) the eastern U-shaped valleys separated by elongated and sharp ridges, locally named '*cuchillos*' (knives).

GEOLOGICAL STRUCTURE AND VOLCANIC STRATIGRAPHY
THE BASAL COMPLEX CONCEPT

Since the early geological study of Hartung (1857), which described a basal "syenite and trapp formation, crossed by an extraordinarily dense swarm of dykes", most of the research on Fuerteventura has been focused on this peculiar and intriguing geological feature. The oldest sequence of the island comprises mafic plutonic rocks, submarine sediments, and volcanics that act as the host rock to a dyke swarm in which the dyke intensity reaches 90% or more in places. This type of formation, which also crops out on La Palma and possibly on La Gomera, was formerly considered to mark a common basement in the archipelago (hence, the term basal complex; Fig. 8.2C,D). Following this concept, the basal complex was defined as a formation comprising all the rocks that lie unconformably below the subaerial volcanic sequences. However, part of the subaerial rocks are frequently included in the basal complex too (eg, Gutiérrez et al., 2006), rendering this stratigraphy confusing (Fig. 8.2C).

The finding of Cretaceous fauna in the marine sediments of Fuerteventura's basal complex (Rothe, 1968) suggested some similarities with the sequence characteristic of constructive plate margins, such as the ophiolite complex of the Troodos Massif in Cyprus (Gastesi, 1973). However, as documented by Robertson and Stillman (1979), these early interpretations of uplifted blocks of oceanic basement were problematic because the igneous rocks are considerably younger than the oceanic sediments and a direct link is not evident.

As Staudigel and Schmincke (1984) demonstrated, the "basal complexes" likely represent the seamount stage of growth of the islands and must be present in the entire archipelago, although they only crop out in the islands that went through important uplifting and central erosion, such as Fuerteventura and La Palma. The older (Cretaceous) submarine sediments on Fuerteventura are

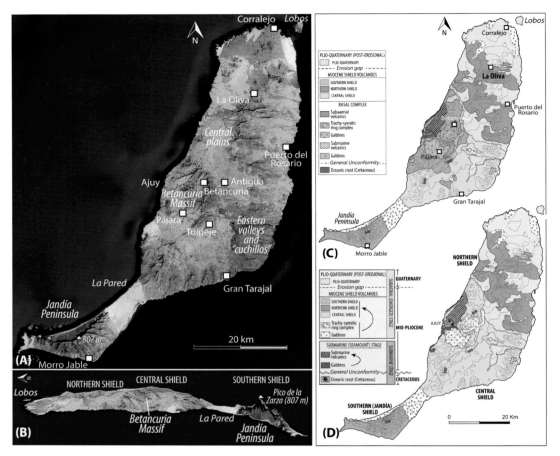

FIGURE 8.2

(A) *Google Earth images* of Fuerteventura. The island is 98 km long and, thus, longer than Tenerife (84 km). Most of the roads are reasonably flat, and it is easy and fast to travel across the island. (B) *Google Earth* profile of Fuerteventura, with vertically exaggerated scale to outline the main geomorphological features. (C). Simplified geological map of Fuerteventura using the concept of a "basal complex." (D) Another approach is to geologically separate the three main temporal units that form the island: (1) the oceanic crust (allochthonous); (2) the submarine (seamount) volcanism of Fuerteventura; and (3) the subaerial volcanism of Fuerteventura. The definition of a "basal complex" has led to confusion in the past, because the subaerial volcanism is then divided into two parts. One part is then included in the "basal complex," whereas the other is not, which creates a "discontinuous" impression, although processes may well have been gradual during the emergence of the island.

part of the oceanic crust and were uplifted by the igneous activity to within the island edifice. In fact, Hansteen and Troll (2003) report contamination signatures in magmas from Gran Canaria that show sedimentary input while also displaying shallow-level crystallization, implying that such an uplifted complex might exist within Gran Canaria island and may even be a feature common to many of the Canary islands.

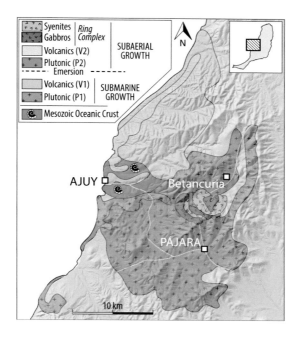

FIGURE 8.3

Geological sketch map of the Betancuria Massif. Note that the older plutonic rocks (P1) correspond to the magma chambers feeding the submarine volcanic suite (V1), whereas the younger plutonic group (P2) fed the subaerial volcanism (V2). Modified from Gutiérrez et al. (2006).

Stillman et al. (1975) synthesized the evolution of Fuerteventura as "a history of transition from normal ocean-floor sedimentation at the foot of the eastern Atlantic continental rise to the build-up of a discrete oceanic island". Therefore, if the "allochthonous" oceanic crust is considered unrelated (see Fig. 8.2D), then what remains are the characteristic formations of construction of an oceanic island, such as the seamount stage and the onset of subaerial volcanism, making the use of the term "basal complex" arbitrary and likely redundant (Carracedo et al., 2001). Consequently, the volcanic stratigraphy of Fuerteventura can be broken down into four progressively younger units: (1) the uplifted (Mesozoic) oceanic crust; (2) the submarine (and transitional) volcanics; (3) the subaerial Miocene shields; and (4) the posterosional volcanic rejuvenation. Note that the older plutonics (P1 in Fig. 8.3) represent the chambers feeding the submarine volcanics (V1), whereas the P2 plutonics relate to the younger surface volcanics (V2).

THE UPLIFTED (MESOZOIC) OCEANIC CRUST

As described by Robertson and Stillman (1979), the composition of this fragment of oceanic crust is tholeiitic and, hence, characteristic of mid-ocean-ridge basalts, which are otherwise very rare in the Canaries. The age has been estimated by the presence of fossils (ammonites and foraminifera) as of Early Jurassic to Cretaceous (Steiner et al., 1998). Therefore, the uplifted crust was formed in the earlier stages of the separation of Africa and America, whereas the volcanic island of Fuerteventura formed more than 100 Ma later.

The basal sequence comprises mid-Atlantic basalts, overlain by a thick (more than 1.5 km) deep-sea sedimentary sequence (Fig. 8.4), which is part of a fan on the rifted African continental margin. Sediments, such as terrigenous and calcareous clastic deposits and black shales with locally heavy mineral concentrations (eg, zircon sands), accumulated probably from the early to mid-Cretaceous (Robertson and Stillman, 1979; Steiner et al., 1998).

FIGURE 8.4

(A) Sequence of overturned Mesozoic ocean floor sediments showing rhythmic alternation of green and white banding (mouth of Barranco de Ajuy). (B) Close-up showing detail of the tilted white (quartz-rich) and green (chlorite-rich) layers with linear striations on the exposed surface. Images courtesy of C. Karsten.

A characteristic sequence of the ocean sediments shows fine-grain rhythmic deposits, with alternating green and white banding (Fig. 8.4). The green layers are mainly formed by chlorite in shale and clay deposits, while the white bands show high quartz contents, indicating a continental origin because quartz is not usually present in volcanic rocks of the Canary archipelago. In places, heavy mineral-enriched layers occur ("zircon sands"), implying highly dynamic transport conditions. These rhythmic deposits were likely formed by many repetitive turbidity events from Africa, and we emphasize that the presence of igneous and sedimentary oceanic crust is unrelated to the Canarian magmatism. Instead, the Canary ocean island volcanism was the cause for the uplift of these older rock sequences.

SUBMARINE GROWTH

An interesting and unusual feature that can be inspected on Fuerteventura is the transition from submarine to the subaerial volcanism.

The earliest volcanic manifestations are thick sequences of basaltic pillow lavas and hyaloclastites resting on an erosive unconformity over the Mesozoic sedimentary crust (Fig. 8.5A,B). Plutonic activity coeval with the generation of the submarine volcanic sequence (see Fig. 8.3) comprises alkaline rocks that crop out along the western coast south of *Ajuy*, such as pyroxenites, nephelinites, and carbonatites (Fúster et al., 1980; Le Bas et al., 1986; Demény et al., 1998). Carbonatites, igneous rocks with more than 50% carbonate minerals, are rare in the Canaries and likely formed on Fuerteventura by liquid immiscibility from a CO_2-rich alkaline silicate magma (Fig. 8.5C).

The submarine succession forms a 2-km-thick sequence (Gutiérrez et al., 2006) that includes intermediate to shallow depth volcanics, as indicated by the vesicularity of pillow lavas (Fig. 8.5B). Units A to F in Fig. 8.5A are progressively shallower, with unit F documenting a shoaling phase through the presence of coral reefs that likely correspond to the first stage of emergence (Fig. 8.5D). Transitional units overlie the submarine sequence and are topped by subaerial volcanic and sedimentary rocks (Miocene to recent, Figs. 8.3 and 8.5A).

THE SUBAERIAL GROWTH OF FUERTEVENTURA

The initial stages of subaerial volcanism erupted large volumes of basaltic effusive lava, constructing an island that was considerably larger and higher than that of today, probably reaching elevations similar to those of present-day Tenerife (eg, Javoy et al., 1986). After this intense eruptive phase (the shield-building stage, lasting only a few Ma, perhaps), in which growth of the edifice through volcanic activity outpaced destruction through erosion and during which most of the volcanic edifice formed, eruptive rates and frequency drastically decreased. Magmas became more evolved (siliceous and thus more viscous), contributing to an increase in volcano height. In fact, Stillman (1999) suggests that a peak as high as the present Mount Teide on Tenerife existed in the central part of the island and was rapidly denuded in less than 2 million years (Ma).

Volcanism then declined and, finally, the island entered extended periods of volcanic inactivity, during which mass wasting through giant landslides and erosion outpaced volcanic growth. This caused the island to become deeply eroded, eventually exposed a window into submarine volcanics

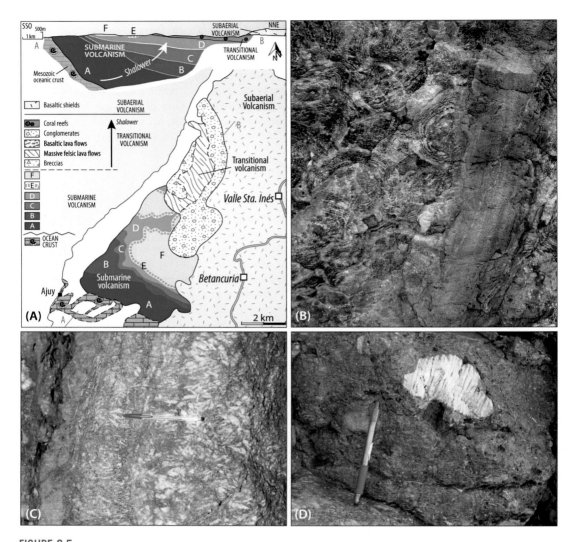

FIGURE 8.5

(A) Geological sketch map showing the submarine and transitional volcanic units defined by Gutiérrez et al. (2006). The northeast—southwest cross-section indicates the gradual emersion of the island. (B) Miocene pillow lavas in Playa del Valle. (C) Carbonatite intrusion in Punta de la Nao. (D) Coral fragment in volcanic deposit, likely xenolith from a coral reef, characterizing the transition from shallow marine to subaerial volcanism in Barranco de la Fuente Blanca. Photos (C) and (D) courtesy of R. Casillas.

FIGURE 8.6

Approximately 2 km south of *Pájara*, in the region east of the km 11 road sign on FV-605, the transitional and subaerial volcanic units unconformably overlie the submarine sequence crossed by the feeder dykes of thousands of submarine and subaerial eruptions. These now form an extremely dense northeast—southwest striking dyke swarm (see also Figs. 8.18, 8.33 and stops 2.2 and 2.3). Google Earth image.

and the roots of the subaerial volcanoes, particularly in the central and northern parts of the island. There, a succession of plutonic intrusions (pyroxenites, gabbros, and syenites) represents the frozen magma reservoirs that fed the Miocene volcanoes of Fuerteventura (see Fig. 8.3).

For instance, the Miocene gabbro-pyroxenite pluton near *Mézquez* (P2 in Fig. 8.3) is interpreted as part of the feeder zone to the subaerial volcanism and is a 3.5×5.5-km shallow-level intrusion. The intrusion was, to some degree, controlled by the regional east—west extensional tectonic setting that affected Fuerteventura during the Miocene (eg, Stillman et al., 1975; Stillman, 1987; Stillman and Robertson, 1977; Fernández et al., 1997), which may have facilitated rapid magma production and consequent ascent and eruption.

The massive and successive plutonic intrusions (Fig. 8.7) and the extraordinary number of dykes caused intense alteration, analyzed in detail by Le Bas et al. (1986) and Hobson et al. (1998). Emplacement of the plutons at high levels in the crust caused hydrothermal metamorphism, likely due to fluid circulation along the dykes because the intensity of metamorphism correlates with the density of dykes. Hence, most of the rocks of the submarine complex were affected by fluid over-print, but notably to variable degrees (Javoy et al. 1986).

Migmatites crop out in a 200-m-wide aureole at the contact between two gabbroic intrusions and the basaltic dyke swarm (Figs. 8.7 and 8.8). Migmatites form under extreme temperature and/or pressure conditions during progressive metamorphism, where partial melting occurs in pre-existing rocks. Incipient melting leads to segregation and crystallization of a leucosome, the lighter-colored part of migmatite, which on Fuerteventura displays a banding of syenitic veins alternating with and migrating through basaltic or pyroxenitic host rocks (zebra structure, Fig. 8.8B).

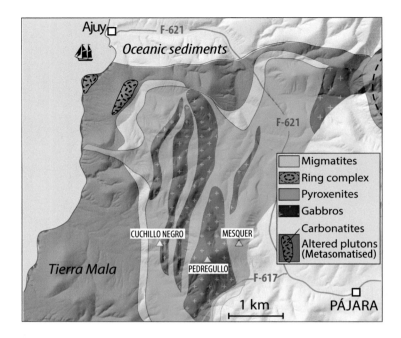

FIGURE 8.7

Geological sketch map of the Betancuria Massif showing a suite of plutonic intrusions (gabbros, pyroxenites, and syenites) in the area between *Pájara* and *Mesquer*. The migmatites (zebra rock see Fig. 8.8) probably formed through the heating and partial melting, and associated internal segregation within the basalt country rocks during intrusion of the associated pyroxenite plutons. Melting in the aureole likely occurred in the presence of abundant fluids (after Le Bas et al., 1986; Hobson et al., 1998).

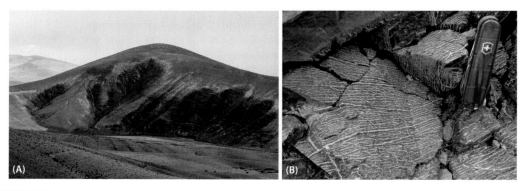

FIGURE 8.8

(A) The black pyroxene pluton of *Mesquer*. (B) Veining in migmatite (zebra rock) in the aureole to the pyroxenite in Bco. de La Arena. Partial melting with concurrent deformation of the basaltic protolith has resulted in feldspathic leucosomes that are present as felsic segregation veins.

THE MIOCENE VEGA DE RIO PALMAS RING COMPLEX

This trachytic—syenitic ring complex, the last stage of the preserved plutonic history of Fuerteventura, intrudes the submarine and transitional volcanic groups and its top cuts across sub-aerial lavas (Figs. 8.9 and 8.10). Thus, the felsic ring complex likely represents the magmatic

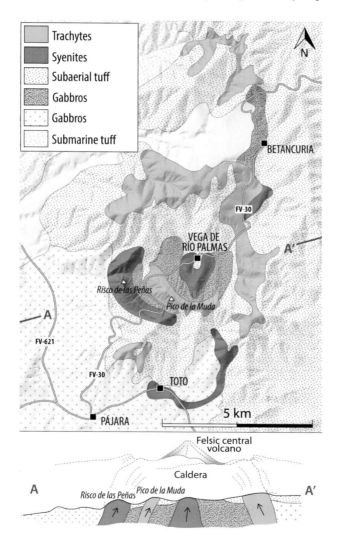

FIGURE 8.9

Geological map and west—east cross-section of the ring dyke complex of *Vega de Río Palmas*. Note the central intrusion is encircled by three concentric ring intrusions (modified after Muñoz, 1969). The upper volcanic structure in the cross-section is inferred from geological evidences (see text for details).

FIGURE 8.10

(A) *Google Earth* image of the ring complex of *Vega de Río Palmas*. (B) Outcrop along the raod at *Risco de la Muda*, a semi-circular trachyte intrusion of the central ring complex.

conduit to a felsic volcanic complex similar to the Las Cañadas—Teide complex on Tenerife, or the Vallehermoso complex on La Gomera, or indeed the Tejeda Caldera on Gran Canaria (see cross-section in Fig. 8.9). The Fuerteventura felsic volcano may have reached elevations of 3000 m or higher above sea level, as suggested by stable isotope data from the dykes (Javoy et al., 1986; Stillman, 1999). Subsequent erosion, probably including massive landslides, removed approximately 3500 km^3 of lavas and volcaniclastics, stripping the western sectors of Fuerteventura down to the rocks of the pre-shield phase (Stillman, 1999).

THE MIOCENE SHIELD VOLCANOES

The subaerial growth of the island gave rise to three large, overlapping shield volcanoes, the central, northern, and southern (Jandía) shields (Figs. 8.11 and 8.12). This volcanic disposition seems to be characteristic of ocean islands, and is seen on the island of Hawaii (Mauna Kea, Mauna Loa, Hualalai, and Kilauea), Tenerife (Teno, Central, Anaga), La Palma (Taburiente, Cumbre Vieja), and Lanzarote (Ajaches, Famara). The explanation for such a frequent arrangement of overlapping

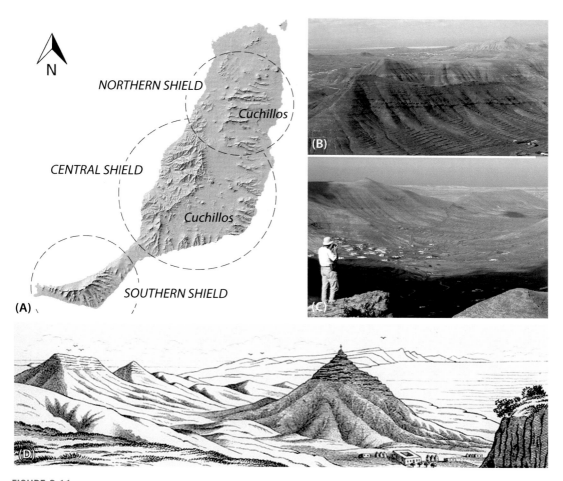

FIGURE 8.11

(A) Three large shield volcanoes constructed Fuerteventura during the Miocene (from Ancochea et al., 1996). The chain of volcanoes continued to the north with the Los Ajaches and Famara shield volcanoes on Lanzarote. (B) Deep erosion of the northern shield of Fuerteventura led to the formation of sharp interfluves ("cuchillos") and wide U-shaped valleys (C). (D) The cuchillos of Fuerteventura attracted the interest of *Hartung* (D), who depicted the phenomenon in this illustration from 1857. Image courtesy of Cabildo Insular de Fuerteventura.

shields may lie in the fact that basaltic magmas are dense, and as a basaltic volcano grows, the excess pressure required for magma to ascend to the surface must surpass the increasing lithostatic pressure of a growing edifice (see Fig. 5.38), thus eventually establishing a "density filter" that may force the magma to migrate laterally to start building a new volcano (Davidson and de Silva, 2000; Pinel and Jaupart, 2000).

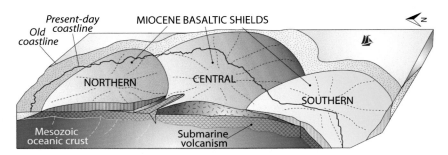

FIGURE 8.12

Sketch of the three shield volcanoes of Fuerteventura illustrating the relationships with each other and with the submarine substratum of the island.

The building of these huge shield volcanoes is relatively fast compared with the long postshield periods in which eruptions decline in number and volume. Then, erosion gradually outpaces volcanic activity. Late rejuvenation cycles of volcanism generally erupt only a minor fraction compared to the volume of the shield stage and are usually separated from the shield stage rocks by a long period of eruptive quiescence. Thus, the configuration reached a zenith during the shield stage, and it has been exposed for many Ma to mass wasting and erosion. Only a small fraction of these shield volcanoes remains exposed at present, a landscape of sharp interfluves (*cuchillos,* meaning knives) separated by wide U-shaped valleys (Fig. 8.11B–D).

Fúster et al. (1968a) included in their "Basaltic Series I" the several hundred meters thick lava sequence that formed the bulk of the emerged island of Fuerteventura during the Miocene (eg, 800 m in the Jandía peninsula). These authors labeled the lower horizontal basaltic lava sequence with a slight seaward dip as the "Series I." Ancochea et al. (2003) applied stratigraphic and structural criteria to separate this formation into three large volcanoes, the central, northern, and southern shield volcanoes (Figs. 8.11–8.13), a disposition that is notably consistent with the published radiometric ages (Coello et al., 1992).

Progress in radiometric dating of the oldest formations of the island met with significant problems, leading to the initial publication of radiometric ages of more than 40 Ma (Eocene) for the older subaerial lavas (eg, Abdel-Monem et al., 1971). More recent dating efforts (eg, Coello et al., 1992) on the Miocene basaltic sequences of the shield volcanoes obtained an age of 20.4 Ma for the oldest rocks on the island (Fig. 8.14).

It is worth noting that the duration of the shield stage still seems to be rather long (up to 7 Ma in the central shield). This could be because the separation of the shield and postshield phases is difficult to determine exactly in this old island. Lavas of the older formations are so deeply weathered that they are frequently unsuitable for radiometric dating, whereas fresh samples likely correspond to the latest stages of the postshield growth, thus 'artificially' extending the duration of the shield stage due to limitations in radioisotopic approaches (see MacDougall and Schmincke, 1976; Guillou et al., 2004a). Moreover, ongoing radiometric and palaeomagnetic (geomagnetic reversal) dating of the Fuerteventura shield volcanoes now yielded a similar age of 20.19 ± 0.30 for the oldest basalts of the island (Guillou et al., in preparation).

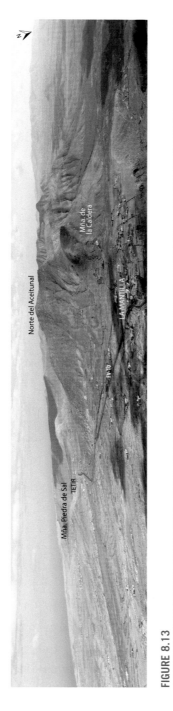

FIGURE 8.13

View from Mña. de la Muda onto El Aceitunal. This part of the island belongs geologically to the northern shield. In the foreground, the small village of *La Mantilla*. To the right, one can see the central plain, and in the far distance the central shield can be seen.

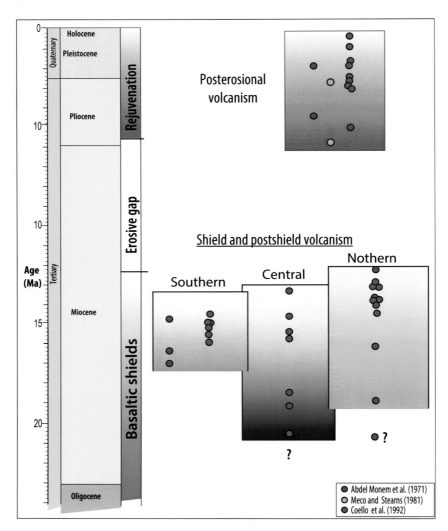

FIGURE 8.14

Simplified stratigraphy of Fuerteventura, based on radiometric ages from Abdel-Monem et al. (1971), Meco and Stearns (1981), and Coello et al. (1992).

THE PLIOCENE TO RECENT POSTEROSIONAL REJUVENATION

After a long period without volcanism, basaltic eruptions resumed in the northern half of the island approximately 5 Ma ago. The considerable duration of the erosional hiatus (Fig. 8.14) produced an extensive coastal abrasion platform, now seen as the unconformable contact of the Miocene to

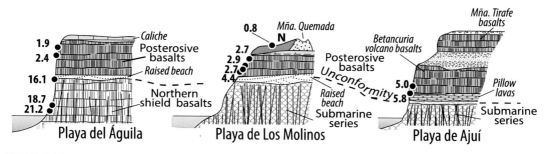

FIGURE 8.15

Geological sections along the western coast of Fuerteventura, indicating the age, polarity, and relationships between the Miocene submarine and subaerial volcanism (after Fúster and Carracedo. Ages from Abdel Monem et al., 1971; Meco and Stearns, 1981 and Coello et al., 1992).

Pliocene volcanism (Fúster and Carracedo, 1979). Marine erosion and correlative coast retreat formed a cliff that extended along most of the western coast, where post-erosional volcanism is overlying the 20-m raised beach and the older Miocene and pre-Miocene formations (Fig. 8.15).

PRE-BRUNHES ERUPTIONS

The initial Pliocene eruptions are scattered in the central part of the island and along the coast around *Ajuy*. This episode of rejuvenation volcanism has been divided according to the Brunhes/Matuyama boundary at approximately 0.78 Ma (Fúster and Carracedo, 1979). Initially, Fúster et al. (1968a) divided the pre-Brunhes eruptions (their "Series II") into two subseries (B1 and B2). The former corresponded to exclusively effusive eruptions forming small shield volcanoes and lavas that were flowing even in very gentle slopes, adapting to the post-erosive topography, such as the Cercado Viejo volcano near *Casillas del Ángel* (Figs. 8.16 and 8.17A), and the still preserved cinder cones, such as the "Volcanes de Tetir" at the center of the island (Figs. 8.16 and 8.17B). The younger and more explosive subseries B2 occurs mainly in the southeast of the island, east of *Pájara* (Figs. 8.16 and 8.17C), and in the north near *Corralejo*, forming the "Malpaís de la Arena" and "Malpaís del Bayuyo" (Figs. 8.16 and 8.17D).

BRUNHES ERUPTIONS

These recent volcanoes are well preserved, frequently forming extensive lava fields, impractical for farming, and locally known as *malpaíses* (badlands). The greatest concentration of recent volcanoes occurs in the northern part of the island (see Figs. 8.16 and 8.17D).

THE INTRUSIVE FACIES OF FUERTEVENTURA

Dykes on Fuerteventura are extraordinarily abundant, particularly in the Betancuria massif, where they reach more than 90% of the rock mass in places (eg, Stillman 1987). Intrusions are so tightly

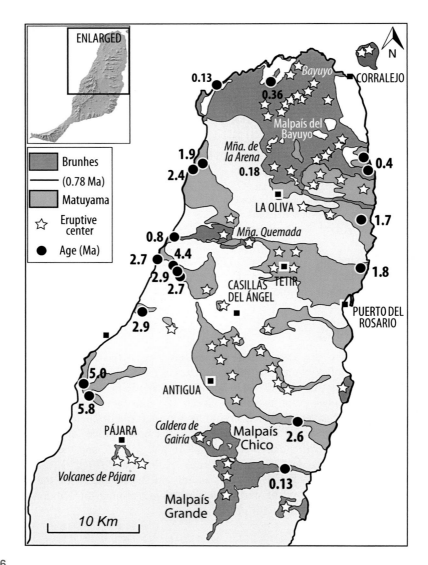

FIGURE 8.16

Post-erosional volcanism in Fuerteventura can be separated into two age groups on the basis of geomagnetic polarities by using the Matuyama/Brunhes limit at 0.78 Ma (ages as in Fig. 8.15).

packed there (see Fig. 8.6) that only small screens of the host rock can be observed in many localities (eg, Figs. 8.18 and 8.19). Most dykes are approximately 0.5–1 m thick; the greater number is basaltic and a small fraction is trachytic to phonolitic. Usually they are intensely overprinted and original minerals have frequently disappeared.

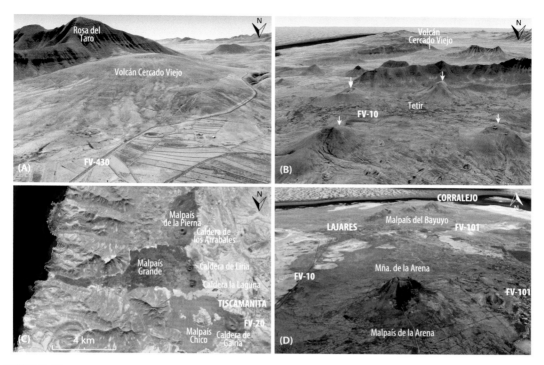

FIGURE 8.17

(A) Small shield volcanoes characterize the stage of rejuvenation volcanism on Fuerteventura, like Cercado Viejo, near *Casillas del Ángel*. (B) The first stage of rejuvenation, Volcanes de Tetir, at the center of the island (shown with arrows) is still preserved. (C) The most recent eruptions (approximately 130 ka) grouped mainly at the southeast of the island (eg, around the Malpaís Grande and Malpaís Chico volcanoes and lava fields) and (D) in the northern part of Fuerteventura, near *Corralejos*, forming the Malpaís de la Arena and the Malpaís del Bayuyo. Images modified from Google Earth.

Because dyke density reaches a maximum in the submarine formation of the Betancuria massif, and because of the presence of Cretaceous marine sediments, some fundamental errors arose in early interpretations of the basal rocks of Fuerteventura. Hartung (1857) interpreted this intrusive dyke complex as a stratigraphic unit (his "Trapp formation"), implying they were tilted lavas. Hausen (1958) consequently described the dyke swarm as lava flows (spilites), tilted from their original horizontal emplacement by the Hercynian orogeny, but also suggested some similarities with ophiolite complexes. Subsequently, several authors considered this formation as a "sheeted dyke complex," a frequent component of ophiolite complexes worldwide. Detailed comparisons between the Betancuria massif and accepted ophiolite complexes that are generally held as representative of oceanic crust in constructive plate margins, such as the Troodos Massif of Cyprus, were then made (eg, Gastesi, 1973).

FIGURE 8.18

The extremely dense dyke swarm in the Betancuria Massif (pink) represents the former feeder system to the submarine and early subaerial volcanism of the island. Dyke swarms in blue correspond to the later subaerial shield-stage volcanism. The green rectangle refers to Fig. 8.6. Modified from Fúster et al. (1968).

However, Fúster et al. (1968a) already pointed out that the "dyke complex" is not a stratigraphic unit *per se*, because the dykes cross different stratigraphic formations simultaneously. Today, we know that similar dyke swarms occur in the active rifts of ocean islands accessible through *galerías* or in the deeply eroded structures of rift zones, such as the northeast rift zone on Tenerife (eg, Carracedo et al., 2011b; Delcamp et al. 2012) or in the Taburiente caldera on La Palma (Staudigel et al., 1986).

Such a high dyke density as seen on Fuerteventura is nevertheless best achieved in an extensional regime. Otherwise, the successive intrusions will progressively increase compressive stresses, hindering and eventually preventing new injections. Because it is very likely that the dyke complex on Fuerteventura fed the numerous eruptions that constructed the submarine and subaerial structure of the island, the origin of the required extensional regime is likely different from a constructive plate margin. Two different options have been considered, which are a fracture propagating from the Atlas to the Canaries (Anguita and Hernán, 1975, 2000) or doming from a mantle plume (Holik et al., 1991; Carracedo et al., 1998).

FIGURE 8.19

(A) Dyke swarm intruding Miocene submarine volcanics, Playa del Valle. (B) Pliocene lavas resting unconformably on the Miocene dyke complex, Bco. de Malpaso, Playa de Ajuy. (C) Densely packed dykes leaving only small screens of the submarine host rock near *Pájara*. (D) Sills in steeply dipping oceanic sediments. These must have intruded prior to uplift and deformation of the oceanic sedimentary rocks in Cueva de Caleta Negra, near *Ajuy*. Images in (A), (B), and (D) courtesy of S. Wiesmaier.

Although there is no significant evidence against a mantle plume accounting for the origin for the Canaries, there are fundamental constraints against the tectonic influence of the nearby Atlas region since the early Miocene. As pointed out by Gutiérrez et al. (2006), the Atlas deformation postdates the construction of Fuerteventura. From the Oligocene to early Miocene, the Africa–Eurasia convergence was mainly absorbed across the ALKAPECA domain (Alborán, Kabylies, Peloritan, and Calabria) and the Iberian–Balearic margin. The Atlas system deformation occurred late in the Miocene and, more specifically, during the Pleistocene and early Quaternary (Frizon de Lamotte et al., 2000), and therefore most probably after the formation of the intrusive complex of Fuerteventura.

FELSIC INTRUSIONS

Outcrops of trachyte and phonolite intrusions are scant on Fuerteventura and generally present as plugs or domes. The most spectacular one is the Tindaya mountain, a Miocene 400-m-high elongated and eroded ridge towering over the Esquinzo plain. Tindaya is a classic case of relief inversion that results from millions of years of differential erosion (Fig. 8.20). The trachyte of the Tindaya quarry is a popular ornamental building stone in the Canaries, and its decorative value is based on the abundant and colorful "Liesegang rings" (see Day 4, Stop 4.1).

Another noteworthy trachyte intrusion crops out in a cliff of Cuchillo del Palo, on the Jandía peninsula. There, a trachytic dyke intrudes the basaltic lavas of the southern shield, expanding into a dome as the intrusion approaches the surface (Fig. 8.20B).

FIGURE 8.20

(A) Montaña Tindaya is a large eroded Miocene trachyte plug (image courtesy of *Foto aérea de Canarias*).

(B) Trachyte dome with its feeder dyke exposed at Cuchillo del Palo on the Jandía peninsula.

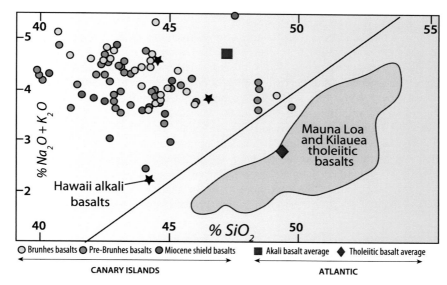

FIGURE 8.21

Simplified TAS diagram of rocks of different basaltic units of Fuerteventura (after Fúster et al., 1968) and comparative data from other Atlantic basalts and of Hawaii alkali and tholeiitic basalt compositions (after Clague and Sherrod, 2014). The data demonstrate the ocean island character of the basaltic rocks of Fuerteventura.

ROCK COMPOSITION

Lavas of Fuerteventura are monotonous from a petrological point of view. Both shield and post-erosive stage rocks are basalts of alkaline character (Fig. 8.21). The more silicic types plot near the boundary with tholeiitic basalts in a TAS diagram and recent eruptions (Series IV of Fúster et al., 1968a) also comprise alkali basalts, very similar in chemical composition to the older formations of the island.

As a general remark, the majority of lavas on Fuerteventura show a low degree of silica saturation and high alkali content. These lavas are even less silica-saturated than the Hawaiian lavas or average Atlantic alkali basalts in general (Fig. 8.21).

YOUNG SEDIMENTARY FORMATIONS

Sedimentary formations, not commonly exposed in the western islands, are a significant geological feature in the eastern islands, particularly on Fuerteventura, Lanzarote, and the iconic Maspalomas sand dunes on Gran Canaria. Specifically, the extensive aeolian sand deposits are relevant for their climatic implications and the definition of raised beaches, which helped enormously in defining the volcano-stratigraphy of the island.

AEOLIC SANDS (JABLES)

Deposits of light-colored sand (locally known as *jables*, from the French *sable*) are frequent in the eastern Canaries (see also Chapter: The Geology of Lanzarote). Thick sand deposits cover extensive areas of Fuerteventura (Fig. 8.22). The sand is formed by a complex aeolian system supplied by Pliocene and Pleistocene marine deposits, partially covered by carbonate crusts (calcrete). Sands are formed mainly by a Mio-Pliocene biogenic component comprising skeletal calcareous algae, shell fragments, and foraminifera that thrived on Fuerteventura during a period of equatorial climate (ie, lacking annual seasonal cycles) that extended from approximately 9−4 Ma (Meco et al., 2007). The closing of the Panama isthmus at approximately 4.6 Ma then initiated the cold Canary current that brought a drastic change in climate to the Canaries. Climate became colder, arid, and with marked annual seasonal cycles (see also Chapter: The Geology of Lanzarote, Fig. 7.9). This led to "mass extinction" of the marine equatorial fauna in the Canaries and provides the source of the *jables* and, hence, the spectacular beaches of Fuerteventura.

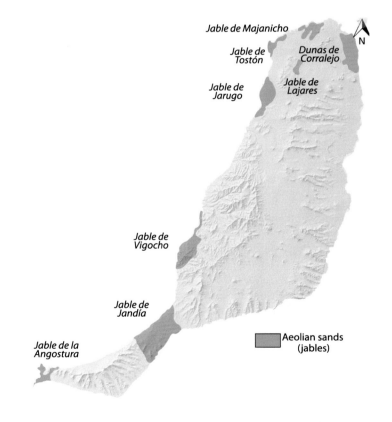

FIGURE 8.22

Distribution of aeolian sand deposits (jables) on Fuerteventura. Notably, the eastern (leeward) coast largely lacks this type of deposit.

A related feature is the *calcrete*, or *caliche*, a hard carbonate crust that partially covers the aeolian sands and is several meters thick in some places, reaching elevations of up to 400 meters above sea level (masl). The calcrete was initially interpreted as a result of dissolution of plagioclase from basalts (Hausen, 1967). This, however, causes difficulties to explain encrusted sand dunes, for example. A more consistent explanation is the encrusting of the sands and sand dunes by repeated dissolution and precipitation (evaporation) events of the biogenic calcium carbonate particles, gradually forming a hard crust that protects the underlying sand from wind erosion (Meco et al., 2007).

This type of "calcrete" mantled large parts of the island of Fuerteventura, although in some areas only vestiges are preserved, implying that during the late Miocene and part of the Pliocene the island was probably completely covered with white sand, likely with only the highest peaks of the Miocene basaltic shields forming black "islets" or "nunataks" in a sea of sand.

A leading industry of the island since the 17th century was therefore obtaining lime (CaO) from the abundant calcareous crusts. Several lime kilns were in operation until the mid-20th century, and one of these will be inspected on Day 4 (Stop 4.3).

RAISED BEACHES

The raised marine terraces on Fuerteventura (and Lanzarote) have long attracted scientific interest. Crofts (1967) recognized seven terraces on Fuerteventura, whereas Fúster et al. (1968a) observed that some of them were stratigraphically related to volcanic formations, using this relationship to define different subaerial volcano-stratigraphic units. This stratigraphic division proved useful on Fuerteventura and Lanzarote because both islands are in an advanced stage of posterosional volcanic rejuvenation and the raised beaches distinctly separate the shield and posterosional volcanic formations and, moreover, different volcanic phases during later episodes (Fig. 8.23).

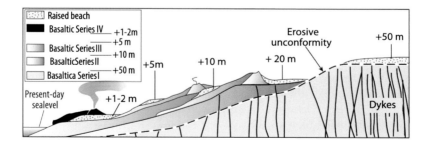

FIGURE 8.23

The first stratigraphic approaches on Fuerteventura used the existence of raised beaches to establish two main units, the Old Basaltic Series or Basaltic Series I, and the Recent Basaltic Series that included Basaltic Series II, III, and IV (after Fúster et al., 1968a). However, an attempt to export these concepts to the other islands, particularly to the western Canaries, proved unfeasible because the "old" Basaltic Series of La Palma and El Hierro were chronologically younger than the "recent" series of the eastern Canaries, thus hinting at a progressive evolution of islands within the archipelago rather than at a simultaneous one.

However, this volcanic stratigraphy proved nontransferable to the western Canaries that are still in the shield stage of evolution and where equivalent raised beaches are lacking. Therefore, Carracedo et al. (1998) proposed applying the same volcano-stratigraphic units used in the Hawaiian Islands, separating the shield and post-erosional stages. Important advantages of this classification are that it is applicable to the entire archipelago, has a genetic connotation, and allows correlation of similar volcanic processes among the different islands and thus across geological time (see 'La Canaria' concept in chapter: The Canary Islands: An Introduction, Fig. 1.34).

GEOLOGICAL ROUTES

Fuerteventura presents a number of features that are unique within the Canarian archipelago. First, it is the oldest island; in its western part, large sections of uplifted oceanic sediments and the submarine seamount stage of growth are exposed. Moreover, erosion and landslides removed the greater part of the Miocene island, exposing the deeper plutonic structure of the Miocene Fuerteventura edifice, such as the different magma reservoirs now preserved as plutonic intrusions and the feeder dykes of thousands of submarine and subaerial eruptions that formed very dense dyke swarms in a number of places.

Because of the low and smooth topography, roads on Fuerteventura are mostly flat, making traveling across the nearly 100-km-long island easy and fast (note that Tenerife is only 98 km long). Tourist resorts are disperse, with the northern *Corralejo* area and the southern Jandía peninsula (*Morro Jable*) being the main tourist hot spots. For this reason, the starting point for the day trips on Fuerteventura are not fixed (Fig. 8.24).

DAY 1: OCEANIC SEDIMENTS AND PLUTONS

Drive to *Pájara* and then follow FV-621 to *Ajuy* (Fig. 8.25). Approximately 1—2 km before reaching the coast, you will see numerous dykes on your left in the road cuts that show striped sedimentary rocks between the dyke intrusion. Note the ratio of dykes to sedimentary country rock. More than 60% of the rock mass here are dyke intrusions. From here, make your way to *Ajuy* and to the larger car park near the beach.

Stop 1.1. Playa de Ajuy (N 28.3991 W 14.1545)

At the northern end of the beach, uplifted sediments are intruded by dykes (Fig. 8.26). The sedimentary rocks comprise dominantly siliciclastic lithologies here and show rhythmic banding for most parts. These sedimentary rocks are, in turn, overlain unconformably by Pliocene beach conglomerates, alluvium, and wind-blown materials, indicating a drop in sea level since the Pliocene and a rather stable situation thereafter. Note the massive gray phonolite dyke that bends within the sediments but is truncated by the younger (Pliocene) lava flows.

Stop 1.2. Zebra rocks in Bco. del Aulagar, Las Arenas (N 28.3841 W 14.1503)

This first outcrop should be followed by a hike up the barranco (approximately 40 min to an hour each way) to inspect dykes and a pyroxenite intrusion with spectacular partial melting textures in their aureoles (zebra rock, Fig. 8.27). For this, walk up the Bco. del Aulagar, which you can access at the southern end of the beach via a dry river bed.

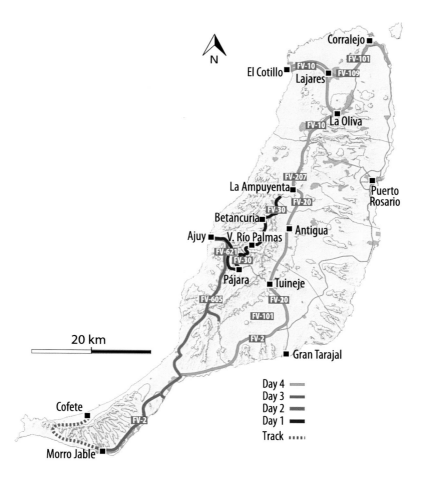

FIGURE 8.24

Map showing the different geological routes offered (Days 1−4). Because of the elongated shape of the island, there is no fixed starting point for the days/trips, but a locality close to the first stop is given for the start of each day. Map modified from IDE Canarias visor 3.0, GRAFCAN.

Note, the coastal walk along the Pliocene unconformity here will follow on Day 2.

After approximately 10 min of walking into the barranco, you will encounter exposure of an old beach with sand, calcite fills, and pillow lava structures intermixed with some hyaloclastites. This must have been a very shallow marine environment at that time, probably marking the transition from submarine to subaerial island growth.

Continue up the stream bed through shallow marine lavas and beach deposits, increasingly cut by dykes. Some of these dykes are zoned, some even with central parts of pyroxenite nodules, likely picked up from the pyroxenite intrusion that we will encounter a little later today. This segregation

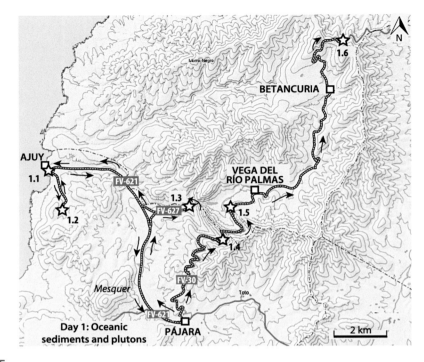

FIGURE 8.25

Map of the central western part of Fuerteventura (the Betancuria massif), with the stops for Day 1 marked. The day is mostly dedicated to the uplifted oceanic sediments and the plutonic core of the island that is exposed in the *Ajuy*, *Pájara*, and *Betancuria* area. Map modified from IDE Canarias visor 3.0, GRAFCAN.

is probably a function of flow differentiation that moves solid inclusions to the center of the active flow during magma transport in a confined fracture (see also Day 5, chapter: The geology of Tenerife).

Continue upstream until you reach a natural terrace and some young trees above the terrace. Keep on the southern track (right) and continue for approximately 10 min through basaltic dykes until you reach the first zebra rock exposure (Fig. 8.27). The zebra rock is in part white to cream-colored in the contact aureoles to the large pyroxenite intrusions that form the dark hills to your left (east). The zebra veins represent frozen partial melt and are made up of evolved syenitic/phonolitic compositions, providing an elegant mechanism for the generation of felsic ocean island magmas. Continue in a southerly direction along the stream bed.

At the junction with Bco. de Las Arenas (N 28.3891 W 14.1498), you can either continue southward to gradually leave the contact aureole again or turn toward the southeast to approach the pyroxenite intrusions. Unfortunately, the pyroxenite is not very fresh, but the large pyroxene crystals are readily visible. A bit further up this southeastern arm of the stream system, a cream-colored "kinky" dyke of felsic composition occurs in the pyroxenite, likely exploiting cooling fractures in the pyroxenite host rock.

FIGURE 8.26

Uplifted oceanic sediments intruded by dykes and, in turn, unconformably overlain by Pliocene beach sands and gravels and by Pliocene basaltic pillow lavas, indicating a relative drop of sea-level since the Pliocene.

Turn back when you feel you have seen sufficient zebra rock and pyroxenite. Remember your return hike will take approximately 40 min, so consider a lunch break here or at the beach after your hike.

From the beach, drive back toward *Pájara* and turn left just before the km 4 road sign on FV-621 to see the Vega de Río Palmas ring dyke. Take the little dirt track at N 28.3931 W14.1273 and park in or near the usually dry river bed at the end of the track.

Stop 1.3. Vega de Río Palmas ring dyke and Las Peñitas dam (N 28.3859 W 14.1061)

Walk up to the little track into the gap in the light-colored syenitic rocks in front of you. The track is well laid out but goes up the next valley after crossing the riverbed for a few minutes before it bends and leads up into the syenite gorge. Please walk carefully on the higher part of the track.

Bulbous (onion skin) weathering is locally observed. Where fresh, the syenite is massive, with abundant K-feldspar, plagioclase, and dark pyroxene prisms, although modal proportions vary considerably across the intrusion (Fig. 8.28A).

Continue on the track up-hill toward the dam (Fig. 8.28B). Note the beige dyke (~1 m wide) on your left just before the dam. The host rock here near the dam is heavily altered and comprises tectonized diorite and gabbro, now beige and locally greenish in color. This phenomenon is likely

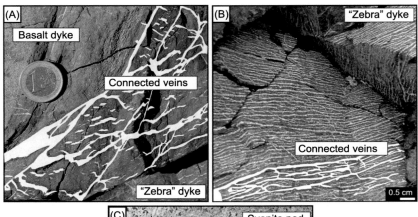

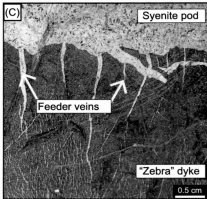

FIGURE 8.27

(A) Annotated field photographs of migmatized basaltic dykes in the contact aureole of pyroxenite PX1 with felsic veins highlighted (see Stillman, 1987; Hobson et al., 1998; Holloway and Bussy, 2007). Segregation veins are seen in isolated as well as merged state and collect into larger syenitic melt pockets that eventually form dyke-sized intrusive bodies. Many isolated leucosomes likely formed by in situ melting of the host basalt. Coin is 2.3 cm in diameter. (B) A partially migmatized basaltic dyke with connected felsic veins that feed into a larger leucocratic vein system. A portion of the connected vein system is highlighted in plain white. (C) A large syenitic melt pocket fed by smaller veins within a migmatized dyke. Images courtesy of F.M. Deegan.

akin to the *Los Azulejos* on Tenerife and Gran Canaria, a near-surface hydrothermal alteration feature along a caldera structure (eg, a ring dyke within a caldera ring fault), implying an eroded caldera complex was likely above this intrusion prior to intense erosion of the island.

Walk a little further and look back onto the Río de la Vega ring dyke. More altered and partly tectonized rocks will appear at the edge of the ring dyke, reminding us of the enormous forces at work during ring faulting and ring dyke emplacement (see Figs. 8.9 and 8.10).

Return to the track on which you came. Once at the car, drive to *Pájara*. From there, follow signs for *Betancuria* (FV-30, km 25) until you reach a parking opportunity at a large view point and car park area to your left, known as *Risco de las Peñas*.

FIGURE 8.28

(A) The outermost (syenitic) ring of the Vega de Río Palmas intrusive complex. (B) Las Peñitas dam is now completely filled with sediment. In the far distance, the darker pyroxenite pluton of Mesquer is visible.

Stop 1.4. Mirador Risco de las Peñas (N 28.3797 W 14.0953)

You are now on top of the Vega de Río Palmas ring dyke. Here, you can inspect the syenite again and take a stroll onto its upper exposures. Locally, pegmatitic domains are present. At this site, squirrels are plentiful, but they should not be fed.

Walk toward the exposed rocks along the available path for a few hundred meters to inspect a larger gabbroic mass that is resting in the syenite and hosts many syenite veins. This is a fallen block from the roof, consistent with the notion of a previous caldera collapse. Assuming these gabbro blocks are from the gabbroic Toto ring dyke, then the Vega de Río Palmas ring dyke is the younger of the two ring intrusions (see Fig. 8.9).

Return to your car and continue northward through mainly dykes of the basal complex. Continue for approximately 2 km to *Mirador las Peñitas* for our next stop.

Stop 1.5. Mirador las Peñitas (N 28.3870 W 14.0924)

Looking west, you can now view the Rio de la Vega ring dyke from above (and from inside)! Note the path taken earlier in the day is visible from here (see Figs. 8.10; 8.28B). The curvature of the intrusion is quite apparent from here also, and so is the geometry of a steep arcute sheet (ie, the bell jar—shaped geometry of the ring dyke).

Note the squirrels! The Barbary ground squirrel (*Atlantoxerus getulus*) was introduced to Fuerteventura from North Africa as a household pet in 1965. After some escaped, they found a natural environment similar to their homeland but without predators. Their numbers have increased to an estimated 300,000 and they are now a threat to the island's endemic flora and fauna.

Continue on FV-30 for approximately 11 km to *Mirador de Corrales de Guise*.

Stop 1.6. Mirador de Corrales de Guise (600 masl, at Guanche statues) (N 28.4407 W 14.0561)

This vantage point is at an altitude of 600 m on FV-30 and offers spectacular views onto the beautiful scenery of Betancuria's landscape to the south and also over northwest Fuerteventura and all the way to Lanzarote (Fig. 8.29).

A pronounced dyke in the near hillside is visible (strike approximately 40° and with a dip to the west). Looking north, in the flat part in the distance beyond the near hillside, note the reddish cinder cone (Mña. Bermeja, from the Spanish *Bermejo*, a type of bright red color). Moreover, in the far distance, a ridge-like trachyte intrusion (Tindaya) and, to the east, lava remnants of the northern shield complex (Miocene 11−9 Ma) are visible (eg, Rosa del Taro, Mña. del Campo; Fig. 8.29).

Looking southeast toward *Betancuria*, you can see rounded hills made up mainly of basal complex dyke intrusions that are also partly exposed in the lower ground.

The two giant bronze statues are a tribute to Guise and Ayose, the native kings of Maxorata and Jandía, the two separate kingdoms of Fuerteventura in the pre-Hispanic era. The larger north of the island was known as Maxorata and the smaller south was known as Jandía, with a dividing wall that was erected on the isthmus of *La Pared*.

This is the last stop of the day. Make your way back to your base, either in the north or south of the island. Note that we will be back in *Ajuy* tomorrow morning.

END OF DAY 1.

DAY 2: THE SOUTH

Make your way to *Pájara* and then again to *Ajuy* (*Puerto de la Peña*), and park again at the beach.

Day 2 will be dedicated mainly to inspect the transition from submarine to subaerial volcanism and the plutonic intrusions of that period (Fig. 8.30).

Stop 2.1. Ajuy, Caleta Negra (N 28.4001 W 14.1572)

We shall now inspect the post-basal complex sequence in the cliffs on the northern end of the beach. Follow the steps up the cliff (Fig. 8.31). Above the steep dipping basal complex sediments and dykes, there is a semihorizontal unit of beach conglomerates. On the steps, the large phonolite dyke from Day 1 appears here as a sill, exploiting the boundary between these major rock units. The beach is approximately 14 masl and has calcareous organic sediments resting on its top section, which are dominantly of aeolian origin. The fossils here are approximately 3 Ma old (ie, Pliocene in age).

There are gastropods and shells to be seen, as well as insect nests (from wasps for instance) mixed with volcanic detritus and rock pebbles of various sources.

Continue through the hardened sand deposits along the coastal walk. After a few minutes you will encounter basaltic rocks on top of the aeolian sandstones, which are Pliocene lavas that were flowing toward the sea. The pillow lavas observed a little further along the path imply that at this time, this site was at sea level (Figs. 8.31 and 8.32). They are, in turn, overlain by regular subaerial lavas.

Continue on the path until it descends. You will soon encounter hyaloclastite breccias and pillow lavas mixed with sand, suggesting a lava delta of some definition. Here, please take a look

FIGURE 8.29

Panorama view from *Mirador los Corrales de Guise* onto the northern half of Fuerteventura. In the foreground is the "smooth" relief of the Betancuria massif. In the distance and in darker color are the remains of the northern shield volcano.

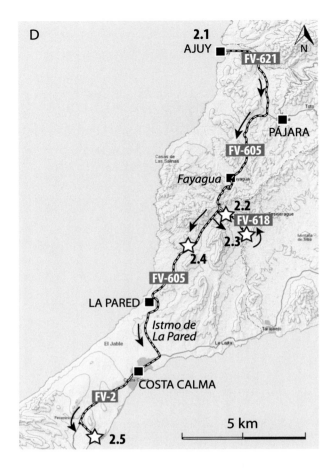

FIGURE 8.30

Map of the central and southern part of Fuerteventura (Betancuria massif to *Costa Calma*) with the stops of Day 2 indicated. This day is dedicated to the transition from submarine to subaerial volcanism and to recent sedimentary deposits. Map modified from IDE Canarias visor 3.0, GRAFCAN.

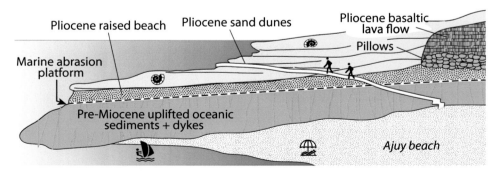

FIGURE 8.31

Sketch of the northern wall of Bco. de Ajuy showing the Mesozoic oceanic sediments and dykes overlain by the Pliocene fossil beach and later basaltic flows. The basaltic pillows indicate the sea level at that time (ie, that before the post-Pliocene sea level drop).

FIGURE 8.32

Pillow lavas formed as the lava flows entered the sea, thus indicating the position of the Pliocene coastline at that time. The basaltic flows overlie a marine abrasion platform (indicated with white arrows) and oceanic sediments that are densely intruded by dykes. Note that the basalt lava created new land at that time by extending the island into the sea.

across the bay for a superb cross-section and more pillows at virtually the same horizon as the one on which you are standing (Fig. 8.32).

Continue down the steps until you can look into the "cueva." Be careful when entering the cueva (especially when entering at high tide!). Here, we can see the approximate north—south trending dykes again that intrude steeply dipping (Mesozoic) sedimentary rocks of the basal complex. Note the sill intrusions within the sediments that are cut by the north—south trending dykes and are locally faulted with the sediments (see Fig. 8.19D). These are likely some of the earliest manifestations of magmatism on Fuerteventura, and they may even precede the uplift stage of the basal complex sediments.

Once you have inspected the cave exposures, return to your vehicle and drive inland toward the T-junction with FV-605 (yellow). Turn right toward *La Pared* (FV-605). Soon the back of the pyroxenite intrusion will appear to your right, and a little further you are again driving through the dykes of the basal complex. A short distance after *Fayagua* (near km 11 on FV-605 road sign), the road climbs up-hill and a noteworthy view of the dykes of the basal complex appears on your left with an approximately north—south strike (of ∼010, making the hillside look striped).

Continue southward on FV-605. A short stretch further, a right turn onto FV-618 is recommended. Drive along this road for approximately 1 km until open ground appears next to the road. Park here for an overview stop.

Stop 2.2. The unconformity at Mña. Melindraga (N 28.2817 W 14.1519)

Here, the transition between submarine and subaerial deposits is well displayed (Fig. 8.33), with semihorizontal lavas and pyroclastics showing prominent layers in the hillside to the north (Mña. Melindraga). On the right side of the hillside, a palaeovalley filled with semihorizontal post-unconformity material becomes apparent, suggesting rugged topography on the basal complex prior to deposition of the younger materials that make up the top of the hillside.

Continue up-hill for approximately 1 km and park at a small lay-by on your left. Park your car so that you can return the way you came.

Stop 2.3. Mña. Redonda (N 28.2775 W 14.1482)

Here you are at the northwest flank of Mña. Redonda, at the uncomformity level. After a walk of a few meters, you can inspect the transition from submarine to the first subaerial lavas again. Basal complex dykes here cut through pyroclasitcs and regular lavas in outcrop.

The ridge toward your right (south) is named El Cardón. The subhorizontal lavas on top of El Cardón are dated at approximately 14 Ma (Hervé Guillou, personal communication), and the entire mountain corresponds to the central shield.

Return to FV-605 and continue southward (to your left), toward *La Pared*. Pass semihorizontal lavas of Miocene age (shield lavas) in the steep hillside for several kilometers while driving downhill. Note a reddish batch in the hillside that is a former cinder cone that took part in the Miocene eruptive episode.

A little further down-hill, the dyke swarm disappears and you are now driving in Miocene shield basalts. A brief inspection is possible at the lower slopes during low tide by taking a little dirt track to your right.

FIGURE 8.33

Subaerial Miocene basaltic flows that rest on sediments. Both units rest unconformably over Miocene submarine volcanics that are intruded by a dense swarm of tilted dykes, in some places forming close to 100% of the exposed rock mass.

Stop 2.4. First view of Playas de Jandía *on the Jandía peninsula* *(N 28.2511 W 14.19200)*

Looking south, you will see the crescent-shaped coastline of the "Peninsula de Jandía," with the Cofete cliffs on the seaside (Fig. 8.34). This is considered to be the result of a gigantic landslide in the late Miocene (see Stillman, 1999).

Continue on FV-605 toward the south. Ahead of you, the sand-covered isthmus between the peninsula and the main part of Fuerteventura is visible now. The sand is a mix of organic remains (tropical fauna, eg, corals) and wind-blown dust from Africa.

Further down-hill, you will drive through sand-covered basalt rocks. The sand is getting increasingly thicker toward *La Pared* and beyond. Drive through *La Pared* and continue further south. Soon you will come to a roundabout. Join FV-2 toward *Costa Calma* and *Morro Jable*.

Drive through *Costa Calma* and into thick sand-covered shield basalts once you leave *Costa Calma*.

Follow the road south, now on FV-2, until you see signs for *Playas de Jandía*. Continue for a short while until you see new signs for *Playas de Jandía*. Turn off FV-2 in an exit to your left just at the km 72 signpost toward *Playas de Jandía/Risco del Paso* (~2.5 km) after the second set of signs for *Playas de Jandía*, which brings you to the southern end of the beach.

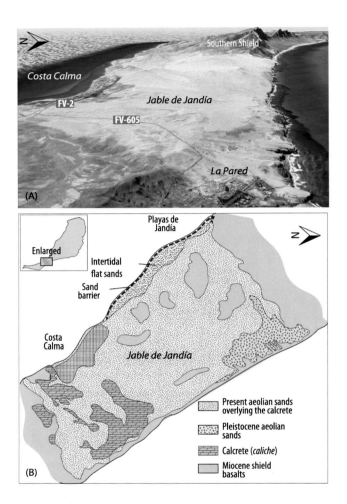

FIGURE 8.34

(A). View in southwest direction onto the isthmus of Jandía that is covered with aeolian and organic sands. In the far distance, the southern shield volcano shows the scarp of a northbound Miocene giant landslide (*Google Earth image*). (B) Sketch map showing the distribution of aeolian sands and caliche on the isthmus of Jandía (simplified after Meco, 1993).

Stop 2.5. Playas de Jandía *(N 28.1117 W 14.2639)*

Playas de Jandía is one of the legendary beaches of Fuerteventura, and it is especially popular with surfers (Fig. 8.35). This is the last stop of the day. Enjoy the beach, possibly some refreshments, and note the dark, fine components in the calcareous sand. This is a mixture of dust from Africa and local detritus and aeolian fossil material of dominantly local derivation again. The calcareous materials are largely organic fossil remains dating back to the time of the tropical climate that

FIGURE 8.35

Playas de Jandía represents an extensive intertidal flat that is protected by a natural sand barrier and can be a very popular site during the high season. Image courtesy of Foto aérea de Canarias.

dominated the older islands of the Canaries prior to the closure of the Panama isthmus (approximately 4 million years ago), which caused the Atlantic currents, and thus also the climate of the Canaries, to change (Meco et al., 2003). This type of beach is not seen on the younger islands (eg, La Palma, El Hierro) because these formed only after this major change in climate had occurred (see chapters: The Geology of El Hierro and The Geology of La Palma, Fig. 7.9).

Once you feel you have sampled sufficient beach impressions, return to FV-2, and from there to your base in either southern, central, or northern Fuerteventura, or, if time is plentiful, consider adding the excursions to the Jandía Peninsula (Day 3).

END OF DAY 2.

DAY 3: THE JANDÍA PENINSULA (THE SOUTHERN SHIELD)

Day 3 is dedicated to inspecting the Jandía peninsula (Fig. 8.36). Start at km 72 on FV-2, where you took the turn off toward *Playas de Jandía/Risco del Paso* for the last Stop of Day 2. Instead of leaving FV-2, continue in a southerly direction on this road along the leeward coast of Jandía. The road crosses a succession of wide barrancos and sharp interfluves carved into the flank of the Miocene southern shield. These have their head walls beheaded by the Jandía landslide (Fig. 8.37).

At FV-2 km 83, the view opens into the wide, crescent-shaped Playa del Matorral. Go straight toward *Morro Jable*, past two roundabouts, and you are in a wide avenue (*Avenida del Saladar*) with plenty of parking space on both sides of the street. Park on the right side and walk to Matorral beach right in front of you.

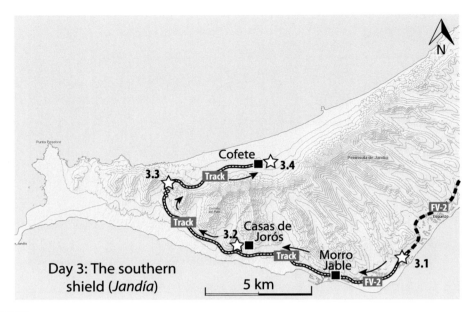

FIGURE 8.36

Map of the southern part of Fuerteventura (Jandía peninsula), with the stops for Day 3. This day is dedicated to the southern shield and the Cofete giant landslide and will close at the beach, so be prepared. Map modified from IDE Canarias visor 3.0, GRAFCAN.

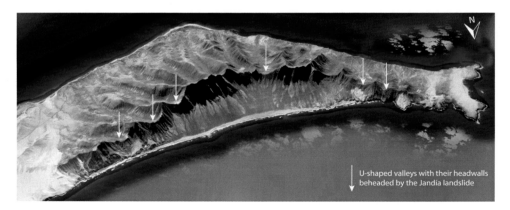

FIGURE 8.37

View of the Jandia jable (in the foreground) and the basalt remnants of the southern shield further to the south. The white arrows indicate the U-shaped valleys that had their headwalls beheaded by the Cofete landslide (after Stillman, 1999). The width of the image covers ∼30 km.

Stop 3.1. Playa del Matorral (N 28.0528 W 14.3239)

The El Matorral beach is a long white sand beach bordered by a salt marsh (saladar), a coastal ecosystem that is regularly flooded by the tides, colonized by salt-tolerant vegetation and plants able to endure periods of submergence. This site is a protected area because of its unique endemic flora.

Continue toward the south on FV-2 to *Morro Jable* and, after passing through the village, take a "busy" dirt track that leads to the *Faro de Jandía*, the westernmost end of the island.

The track crosses valleys and ridges of the southern shield, which is covered with white spots of indurate aeolian sands.

After approximately 5 km from *Morro Jable*, you will see a small cluster of houses (*Casas de Jorós*). Continue for another 300 m to Bco. de Jorós and park the car next to a group of palm trees to inspect a succession of marine terraces.

Stop 3.2. Bco. de Jorós (N 28.0610 W 14.4042)

Deposits of white sand partially cover the Miocene lavas on the right of the track. On closer inspection, you will see layers of rounded conglomerates with marine fossils corresponding to raised marine terraces that range from +62 to 0 m and that correspond to "marine episodes I to XII" (Zazo et al., 2002). These authors performed a systematic study of marine terraces on Fuerteventura, and of this site in particular (Fig. 8.38).

Continue west for approximately 4 km and, on arriving at a major crossroad, turn to the north toward *Cofete*. More deposits of aeolian sands appear attached to the walls of the barranco. After approximately

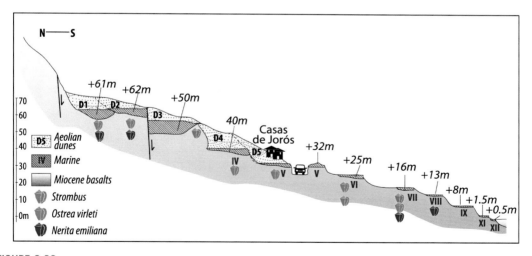

FIGURE 8.38

Raised marine terraces and warm faunas, corresponding to marine Episodes I to XII, form a 12-step "staircase" near *Casas de Jorós*, Jandia. The climbing aeolian dunes overlie fossiliferous marine deposits. Tropical and warm-water faunas are present in the older episodes, whereas in the youngest ones (episodes IX to XII) they are lacking. Modified after Zazo et al. (2002).

2 km of up-hill driving, the track arrives at a mirador at the crest of the cliff above *Cofete*, with a superb view of a crescent-shaped scarp of the entire northern face of the Jandía peninsula.

Stop 3.3. The Jandía landslide scarp (N 28.0933 W 14.4334)

We have described the giant landslide scarp as a relatively frequent feature in the Canaries, particularly on Tenerife, La Palma, and El Hierro. Similar structures are less evident in the eastern Canaries, mainly because erosion has removed many of the few on-land diagnostic features, and marine sedimentation conceals off-shore evidence such as debris-avalanche deposits.

Nevertheless, Stillman (1999) points to the generation of giant landslides from Fuerteventura, especially from this site. He argues that these were analogous to those seen on the younger islands and a principal mechanism of erosion on Fuerteventura. These landslides appear to have removed approximately 3500 km^3 of lavas and volcaniclastics, stripping the western sectors of the Miocene central and northern shields down to the preshield phase rocks. Features characteristic of giant landslides are more evident in the Jandía peninsula than elsewhere on the island, preserving a 12-km escarpment and wide depression, which probably originated as a consequence of a northbound slide approximately 15 Ma ago (Fig. 8.39). Ancochea et al. (1996), for example, describe that, in the north, the submarine floor of the southern shield descends rapidly, whereas in the south the submarine slope is gentle (see Fig. 8.1), consistent with large-volume deposits derived from the island.

Other evidence that indicates that the southern shield was considerably larger and higher before losing the northern half of the edifice is the preserved fossil drainage pattern that was initiated on these former peaks but was "beheaded" by the landslide (see also Fig. 8.37).

Drive down the track for approximately 5 km. The track progressively cuts older basaltic flows from approximately 15 Ma to more than 13 Ma (Fig. 8.40). Further down, well-developed piedemonts cover the basaltic flows at the foot of the cliff. The entire succession of basaltic flows forming the landslide scarp has only two different geomagnetic polarities (see inset in Fig. 8.40), consistent with the fast growth of the southern shield. The track finally arrives at the hamlet of *Cofete*, where you can enjoy a full view of the giant landslide embayment. A question remains, however, regarding why the landslide was not followed, as occurred in the majority of the giant landslides in the Canaries, by postcollapse eruptive activity, which seems entirely absent in this particular landslide embayment here.

Stop 3.4. Cofete (N 28.1027 W 14.3889)

From *Cofete*, you can walk to the beach (less than 1 km). The remoteness and the imposing silence of this place are sometimes broken by brays of feral donkeys that were abandoned and roam freely in some of the more remote parts of the island, particularly here in *Cofete*.

END OF DAY 3.

DAY 4: THE NORTH

Day 4 will include a tour in the northern part of Fuerteventura (Fig. 8.41).

Join FV-2 and after *Costa Calma*, continue to *Tarajalejo*. After a long drive, you will arrive at a roundabout where you can join FV-20 in the direction toward *Tuineje*. There, you will find *malpais* (badlands) on your right, made of extensive "aa lavas" that originate from the cinder cones that rise from the plain. These are dated at approximately 136 kiloyears (ky) (Hervé Guillou personal communication); therefore, they are among the youngest rocks on the island (Fig. 8.42).

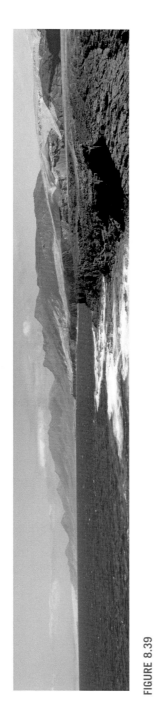

FIGURE 8.39

Impression of the windward coast of the southern shield viewed from the southern end of the island. Aeolian dunes have been deposited in the collapse scarp. The jable de Jandía appears faintly in the far distance. Note, the dunes in the central part of the picture are trying to "climb" the slopes of the collapse scarp.

FIGURE 8.40

The track to *Cofete* cuts the entire sequence of basaltic flows of the southern shield. This succession of basaltic flows was erupted between approximately 15 and 13 Ma (*H. Guillou pers.com.*). These ages and the presence of only two geomagnetic polarities (see inset) are consistent with the construction of basaltic shields in a relatively short period of time, maybe 1−2 Ma.

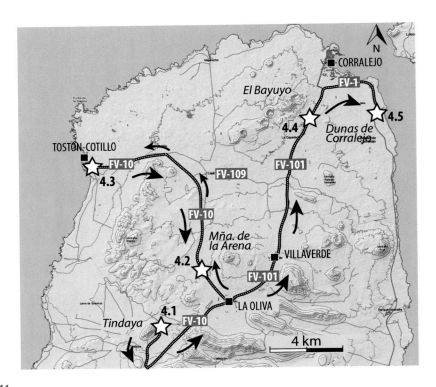

FIGURE 8.41

Map of the northern part of Fuerteventura with the stops for Day 4 indicated. Map modified from IDE Canarias visor 3.0, GRAFCAN.

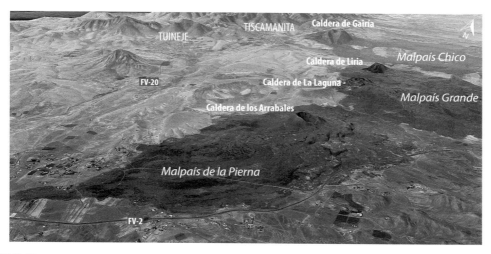

FIGURE 8.42

Fuerteventura lacks Holocene eruptions. The youngest volcanism is older than 100 ka. The volcanic cones of central Fuerteventura (*Google Earth image*) have been dated at ~139 ky (*H. Guillou, personal communication*).

Once you reach *Tuineje*, please turn right, but stay on FV-20 in the direction of *Antigua*. Drive through *Tiscamanita*. A cinder cone rises to your right just behind the village (Caldera de Gairía), which also belongs to the posterosional eruptive episode and is dated at 136 ky (Hervé Guillou personal communication).

In *Tiscamanita* you also have the opportunity to visit the *Centro Interpretativo de los Molinos* if you are interested in the windmill tradition of the island. Otherwise, continue on FV-20 toward *Antigua*. Just before the village of *Agua de Bueyes*, you will be driving through dykes of the basal complex exposed in the road cuts to your right.

Continue north, passing large flat stretches with red caliche (calcrete) that is in part from organic aeolian sands that covered almost the entire island. Drive through *La Ampuyenta* and turn left after the roundabout upon leaving the village.

Note the lavas of the north shield on the right side. These lavas are approximately 14 Ma old. Continue for a few kilometers on this small road. When you reach a fork, take a left turn, and then soon after a right turn, which will get you on FV-207 in the direction of *Tindaya* and *La Oliva*.

On FV-207 you will soon pass a volcanic edifice on your left. This cone is approximately 2.8 Ma old and unusually reddish (Mña. Bermeja, meaning an earthy red in Spanish). Note the quarry on the south side, which delivers a popular building stone in the Canaries (*toba*).

Continue northward and pass another young cone called Mt. Quemada (~1 Ma; the burned hill) on your left. The road will soon join FV-10. Turn left toward *La Oliva*, and you will soon drive within the basalts of the northern shield.

After a short drive further north on FV-10, you will get a view of Mt. Tindaya on your left (see also Fig. 8.20A). A small quarry is visible on the south end, which can be visited. Drive through the village to the quarry. The quarry is behind private land, and you will have to ask the owner for permission to access the site. If you succeed, then walk up to the quarry from the track to inspect the trachyte and the colorful and pronounced "Liesegang rings."

Stop 4.1. Montaña Tindaya (N 28.5918 W 13.9783)

Mña. Tindaya is a trachyte intrusion, likely of Upper Miocene age, that shows the characteristic fractures of a felsic dome (eg, concentric and radial joints), which formed after undergoing volume retraction upon cooling (see also chapter: The Geology of La Gomera, Figs. 4.28 and 4.29).

Here, you will also see intense Lisengang alteration originating from pervasive fluid percolation through joints and especially through porous flow (Fig. 8.43A,B). This feature popularized the use

FIGURE 8.43

(A) Quarry at Mña. Tindaya. The abundant and colorful concentric features are "Liesegang" rings, secondary structures that are found in permeable igneous rocks that have been "chemically" weathered by fluid circulation. (B) Close-up of the Liesegang rings and internal fractures in the Tindaya trachyte. (C) A costly project was proposed in the late 1990s aiming to excavate a hollow cube with a 50-m side length inside the mountain and leave a conduit to allow sunlight into the cube. The project was abandoned because of the enormous costs and the possible instability caused by the internal fractures in the trachytic rock.

of this rock to embellish the facades of prominent edificies, for example, several banks and administrative buildings in *Puerto del Rosario* have it on display.

A gigantic art project that did cost approximately 12 million Euro for planning and testing alone (ie, before even starting the job), with a much greater total anticipated budget of more than 100 million, proposed to excavate a $50 \times 50 \times 50$-m hollow chamber inside the mountain and a vertical tunnel in the roof to allow sunlight into the chamber (Fig. 8.43C). This project did not receive positive evaluations by the geologists and engineers who work on the island due to the extraordinary cost and also because of the presence of internal fractures that cause instability inside the mountain and likely make such a project technically unfeasible and even dangerous. Hence, this art project was eventually abandoned.

Return to your vehicle and continue on FV-10. Soon you will pass a spectacular caliche-covered fan in the hillside on your right. This is a fossilized aeolian deposit that was laid down on the slopes of the hill and solidified due to reprecipitation of calcite, which is soluble in cold water (colder than $\sim 28°C$) but precipitates as aragonite in warmer waters.

Approaching *La Oliva*, you will see a recent cinder cone that has an eruption dated at 185 ky BP (Meco et al., 2007) and is known as Mña. de la Arena (arena = lapilli in local Canary terminology) (Fig. 8.44).

In *La Oliva*, you may want to visit the *Casa de los Coroneles*, which was built in the 18th century and was the former seat of the military governor of Fuerteventura. Also, note the church in *La Oliva*, which was built from local stone in early colonial style. Lanzarote and Fuerteventura were the first two of the Canary Islands to be conquered by the Spanish, dating back to approximately 1420 AD.

From *La Oliva*, turn left (West) toward *El Cotillo* (FV-10). Drive through the malpaís of Mña. de la Arena, which covers large tracts toward the coast, and park the car in a lay-by at the km 25 signpost for a close-up view of the vent and lava flows.

Stop 4.2. Mña. de la Arena *(N 28.6289 W 13.9451)*

Mña. de la Arena is a multicrater basaltic cinder cone dated at 185 Ma (Meco et al., 2007) that poured short lava flows in all directions (Fig. 8.44A). The volcanic cone and lava flows rest on a bed of aeolian organic sands topped with a thin layer of calcrete, which overlie the Miocene basalts of the northern shield (see cross-section in Fig. 8.44B).

Continue toward *El Cotillo*. When reaching the settlement, keep left and continue toward the coast. The coastal recession has formed a cliff that extends along the northern coast of the island, from *El Cotillo* all the way to the Betancuria massif. Park your car near the *Castillo de El Tostón*, a small round fort that was built in the late 1790s to defend and protect against pirates. From there, you have a superb panoramic view of the coast, and it is even better from the top of the watchtower, but a small entrance fee applies.

Stop 4.3. Castillo de El Tostón *(N 28.6801 W 14.0103)*

Inside the tower, you can inspect the 18th century fortification made of local basalt. Nearby, two chalk-burning ovens (kilns) are visible, which were used in the past to produce lime (CaO) from the caliche (calcrete) layers and deposits, which was a highly valuable commodity at the time (in simple terms: $CaCO_3 + heat \rightarrow CaO + CO_2 \uparrow$).

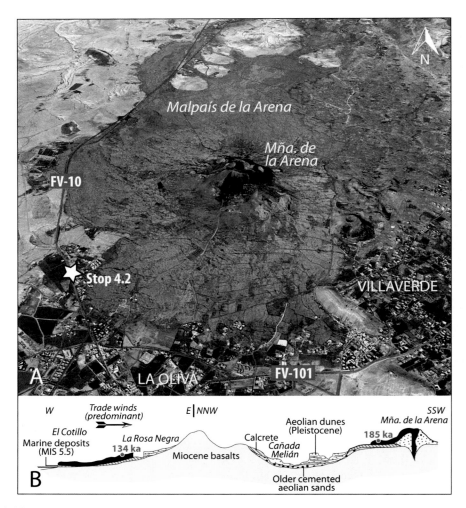

FIGURE 8.44

(A) Mña. de la Arena, a multicrater basaltic cinder cone encircled by its own lava flows (*Google Earth image*). This eruption was dated to ~185 ky (*H. Guillou, personal communication*). Note the quarry in the volcanic cone to extract lapilli ("arena" is sand in the local dialect). (B) Cross-section showing the relationship between the recent eruption deposits and the marine deposits and aeolian sand dunes of the area.

To the south, a large platform opens up that is formed by basalt lavas with an age of approximately 2 Ma and is associated with Mña. Ventosilla, a vent near Tindaya (Fig. 8.45). Note these lavas form a platform, implying that new land was created during this event. Enjoy the view and the breeze before returning to *La Oliva* on FV-10.

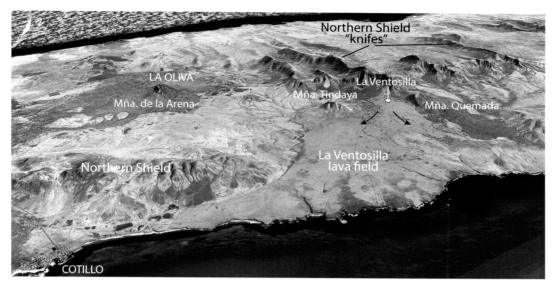

FIGURE 8.45

View of the western coast of Fuerteventura, south of *El Cotillo* (Google Earth image), showing Miocene cuchillos and U-shaped valleys in the remains of the northern shield. Mña. de la Arena and Mña. Quemada are recent volcanic cones (Upper Pleistocene); together with the extensive lava field of the Pliocene La Ventosilla eruption, they are overlying a Miocene erosional landscape in this part of the island.

On the return journey you will pass a sand quarry on your right, which shows several meters of aeolian sand deposits that form a rather thick cover in the north of Fuerteventura. This may be worth a stop if you have the time.

Otherwise, continue to *La Oliva* and join FV-101 in the direction toward *Corralejo*. Continue through *Villaverde*, but go slowly at the end of the village and follow signs for *Cueva del Llano*, which gives you access to a real lava tunnel of Pleistocene age (Fig. 8.46). Follow the small road for approximately 700 m. A visit will take maybe half an hour and will cost approximately €5. Guided tours run hourly.

Return to the main road (FV-101) and continue toward *Corralejo*. Approximately 3 km before *Corralejo*, you will be driving through the extensive northern malpais (age ~100–400 ky; see Fig. 8.16), displaying beautiful "aa" flow features that are characteristic of young basaltic lava flows in the distal parts of such deposits. Notably, these lavas here also rest on a thick layer of aeolian sands.

On your left, the volcanic alignment of El Bayuyo will now become visible. Lapilli mining and domestic refuse deposits have unfortunately degraded this spectacular volcanic group considerably.

Drive through *Corralejo* and on a major roundabout join FV-1 to the right, which will bring you into the sand dunes of the north. Note the small island on your left, "Isla de Lobos" (lobos de mar = sea lions or seals), which hosts a Pleistocene cinder cone with lavas and hornito-type blisters that can be seen from a considerable distance.

FIGURE 8.46

Interpretation center and entrance to *Cueva del Llano* lava tube that formed from an eruption approximately 800,000 years ago. The 680-m-long tube preserves a record of layers of sediment and small fossils that were deposited inside the lava tube over many thousands of years, during which Fuerteventura changed from a humid island to the present-day desert-type landscape. The cave has a noteworthy collection of fossils, and some of them are by now long extinct on the island.

Stop 4.4. Dunas de Corralejo (N 28.6873 W 13.8333)

Here, you can imagine yourself in a desert. Just park along the roadside and walk west for a few hundred meters through the dunes of Corralejo to get a feeling for the extensive area covered by the dunes (Figs. 8.47 and 8.48).

Sand dunes form an important and unique system that can be mobile or fixed by vegetation (Fig. 8.48A). Dry sand particles lack cohesion, and wind erosion is the main limiting factor for vegetation. Dunes can become vegetated when the wind power is sufficiently low. Once vegetated, a much higher wind stress is needed to destroy the plants and reactivate the dunes. The Corralejo field of dunes is largely devoid of vegetation, and wind transport of sand frequently shifts the detailed dune arrangement (Fig. 8.48B).

A close inspection of the sand will show the presence of a fine fraction with quartz crystals, a component that does not exist in the lavas of the Canaries. In fact, the quartz comes from Africa with relatively frequent strong hot winds (eg, the *Calima*) that bring massive dust invasions to the Canary Islands (Fig. 8.48C), particularly on Fuerteventura and Lanzarote, sometimes leading to "blood rain" (Criado and Dorta, 2003), which is when red dust falls with coinciding precipitation. Microscopic observation of the aeolian sand shows foraminifera and shell fragments of the extinct tropical Pliocene fauna, the main source of "white" beaches in the archipelago (Fig. 8.48D).

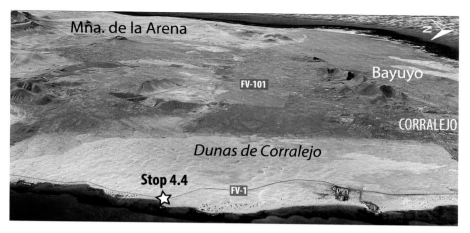

FIGURE 8.47

View of the northeast coast of Fuerteventura, south of *Corralejo* (Google Earth image), showing the aeolian sands of the "Dunas de Corralejo," a 2.7 × 7-km field of aeolian sand deposits

FIGURE 8.48

(A) and (B) Wind-blown aeolian sand also blows over highway FV-1, which has to be cleaned frequently. (C) Massive intrusion of dust from Africa into the Canaries (image courtesy of *NASA*). Vegetation reduces the impact of wind and water and helps to stabilize sand dunes (A). (D) Photomicrograph of the aeolian sand showing foraminifera and shell fragments of an extinct tropical Pliocene fauna.

The day ends here. If you are staying in *Corralejo*, then your return journey is short today.

Please note that a car ferrry operates between *Corralejo* and *Playa Blanca* on Lanzarote, returning within the day (approximately €80). If you wish to explore another island while based on Fuerteventura, then this offers a practical, albeit slightly costly, solution (see chapter The Geology of Lanzarote).

END OF DAY 4.

GLOSSARY

Aa lava a type of basaltic lava featuring a jagged blocky surface. The term originates from the Polynesian language of Hawaii.

Alkali basalt type of *basalt* containing greater amounts of sodium and potassium and fewer amount of silica relative to the average *tholeiitic* basalt. Phenocrysts of *olivine, augite*, and *plagioclase* may occur.

Alkaline magma series a sequence of *magma* evolution (alkali basalt → trachyte → phonolite) that produces progressively higher alkali elements, like sodium and potassium, with higher silica content (See also *differentiation*).

Amphibole a *rock-forming mineral* that possesses a double-chain silicate structure and various elements such as calcium, iron, sodium, magnesium, aluminum, and OH (water) between the silica-based double chains. It occurs in intermediate stages of the discontinuous strand of *Bowen's reaction series*.

Andesite a type of igneous rock with more silica than basalt and less than rhyolite. It has similar amounts of silica as trachyte; however, it has lower concentrations of alkali elements. Andesite derives its name from the most common rock type of the Andes and commonly contains pyroxene, plagioclase feldspar, and amphibole.

Ankaramite a *pyroxene*-rich and *olivine*-rich (*pyroxene > olivine*) basalt often found in *ocean islands* and some island arcs. Ankaramites often erupt from deep magma reservoirs, hence the high *pyroxene* and *olivine* content, and are sometimes referred to as "oceanites" in older accounts.

Ash volcanic rock particles with a typical size of <2.65 mm. It is ejected from a volcanic vent and often protrudes high up toward the atmosphere. When it settles, an "ash-fall" deposit is formed (see also *tephra*).

Biotite a *rock-forming mineral*, similar to muscovite in terms of structure (*sheet silicate*), although considerably darker in color, ranging from dark brown or greenish-black to black. It contains iron (Fe) in place of potassium (K). When euhedral, it often appears six-sided (pseudohexagonal).

Bomb volcanic ejecta (*tephra*) that is more than 65 mm (6.5 cm) across.

Bowen's reaction series gives the sequence by which minerals crystallize from cooling magma. The series is divided in two series: the discontinuous series (olivine → pyroxene → amphibole → biotite) and the continuous series (*calcium-rich* to *sodium-rich plagioclase*). The final minerals to crystallize as *magma* cools, and the final meeting point of the reaction series, are *quartz, white mica*, and *potassium-rich feldspar*, but this stage is not reached in the Canaries.

Breccia is a fragmented rock that contains larger angular clasts in a finer-grain cement or matrix. Breccias form from sedimentary, volcanic, tectonic, or impact processes. They all require dynamic motion to achieve rock formation.

Calcite a *rock-forming mineral* with chemical composition $CaCO_3$ (calcium carbonate). It can be precipitated from evaporating sea water or from warm aqueous solutions. It is essential in the making of calcareous sedimentary rocks and is frequent in hydrothermal vein and crack fillings.

Caldera volcanic depression typically more than 1 km in diameter, often sub-circular, and can be >10 km in diameter. It is produced as a result of a collapse of a *magma chamber* roof into an emptying *magma chamber* (collapse caldera). Explosive eruptions are frequently associated with this collapse, leading to intra-caldera and extra-caldera ignimbrite (ash-flow) deposits. Note, however, giant landslide embayments and wide craters are also frequently referred to as "Calderas" in the Canary Islands.

Chalcedony a variety of fracture and vein *quartz*, but in a microcrystalline form. Usually it has a light blue color. It occurs in geodes, as fill in former vesicles, and in fractures together with other secondary fillings.

Chlorite a sheet silicate related to the mica group of minerals and that is varying in shades of green. Chlorite is frequently found in low-grade metamorphic rocks and in diagenetically affected sedimentary rocks, but it is a sign of alteration when seen in igneous rocks.

Cinder cone a volcanic cone built up of basaltic ejecta and pyroclastic material (scoria or spatter). Deposits usually fluctuate between moderately and poorly consolidated, with straight outer slopes of ≤30 degree. Cinder cones generally form due to a single eruptive episode (*monogenetic*).

Clinopyroxene a *rock-forming mineral*, usually black and prism-shaped. Rich in iron and magnesium, it belongs to the single-chain silicates. Pyroxene comes second, just after *olivine*, in Bowen's discontinuous reaction series and is therefore frequent in mafic and intermediate igneous rocks.

Cone sheet curved or circular *dyke* intrusion with the shape of an inverted cone (ie, the intrusion dips inward, toward a common center). Cone sheets usually occur in swarms of many hundreds of individual sheets.

Conglomerate rock type with coarse-grain rounded clasts (more than 2 mm in size) within a finer-grain cement or matrix. The characteristic roundness of the clasts suggests dynamic transport prior to lithification.

Dacite an igneous rock, rich in silica, with a composition between *andesite* and *rhyolite*. Dacite is a member of the *subalkaline* or *tholeiitic magma series*.

Debris avalanche a large and speedy mass movement of mixed debris driven by gravity. Debris avalanches form from large landslides and usually contain a combination of debris types.

Debris avalanche deposit Product of a *debris avalanche*.

DRE (dense rock equivalent) a measure to describe the volume of erupted pyroclastic products of an eruption. It is usually given in km^3. The volume of ejected tephra during an explosive eruption can be calculated by mapping the volume of deposited material and correct the result via the density of the deposit to the volume of massive bulk rock (ie, equivalent to lava or dense rock).

Differentiation (magmatic) a collection of processes that can produce a series of daughter magmas from one parental *magma*. Although several factors affect differentiation (eg, *magma mixing*, *crustal assimilation*), it is widely assumed to dominantly work through *fractional crystallization*, that is, the progressive change of a parental magma's composition as a result of the successive removal of mineral species following *Bowen's reaction series*.

Diorite an intrusive rock made up of black and white speckles of feldspar, pyroxene, and amphibole. It is of intermediate silica content and belongs to the *subalkaline magma series*.

Dolerite is medium-grain intrusive rock that is compositionally equivalent to basalt but is of coarser grain size. It can contain *clinopyroxene* and/or *orthopyroxene*, *plagioclase*, and *olivine*. Dolerite is usually found in dykes, as well as sills, lopoliths, laccoliths, and other smaller-scale intrusions.

Dyke solidified magma that fills a fracture or fissure in surrounding rock. Dykes are overall penny-shaped and therefore usually appear as linear features in outcrop. Dykes are often approximately 1 m wide when basaltic, but rhyolitic dykes can reach tens of meters in thickness.

Enclave (in magma mixing) usually consist of globular volumes of rock surrounded by the intruded host rock. This texture records the temperature-related viscosity differences of the component magmas. The end-members of a mixed magma may be broadly related or may have fundamentally different origins.

Feldspar a *rock-forming mineral* that is the most widely recognized mineral in the Earth's crust. Feldspar comes in a number of forms, including potassium feldspar (K-feldspar) and plagioclase feldspar (Ca-feldspar). All feldspars have a complex silicate skeleton structure (tecto-silicate). The continuous series of *Bowen's reaction series* advances from plagioclase to K-feldspar. This frequently causes feldspar crystals to show internal zoning that reflects progressive magma development (like tree rings, for example).

Felsic an igneous rock that is light in color and holds a high proportion of feldspar, muscovite, and quartz (silica) is referred to as felsic (felsic = feldspar + silica).

Fractional crystallization the process that causes a change in the general chemistry of a *magma* through means of removing crystals of lower silica content than the melt. Crystals are formed at distinctive temperatures and are afterward consecutively separated from the melt because of settling or various other physical process as *magma* progressively cools (See also *differentiation*).

Gabbro a coarse-grain igneous rock that is medium to dark in color and holds *pyroxene*, *plagioclase*, and at times *olivine*. It is chemically comparable to *basalt*, but it formed at a greater depth (slower cooling) to allow the growth of larger minerals than in *basalt* or *dolerite*.

Geomagnetic Polarity Time Scale (GPTS) a time scale that has been constructed from analysis of seafloor magnetic anomalies and reversal sequences on land. Positive anomalies are periods when the magnetic field was pointing toward the north (as at present), whereas negative anomalies are periods when the magnetic field reversed to the opposite direction. The current GPTS time scale reaches back to the oldest oceanic crust (Upper Jurassic).

Graben a down-faulted area of land between two bounding faults, making an elongate basin. A graben is caused by extensional stresses in the Earth's crust.

Hornito a rootless spatter cone that forms from a basaltic lava flow (normally pahoehoe). A hornito is created when lava is pushed up through an opening in the cooled top surface of a lava flow, after which lava collects around the opening. Hornitos are typically steep-sided and form noticeable pinnacles or stacks. They are "rootless" because they are supplied from the interior of the underlying lava flow, rather than from a deeper magmatic fissure or conduit system.

Hot spot an uncharacteristically hot portion of the Earth's *mantle* that allows buoyant mantle material to rise toward the Earth's surface. Approximately 100 km beneath the surface, the mantle material (*peridodite* or *pyroxenite*) starts to melt from decompression. Rising *magma* (the melt) then feeds eruptions on the surface of the Earth. Hot spots are assumed to feed many ocean island volcanoes and are generally considered stationary.

Hyaloclasite an aggregated rock containing fine fragments of volcanic materials that was created from the explosive reaction of hot *lava* and water. This rock type occurs frequently in shield successions of ocean islands where lava runs into the sea or where submarine facies merge into nonmarine units.

Hydrothermal alteration a chemical process in rocks and minerals brought about by contact with hydrothermal solutions (hot waters). This is a widespread process in volcanic rocks, especially where ground or rainwater gets heated from magmatic activity. Fluid flow can be in veins or cracks or pervasive.

Ignimbrite (deposit) a rock body that is poorly sorted and was assembled from pyroclastic (pumice-rich) flow currents. Ignimbrites (ie, ignimbrite deposits) are regularly sheet-like or fill valleys and have a tendency to "smooth" the topography. Welded and nonwelded varieties exist, which reflect various eruptive temperatures and flow behavior.

Isopach contour that connects points of the same thickness of a stratigraphic unit. An isopach is measured perpendicular to the bedding and therefore shows the true thickness of a stratigraphic unit.

Jasper a common and impure variety of fracture and vein *quartz*. Color is typically red, reddish-brown, or yellow. It is the result of silica remobilization from sediments and volcanic deposits and involves heat and hydrothermal fluids. Mineral impurities usually cause the reddish color.

Juvenile magma describes magma of an ongoing eruption, as opposed to recycled fragments of previous eruptions that are often erupted in phreatomagmatic or exhalative events.

Lahar mix of mud and water enriched in volcanic material, produced by heavy precipitation on a volcano or ice melting from heat flow caused by a volcano. Lahars can be hot and very devastating. The term is originally from Indonesia.

Landslide a large movement of mass that occurs when Earth material is destabilized and moves down-hill. Landslides produce a *debris avalanche* or *debris flow deposit*.

Lava flow molten rock that moves on the surface of the Earth in a coherent fashion and was emitted during a volcanic eruption. Thick, viscous lava flows are termed "*aa lava*" and very runny, usually thin lava flows are termed "*pahoehoe*" lava, following Hawaiian terminology. Geologists frequently utilize the expression "*lava*" for solidified lava flows, whereas a "*lava flow*" would be actively flowing lava.

Maar a form of volcanic crater, frequently occupied by a lake or sediment, encompassed by a ring of tuffaceous deposits produced by often ferocious explosions of *magma* in contact with water (any source). This often prompts high proportions of fragments of country rock in the lower maar deposits that give way to more juvenile magmatic materials in the upper layers.

Mafic *magma*, rock or minerals that hold high proportions of magnesium and iron (mafic = magnesium and iron), such as an igneous rock that holds adequate proportions of ferromagnesian minerals such as *olivine* and *pyroxene* coupled with low concentrations of *feldspar* and *quartz*.

Magma molten rock at high temperatures beneath the Earth's surface. Magma contains crystals, liquid melt, and dissolved volatiles that become gas bubbles when pressure is released (eg, on ascent).

Magma chamber (or magma reservoir) zone within the Earth where magma stalls in equilibrium with the crustal rocks before traveling further or solidifying. Magma chambers may take the form of a sponge-like mush of solid minerals and liquid magma and can be several kilometers in width. They provide the location for *magma mixing* and *differentiation*.

Magma Mixing is the process by which two magmas interact and attempt to hybridize, resulting frequently in a mix of two magma types in rocks or deposits (composite). Note, complete blending of two magmas is not common and incomplete mixing is what is frequently observed.

Mantle zone above the Earth's core and underneath the Earth's crust, approximately 2300 km thick, and most likely of dominantly *garnet peridotite* with regard to composition, but with various internal inhomogeneities. The mantle makes up 84% of the Earth's volume and 68% of its mass. Different *magma* types (see *alkaline series*, *tholeiitic series*) are produced from melting the mantle at different depths and to variable degrees.

Monogenetic rocks or rock formations that are created from one particular source or process, for example, through a single eruptive event.

MORB (mid-ocean ridge-type *basalt*) a broad term for the *tholeiitic basalt series* along mid-oceanic spreading centers (ridges). Tholeiitic rocks are *sub-alkaline* in the *TAS diagram* and can occur on ocean islands also, usually indicating large degrees of mantle melting relative to OIB basalts.

Muscovite *rock-forming mineral*, of the mica group, that forms thin flaky sheets that range from colorless to exceptionally pale green or light brown, but usually with a silvery sheen. It incorporates K rather than Fe (*biotite*, black) or Mg (*phlogopite*, brown).

Oceanic crust made of *mafic* basaltic and gabbroic rocks (MORB *type*) that cover approximately 60% of the Earth's surface. Oceanic crust is approximately 6−8 km thick and has a mean density of approximately 3.2−3.3 g/cm^3 (see also *mafic* and *tholeiitic magma series*). A layer of pillow basalts intermingled with sediment grades downward through feeder dykes into what used to be the magma chamber beneath the ridge, the *gabbros*.

Ocean island an area of volcanic rock, volcanic detritus, and volcanic soil that is totally surrounded by water and rises to the surface from the floor of the ocean.

OIB (ocean island basalt) type of rock occurring on many volcanic intraplate *ocean islands* (ie, away from tectonic plate boundaries). OIB is usually associated with *hot spots* (see also *alkaline magma series*).

Olivine *rock-forming mineral* typically olive green in color and found in silica-poor igneous rocks that are *mafic* (ie, rich in Mg and Fe), such as in *basalts* and *gabbros*. Olivine contains isolated Si-tetrahedra (inosilicate).

Phlogopite a *rock-forming mineral*, part of the mica group, similar to *muscovite* and *biotite*, although it contains more magnesium. Color ranges from brown to dark golden, with shiny or pearly luster. It is quite often found in the *felsic* volcanic and plutonic rocks of the Canary Islands.

Pahoehoe type of *lava flow* with a smooth surface texture, in contrast to the sharp blocky texture of "*aa* lava." The term is derived from the Polynesian language of Hawaii.

Peridotite ultra-*mafic* coarse-grain rock, with *olivine* and *pyroxenes*, with or without garnet. Peridotite possesses a rather low content of silica and is usually very dense. Peridotite seems, by all accounts, to be the dominant rock of the Earth's *mantle* and produces a basaltic *magma* on partial melting. Peridotite fragments coming from the mantle do occasionally occur in ocean island lavas as mantle nodules or xenoliths.

Phonolite a medium to light grey to green igneous rock of a fine-grain nature and of evolved alkaline composition. It is the highest differentiation product of the *alkaline magma series* and is frequent on Atlantic ocean islands. Phonolite is also found on the continents where it forms during anorogenic upwarping and rifting. When struck, *phonolite* makes a distinctive sound, hence the name (from Greek soundstone).

Picrite a coarse-grain type of basaltic igneous rock holding abundant *olivine* over *pyroxene*, and maybe small portions of plagioclase feldspar (usually >18 wt% MgO in the rock). Picrites indicate high degrees of mantle melting.

Piedmont(s) a landform at the foot of a mountain made from the accumulation of debris due to shifting rivers and streams.

Pillow lava a lava flow that solidified in or under water and that formed pillow structures. Pillow lava is characterized by a glassy rind enclosing a coarser center. Its bulbous pillow and tube-like shapes are brought about by chilling of *magma* against water, creating a solid rind that enables further *magma* flow inside the tubes. These tubes regularly stack up in heaps and, when cut at a high angle, they appear like a pillow stack.

Plagioclase (feldspar) a *rock-forming mineral* that is formed by a silicate framework (tecto-silicate). Plagioclase is rich in calcium and aluminum, in addition to holding sodium. Plagioclase feldspar is the most abundant mineral in the Earth's crust and is frequently found in mafic and intermediate igneous rocks of plutonic and volcanic derivation. Ca-rich plagioclase forms the beginning in *Bowen's continuous reaction series.*

Plinian eruption an explosive volcanic eruption of pyroclastic ejecta characterised by a plume of ash and gas, which because of its high gas content, rises high into the atmosphere. Named after Plini's the Younger's description of the eruption of Mt. Vesuvius in AD 79, which took his uncles' life (Plini the Elder).

Polygenetic volcano volcano that originates from more than one period of activity, usually from a long history or succession of constructive events.

Pyroclastic flow a flow made up of hot volcanic gas and rock fragments (tephra), either originating from gravity or from a blast down the flank of an erupting volcano. Pyroclastic flows travel at speeds of ≤500 km/h and are as hot as 600°C in temperature at distances of up to 10 km away from their vent. An *ignimbrite deposit* often results from a pyroclastic flow event.

Pyroxene a *rock-forming mineral* that is black, prismatic, and rich in iron and magnesium. It possesses a single-chain silicate structure and comes second in Bowen's discontinuous reaction series (after *olivine* and before *amphibole*). *Pyroxene* occurs as clino-pyroxene and ortho-pyroxene, which differ on a crystallographic level.

Quartz a *rock-forming mineral* that is usually colorless or white but can be found in an extensive variety of colors depending on its composition (eg, blue chalcedony or red carneol and jasper). If crystalline, crystals are typically six-sided prisms terminated by a six-faced pyramid. Quartz has a complex framework structure (tecto-silicate) and is last in Bowen's reaction series. Note, however, that primary igneous quartz is not present in the Canary Islands. This being said, secondary (hydrothermal) quartz is common (eg, as vein, fracture, or vesicle fills).

Radiometric dating a technique used to date natural materials such as rocks or plant remains. Usually the ratio of a naturally occurring radioactive isotope relative to their decay product is measured, which allows calculation of the age of an object if the decay constant for the isotope system is known.

Rheomorphic flow postdepositional (downslope) movement of hot, plastic material (eg, an *ignimbrite deposit*) prompting fold and thrust structures, for example, in some of the ignimbrites in southern Gran Canaria.

Rhyolite a fine-grain, normally light-colored (white, gray, reddish) igneous rock that is usually very high in silica content (≥70%), often containing *feldspar* (K-*feldspar* and perhaps *plagioclase*), *amphibole*, biotite, and possibly *quartz*. Rhyolite is the highest differentiation product of the *subalkaline (tholeiitic) magma series* or the melting product of crustal rocks like silica-rich sediments.

Rift zone (volcanic) an alignment of *cinder cones* and other vent types at the surface fed by a swarm of *dykes* at depth. In the Canary Islands, volcanic rift zones typically take the form of long and narrow ridges. These volcanic rifts are not identical to mid-ocean rift situations, however.

Rock-forming minerals the most widely recognized minerals that make up 95% of all rocks in the Earth's crust and *mantle*. Silicates are the most fundamental rock-forming minerals in magmatic rocks and are subdivided into two distinct groups, *mafic* and *felsic*. The mafic ones are *pyroxene, amphibole, biotite*, and *olivine*, whereas the felsic minerals are *potassium K-feldspar, plagioclase feldspar, muscovite*, and *quartz*.

Scoria basaltic ejecta with sizes ranging from pebble to fist size, encompassing a rough texture and frequent gas pockets or vesicles. Scoria usually forms from *lava* fountains, eruptive clouds, or by the grinding and crunching of materials by an oncoming yet cooling flow of lava.

Scoria cone (see *cinder cone*)

Seamount large submarine volcanic mountain that rises, at minimum, 1000 m above the ambient ocean floor. Once a volcanic seamount breaks the surface of the ocean, it becomes an *ocean island*.

Shield volcano large volcanic landform that comprises shallow slopes and is often crowned by a caldera or pit crater (eg, Hawaii). Shield volcanoes look like a warrior's shield lying on its back and thus with a convex upward shape, hence the name.

Sill a tabular intrusion, igneous in nature, with semiconcordant contact surfaces with the bedding or layering of the intruded rock.

Spatter cone a small volcanic cone that formed from *tephra* propelled out from a central vent as lapilli or bombs of relatively fluid basaltic lava (similar to *cinder* or *scoria cones*).

Strombolian eruption the kind of eruption in which pockets of gas periodically burst at the highest point of a basaltic *magma* column, tossing *tephra* into the air. Very common in *cinder cone* eruptions.

Sub-alkaline magma series magmatic evolutionary series that comprises the majority of igneous rocks on the globe (>80%). The *subalkaline series* arises from partial melting of mantle rocks at shallow depths and/or larger degrees of partial melting relative to alkaline basaltic magmas (see also *tholeiitic magma series*).

Syenite a coarse-grain felsic plutonic rock of the same general composition as *trachyte* or *phonolite*; however, slow cooling at depth permitted the growth and development of larger crystals. Syenite is found as discrete intrusions in the stable continental crust and in the cores of some ocean island volcanoes, for example, on Gran Canaria and Fuerteventura.

Tephra any particle or fragment, of any size, expelled from a volcano (eg, *ash, lapilli*, and *bombs*).

Tholeiite fine-grain basaltic igneous rock. A *subalkaline* type of basalt commonly found on the ocean floor (see *MORB*). It usually reflects large degrees of mantle melting.

Tholeiitic magma series a *subalkaline magma* evolutionary series (basalt → andesite → dacite → rhyolite). It is the most common magmatic series at mid-ocean ridge settings (see *MORB*).

Trachyte an intermediate volcanic rock of alkaline affinity. Similar to *andesite* in silica concentration, but sodium and potassium are both higher than in *andesite*. Trachyte largely derives from alkaline basalt *magma* by *differentiation* and occurs frequently in hot spot settings. Trachyte is the volcanic equivalent of *syenite* and often erupts as domes or explosively.

Viscosity the internal resistance to flow of a liquid, such as in *magma*. Viscosity is primarily a function of composition and temperature and is usually higher in felsic magmas than in mafic ones.

Vitrophyre a glassy volcanic rock that could conceivably hold larger crystals (phenocrysts) implanted in the glassy ground mass. A vitrophyre is formed when molten rock (magma) cools rapidly and solidifies on contact with a cold substrate, such as the Earth's surface.

Volatile a substance that melts or vaporizes at medium to low pressures and temperatures. As an example, volatiles dissolved in *magma* (eg, H_2O, CO_2, H_2S, etc.) are discharged as a free gas phase (via bubbles) when the temperature and/or pressure on the *magma* is lowered to the point where the magma's volatile solubility falls below its volatile content.

References

Abdel Monem, A., Watkins, N.D., Gast, P.W., 1971. Potassium-argon ages, volcanic stratigraphy, and geomagnetic polarity history of the Canary Islands: Lanzarote, Fuerteventura, Gran Canaria, and La Gomera. Am. J. Sci. 271, 490−521.

Abdel Monem, A., Watkins, N.D., Gast, P.W., 1972. Potassium-argon ages, volcanic stratigraphy, and geomagnetic polarity history of the Canary Islands: Tenerife; La Palma, and Hierro. Am. J. Sci. 272, 805−825.

Ablay, G.J., Marti, J., 2000. Structure, stratigraphy and volcanic evolution of the Pico Teide-Pico Viejo formation, Tenerife, Canary Islands. J. Volcanol. Geoth. Res. 103, 175−208.

Ablay, G.J., Ernst, G.G.J., Martí, J., Sparls, R.S.J., 1995. The 2 Ka subplinian eruption of Montaña Blanca, Tenerife. Bull. Volcanol. 57 (5), 337−355.

Ablay, G.J., Carroll, M.R., Palmer, M.R., Martí, J., Sparks, R.S.J., 1998. Basanite-phonolite lineages of the Teide Pico Viejo volcanic complex, Tenerife, Canary Islands. J. Petrol. 39 (5), 905−936.

Afonso, A., Aparicio, A., Hernández −Pacheco, A., Badiola, E.R., 1974. Morphology evolution of Teneguía volcano area. Estud. Geol. 19−26, Vol. Teneguía.

Albert-Beltran, J.F., Araña, V., Diez, J.L., Valentin, A., 1990. Physical-chemical conditions of the Teide Volcanic System (Tenerife, Canary Islands). J. Volcanol. Geoth. Res. 43, 321−332.

Allibon, J., Bussy, F., Lewin, É., Darbellay, B., 2011. The tectonically controlled emplacement of a vertically sheeted gabbro-pyroxenite intrusion: feeder-zone of an ocean-island volcano (Fuerteventura, Canary Islands). Tectonophysics 500, 78−97.

Alonso, J.J., Araña, V., Martí, J., 1988. La ignimbrita d Arico (Tenerife). Mecanismos de emission y de desplazamiento. Revista de la Sociedad Geológica de España 1 (1−2), 15−24.

Ancochea, E., Fúster, J.M., Ibarrola, E., Cendrero, A., Coello, J., Hernán, F., et al., 1990. Volcanic evolution of the island of Tenerife (Canary Islands) in the light of new K-Ar data. J. Volcanol. Geoth. Res. 44, 231−249.

Ancochea, E., Hernán, F., Cendrero, A., Cantagrel, J.M., Fúster, J.M., Ibarrola, E., et al., 1994. Constructive and destructive episodes in the building of a young oceanic island, La Palma, Canary Islands, and genesis of the Caldera de Taburiente. J. Volcanol. Geoth. Res. 60, 243−262.

Ancochea, E., Brandle, J.L., Cubas, C.R., Hernán, F., Huertas, M.J., 1996. Volcanic complexes in the eastern ridge of the Canary Islands: the Miocene activity of the island of Fuerteventura. J. Volcanol. Geoth. Res. 70, 183−204.

Ancochea, E., Huertas, M., Cantagrel, J.M., Coello, J., Fúster, J.M., Arnaud, N., et al., 1999. Evolution of the Cañadas edifice and its implications for the origin of the Las Cañadas Caldera (Tenerife, Canary Islands). J. Volcanol. Geoth. Res. 88 (3), 177−199.

Ancochea, E., Brandle, J.L., Huertas, M.J., Cubas, C.R., Hernán, F., 2003. The felsic dikes of La Gomera (Canary Islands): identification of cone sheet and radial dike swarms. J. Volcanol. Geoth. Res. 120, 197−206.

Ancochea, E., Hernán, F., Huertas, M.J., Brändle, J.L., Herrera, R., 2006. A new chronostratigraphical and evolutionary model for La Gomera: implications for the overall evolution of the Canarian Archipelago. J. Volcanol. Geoth. Res. 157 (4), 271−293.

Ancochea, E., Brändle, J.L., Huertas, M.J., Hernán, F., Herrera, R., 2008. Dike-swarms, key to the reconstruction of major volcanic edifices: the basic dikes of La Gomera (Canary Islands). J. Volcanol. Geoth. Res. 173, 207−216.

Anguita, F., Hernán, F., 1975. A propagating fracture model versus a hot spot origin for the Canary Islands. Earth Planet. Sci. Lett. 27, 11−19.

Anguita, F., Hernán, F., 2000. The Canary Islands origin: a unifying model. J. Volcanol. Geoth. Res. 103, 1−26.

Aparicio, A., Araña, V., Díez-Gil, J.L., 1994. Una erupción hidromagmática en la isla de Lanzarote: La caldera de El Cuchillo. In: García, A., Felpeto, A. (Eds.), *In memoriam Dr. José Luis Díez Gil*. Serie Casa de Los Volcanes. Cabildo Insular de Lanzarote, pp. 109−120.

589

Aparicio, A., Bustillo, M.A., Garcia, R., Araña, V., 2006. Metasedimentary xenoliths in the lavas of the timanfaya eruption (1730–1736, Lanzarote, Canary Islands): metamorphism and contamination processes. Geol. Mag. 143, 181–193.

Aparicio, A., Tassinari, C.C.G., Garcia, R., Araña, V., 2010. Sr and Nd isotope composition of the metamorphic sedimentary and ultramafic xenoliths of Lanzarote (Canary Islands): implications for magma sources. J. Volcanol. Geoth. Res. 189, 143–150.

Araña, V., Ibarrola, E., 1973. Rhyolitic pumice in the basaltic pycroclasts from the 1971 eruption of Teneguía volcano, Canary Islands. Lithos 6 (3), 273–278.

Araña, V., Carracedo, J.C., 1980. Los Volcanes de Las Islas Canarias – Canarian Volcanoes (Vol. 3). Editorial Rueda, Madrid, p. 175.

Arnaud, N., Huertas, M.J., Cantagrel, J.M., Ancochea, E., Fúster, J.M., 2001. Edades 39 Ar/40 Ar de los depósitos de Roques de García (Las Cañadas, Tenerife). Ceogaceta 29, 19–22.

Balcells, R., Barrera, J.L., Gómez, J.A., 1992. Geological Map 21-21/21-22 Isla de Gran Canaria.

Barker, A., Troll, V., Carracedo, J., Nicholls, P., 2015. The magma plumbing system for the 1971 Teneguía eruption on La Palma, Canary Islands. Contributions to Mineralogy and Petrology, 170 (54).

Berggren, W.A., Kent, D.V., Swisher III, C.C., Aubry, M.-P., 1995. A revised Cenozoic geochronology and chronostratigraphy. Soc. Sediment. Geol. Spec. Publ. 54, 129–212.

Bethéncourt-Massieu, A., 1982. Los terremotos de 1793 en El Hierro. Homenaje a Alfonso Trujillo 2, 13–28.

Blake, S., 1990. Viscoplastic models of lava domes. *IAVCEI Procedings in Volcanology*, Vol.2. Lava Flows and Domes. Springer Verlag, Heidelberg, pp. 88–126.

Blumenthal, M.H., 1961. Rasgos principales de la geologia de las Islas Canarias, con datos sobre Madeira. Bolletin del Instituto Geológico y Minero de España tomo LXXII, 1–130.

Bonelly Rubio, J.M., 1950. Contribución al estudio de la erupción del volcán del Nambroque o San Juan (Isla de La Palma), del 24 de junio al 4 de agosto de 1949. Instituto Geográfico y Catastral, Madrid, p. 34.

Booth, B., 1973. The Granadilla pumice deposit of southern Tenerife, Canary Islands. Proc. Geol. Assoc. 84, 353–370.

Bory de St. Vincent, J.B.G.M., 1803. Essais sur les isles Fortunées et l'antique Atlantide, ou, Précis de l'histoire générale de l'archipel des Canaries. Baudouin, Paris, p. 522. an XI.

Bosshard, E., MacFarlane, D.J., 1970. Crustal structure of the Western Canary Islands from seismic refraction and gravity data. J. Geophys. Res. 75, 4901–4918.

Bourcart, J., Jérémine, E., 1937. La Grande Canarie. Bull. Volcanol. Ser. II 2, 1–77.

Bravo, T., 1964. Estudio Geológico y Petrográfico de la Isla de La Gomera (Tesis Doctoral). Estudios Geológicos. Vol. 20. Madrid: Instituto Lucas Mallada (CSIC): 1–56.

Brey, G., Schmincke, H.-U., 1980. Origin and diagenesis of the Roque Nublo volcanics II. Bull. Volcanol. 43–1, 15–33.

Brown, R.J., Branney, M.J., 2004. Event-stratigraphy of a caldera-forming ignimbrite eruption on Tenerife. The 273 ka Poris formation. Bull. Volcanol. 66, 392–416.

Brown, R.J., Barry, T.L., Branney, J.J., Pringle, M.S., Bryan, S.E., 2003. The quaternary pyroclastic succession of southeast Tenerife, Canary Islands, explosive eruptions, related caldera subsidence, and sector collapse. Geol. Mag. 140, 265–288.

Bryan, S.E., Martí, J., Cas, R.A.F., 1998. Stratigraphy of the bandas del Sur formation. An extracaldera record of quaternary phonolitic ex-plosive eruptions from the Las Cañadas edifice, Tenerife (Canary Islands). Geol. Mag. 135, 605–636.

Bryan, S.E., Cas, R.A.F., Martí, J., 2000. The 0.57 Ma plinian eruption of the Granadilla member, Tenerife (Canary Islands): an example of complexity in eruption dynamics and evolution. J. Volcanol. Geoth. Res. 103, 209–238.

Burchardt, S., Troll, V.R., Mathieu, L., Emeleus, H., Donaldson, C.H., 2013. Ardnamurchan 3D cone-sheet architecture explained by a single elongate magma chamber. Sci. Rep. 3, 2891.

Calamai, A., Ceron, P., 1970. Air convection within "Montaña del Fuego" (Lanzarote, Canary Archipelago). Geothermics 2, 611—614.

Canales, J.P., Dañobeitia, J.J., 1998. The Canary Islands swell: a coherence analysis of bathymetry. Geophys. J. Int. 132, 479—488.

Cantagrel, J.M., Cendrero, A., Fúster, J.M., Ibarrola, E., Jamond, C., 1984. K-Ar chronology of the volcanic eruptions in the Canarian archipelago: Island of La Gomera. Bull. Volcanol. 47, 597—609.

Carracedo, J.C., 1979. Paleomagnetismo e historia volcánica de Tenerife. Aula Cultura Cabildo Insular de Tenerife, Santa Cruz de Tenerife, p. 81.

Carracedo, J.C., 1984. Marco Geográfico, En: Geografía Física de Canarias, vol. 1. Editorial Interinsular Canaria, Santa Cruz de Tenerife, pp. 10—16.

Carracedo, J.C., 1994. The Canary Islands: an example of structural control on the growth of large oceanic-island volcanoes. J. Volcanol. Geoth. Res. 60 (3—4), 225—242.

Carracedo, J.C., 1996a. Morphological and structural evolution of the western Canary Islands: hotspot-induced three-armed rifts or regional tectonic trends? J. Volcanol. Geoth. Res. 72, 151—162.

Carracedo, J.C., 1996b. A simple model for the genesis of large gravitational landslide hazards in the Canary Islands. In: McGuire, W., Neuberg, J., Jones, A. (Eds.), Volcano Instability on the Earth and Other Planets, *Geological Society of London, Special Publications*, London, 110. pp. 125—135.

Carracedo, J.C., 1999. Growth, structure, instability and collapse of Canarian volcanoes and comparisons with Hawaiian volcanoes. J. Volcanol. Geoth. Res. 94, 1—19.

Carracedo, J.C., 2008a. Outstanding geological values: the basis of Mt Teide's World Heritage nomination. Geol. Today 24 (3), 104—111.

Carracedo, J.C., 2008b. Canarian Volcanoes IV: La Palma, La Gomera y El Hierro. Rueda, Madrid, p. 213.

Carracedo, J.C., 2011. Geología de Canarias I: Origen, evolución, edad y volcanismo. Editorial Rueda, Madrid, p. 398.

Carracedo, J.C., Day, S.J., 2002. Canary Islands, Series Classic Geology in Europe, vol. 4. Terra Publishing, Harpenden, p. 249.

Carracedo, J.C., Troll, V.R., 2013. Teide Volcano — Geology and Eruptions of a Highly Differentiated Oceanic Stratovolcano. Springer-Verlag, Berlin Heidelberg, p. 234.

Carracedo, J.C., Rodríguez Badiola, E., 1991. Lanzarote. La erupción volcánica de 1730: estudio volcanológico de una de las erupciones basálticas fisurales de mayor duración y magnitud de la historia. El Museo Canario, Las Palmas de Gran Canaria. Cabildo Insular de Lanzarote, Arrecife de Lanzarote, p. 183.

Carracedo, J.C., Rodríguez Badiola, E., 1993. Evolución geológica y magmática de la isla de Lanzarote, Islas Canarias. Revista Academia Canaria Ciencias 4, 25—58.

Carracedo, J.C., Heller, F., Soler, V., 1986. Evidencia de un fenómeno de autoinversión de la RMN en piroclastos de la erupción de 1985 del Nevado del Ruiz (Colombia). Geogaceta 1, 11—13.

Carracedo, J.C., Rodríguez Badiola, E., Soler, V., 1990. Aspectos volcanológicos y estructurales, evolución petrológica e implicaciones en riesgo volcánico de la erupción de 1730 en Lanzarote, Islas Canarias. Estud. Geol. 46, 25—55.

Carracedo, J.C., Rodríguez Badiola, E.R., Soler, V., 1992. The 1730—1736 eruption of Lanzarote: an unusually long, high magnitude fissural basaltic eruption in the recent volcanism of the Canary Islands. J. Volcanol. Geoth. Res. 53, 239—250.

Carracedo, J.C., Day, S., Guillou, H., Rodríguez Badiola, E., 1996. The 1677 eruption of La Palma, Canary Islands. Estud. Geol. 52, 345—357.

Carracedo, J.C., Day, S.J., Guillou, H., Rodríguez Badiola, E., Canas, J.A., Pérez Torrado, F.J., 1998. Hotspot volcanism close to a passive continental margin: the Canary Islands. Geol. Mag. 135 (5), 591—604.

Carracedo, J.C., Day, S.J., Guillou, H., Gravestock, P., 1999. Later stages of volcanic evolution of La Palma, Canary Islands: rift evolution, giant landslides, and the genesis of the Caldera de Taburiente. Geol. Soc. Am. Bull. 111, 755–768.

Carracedo, J.C., Rodríguez Badiola, E., Guillou, H., De La Nuez, J., Pérez Torrado, F.J., 2001. Geology and volcanology of La Palma and El Hierro, Western Canaries., Volumen Especial de Estud. Geol. 57, 124.

Carracedo, J.C., Pérez-Torrado, F.J., Ancochea, E., Meco, J., Hernán, F., Cubas, C.R., et al., 2002. Cenozoic volcanism II: the Canary Islands. In: Gibbons, W., Moreno, T. (Eds.), The Geology of Spain. The Geological Society of London, London, pp. 439–472.

Carracedo, J.C., Guillou, H., Paterne, M., Scaillet, S., Rodríguez Badiola, E., Paris, R., et al., 2003. La Erupción y el tubo volcánico del volcán Corona (Lanzarote, Islas Canarias). Estud. Geol. 59, 277–302.

Carracedo, J.C., Guillou, H., Paterne, M., Scaillet, S., Rodríguez Badiola, E., Paris, R., et al., 2004a. Análisis del riesgo volcánico asociado al flujo de lavas en Tenerife (Islas Canarias): escenarios previsibles para una futura erupción en la isla. Estud. Geol. 60, 63–93.

Carracedo, J.C., Pestano Pérez, G., García Martínez, A., Rodríguez Valdés, E., Cabrera Peraza, J.M., Bermejo Dominguez, J.A., et al., Avance de un mapa de peligrosidad volcánica de Tenerife (escenarios previsibles para una futura erupción en la isla). Servicio Publicaciones de la Caja Gral, Ahorros de Canarias, p. 47.

Carracedo, J.C., Rodríguez Badiola, E., Pérez Torrado, F.J., Hansen, A., Rodríguez-González, A., Scaillet, S., et al., 2007a. La erupción que Cristóbal Colón vió en la isla de Tenerife (Islas Canarias). Geogaceta 41, 39–42.

Carracedo, J.C., Rodríguez Badiola, E., Guillou, H., Paterne, M., Scaillet, S., Pérez Torrado, F.J., et al., 2007b. Eruptive and structural history of Teide volcano and rift zones of Tenerife, Canary Islands. Geol. Soc. Am. Bull. 19, 1027–1051.

Carracedo, J.C., Fernandez-Turiel, Gimeno, D., Guillou, H., Klügel, A., Krastel, S., et al., 2011a. Comment on "The distribution of basaltic volcanism on Tenerife, Canary Islands: implications on the origin and dynamics of the rift systems" by A. Geyer and J. Martí (2010). Tectonophysics 483, 310–326, Tectonophysics 503 (3–4), 239–241.

Carracedo, J.C., Guillou, H., Nomade, S., Rodríguez-Badiola, E., Pérez-Torrado, F.J., Rodríguez-González, A., et al., 2011b. Evolution of ocean-island rifts: the northeast rift zone of Tenerife, Canary Islands. Geol. Soc. Am. Bull. 123, 562–584.

Carracedo, J.C., Pérez-Torrado, F.-J., Rodríguez-González, A., Fernandez-Turiel, J.-L., Klügel, A., Troll, V.R., et al., 2012a. The ongoing volcanic eruption of El Hierro, Canary Islands. EOS Trans. Am. Geophys. Union 93 (9), 89–90.

Carracedo, J.C., Pérez Torrado, F., Rodríguez-González, Soler, V., Fernández Turiel, J.L., Troll, V.R., et al., 2012b. The 2011 submarine volcanic eruption in El Hierro (Canary Islands). Geol. Today 28 (2), 53–58.

Carracedo, J.C., Troll, V.R., Zaczek, K., Rodríguez-González, A., Soler, V., Deegan, F.M., 2015a. The 2011/2012 submarine eruption off El Hierro, Canary Islands: new lessons in oceanic island growth and volcanic crisis management. Earth Sci. Rev. 150, 168–200.

Carracedo, J.C., Pérez-Torrado, F.J., Rodríguez-Gonzalez, A., Paris, R., Troll, V.R., Barker, A.K., 2015b. Volcanic and structural evolution of Pico do Fogo, Cape Verde. Geol. Today 31 (4), 146–152.

Cendrero, A., 1967. Nota previa sobre la geología del complejo basal de la isla de La Gomera. Estud. Geol. 23, 71–79.

Cendrero, A., 1971. Estudio geológico y petrológico del complejo basal de la isla de La Gomera (Canarias). Estud. Geol. 27, 3–73.

Clague, D., Sherrod, D.R., 2014. Growth and degradation of Hawaiian volcanoes. In: Poland, M.P., Takahasi, T.J., Landowski, C.M. (Eds.), Characteristics of Hawaiian Volcanoes, vol. 1801. U.S. Geological Survey, Reston, VA, pp. 97–146, Professional Paper.

Clarke, H., Troll, V.R., Carracedo, J.C., Byrne, K., Gould, R., 2005. Changing eruptive styles and textural features from phreatomagmatic to strombolian activity of basaltic littoral cones: Los Erales cinder cone, Tenerife, Canary Islands. Estud. Geol. 61 (3–6), 121–134.

Clarke, H., Troll, V.R., Carracedo, J.C., 2009. Phreatomagmatic to Strombolian eruptive activity of basaltic cinder cones: Montana Los Erales, Tenerife, Canary Islands. J. Volcanol. Geoth. Res. 180, 225–245.

Clarke, S.C.L., Spera, F.J., 1990. Evolution of the Miocene Tejeda magmatic system, Gran Canaria, Canary Islands. Contrib. Mineral. Petrol. 104, 681–699.

Coello, J., 1987. Las aguas subterráneas en las formaciones volcánicas del Norte de La Palma (Islas Canarias), Com. Simp. Int. Rec, 2000. Hidraulica, Canarias Agua, p. 19.

Coello, J., Cantagrel, J.M., Hernán, F., Fúster, J.M., Ibarrola, E., Anochea, E., et al., 1992. Evolution of the eastern volcanic ridge of the Canary Islands based on new K-Ar data. J. Volcanol. Geoth. Res. 53, 251–274.

Collier, J.S., Watts, A.B., 2001. Lithospheric response to volcanic loading by the Canary Islands: constraints from seismic reflection data in their flexural moat. Geophys. J. Int. 147, 660–676.

Coppo, N., Schnegg, P.A., Falco, P., Costa, R., Burkhard, M., 2008. Structural pattern of the Western Las Cañadas caldera (Tenerife, Ca-nary Islands) revealed by audiomag-netotellurics. Swiss J. Geosci. 101, 409–413.

Cousens, B.L., Spera, F.J., Tilton, G.R., 1990. Isotopic patterns in silicic ignimbrites and lava flows of the Mogán and lower Fataga Formations, Gran Canaria, Canary Islands: temporal changes in mantle source composition. Earth Planet. Sci. Lett. 96, 319–335.

Criado, C., Dorta, P., 2003. An unusual 'blood rain' over the Canary Islands (Spain). The storm of January 1999. J. Arid Environ. 55, 765–783.

Crisp, J.A., Spera, F.J., 1987. Pyroclastic flows and lavas of the Mogán and Fataga formations, Tejeda volcano, Gran Canaria, Canary Islands: mineral chemistry, intensive parameters, and magma chamber evolution. Contrib Mineral Petrol 96, 503–518.

Crofts, R., 1967. Raised beaches and chronology in northwest Fuerteventura, Canary Islands. Quaternaria 9, 247–260.

Cubas, C.R., 1978. Estudio de los domos sálicos de la Isla de La Gomera (Islas Canarias). I. Vulcanología. Estud. Geol. 34, 53–70.

Cubas, C.R., Ancochea, E., Hernán, F., Huertas, M.J., Brändle, J.L., 2002. Edad de los domos sálicos de la isla de La Gomera. Geogaceta 32, 71–74.

Darias y Padrón, D.V., 1929. Noticias generales históricas sobre la Isla del Hierro, una de las Canarias. Imprenta Curbelo, San Cristóbal de La Laguna, p. 407.

Davidson, I., 2005. Central Atlantic margin basins of North West Africa: geology and hydrocarbon potential (Morocco to Guinea). J. Afr. Earth Sci. 43, 254–274.

Davidson, J.P., de Silva, S., 2000. Composite volcanoes. In: Sigurdsson, H. (Ed.), Encyclopedia of Volcanoes. Academic Press, San Diego, pp. 663–681.

Dávila-Harris, P., 2009. Explosive ocean-island volcanism: the 1.8–0.7 Ma explosive eruption history o Cañadas volcano recorded by the pyroclastic successions around Adeje and Abona, southern Tenerife, Canary Islands. PhD Thesis, University of Leicester, Leicester, p. 170.

Dávila-Harris, P., Branney, M.J., Storey, M., 2011. Large eruption-triggered ocean-island landslide at Tenerife: onshore record and long-term effects on hazardous pyroclastic dispersal. Geology 39–10, 951–954.

Day, S.J., Carracedo, J.C., Guillou, H., 1997. Age and geometry of an aborted rift collapse: the San Andres fault system, El Hierro, Canary Islands. Geol. Mag. 134, 523–537.

Day, S.J., Carracedo, J.C., Guillou, H., Gravestock, P., 1999. Recent structural evolution of the Cumbre Vieja volcano, La Palma, Canary Islands: volcanic rift zone reconfiguration as a precursor to volcanic flank instability? J. Volcanol. Geoth. Res. 94, 135–167.

Decker, R.W., Klein, F.W., Okamura, A.T., Okubo, P.G., 1995. Forecasting eruptions of Mauna Loa Volcano, Hawaii. In: Rhodes, J.M., Lockwood, J.P. (Eds.), Mauna Loa revealed: Structure, Composition, History, and Hazards. Geophysical Monograph Series American Geophysical Union, Washington, DC, pp. 337−348. (Book 92)

Deegan, F.M., Troll, V.R., Barker, A.K., Harris, C., Chadwick, J.P., Carracedo, J.C., et al., 2012. Crustal versus source processes recorded in dykes from the Northeast volcanic rift zone of Tenerife, Canary Islands. Chem. Geol. 334, 324−344.

De la Nuez, J., Quesada, M.L., Alonso, J.J., 1997. Los volcanes de los islotes al norte de Lanzarote. Fundación César Manrique 233.

Delcamp, A., Petronis, M., Troll, V.R., van Wyk de Vries, B., Carracedo, J.C., Pérez-Torrado, F.J., 2010. Paleomagnetic evidence for rotation of the NE-rift on Tenerife, Canary Islands. Tectonophysics 492, 40−59.

Delcamp, A., Troll, V.R., van Wyk de Vries, B., Carracedo, J.C., Petronis, M.S., Pérez-Torrado, F.J., et al., 2012. Dykes and structures of the NE rift of Tenerife, Canary Islands: a record of stabilization and destabilization of ocean island rift zones. Bull. Volcanol. 74 (5), 963−980.

Delcamp, A., Petronis, M., Troll, V.R., 2014. Discerning magmatic flow patterns in shallow-level basaltic dykes from the NE rift zone of Tenerife, Spain, using the Anisotropy of Magnetic Susceptibility (AMS) technique. In: Ort, M.H., Porreca, M., Geissmann, J.W. (Eds.), The Use of Palaeomagnetism and Rock Magnetism to Understand Volcanic Processes, 396. Geological Society of London, Special Publications, London, pp. 87−106.

Demény, A., Ahijado, A., Casillas, R., Vennemann, T.W., 1998. Crustal contamination and fluid/rock interaction in the carbonatites of Fuerteventura (Canary Islands, Spain): a C, O, H isotope study. Lithos 44, 101−115.

Dietz, R.S., Sproll, W.P., 1979. East Canary Islands as a microcontinent within the Africa-North America continental drift fit. Nature 226, 1043−1045.

Donoghue, E., Troll, V.R., Harris, C., O'Halloran, A., Pérez-Torado, J.F., Walter, T.R., 2008. Low-temperature hydrothermal alteration of intra-caldera tuffs, Tejeda caldera, Gran Canaria: mineralogical and isotopic changes. J. Volcanol. Geoth. Res. 176, 551−564.

Donoghue, E., Troll, V.R., Harris, C., 2010. Hydrothermal alteration of the Miocene Tejeda Intrusive Complex, Gran Canaria, Canary Islands: insights from petrography, mineralogy and O- and H- isotope geochemistry. J. Petrol. 51, 2149−2176.

Edgar, C.J., 2003. The stratigraphy and eruption dynamics of a quaternary phonolitic plinian eruption sequence. The Diego Hernandez Formation, Tenerife, Canary Islands (Spain) (PhD Thesis). Monash University, Clayton, p. 264.

Edgar, C.J., Wolff, J.A., Nichols, H.J., Cas, R.A.F., Martí, J., 2002. A complex quaternary ignimbrite-forming phonolitic eruption: the Porís Member of the Diego Hernández Formation (Tenerife, Canary Islands). J. Volcanol. Geoth. Res. 118, 99−130.

Edgar, C.J., Wolff, J.A., Olin, P.H., Nichols, H.J., Pittari, A., Cas, R.A.F., et al., 2007. The late quaternary Diego Hernández formation, Tenerife: volcanology of a complex cycle of voluminous explosive phonolitic eruptions. J. Volcanol. Geoth. Res. 160, 59−85.

Feraud, G., Kaneoka, I., Allègre, C.J., 1980. K/Ar ages and stress pattern in the Azores: geodynamic implications. Earth Planet. Sci. Lett. 46, 275−286.

Fernández, C., Casillas, R., Ahijado, A., Perello, V., Hernández-Pacheco, A., 1997. Shear zones as a result of intraplate tectonics in oceanic crust: the example of the Basal Complex of Fuerteventure (Canary Islands). J. Struct. Geol. 19 (1), 41−57.

Fernández Palacios, J.M., Nascimento, L., Otto, R., Delgado, J.D., Garcia-del-Rey, E., Arevalo, J.R., et al., 2010. A reconstruction of Palaeo-Macaronesia, with particular reference to the long-term bio-geography of the Atlantic island laurel forests. J. Biogeogr. 38, 226−246.

Fiske, R.S., Jackson, E.D., 1972. Orientation and growth of Hawaiian volcanic rifts. Proc. R Soc. A 329, 299—326.

Freundt, A., 2003. Entrance of hot pyroclastic flows into the sea: experimental observations. Bull. Volcanol. 65, 144—164.

Freundt, A., Schmincke, H.-U., 1992. Mixing of rhyolite, trachyte and basalt magma erupted from a vertically and laterally zoned reservoir, composite flow P1, Gran Canaria. Contrib. Mineral. Petrol. 112, 1—19.

Freundt, A., Schmincke, H.-U., 1995. Petrogenesis of rhyolite—trachyte—basalt composite ignimbrite P1, Gran Canaria, Canary Islands. J. Geophys. Res. 100, 455—474.

Freundt, A., Schmincke, H.-U., 1998. Emplacement of ash-layers related to high-grade ignimbrite P1 in the sea around Gran Canaria. In: Weaver, P.P.E., Schmincke, H.-U., Firth, J.V., Duffield, W.A. (Eds.), Proc. ODP Sci. Results. 157. Ocean Drilling Program, College Station, TX, pp. 201—218.

Freundt, A., Tait, S.R., 1986. The entrainment of high viscosity magma into low-viscosity magma in eruption conduits. Bull. Volcanol. 48, 325—339.

Fritsch, K.W.G., Reiss, W., 1868. Geologische beschreibung der insel Tenerife: ein Beitrag zur kenntnis vulkanischer gebirge. Verlag Wurster and Co, Winterthur, p. 494.

Frisch, T., Schmincke, H.-U., 1969. Petrology of clinopyroxene-amphibiole inclusions from the Roque Nublo volcanics, Gran Canaria, Canary Islands (Petrology of Roque Nublo volcanics I). Bull. Volcanol. 33, 1073—1088.

Frizon de Lamotte, D., Saint Bezar, B., Bracene, R., Mercier, E., 2000. The two main steps of the Atlas building and geodynamics of the western Mediterranean. Tectonics 19 (4), 740—761.

Funck, T., Schmincke, H.-J., 1998. Growth and destruction of Gran Canaria deduced from seismic reflection and bathymetric data. J. Geophys. Res. 103 (B7), 15393—15407.

Fúster, J.M., Carracedo, J.C., 1979. Magnetic polarity mapping of quarternary volcanic activity of Fuerteventura and Lanzarote (Canary Islands). Estud. Geol. 35, 59—65.

Fúster, J.M., Cendrero, A., Gastesi, P., Ibarrola, E., Ruiz, J.L., 1968a. Geology and volcanology of the Canary Islands, Fuerteventura. In: Instituto Lucas Mallada, Madrid, International Symposium Volcanology, Tenerife September 1968, Special Publications, p. 239.

Fúster, J.M., Santin, S.F., Sagredo, A., 1968b. Geology and volcanology of the Canary Islands, Lanzarote. In: Institut Lucas Mallada, Madrid, International Symposium Volcanology, Tenerife September 1968, Special Publications, p. 177.

Fúster, J.M., Hernández-Pacheco, A., Muñoz, M., Badiola, E.R., Cacho, L.G., 1968c. Geology and volcanology of the Canary Islands, Gran Canaria. In: Institut Lucas Mallada, Madrid, International Symposium Volcanology, Tenerife September 1968, Special Publications, p. 243.

Fúster, J.M., Arana, V., Brändle, J.L., Navarro, M., Alonso, U., Aparicio, A., 1968d. Geology and volcanology of the Canary Islands, Tenerife. In: Institut Lucas Mallada, Madrid, International Symposium Volcanology, Tenerife September 1968, Special Publications, p. 218.

Fúster, J.M., Muñoz, M., Sagredo, J., Yebenes, A., Bravo, T., Hernández-Pacheco, A., 1980. Islas Canarias: Fuerteventura y Tenerife (excursión 121 A + C). Boletín Geológico y Minero 91 (2), 103—130.

Fúster, J.M., Ibarrola, E., Snelling, N.J., Cantagrel, J.M., Huertas, M.J., Coello, J., et al., 1994. Cronología K-Ar de la Formación Cañadas en el sector suroeste de Tenerife: implicaciones de los episodios piroclásticos en la evolución volcánica. Boletín Real Sociedad Espannóla de Historia Natural (Section Geológia) 89, 25—41.

Gabaldón, V., Cabrera, M.C., Cueto, L.A., 1989. Formación detrítica de Las Palmas. Sus facies y evolución sedimentológica. ESF Meeting on Canarian Volcanism, Lanzarote, 210—215.

Galipp, K., Klügel, A., Hansteen, T.H., 2006. Changing depths of magma fractionation and stagnation during the evolution of an oceanic island volcano: La Palma (Canary Islands). J. Volcanol. Geoth. Res. 155, 285—306.

García Cacho, L., Díez-Gil, J.L., Araña, V., 1994. A large volcanic debris avalanche in the Pliocene Roque Nublo Stratovolcano, Gran Canaria, Canary Islands. J. Volcanol. Geoth. Res. 63, 217—229.

García-Talavera, F., 1990. Aves gigantes en el Mioceno de Famara. Revista de la Academia Canaria de Ciencias 2, 71—79.

Gastesi, P., 1973. Is Betancuria Massif, Fuerteventura, Canary Islands, an uplifted piece of oceanic crust? Nat. Phys. Sci. 246 (155), 102—104.

Gee, M.J.R., Masson, D.G., Watts, A.B., Mitchell, N.C., 2001. Offshore continuation of volcanic rift zones, El Hierro, Canary Islands. J. Volcanol. Geoth. Res. 105, 107—119.

Geldmacher, J., Hoernle, K., Van den Bogaard, P., Zankl, G., Garbe-Schönberg, D., 2001. Earlier history of the C70-Ma-old Canary hotspot based on the temporal and geochemical evolution of the Selvagen archipelago and neighbouring seamounts in the eastern north Atlantic. J. Volcanol. Geoth. Res. 111, 55—87.

Geldmacher, J., Hoernle, K., Van den Bogaard, P., Duggen, S., Werner, R., 2005. New Ar-40/Ar-39 age and geochemical data from seamounts in the Canary and Madeira volcanic provinces: support for the mantle plume hypothesis. Earth Planet. Sci. Lett. 237, 85—101.

Geyer, A., Martí, J., 2011. The distribution of basaltic volcanism on Tenerife, Canary Islands: implications on the origin and dynamics of the rift system, reply to the comment by Carracedo et al. Tectonophysics 503 (33—4), 234—238.

Giachetti, T., Paris, R., Kelfoun, K., Pérez-Torrado, F.J., 2011. Numberical modelling of the tsunami triggered by the Güìmar debris avalanche, Tenerife (Canary Islands): comparison with field-based data. Mar. Geol. 284 (1—4), 189—202.

González, P.J., Samsonov, S.V., Pepe, S., Tiampo, K.F., Tizzani, P., Casu, F., et al., 2013. Magma storage and migration associated with the 2011—2012 El Hierro eruption: implications for crustal magmatic systems at oceanic island volcanoes. J. Geophys. Res. Solid Earth 118 (8), 4361—4377.

Gudmundsson, A., 1990. Emplacement of dikes, sills and crustal magma chambers at divergent plate boundaries. Tectonophysics 176, 257—275.

Guillou, H., Carracedo, J.C., Pérez-Torrado, F.P., Badiola, E.R., 1996. K-Ar ages and magnetic stratigraphy of a hotspot-induced, fast grown oceanic island: El Hierro, Canary Islands. J. Volcanol. Geoth. Res. 73, 141—155.

Guillou, H., Carracedo, J.C., Day, S.J., 1998. Dating the upper Pleistocene-Holocene volcanic activity of La Palma using the unspiked K-Ar technique. J. Volcanol. Geoth. Res. 86, 137—149.

Guillou, H., Carracedo, J.C., Duncan, R., 2001. K-Ar, 40Ar/39Ar Ages and magnetostratigraphy of Brunhes and Matuyama lava sequences from La Palma Island. J. Volcanol. Geoth. Res. 106, 175—194.

Guillou, H., Carracedo, J.C., Paris, R., Pérez-Torrado, F.J., 2004a. Implications for the early shield-stage evolution of Tenerife from K/Ar ages and magnetic stratigraphy. Earth Planet. Sci. Lett. 222, 599—614.

Guillou, H., Pérez Torrado, F.J., Hansen Machin, A.R., Carracedo, J.C.C., Gimeno, D., 2004b. The Plio-Quarternary volcanic evolution of Gran Canaria based on new K-Ar ages and magnetostratigraphy. J. Volcanol. Geoth. Res. 135 (3), 221—246.

Gurenko, A.A., Chaussidon, M., Schmincke, H.-U., 2001. Magma ascent and contamination beneath one intraplate volcano: evidence from S and O isotopes in glass inclusions and their host clinopyroxenes form Miocene shield basaltic hyaloclastites southwest of Gran Canaria (Canary Islands). Geochemica et Cosmochimica Acta 65 (23), 4359—4374.

Gurenko, A.A., Sobolev, A.V., Hoernle, K.A., Hauff, F., Schmincke, H.-U., 2009. Enriched, HIMU-type peridotite and depleted recycled pyroxenite in the Canary plume: a mixed-up mantle. Earth Planet. Sci. Lett. 277 (3—4), 514—524.

Gurenko, A.A., Hoernle, K.A., Sobolev, A.V., Hauff, F., Schmincke, H.-U., 2010. Source components of the Gran Canaria (Canary Islands) shield stage magmas: evidence from olivine composition and Sr-Nd-Pb isotopes. Contrib. Mineral. Petrol. 159, 689—702.

Gutiérrez, M., Casillas, R., Fernández, C., Balogh, K., Ahijado, A., Castillo, C., et al., 2006. The submarine volcanic succession of the basal complex of Fuerteventura, Canary Islands: a model of submarine growth and emergence of tectonic volcanic islands. Geol. Soc. Am. Bull. 118, 785–804.

Hansen, A., 1993. Bandama: paisaje y evolución. Cabildo Insular, Las Palmas de Gran Canaria, p. 127.

Hansen, A.M., Rodríguez-González, A., Pérez-Torrado, F.J., 2008. Vulcanismo Holoceno: Bandama y su entorno. In: Pérez-Torrado, F.J., Cabrera, M.C. (Eds.), Itinerarios geológicos por las Islas Canarias, 5. Sociedad Geológica de España, Gran Canaria, pp. 89–103. , Vol. Geo-Guías.

Hansteen, T.H., Troll, V.R., 2003. Oxygen isotope composition of xenoliths from the oceanic crust and volcanic edifice beneath Gran Canaria (Canary Islands): consequences for crustal contamination of ascending magmas. Chem. Geol. 193, 181–193.

Hansteen, T.H., Klügel, A., Schmincke, H.-U., 1998. Multi-stage magma ascent beneath the Canary Islands: evidence from fluid inclusions. Contrib. Mineral. Petrol. 132 (1), 48–64.

Hartung, G. (1857) Die geologischen Verhältnisse der Inseln Lanzarote und Fuerteventura. Zürich: Neue Denkschriften der Schweizerischen Gesellschaft für allgemeine Naturwissenschaften, Bd. XV., p. 166.

Hausen, H., 1958. On the geology of Fuerteventura (Canary Islands). Societas Scientiarum Fennica, Commentationes Physico-Mathematicae 22, 211.

Hausen, H., 1971. Outlines of the geology of Gomera. Soc. Sci. Fenn. Comment. Phys. Math. 41, 1–53.

Hausen, H.M., 1967. Sobre el desarrollo geológico de Fuerteventura (Islas Canarias). Una breve reseña. Anuario de Estudios Atlánticos 1 (13), 11–37.

Hernán, F., Vélez, R., 1980. El sistema de diques cónicos de Gran Canaria y la estimación estadística de sus características. Estud. Geol. 36, 65–73.

Hernández-Pacheco, A., 1982. Sobre una posible erupción en 1973 en la isla de El Hierro (Canarias). Estud. Geol. 38, 15–25.

Hernández-Pacheco, A., Valls, C., 1982. The historic eruptions of La Palma island (Canaries). Revista Universidad dos Acores 3, 83–94.

Hernández-Pacheco, E., 1910. Estudio geológico de Lanzarote y de las Isletas Canarias. Memorias de l Real Sociedad Española de Historia Natural VI 4, 331.

Herrera, R., Huertas, M.J., Ancochea, E., 2008. Edades 40Ar-39Ar del Complejo Basal de la isla de La Gomera (40Ar-39Ar ages of the Basal Complex of La Gomera Island). Geogaceta 44, 7–10.

Hobson, A., Bussy, F., Hernandez, J., 1998. Shallow-level migmatization of gabbros in a metamorphic contact aureole, Fuerteventura Basal Complex, Canary Islands. J. Petrol. 39, 1025–1037.

Hoefs, J., 1997. Stable Isotope Geochemistry. Springer-Verlag, Berlin, Heidelberg, p. 210.

Hoernle, K., 1998. Geochemistry of Jurassic ocean crust beneath Gran Canaria, Canary Islands: implication for crustal recycling and assimilation. J. Petrol. 39, 859–880.

Hoernle, K., Schmincke, H.U., 1993. The role of partial melting in the 15-Ma geochemical evolution of Gran Canaria: a blob model for the Canary Hotspot. J. Petrol. 34 (Part 3), 599–626.

Holik, J.S., Rabinowitz, P.D., Austin, J.A., 1991. Effects of Canary hotspot volcanism on structure of oceanic-crust off Morocco. J. Geophys. Res. Solid Earth Planets 96 (B7), 12039–12067.

Holloway, M.I., Bussy, F., 2007. Trace element distribution among rock-forming minerals from metamorphosed to partially molten basic igneous rocks in a contact aureole (Fuerteventura, Canaries). Lithos 102, 616–639.

Holloway, M.I., Bussy, F., Vennemann, T.W., 2008. Low-pressure, water-assisted anatexis of basic dykes in a contact metamorphic aureole, Fuerteventura (Canary Islands): oxygen isotope evidence for a meteoric fluid origin. Contrib. Mineral. Petrol. 155, 111–121.

Huertas, M.J., Arnaud, N.O., Ancochea, E., Cantagrel, J.M., Fúster, J.M., 2002. Ar-40/Ar-39 stratigraphy of pyroclastic units from the Cañadas Volcanic Edifice (Tenerife, Canary Islands) and their bearing on the structural evolution. J. Volcanol. Geoth. Res. 115 (3–4), 351–365.

Hunt, J.E., Wynn, R.B., Talling, P.J., Masson, D.G., 2013. Multistage collapse of eight western Canary Island landslides in the last 1.5 Ma: sedimentological and geochemical evidence from subunits in submarine flow deposits. Geochem. Geophys. Geosyst. 14 (7), 2159−2181.

IGME (Instituto Geológico y Minero de España), 1992. Proyecto MAGNA. Memoria y mapa geológico de España a escala 1:100.000. Hoja no 21-21/21-22. Isla de Gran Canaria. IGME, Madrid.

Izquierdo, T., 2014. Conceptual hydrogeological model and aquifer system classification of a small volcanic island (La Gomera; Canary Islands). Catena 114, 119−128.

Javoy, M., Stillman, C.J., Pineau, F., 1986. Oxygen and hydrogen isotope studies on the basal complexes of the Canary Islands: implications on the conditions of their genesis. Contrib. Mineral. Petrol. 92, 225−235.

Johansen, T.S., Hauff, F., Hoernle, K., Klügel, A., Kokfelt, T.F., 2005. Basanite to phonolite differentiation within 1550-1750 yr: U-Th-Ra isotopic evidence from the A.D. 1585 eruption on La Palma, Canary Islands. Geology 33, 897−900.

Johnson, S.E., Tate, M.C., Fanning, C.M., 1999. New geological mapping and SHRIMP U-Pb zircon data in the Peninsular ranges batholith, Baja California, Mexico: evidence for a suture? Geology 27, 743−746.

King, S.D., 2007. Hotspots and edge-driven convection. Geology 35 (3), 223−226.

Kissel, C., Guillou, H., Laj, C., Carracedo, J.C., Nomade, S., Pérez-Torrado, F.J., et al., 2011. The Mono Lake excursion recorded in phonolitic lavas from Tenerife (Canary Islands): paleomagnetic analyses and coupled K/Ar and Ar/Ar dating. Phys. Earth Planet. Interiors 187 (3−4), 232−244.

Kissel, C., Guillou, H., Laj, C., Carracedo, J.C., Pérez-Torrado, F., Wandres, C., et al., 2014. A combined paleomagnetic/dating investigation of the upper Jaramillo transition from a volcanic section at Tenerife (Canary Islands). Earth Planet. Sci. Lett. 406, 59−71.

Kissel, C., Rodríguez-Gonzalez, A., Laj, C., Carracedo, J.C., Pérez-Torrado, F.J., Wandres, C., et al., 2015. Paleosecular variation of the earth magnetic field at the Canary Islands over the last 15 ka. Earth Planet. Sci. Lett. 412, 52−60.

Klügel, A., 1998. Reactions between mantle xenoliths and host magma beneath La Palma (Canary Islands): constraints on magma ascent rates and crustal reservoirs. Contrib. Mineral. Petrol. 131, 237−257.

Klügel, A., Schmincke, H.U., White, J.D.L., Hoernle, K.A., 1999. Chronology and volcanology of the 1949 multi-vent rift-zone eruption on La Palma (Canary Islands). J. Volcanol. Geoth. Res. 94, 267−282.

Klügel, A., Hoernle, K.A., Schmincke, H.U., White, J.D.L., 2000. The chemically zoned 1949 eruption on La Palma (Canary Islands): petrologic evolution and magma supply dynamics of a rift-zone eruption. J. Geophys. Res. 105 (B3), 5997−6016.

Klügel, A., Hansteen, T.H., Galipp, K., 2005. Magma storage and underplating beneath Cumbre Vieja volcano, La Palma (Canary Islands). Earth Planet. Sci. Lett. 236, 211−226.

Klügel, A., Hansteen, T.H., van den Bogaard, P., Strauss, H., Hauff, F., 2011. Holocene fluid venting at an extinct Cretaceous seamount, Canary archipelago. Geology 39, 855−858.

Kobberger, G., Schmincke, H.-U., 1999. Deposition of rheomorphic ignimbrite D (Mogán Formation), Gran Canaria, Canary Islands, Spain. Bull. Volcanol. 60, 465−485.

Krastel, S., Schmincke, H.-U., 2002a. Crustal structure of northern Gran Canaria deduces from active seismic tomography. J. Volcanol. Geoth. Res. 115, 153−177.

Krastel, S., Schmincke, H.-U., 2002b. The channel between Gran Canaria and Tenerife: constructive processes and destructive events during the evolution of volcanic islands. Int. J. Earth Sci. 91, 629−641.

Krastel, S., Schmincke, H.-U., Jacobs, C.L., Rihm, R., Le Bas, T.P., Alibés, B., 2001. Submarine landslides around the Canary Islands. J. Geophys. Res. Solid Earth 106 (B3), 3977−3997.

Kröchert, J., Buchner, E., 2009. Age distribution of cinder cones within the Bandas del Sur Formation, southern Tenerife, Canary Islands. Geol. Mag. 146 (2), 161−172.

Krumbholz, M., Hieronymus, C.H., Burchardt, S., Troll, V.R., Tanner, D.C., Friese, N., 2014. Weibull-distributed dyke thickness reflects probabilistic character of host-rock strength. Nat. Commun. 5, Article number: 3272.

Leal, M., Lillo, J., Márquez, A., 2014. Assessment of groundwater circulation in La Gomera aquifers (Canary Islands, Spain) from their hydrochemical features. Environ. Earth Sci. 71, 23–30.

Leat, P.T., Schmincke, H.-U., 1993. Large-scale rheomorphic shear deformation in Miocene peralkaline ignimbrite E, Gran Canaria. Bull. Volcanol. 55, 155–165.

Le Bas, M.J., Rex, D.C., Stillman, C.J., 1986. The early magmatic chronology of Fuerteventura, Canary Islands. Geol. Mag. 123, 287–298.

Lietz, J., Schmincke, H.U., 1975. Miocene-Pliocene sea level changes and volcanic episodes on Gran Canaria (Canary Islands) in the light of new K-AR ages. Palaeogeogr. Palaeoclimatol. Palaeoecol. 18, 213–239.

Lomoschitz, A., Corominas, J., 1997. La depresión de Tirajana, Gran Canaria. Una macroforma erosive producida por grandes desilizamientos. Cuarternario y Geomorfología 11 (3–4), 75–92.

Longpré, M.-A., Troll, V.R., Hansteen, T.H., 2008a. Upper mantle magma storage under a Canarian shield volcano: teno massif, Tenerife, Spain. J. Geophys. Res. 113, B08203.

Longpré, M.-A., del Porto, R., Troll, V.R., Nicoll, G., 2008b. Engineering geology and future stability of the El Risco landslide, NW-Gran Canaria, Spain. Bull. Eng. Geol. Environ. 67, 165–172.

Longpré, M.-A., Troll, V.R., Walter, T.R., Hansteen, T.H., 2009. Volcanic and geochemical evolution of the Teno massif, Tenerife, Canary Islands: some repercussions of giant landslides on ocean island magmatism. Geochem. Geophys. Geosyst. 10 (12), Q12017.

Longpré, M.A., Chadwick, J.P., Wijbrans, J., Iping, R., 2011. Age of the El Golfo debris avalanche, El Hierro (Canary Islands): new constraints from laser and furnace 40Ar/39Ar dating. J. Volcanol. Geoth. Res. 203, 76–80.

López, C., Blanco, M.J., Abella, R., Brenes, B., Cabrera-Rodríguez, V.M., Casas, B., et al., 2012. Monitoring the volcanic unrest of El Hierro (Canary Islands) before the onset of the 2011–2012 submarine eruption. Geophys. Res. Lett. 39, L13303.

Lundstrom, C.C., Hoernle, K., Gill, J., 2003. U-series disequilibria in volcanic rocks from the Canary Islands: plume versus lithospheric melting. Geochim. Cosmochim. Acta 67, 4153–4177.

Luongo, G., Cubellis, R., Obrizzo, R., Petrazzuoli, S.M., 1991. A physical model for the origin of volcanism of the Tyrrhenian margin: the case of the Neapolitan area. J. Volcanol. Geoth. Res. 48, 173–185.

Lyell, C., 1854. Notes and correspondence on the geology of the Madeiran and Canary Islands.

Macdonald, G.A., 1972. Volcanoes. Prentice Hall PTR, New Jersey, p. 510.

MacFarlane, D.J., Ridley, W.I., 1968. An interpretation of gravity data for Tenerife, Canary Islands. Earth Planet. Sci. Lett. 4 (6), 481–486.

MacFarlane, D.J., Ridely, W.I., 1969. An interpretation of gravity data for Lanzarote, Canary Islands. Earth Planet. Sci. Lett. 6, 431–436.

Mader, C., 2001. Modeling the La Palma landslide tsunami. Sci. Tsunami Hazards 19, 150–170.

Manconi, A., Longpré, M.-A., Walter, T.R., Troll, V.R., Hansteen, T.H., 2009. The effects of flank collapses on volcano plumbing systems. Geology 37, 1099–1102.

Márquez, A., López, I., Herrera, R., Martín-González, F., Izquierdo, T., Carreño, F., 2008. Spreading and potential instability of Teide volcano, Tenerife, Canary Islands. Geophys. Res. Lett. 35, L05305.

Martí, J., Gudmundsson, A., 2000. The Las Cañadas caldera (Tenerife, Ca-nary Islands): an overlapping collapse caldera generated by magma-chamber migration. J. Volcanol. Geoth. Res. 103, 161–173.

Martí, J., Mitjavila, J., Araña, V., 1994. Stratigraphy, structure and geo-chronology of the Las Cafiadas cal-dera (Tenerife, Canary Islands). Geol. Mag. 131, 715–727.

Martí, J., Geyer, A., Andujar, J., Teixidó, F., Costa, F., 2008. Assessing the potential for future explosive activity from Teide-Pico Viejo stratovolcanoes (Tenerife, Canary Islands). J. Volcanol. Geoth. Res. 178, 529−542.

Martín, J.L., Díaz, M., 1984. El Tubo volcánico de Los Naturalistas (Lanzarote − Islas Canarias). Lapiaz 13, 51−53.

Martínez del Olmo, W., Buitrago, J., 2002. Sedimentación y volcanismo al este de las islas de Fuerteventura y Lanzarote (Surco de Fúster Casas). Geogaceta 32, 51−54.

Masson, D.G., 1996. Catastrophic collapse of the volcanic island of Hierro 15 ka ago and the history of landslides in the Canary Islands. Geology 24, 231−234.

Masson, D.G., Watts, A.B., Gee, M.J.R., Urgelés, R., Mitchell, N.C., Le Bas, T.P., et al., 2002. Slope failures on the flanks of the western Canary Islands. Earth Sci. Rev. 57, 1−35.

Mathieu, L., van Wyk de Vries, B., Holohan, E.P., Troll, V.R., 2008. Dykes, cups, saucers and sills: analogue experiments on magma intrusion into brittle rocks. Earth Planet. Sci. Lett. 271, 1−13.

McDougall, I., Schmincke, H.U., 1976. Geochronology of Gran Canaria, Canary Islands: age of shield building volcanism and other magmatic phases. Bull. Volcanol. 40 (1), 57−77.

McHone, J.G., 2000. Non-plume magmatism and rifting during the opening of the central Atlantic Ocean. Tectonophysics 316, 287−296.

McKenzie, D., Bickle, M.J., 1988. The volume and composition of melt generated by extension of the lithosphere. J. Petrol. 29, 625−679.

Meco, J., Stearns, C.E., 1981. Emergent littoral deposits in the eastern Canary Islands. Quart. Res. 15 (2), 199−208.

Meco, J., Petit-Maire, N., Guillou, H., Carracedo, J.C., Lomoschitz, A., Ramos, A.J.G., et al., 2003. Climatic changes over the last 5,000,000 years as recorded in the Canary Islands. Episodes 26, 133−134.

Meco, J., Scaillet, S., Guillou, H., Lomoschitz, A., Carracedo, J.C., Ballester, J., et al., 2007. Evidence for long-term uplift on the Canary Islands from emergent Mio-Pliocene littoral deposits. Glob. Planet. Change. 57, 222−234.

Mehl, K.W., Schmincke, H.-U., 1999. Structure and emplacement of the Pliocene Roque Nublo debris avalanche deposit, Gran Canaria, Spain. J. Volcanol. Geoth. Res. 94, 105−134.

Meletlidis, S., Di Roberto, A., Pompilio, A., Bertagnini, M., Iribarren, A., Felpeto, I., et al., 2012. Xenopumices from the 2011−2012 submarine eruption of El Hierro (Canary Islands, Spain): constraints on the plumbing system and magma ascent. Geophys. Res. Lett. 39, L17302.

Menéndez, I., Silva, P.G., Martín-Betancor, M., Pérez-Torrado, F.J., Guillou, H., Scaillet, S., 2008. Fluvial dissection, isostatic uplift, and geomorphological evolution of volcanic islands (Gran Canaria, Canary Islands, Spain). Geomorphology 102, 189−203.

Mitjavila, J., Villa, I., 1993. Temporal evo-lution of Diego Hernández formation Las Cañadas, Tenerife and using the 40Ar-39Ar method. Revista de la Sociedad Geológica de España 6, 61−65.

Montelli, R., Nolet, G., Dahlen, F.A., Masters, G., Engdahl, E.R., Hung, S.H., 2004. Finite-frequency tomography reveals a variety of plumes in the mantle. Science 303, 338−343.

Montelli, R., Nolet, G., Dahlen, F.A., Masters, G., 2006. A catalogue of deep mantle plumes: new results from finite-frequency tomography. Geochem. Geophys. Geosyst. 7, Q11007.

Moore, J.G., 1964. Giant submarine landslides on the Hawaiian Ridge. US Geol. Surv. Prof. Paper 501, D95−D98.

Moore, J.G., Clague, D.A., Holcomb, R.T., Lipman, P.W., Normark, W.R., Torresan, M.E., 1989. Prodigious submarine landslides on the Hawaiian Ridge. J. Geophys. Res. 94, 17465−17484.

Moss, J.L., McGuire, W.J., Page, D., 1999. Ground deformation monitoring of a potential landslide at La Palma, Canary Islands. J. Volcanol. Geoth. Res. 94, 251−265.

Muñoz, M., 1969. Ring complexes of Pajara in Fuerteventura island. Bull. Volcanol. 33, 840−861.

Navarro Latorre, J.M., Coello, J., 1989. Depressions originated by landslide processes in Tenerife. In: ESF meeting on Canarian Volcanism, Lanzarote. ESF, Strasbourg, pp. 150–152.

Newhall, C.G., Melson, W.G., 1983. Explosive activity associated with the growth of volcanic domes. J. Volcanol. Geoth. Res. 1, 111–131.

Nichols, H.J., 2001. Petrologic and geochemical variation of a caldera-forming ignimbrite: the Abrigo Member, Diego Hernández Formation, Tenerife, Canary Islands (Spain). MSc Thesis, Washington State University, Pullman, p. 123.

Pais, J., 1997. La Cucaracha II: Una Necrópolis única. In: eldiario.es, 2015. <http://www.eldiario.es/temas/cucaracha/>.

Paradas Herrero, A., Fernández Santín, S., 1984. Estudio vulcanológico y geoquímico del maar de la Caldera del Rey. Tenerife (Canarias). Estud. Geol. 40 (5–6), 285–313.

Pararas-Carayannis, G., 2002. Evaluation of the threat of mega tsunami generation from postulated massive slope failures of island stratovolcanoes on La Palma, Canary Islands, and on the island of Hawaii. Sci. Tsunami Hazards 20 (5), 251–277.

Paris, R., Carracedo, J.C., Pérez-Torrado, F.J., 2005a. Massive flank failures and tsunamis in the Canary Islands: past, present, future. Zeitschrift für Geomorphologie Supplement Series 140, 37–54.

Paris, R., Guillou, H., Carracedo, J.C., Pérez Torrado, F.J., 2005b. Volcanic and morphological evolution of La Gomera (Canary Islands), based on new K-Ar ages and magnetic stratigraphy: implications for oceanic island evolution. J. Geol. Soc. Lond. 162, 501–512.

Pedrazzi, D., Martí, J., Geyer, A., 2013. Stratigraphy, sedimentology and eruptive mechanisms in the tuff cone of El Golfo (Lanzarote, Canary Islands). Bull. Volcanol. 75, 740.

Pedrazzi, D., Becerril, L., Martí, J., Meletlidis, S., Galindo, I., 2014. Explosive felsic volcanism on El Hierro (Canary Islands). Bull. Volcanol. 76, 863.

Pellicer, M.J., 1977. Estudio volcanológico de la Isla de El Hierro (Islas Canarias) (Volcanological study of the island of El Hierro (Canary Islands). Estud. Geol. 33, 181–197.

Pérez-Torrado, F.J., 2000. Volcanostratigrafía del grupo Roque Nublo, Gran Canaria. Universidad de Las Palmas de Gran Canaria. Servicio de Publicaciones: Ediciones del Cabildo de Gran Canaria, Las Palmas, p. 459.

Pérez-Torrado, F.J., Carracedo, J.C., Mangas, J., 1995. Geochronology and stratigraphy of the Roque Nublo Cycle, Gran Canaria, Canary Islands. J. Geol. Soc. Lond. 152, 807–818.

Pérez Torrado, F.J., Paris, R., Cabrera, M.C., Schneider, J.-L., Wassmer, P., Carracedo, J.C., et al., 2006. Tsunami deposits related to flank collapse in oceanic volcanoes: the Agaete valley evidence, Gran Canaria, Canary Islands. Mar. Geol. 227, 135–149.

Pérez-Torrado, F.J., Rodríguez-Gonzalez, A., Carracedo, J.C., Fernández-Turiel, J.L., Guillou, H., et al., 2011. Edades C-14 Del Rift ONO de El Hierro (Islas Canarias). In: Turu, V., Constante, A. (Eds.), El Cuaternario en España y Áreas Afines, Avances en 2011. Asociación Española para el Estudio del Cuaternario (AEQUA), Andorra, pp. 101–104.

Pérez-Torrado, F.J., Carracedo, J.C., Rodríguez-Gonzalez, A., Soler, V., Troll, V.R., Wiesmaier, S., 2012. La erupción submarina de La Restinga en la isla de El Hierro, Canarias: Octubre 2011-Marzo 2012 (The submarine eruption of La Restinga (El Hierro, Canary Islands): October 2011–March 2012. Estud. Geol. 68 (1), 5–27.

Pérez-Torrado, F.J., Gimeno, D., Aulinas, M., Cabrera, M.C., Guillou, H., et al., 2015. Polygonal feeder tubes filled with hydroclasts: a new volcanic lithofacies marking shoreline subaerial-submarine transition. J. Geol. Soc. Lond. 172, 29–43.

Pinel, V., Jaupart, C., 2000. The effect of edifice load on magma ascent beneath a volcano. Philos. Trans. R Soc. Lond. 358, 1515–1532.

Pittari, A., Cas, R.A.F., 2004. Sole marks at the base of the late Pleisto-cene Abrigo Ignimbrite, Tenerife: implications for transport and depositional processes at the base of pyroclastic flows. Bull. Volcanol. 66, 356—363.

Pittari, A., Cas, R.A.F., Martí, J., 2005. The occurrence and origin of prominent massive, pumice-rich ignimbrite lobes within the late pleistocene Abrigo ignimbrite, Tenerife, Canary Islands. J. Volcanol. Geoth. Res. 139, 271—293.

Pittari, A., Cas, R.A.F., Edgar, C.J., Nichols, H.J., Wolff, J.A., Martí, J., 2006. The influence of palaeotopography on facies architecture and pyroclastic flow processes of a lithic-rich ignimbrite in a high gradient setting: the Abrigo Ignimbrite, Tenerife, Canary Islands. J. Volcanol. Geoth. Res. 152, 273—315.

Pittari, A., Cas, R.A.F., Wolff, J.A., Nichols, H.J., Larson, P.B., Martí, J., 2008. The use of lithic clast distributions in pyroclastic deposits to understand pre- and syn-caldera collapse processes: a case study of the Abrigo Ignimbrite, Tenerife, Canary Islands. Developments in Volcanology 10, Caldera Volcanism: Analysis, Modelling and Response, pp. 97—142.

Plan Hidrológico Insular Memoria Walter de La Palma In: Boletin Oficial de Canarias núm. 141, lunes 29 de octubre de 2001, 16232—16557.

Porter, S.C., 1972. Distribution, morphology, and size frequency of cinder cones of Mauna Kea volcano, Hawaii. Geol. Soc. Am. Bull. 83, 3607—3612.

Ramalho, R., Helffrich, G., Cosca, M., Vance, D., Hoffman, D., Schmidt, D.N., 2010. Episodic swell growth inferred from variable uplift of the Cape Verde hotspot island. Nat. Geosci. 2 (11), 774—777.

Real Audiencia de Canarias, 1731. Descripción del estado a que tiene reducida el Volcán la Isla de Lanzarote desde el primer día de Septiembre de 1730 asta el 4 de Abril de 1731. G.yJ., Leg. 89, Archivo de Simancas (Manuscrito, 56 p.).

Ridley, W.I., 1971. The field relations of the Las Cañadas volcanoes, Tenerife, Canary Islands. Bull. Volcanol. 35, 318—334.

Rihm, R., Jacobs, C.L., Krastel, S., Schmincke, H.-U., Alibes, B., 1998. Las Hijas seamounts — the next Canary Island? Terra Nova 10 (3), 121—125.

Rivera, J., Lastras, G., Canals, M., Acosta, J., Arrese, B., Hermida, N., et al., 2013. Construction of an oceanic island: insights from El Hierro 2011-2012 submarine volcanic eruption. Geology 41, 355—358.

Rivera, J., Hermida, N., Arrese, B., González-Aller, D., Sánchez de Lamadrid, J.L., Gutiérrez de la Flor, D., et al., 2014. Bathymetry of a new-born submarine volcano: El Hierro, Canary Islands. J. Maps 10, 82—89.

Robertson, A.H.F., Stillman, C.J., 1979. Submarine volcanic and associated sedimentary rocks of the Fuerteventura Basal complex, Canary Islands. Geol. Mag. 116, 203—214.

Rodríguez-Badiola, E., Pérez-Torrado, F.J., Carracedo, J.C., Guillou, H., 2006. Petrografía y geoquímica del edifico volcánico Teide-Pico Viejo y las dorsales noreste y noroeste de Tenerife. In: Carracedo, J.C. (Ed.), Los volcanes del Parque Nacional del Teide/El Teide, Pico Viejo y las dorsales activas de Tenerife. Naturaleza y Parques Nacionales, Ministerio de Medio Ambiente, Madrid, pp. 129—186.

Rodríguez-Gonzalez, A., Fernandez-Turiel, J.L., Pérez-Torrado, F.J., Hansen, A., Aulinas, M., Carracedo, J.C., et al., 2009. The Holocene volcanic history of Gran Canaria island: implications for volcanic hazards. J. Quat. Sci. 24, 697—709.

Rodríguez-Gonzalez, A., Fernandez-Turiel, J.L., Pérez-Torrado, F.J., Paris, R., Gimeno, D., Carracedo, J.C., et al., 2012. Factors controlling the morphology of monogenetic basaltic volcanoes: the Holocene volcanism of Gran Canaria (Canary Islands, Spain). Volcano Geomorphol. Landforms, Processes Hazards 136, 31—44.

Rodríguez-Losada, J.A., Martínez-Frías, J., 2004. The felsic complex of the Vallehermoso Caldera: interior of an ancient volcanic system (La Gomera, Canary Islands). J. Volcanol. Geoth. Res. 137, 261—284.

Romero Ruiz, C., 1991. Las manifestaciones volcánicas históricas del Archipiélago Canario (2 Tomos). Gobierno de Canarias (Consejería de Política Territorial), Tenerife.

Rothe, P., 1968. Mesozoische Flysch-Ablagerungen auf der Kanareninsel Fuerteventura. Geologische Rundschau 58, 314–322.

Rothe, P., 1974. Canary Islands-origin and evolution. Naturwissenschaften 61 (12), 526–533.

Rothe, P., Schmincke, H.U., 1968. Contrasting origins of the eastern and western islands of the Canarian Archipelago. Nature 218, 1152–1154.

Rumeu de Armas, A., Araña, V., 1982. Diario pormenorizado de la erupción volcánica de Lanzarote en 1824. Anuario de estudios atlánticos 28, 15–61.

Santiago, M., 1960. Los volcanes de La Palma: Datos histórico-descriptivos. El Museo Canario, no 75–76 Enero-Diciembre 1960: 281–346.

Schirnick, C., van den Bogaard, P., Schmincke, H.-U., 1999. Cone sheet formation and intrusive growth of an oceanic island – the Miocene Tejeda complex on Gran Canaria (Canary Islands). Geology 27, 207–210.

Schmincke, H.-U., 1967. Cone sheet swarm, resurgence of Tejeda Caldera, and the early geologic history of Gran Canaria. Bull. Volcanol. 31, 153–162.

Schmincke, H.-U., 1969. Ignimbrite sequence on Gran Canaria. Bull. Volcanol. 33, 1199–1219.

Schmincke, H.-U., 1973. Magmatic evolution and tectonic regime in the Canary, Madeira and Azores Islands Groups. Geol. Soc. Am. Bull. 84, 633–648.

Schmincke, H.-U., 1974. Volcanological aspects of peralkaline silicic welded ash-flow tuffs. Bull. Volcanol. 38, 594–636.

Schmincke, H.-U., 1976. Geology of the Canary Islands. In: Kunkel, G. (Ed.), Biogeography and Ecology in the Canary Islands. W. Junk, The Hague, pp. 67–184.

Schmincke, H.-U., 1979a. Age and crustal structure of the Canary Islands. J. Geophys. Res. 46, 217–224.

Schmincke, H.-U., 1979b. Volcanic and chemical evolution of the Canary Islands. In: von Rad, U., et al., (Eds.), Geology of the Northwest African Continental Margin. Springer Verlag, Berlin Heidelberg New York, pp. 273–306.

Schmincke, H.-U., 1993. Geological Field Guide of Gran Canaria. Pluto Press, Kiel, pp. 1–227.

Schmincke, H.-U., 2004. Volcanism. Springer-Verlag, Berlin Heidelberg, p. 324.

Schmincke, H.-U., Graf, G., 2000. DECOS OMEX II, Cruise No. 43, 25 November 1998–14 January 1999, Metero-Berichte, 00-2, Universität Hamburg, Hamburg, Germany, p. 99.

Schmincke, H.U., Rihm, R., 1994. Ozeanvulkan 1993, Cruise No. 24, 15 April–9 May 1993. Meteor-Bericht 94-2, Universität Hamburg, Hamburg, Germany, p. 88.

Schmincke, H.-U., Segschneider, B., 1998. Shallow submarine to emergent basaltic shield volcanism of Gran Canaria: evidence from drilling into the volcanic apron (ODP Leg 157). In: Weaver, P.P.E., Schmincke, H.-U., Firth, J.V. (Eds.), Proceedings of the Ocean Drilling Program, Scientific Results 157. Ocean Drilling Program, College Station, TX, pp. 141–181.

Schmincke, H.-U., Sumita M., 1998. Volcanic evolution of Gran Canaria reconstructed from apron sediments: synthesis of VICAP project drilling. In: Proceedings of the Ocean Drilling Program, Scientific Results, 157.

Schmincke, H.-U., Sumita, M., 2010. Geological evolution of the Canary Islands: a young volcanic archipelago of the old African continent. Görres Druckerei und Verlag GmbH, Koblenz (Germany), p. 196.

Schmincke, H.-U., Swanson, D.A., 1967. Laminar viscous flowage structures in ashflow tuffs from Gran Canaria, Canary Islands. J. Geol. 75, 641–664.

Schmincke, H.U., Weaver, P.P.E., Firth, J.V., Shipboard Scientific Party, 1995. Proc ODP, Initial Reports, Leg 157, College Station TX, (Ocean Drilling Program), p. 843.

Schmincke, H.U., Park, C., Harms, E., 1999. Evolution and environmental impacts of the eruption of Laacher See Volcano (Germany) 12,900 a BP. Quat. Int. 61 (1), 161–172.

Sigmarsson, O., Carn, S., Carracedo, J.C., 1998. Systematics of U-series nuclides in primitive lavas from the 1730-36 eruption on Lanzarote, Canary Islands, and implications for the role of garnet pyroxenites during oceanic basalt formations. Earth Planet. Sci. Lett. 162, 137–151.

Sigmarsson, O., Laporte, D., Carpentier, M., Devouard, B., Devidal, J.-L., Marti, J., 2013. Formation of U-depleted rhyolite from a basanite at El Hierro, Canary Islands. Contrib. Mineral. Petrol. 165 (3), 601−622.

Singer, B., Brown, L.L., 2002. The Santa Rosa event: Ar-40/Ar-39 and paleomagnetic results from the Valles rhyolite near Jaramillo Creek, Jemez Mountains, New Mexico. Earth Planet. Sci. Lett. 197, 51−64.

Singer, B.S., Relle, M.K., Hoffman, K.A., Battle, A., Laj, C., Guillou, H., et al., 2002. Ar/Ar ages from transitionally magnetized lavas on La Palma, Canary Islands, and the geomagnetic instability timescale. J. Geophys. Res. 107 (B11), 2307.

Solana, M.C., Aparicio, A., 1999. Reconstruction of the 1706 Montaña Negra eruption. Emergency procedures for Garachico and El Tanque, Tenerife, Canary Islands. Geol. Soc. Lond. Spec. Publ. 161, 209−216.

Solana, M.C., Kilburn, C.R.J., 2003. Public awareness of landslide hazards: the Barranco de Tirajana, Gran Canaria, Spain. Geomorphology 54, 39−48.

Solana, M.C., Kilburn, C.R.J., Rodríguez Badiola, E., Aparicio, A., 2004. Fast emplacement of extensive pahoehoe flow-fields: the case of the 1736 flows from Montaña de las Nueces, Lanzarote. J. Volcanol. Geoth. Res. 132, 189−207.

Sparks, R.S.J., 1986. The dimensions and dynamics of volcanic eruption columns. Bull. Volcanol. 48, 3−15.

Sperling, F.N., Washington, R., Whittaker, R.J., 2004. Future climate change of the subtropical North Atlantic: implications for the cloud forests of Tenerife. Clim. Change 65, 103−123.

Staudigel, H., Schmincke, H., 1981. Structural evolution of a seamount: evidence from the uplifted intraplate seamount on the island of La Palma, Canary Islands. Eos 62, 1075.

Staudigel, H., Schmincke, H.U., 1984. The Pliocene seamount series of La Palma, Canary Islands. J. Geophys. Res. 89 (B−13), 11190−11215.

Staudigel, H., Feraud, G., Giannerini, G., 1986. The history of intrusive activity on the island of La Palma (Canary Islands). J. Volcanol. Geoth. Res. 27, 299−322.

Stearns, H.T. (1946). Geology of the Hawaiian islands. U.S. Geological Survey, Division Hydrography, p. 106.

Steiner, C., Hobson, A., Favre, P., Stampfli, G.M., Hernandez, J., 1998. Mesozoic sequence of Fuerteventura (Canary Islands): witness of early Jurassic sea-floor spreading in the central Atlantic. Geol. Soc. Am. Bull. 110 (10), 1304−1317.

Stillman, C.J., 1987. A Canary Island Dyke Swarm: implications for formation of Oceanic Islands extensional fissural volcanism. In: Halls, H.C., Fahrig, W.F. (Eds.), Mafic Dyke Swarms: A Collection of Papers Based on the Proceedings of an International Conference Held at Erindale College. University of Toronto, ON, Canada, June 4 to 7, 1985. Geological Association of Canada Special Publication, vol. 34, International Lithosphere Program 120, St. John's Nfld., Canada, p. 503.

Stillman, C.J., 1999. Giant Miocene landslides and the evolution of Fuerteventura, Canary Islands. J. Volcanol. Geoth. Res. 94 (1−4), 89−104.

Stillman, C.J., Robertson, A.H.F., 1977. The dike swarm of Fuerteventura Basal Complex, Canary Islands. J. Geol. Soc. Conf. Rep. 135, 461.

Stillman, C.J., Bennell-Baker, M.J., Smewing, J.D., Fúster, J.M., Muñoz, M., Sagredo, J., 1975. Basal complex of Fuerteventura (Canary Islands) is an oceanic intrusive complex with rift-system affinities. Nature 257, 469−471.

Stroncik, N.A., Klügel, A., Hansteen, T.H., 2009. The magmatic plumbing system beneath El Hierro (Canary Islands): constraints from phenocrysts and naturally quenched basaltic glasses in submarine rocks. Contrib. Mineral. Petrol. 157 (5), 593−607.

Sumner, J.M., Wolff, J., 2003. Petrogenesis of mixed-mamga, high-grade, peralkaline ignimbrite "TL" (Gran Canaria): diverse styles of mixing in a replenished, zoned magma chamber. J. Volcanol. Geoth. Res. 126, 109−126.

Swanson, D.A., Duffield, W.A., Fiske, R.S., 1976. Displacement of the south flank of Kilauea Volcano: the result of forceful intrusion of magma into the rift zones: interpretation of Geodetic and Geologic information leads to a new model for the structure of Kilauea Volcano. USGS Professional Paper 963, 93.

Thirlwall, M.F., Singer, B.S., Marriner, G.F., 2000. 39Ar-40Ar ages and geochemistry of the basaltic shield stage of Tenerife, Canary Islands, Spain. J. Volcanol. Geoth. Res. 103 (1–4), 247–297.

Thomas, L.E., Hawkesworth, C.J., Van Calsteren, P., Turner, S.P., Rogers, N.W., 1999. Melt generation beneath ocean islands: a U-Th-Ra isotope study from Lanzarote in the Canary Islands. Geochim. Cosmochim. Acta 63, 4081–4099.

Troll, V.R., Schmincke, H.-U., 2002. Magma mixing and crustal recycling recorded in ternary feldspar from compositionally zoned peralkaline ignimbrite 'A', Gran Canaria, Canary Islands. J. Petrol. 43, 243–270.

Troll, V.R., Walter, T.R., Schmincke, H.-U., 2002. Cyclic caldera collapse: piston or piecemeal subsidence? Field and experimental evidence. Geology 30, 135–138.

Troll, V.R., Klügel, A., Longpré, M.-A., Burchardt, S., Deegan, F.M., Carracedo, J.C., et al., 2011. Floating stones off El Hierro (Canary Islands, Spain): the peculiar case of the October 2011 eruption. Solid Earth Discuss. 3, 975–999.

Troll, V.R., Klügel, A., Longpré, M.-A., Burchardt, S., Deegan, F.M., Carracedo, J.C., et al., 2012. Floating stones off El Hierro, Canary Islands: xenoliths of pre-island sedimentary origin in the early products of the October 2011 eruption. Solid Earth 3, 97–110.

Troll, V.R., Deegan, F.M., Burchardt, S., Zaczek, K., Carracedo, J.C., Meade, F.C., et al., 2015. Nannofossils: the smoking gun for the Canarian hotspot. Geol. Today 31 (4), 137–145.

Urgelés, R., Canals, M., Baraza, J., Alonso, B., Masson, D.G., 1997. The last major megalandslides in the Canary Islands: the El Golfo debris avalanche and the Canary debris flow, west Hierro Island. J. Geophys. Res. 102, 305–323.

Urgelés, R., Canals, M., Baraza, J., Alonso, B., 1998. Seismostratigraphy of the western flanks of El Hierro and La Palma (Canary Islands): a record of Canary Island volcanism. Mar. Geol. 146, 225–241.

Urgelés, R., Masson, D.G., Canals, M., Watts, A.B., Le Bas, T., 1999. Recurrent large-scale landsliding on the west flank of La Palma, Canary Islands. J. Geophys. Res. -Solid Earth 104, 25331–25348.

Urgelés, R., Canals, M., Masson, D.G., 2001. Flank stability and processes off the western Canary Islands: a review from El Hierro and La Palma. Scientia Marina 65, 21–31.

van den Bogaard, P., 2013. The origin of the Canary Island seamount province — new ages of old seamounts. Sci. Rep. 3 (2107), 1–7.

van den Bogaard, P., Schmincke, H.-U., 1998. Chronostratigraphy of Gran Canaria. In: Weever, P.P.E., Schmincke, H.U., Firth, J.V., Duffield, W.A. (Eds.), Proceedings of the Ocean Drilling Program, Scientific Results 157. Ocean Drilling Program, College Station, TX, pp. 127–140.

Viera y Clavijo, J., 1776. Noticias de la historia general de las Islas Canarias. Goya Ediciones (publicado en 1951), Santa Cruz de Tenerife, p. 333.

Vogt, P.R., Smoot, N.C., 1984. The Geisha Guyots — Multibeam bathymetry and morphometric interpretation. J. Geophys. Res. 89 (NB13), 1085–1107.

von Buch, L., 1815. Description physique des Iles Canaries: atlas. Imprimerie d'Hippolyte Tillard, Paris.

von Buch, L., 1825. Physikalische Beschreibung der Canarischen Inseln. Berlin, p. 201.

von Humboldt, A., 1814. Voyage de Humboldt et Bonpland. Première Partie. Relation Historique. Tome Premier. G. Dufour et Comp., Rue des Mathurins-Saint-Jaques 7, Paris.

Walker, G.P.L., 1990. Geology and volcanology of the Hawaiian Islands. Pac. Sci. 44 (4), 315–347.

Walker, G.P.L., 1999. Volcanic rift zones and their intrusion swarms. J. Volcanol. Geoth. Res. 94, 21–34.

Walter, T., Schmincke, H.-U., 2002. Rifting, recurrent landsliding and Miocene structural reorganization on NW-Tenerife (Canary Islands). Int. J. Earth Sci. (Geologische Rundschau) 91 (4), 615–628.

Walter, T.R., Troll, V.R., 2001. Formation of caldera periphery faults, an experimental study. Bull. Volcanol. 63, 191–203.

Walter, T.R., Troll, V.R., 2003. Experiments on rift zone formation in unstable volcanic edifices. J. Volcanol. Geoth. Res. 127, 107−120.

Walter, T.R., Troll, V.R., Caileau, B., Schmincke, H.-U., Amelung, F., van den Bogaard, P., 2005. Rift zone reorganization through flank instability on ocean islands − an example from Tenerife, Canary Islands. Bull. Volcanol. 65, 281−291.

Ward, S.N., Day, S., 2001. Cumbre Vieja Volcano − potential collapse and tsunami at La Palma, Canary Islands. Geophys. Res. Lett. 28 (17), 3397−3400.

Watts, A.B., Masson, D.G., 1995. A giant landslide on the north flank of Tenerife, Canary Islands. J. Geophys. Res. 100, 24487−24498.

Watts, A.B., Masson, D.G., 2001. New sonar evidence for recent catastrophic collapses of the north flank of Tenerife, Canary Islands. Bull. Volcanol. 63 (1), 8−19.

White, R., McKenzie, D., 1989. Magmatism at rift zones: the generation of volcanic continental margins and flood basalts. J. Geophys. Res. Solid Earth 94 (B6), 7685−7729.

Wiesmaier, S., Deegan, F.M., Troll, V.R., Carracedo, J.C., Chadwick, J.P., Chew, D.M., 2011. Magma mixing in the 1100 AD Montaña Reventada composite eruption, Tenerife, Canary Islands: interaction between rift zone and central volcano plumbing systems. Contrib. Mineral. Petrol. 162, 651−669.

Wiesmaier, S., Troll, V.R., Carracedo, J.C., Ellam, R.M., Bindeman, I., Wolff, J.A., 2012. Bimodality of Lavas in the Teide−Pico Viejo succession in Tenerife—the role of crustal melting in the origin of recent phonolites. J. Petrol. 53, 2465−2495.

Wiesmaier, S., Troll, V.R., Wolff, J.A., Carracedo, J.C., 2013. Open-system processes in the differentiation of mafic magma in the Teide−Pico Viejo succession, Tenerife. J. Geol. Soc. London. 170, 557−570.

Wilson, J.T., 1963. A possible origin of the Hawaiian Islands. Can. J. Phys. 41, 863−870.

Wolff, J.A., 1983. Petrology of Quaternary pyroclastic deposits from Tenerife, Canary Islands. Ph.D. Thesis, Univerisity of London, London.

Wolff, J.A., Grandy, J.S., Larson, P.B., 2000. Interaction of mantle-derived magma with island crust? Trace element and oxygen isotope data from the Diego Herandez Formation, Las Cañadas, Tenerife. J. Volcanol. Geoth. Res. 103, 343−366.

Zaczek, K., Troll, V.R., Cachao, M., Ferreira, F., Deegan, F.M., Carracedo, J.C., et al., 2015. Nannofossils in 2011 El Hierro eruptive products reinstate plume model for Canary Islands. Sci. Rep. 5 (7945), 5.

Zazo, C., Goy, J.L., Hillaire-Marcel, C., Gillot, P.Y., Soler, V., González, J.A., et al., 2002. Raised marine sequences of Lanzarote and Fuerteventura revisited − a reappraisal of relative sea-level changes and vertical movements in the eastern Canary Islands during quarternary. Quart. Sci. Rev. 21, 2019−2046.

Index

Note: Page numbers followed by "*f*" and "*t*" refer to figures and tables, respectively.

Printed in the United States
By Bookmasters